Advanced Structural Ceramics

Advanced Structural Ceramics

Bikramjit Basu
Department of Materials Science and Engineering
Indian Institute of Technology
Kanpur, India

Currently at the Materials Research Center
Indian Institute of Science
Bangalore, India

Kantesh Balani
Department of Materials Science and Engineering
Indian Institute of Technology
Kanpur, India

The American Ceramic Society

WILEY

A John Wiley & Sons, Inc., Publication

Published by John Wiley & Sons, Inc., Hoboken, New Jersey.
Published simultaneously in Canada.

Library of Congress Cataloging-in-Publication Data:

Basu, Bikramjit.
Advanced Structural Ceramics / Prof. Bikramjit Basu, Dept. of Materials Science and Engineering, Indian Institute of Technology Kanpur, India, & Prof. Kantesh Balani, Dept. of Materials Science and Engineering, Indian Institute of Technology Kanpur, India.
pages cm
Includes index.
ISBN 978-0-470-49711-1 (cloth)
1. Ceramic materials. 2. Ceramic-matrix composites. I. Balani, Kantesh. II. Title.
TA455.C43B375 2011
620.1'4–dc22

2010048280

oBook ISBN: 978-1-118-03730-0
ePDF ISBN: 978-1-118-03728-7
ePub ISBN: 978-1-118-03729-4

Printed in the United States of America.

10 9 8 7 6 5 4 3 2 1

Bikramjit Basu dedicates this book with a great sense of gratitude to his uncle, Mr. Ranjit Mazumdar

Kantesh Balani dedicates this book to his father, the late Mr. Parmanand B. Balani

Contents

Section Five High-Temperature Ceramics

Preface

The field of advanced structural ceramics is widely recognized as an increasingly important area for material scientists, space technologists, mechanical engineers and tribologists, biotechnologists, chemists, and medical professionals. Recent developments in our understanding of fundamental concepts of materials science have enabled impressive progress in the attempts to develop smart and tough structural ceramics. The progress in advanced structural ceramics clearly requires an improved understanding in multiple disciplines as well as the development of new design methodologies in order to obtain better properties in terms of physical, tribological, high-temperature, and even biological performance. From this perspective, this book has been structured into various theme sections, each of which contains a number of chapters. The first section of this book has been designed to facilitate readers who do not have a background in the area of structural ceramics processing and properties. While conceiving the contents of this book, the authors desired to motivate students and young researchers as well as to provide experts in the field with a healthy balance of topics for teaching and academic pursuits. It is expected that this book, if used as a text, would strongly benefit senior undergraduate and postgraduate students.

This unique book illustrates some recent examples of the development of new ceramic compositions or ceramics with refined microstructure and properties for various engineering applications, while covering requisite fundamentals necessary to understand the progress being made in ceramics science. This book aptly describes the fundamentals of mechanical properties and processing, while highlighting some of the recent advances in processing tools for fabricating ceramic-based bulk and coating materials. Further, the authors strongly consider the importance of tough ceramics (MAX phases, zirconia [ZrO_2], and SiAlON-based ceramics) engineered for structural applications. The use of advanced ceramics as coatings for high-temperature applications is also well addressed. Additionally, this book deals pertinently with the newly enticing area of nanoceramic composites and biomaterials. An important feature of this book is that two sections review and refresh the reader's familiarity with the fundamentals of the structure–property correlation as well as basic aspects of processing; subsequent sections covering theme areas also start with an overview chapter for easier understanding of the entire contents of the book. Each theme area also covers the most important ceramic systems.

In Section I, on the *fundamentals*, general properties of various engineering materials are discussed with particular reference to the distinguishable properties of ceramics. Various broad classifications of ceramics are also presented. The bird's-eye view of their temporal growth, applications, and properties is also touched upon. The basic aspects of atomic bonding, structure, and physical properties, and their applications, are discussed in this section. The development of material properties resulting from the fundamental bonding and structure is also elucidated in this section. An important aspect of this section is the science-based discussion on the origin of brittle fracture and strength variability of ceramics. The concept of fracture toughness and measurement of various mechanical properties as well as a brief discussion on toughening mechanisms are also presented in this section.

Section II covers the (1) *synthesis* of high purity powders, which serve as the starting block for consequent sintering and shaping into useful products, (2) *sintering mechanism*, which is detailed with conventional sintering methods, and (3) emergence of advanced sintering techniques (i.e., *thermomechanical processing*), utilizing pressure in addition to temperature for processing ceramics. The synthesis of ZrO_2, titanium diboride (TiB_2), hydroxyapatite (HAp), and tungsten carbide (WC) powders is described to illustrate how some of the important technologically relevant powders can be synthesized.

Section III constitutes the environment and engineering of ceramic materials, associated with the damage that might result upon interaction. Hence, *surface coatings* become an essential component of applying ceramics for protection of surfaces exposed to high temperature, corrosion, wear, oxidation, and so on. Major thermal spraying techniques, such as plasma spraying, high-velocity oxy-fuel (HVOF), detonation-gun (D-gun), and electric arc, are described in this section. Thereupon, the role of coatings in material protection or functionality (such as in body implants), ultra-high-temperature wear resistance, thermal barriers, and so on, is represented.

A major drawback of ceramics is their highly brittle nature, which becomes a key parameter when developing structural components. Hence, Section IV concentrates on a special class of *advanced tough ceramics*, such as ZrO_2-based ceramics, S-SiAlON ceramics, and the MAX phases. Toughness optimization via microstructural tailoring and controlling the processing conditions to achieve optimal performance of structural ceramics is presented herein.

Section V emerges with the classification of *high-temperature ceramics*, followed by the processing requirement of using sintering additives toward achieving full densification. Consequently, the technological aspect of optimizing the sintering conditions to attain uniform or controlled microstructure and, thereby, enhance performance of high-temperature ceramics is discussed in reference to recently developed TiB_2 or ZrB_2 ceramics with silicide additives. Following which, high-temperature oxidation and mechanical properties are detailed for better understanding of stable high-temperature ceramics for advanced applications.

Nanoceramic composites, a newly sustainable arena, are detailed in Section VI, demonstrating their relevance in today's scenario. Processing-related challenges and microstructure development are described to illustrate how to develop strategies to retain their nanograined feature using some advanced sintering routes, for example, spark plasma sintering. The need for tailoring composition and process parameters is also presented in reference to recent development of oxide and non-oxide ceramic nanocomposites.

One of the important areas in which ceramics are receiving more appreciation is *biomedical applications*. After mentioning some of the concepts required for the development of bioceramics and biocomposites, the processing and properties of HAp-based bioceramic composites for hard-tissue replacement applications are addressed in Section VII. Furthermore, this final section concentrates on the *in vitro* and *in vivo* properties of bioceramic composites followed by assessment of their tribological properties, which is essential when evaluating new materials as potential biocomposite ceramics.

Hence, this book provides an entirely new paradigm of visualizing ceramics not only as an isolated category of materials limited to high-temperature wear and corrosion resistance, but also as load-bearing structural materials for advanced and new engineered and/or scientific applications. The previously described layout of the book as well as the succession of the various sections and chapters is primarily meant to provide easy understanding for both students and experts pursuing the field of structural ceramics. In particular, this book has the following major important features: (1) the fundamental structure–processing–property–application pyramid is presented, enabling the book to be used as a textbook for teaching, academics, and research; (2) a broad range of topics is covered, including contemporary and exciting areas such as advanced tough ceramics, high-temperature ceramics, advanced ceramic processing techniques, and the *in vivo* and *in vitro* properties of bioceramics; (3) the basics addressed in this book will appeal to a large number of active researchers from various disciplines—biological sciences, metallurgy and materials science, ceramics, and biotechnology—as well as engineers, manufacturers, dentists, and surgeons.

This book is the outcome of several years of teaching undergraduate- and postgraduate-level courses in the areas of ceramics, composite materials, biomaterials, manufacturing of materials, and other related fundamental courses in the area of materials science offered to students of the Indian Institute of Technology (IIT) Kanpur, India. More important, the research results of many of the postgraduate students from our research groups are also summarized in some chapters. Bikramjit Basu would like to mention the contribution of his past and present students, B. V. Manoj Kumar, G. Brahma Raju, Amartya Mukhopadhyay, Shekhar Nath, Naresh Saha, Shouriya Dutta Gupta, K. Madhav Reddy, Subhodip Bodhak, Srimanta Das Bakshi, D. Sarkar, P. Suresh Babu, Manisha Taneja, Amit S. Sharma, Garima Tripathi, Alok Kumar, Shilpee Jain, Neha Gupta, Ashutosh K. Dubey, Devesh Tiwari, Shibayan Roy, Ravi Kumar, A. Tewari, Prafulla Mallick, R. Tripathy, T. Venkateswaran, U. Raghunandan, Divya Jain, Nitish Kumar, Atiar

R. Molla, and Sushma Kalmodia. The dedication of these students to developing various ceramics and composites is reflected in the research results summarized in many of the chapters. With a great sense of appreciation and gratitude, B. Basu notes past and present research collaboration with a number of researchers and academicians, including Drs. Omer Van Der Biest, Jozef Vleugels, G. Roebben, D. Dierickx, C. Zhao, R. K. Bordia, Dileep Singh, M. Singh, T. Goto, T. J. Webster, Amar S. Bhalla, Ruyan Guo, Mauli Agrawal, Artemis Stamboulis, G. Sundararajan, K. Chattopadhyay, K. Biswas, N. K. Mukhopadhyay, M. Banerjee, R. Gupta, D. Kundu, R. Prasad, S. K. Mishra, Mira Mohanty, P. V. Mohanan, A. K. Suri, Mike H. Lewis, Ender Suvaci, Hasan Mondal, Ferhat Kara, Nurcan Kalis Ackibas, D. Roy, M. C. Chu, S. J. Cho, Doh-Yeon Kim, J. H. Lee, Jo Wook, S. Kang, Alok Pandey, Alok Dhawan, Arvind Sinha, A. Basumallick, and Animesh Bose. The encouragement and collaboration with two of our former colleagues, the late Prof. R. Balasubramaniam and the late Prof. V. S. R. Murty, are also remembered. B. Basu also remembers the constant inspiration of a number of colleagues and former teachers, including Profs. S. Ranganathan, Ashutosh Sharma, Goutam Biswas, Kalyanmoy Deb, Vikram Jayaram, Dipankar Banerjee, Atul Chokshi, and B. S. Murty.

Kantesh Balani thanks the contribution of his students and staff: Dr. Neelima Mahato, Ankur Gupta, Milind R. Joshi, Samir Sharma, S. Ariharan, Anup Patel, and Raja Choudhary. He also expresses his sincere gratitude to long-term colleagues and friends: Dr. Debrupa Lahiri and Profs. Yao Chen, Srinivasa Rao Bakshi, Rajesh Srivastava, Sanjay Mittal, Ashwini Kumar, and Anup Keshri.

The authors also extend sincere gratitude to their colleagues at the Materials Science and Engineering Department at IIT Kanpur, especially Profs. Dipak Mazumdar, S.P. Mehrotra, Anandh Subramaniam, Anish Upadhyaya, Gouthama, Kallol Mondal, Krishanu Biswas, Rajiv Shekhar, Monica Katiyar, Deepak Gupta, and Vivek Verma, for their consistent support. The authors express their sincere thanks to Mr. Divakar Tiwari for his untiring efforts and excellent assistance during the manuscript preparation.

The authors would like to take this opportunity to acknowledge the financial support of various governmental agencies of India, including the Council of Scientific and Industrial Research (CSIR), Department of Atomic Energy (DAE), Department of Biotechnology (DBT), Ministry of Human Resource and Development (MHRD), Defense Research and Development Organization (DRDO), Department of Science and Technology (DST), UK–India Education and Research Initiative (UKIERI), and Indo–US Science and Technology Forum (IUSSTF), which facilitated research in the area of ceramics and composites in our group. We would also like to thank the CARE grant, and the Centre for Development of Technical Education (CDTE), IIT Kanpur, for extending financial support during the writing of this book. Bikramjit Basu expresses gratitude to his long-time friend and collaborator, Dr. Jaydeep Sarkar, for constant inspiration during the writing of this book. Kantesh Balani also extends his sincere gratitude for the tutelage and support extended by Prof. Arvind Agarwal, Florida International University, Miami. We thank Prof. Sir Richard Brook for constructive criticism and comments.

The authors sincerely express their gratitude to Prof. Michel Barsoum for writing the Foreword.

Finally, we acknowledge the moral support extended by our parents, in-laws, and family members during the course of writing this book.

Bikramjit Basu

Laboratory of Biomaterials
Materials Science and Engineering
IIT Kanpur, India

Currently at the Materials Research Center
Indian Institute of Science
Bangalore, India

Kantesh Balani

Biomaterials Processing and Characterization Laboratory
Materials Science and Engineering
IIT Kanpur, India

July 2011

To view color versions of the figures in this book, please visit
ftp://ftp.wiley.com/public/sci_tech_med/advanced_structural_ceramics.

Foreword

Ceramics have long been recognized as brittle materials, which in turn has limited their applications. With the advent of tougher ceramics, however, their utility has increased concomitantly. This book explains how, and why, today advanced structural ceramics represent a multibillion dollar industry that is still growing. Ceramics are increasingly used in both monolithic and composite form in advanced aerospace, automotive, biomedical, industrial, and consumer applications. The vast majority of books dealing with the topic of structural ceramics and their uses are edited compilations or conference proceedings that are of little use for somebody trying to get a better handle on the topic. Since they are geared toward researchers and scientists who are more or less familiar with the topics at hand, these compilations do not attempt to explain the fundamental science behind the topics they discuss. This book tries to bridge the gap from basics to applications.

This book is divided into seven sections. The first introduces ceramics and the basics behind their bonding, as well as their mechanical properties and how they are quantified. The second section deals with the synthesis of ceramics powders and their compaction and sintering. The third reviews coatings and the thermal spray of ceramics. Section IV deals with the toughening of zirconia, SiAlONs, and the MAX phases. Section V considers ultra-high-temperature ceramics and their processing, mechanical properties, and oxidation resistances. The penultimate section reviews work on nanostructured ceramics, in both monolithic and composite form. The last section deals with bioceramics and their uses.

One of the major strengths of this book is the large number of examples and references—many from the authors' own work—used to illustrate the ideas presented. Another advantage of this book is that it is conceived, from the initial stages, as a textbook and is based in part on the authors' class notes, which from my experience is a valuable and almost indispensible requirement for writing a good textbook. This book can be used as a textbook for students—both graduate and senior undergraduate—and academicians, or as a practical guide for industrial researchers and engineers.

Michel W. Barsoum
Grosvenor and Distinguished Professor
Department of Materials Science and Engineering
Drexel University, Philadelphia, PA

About the Authors

Dr. Bikramjit Basu is currently an Associate Professor, Materials Research Center, Indian Institute of Science, Bangalore, India. He is on leave from the Indian Institute of Technology (IIT) Kanpur, India. Bikramjit Basu obtained his undergraduate and postgraduate degrees, both in metallurgical engineering, from National Institute of Technology, Durgapur, and the Indian Institute of Science, Bangalore, in 1995 and 1997, respectively. He earned his PhD in ceramics at Katholieke Universiteit Leuven, Belgium, in 2001. After a brief stint of postdoctoral research at University of California, Santa Barbara, he joined IIT Kanpur, India, in 2001 as assistant professor. He has held visiting positions at University of Warwick, U.K., Seoul National University, South Korea, and University Polytechnic Catalonia, Barcelona. In India, Dr. Basu established vibrant research programs in ceramics and biomaterials with government funding of more than five crores. In the structural ceramics area, he demonstrated the unique capability of spark plasma sintering in developing nanoceramic materials in zirconia (ZrO_2) and tungsten carbide (WC) systems. In biomaterials, his primary focus is on optimizing the physical and biological properties in hydroxyapatite-based biocomposites and glass-ceramics for hard-tissue replacement.

Dr. Basu has authored or co-authored more than 150 peer-reviewed research papers, including 20 papers in *Journal of American Ceramic Society*. He has delivered more than 80 invited lectures, both nationally and internationally, including in the United States, United Kingdom, Germany, Japan, and Canada. He is on the editorial board of five international journals (including *Materials Science and Engineering C* and *International Journal of Biomaterials*) and serves as reviewer of more than 20 Science Citation Index journals in the area of ceramics and biomaterials. He is principal editor of the book *Advanced Biomaterials: Fundamentals, Processing and Applications* (which was published in September 2009 by John Wiley & Sons). He is currently the principal investigator of two major international research programs in biomaterials, funded by UK–India Educational and Research Initiative and Indo–US Science and Technology Forum. In recognition of his contributions to the fields of ceramics, tribology, and biomaterials, Dr. Basu received noteworthy awards from the Indian Ceramic Society (2003), Indian National Academy of Engineering (2004), and Indian National Science Academy (2005), as well as the Metallurgist of the Year award (2010), instituted by Ministry of Steels, Government of India. He is the first Indian from India to receive the prestigious Coble Award for Young Scholars from the American Ceramic Society in 2008. Recently, he received the National Academy of Science, India (NASI)-SCOPUS Young Scientist 2010 award in Engineering Sciences.

Dr. Kantesh Balani joined as an assistant professor in the Department of Materials and Metallurgical Engineering (now Materials Science & Engineering) at the IIT Kanpur in July 2008. He earned his doctorate in mechanical engineering from Florida International University, Miami, in 2007. His research concentrated on the role of carbon nanotube dispersion in enhancing the fracture toughness of alumina (Al_2O_3) nanocomposites. He has also worked on bioceramic hydroxyapatite coatings for biomedical applications. He pursued his post-doctoral research in the Nanomechanics and Nanotribology Laboratory (NMNTL) and Plasma Forming Laboratory (PFL), Florida International University, Miami. He is recipient of several fellowships and awards, such as Young Engineer Award 2010 (Indian National Academy of Engineering), Young Metallurgist Award 2010 (Indian Institute of Metals), Young Scientist Award 2009 (Materials Science Division, Indian Science Congress Association), R.L. Thakur Memorial Prize 2009 (Indian Ceramics Association), David Merchant International Student Achievement Award 2007, Arthur E. Focke LeaderShape Award 2004, Research Challenge Trust Fund (RCTF) Fellowship 2002, Sudharshan Bhat Memorial Prize and S. Ananthramakrishnan Memorial Prize 2001, and Deutscher Akademischer Austausch Dienst (DAAD) Scholarship 2001. He has presented over 25 lectures at international conferences and has over 45 publications in peer-reviewed journals and conference proceedings. His research interests include ab initio molecular modeling, electron microscopy, and nanomechanics and nanotribology of bio/nanocomposites. Currently, he is reviewer of over 20 technical journals from Elsevier, Blackwell Publishing Inc., Wiley, Springer, Hindawi, Highwire, Materials Research Society India/Indian National Science Academy, and American Society of Metals, serves as a key reader for Metallurgical and Materials Transactions A, and is involved as one of the editorial board members of *Recent Patents on Materials Science* (*Bentham*), *Recent Patents on Nanotechnology* (*Bentham*), and *Nanomaterials and Energy* (Institution of Civil Engineers).

Section One

Fundamentals of Nature and Characteristics of Ceramics

Chapter 1

Ceramics: Definition and Characteristics

In this chapter, the general properties of ceramics are discussed in reference to other primary classes of materials. Further, the need for the development of high-toughness ceramics with high hardness, strength, and wear resistance are addressed. The development of ceramic materials for high-temperature applications are also discussed.

1.1 MATERIALS CLASSIFICATION

There is a general consensus that engineering materials can be classified into three primary classes: metals and alloys; ceramics and glasses; and polymers. Among these three primary classes, metals, metallic alloys, and polymers are, by far, more widely used than ceramics and glasses for various structural and engineering applications. Nevertheless, ceramics have attracted attention in the scientific community in the last three decades.[1–4] The widespread use of metallic materials is driven by their high tensile strength and high toughness (crack growth resistance) as well as their ability to be manufactured in various sizes and shapes using reproducible fabrication techniques. Similarly, polymers have distinct advantages in terms of their low density, high flexibility, and ability to be molded into different shapes and sizes. Nevertheless, polymeric materials have low melting point (less than 400°C) as well as very low strength and elastic modulus. Compared with ceramics, metals have much lower hardness and many commonly used metallic materials have a much lower melting point (<2000°C). From this perspective, ceramics and glasses have advantageous properties, including refractoriness (capability to withstand high temperatures), strength retention at high temperature, high melting point, and good mechanical properties (hardness, elastic modulus, and compressive strength). In view of such an attractive combination of properties, ceramics are considered as potential materials for high-temperature structural applications and various tribological applications requiring high hardness and wear resistance. Despite having such

Advanced Structural Ceramics, First Edition. Bikramjit Basu, Kantesh Balani.

potential applications, the widespread use of ceramics has been limited, because of their brittleness (poor fracture toughness) and variability in mechanical properties.

To combine various advantageous properties of the three primary material classes, a derived material class—that is, composites—is being developed. The composites are generally defined as a class of materials that comprise at least two intimately bonded microstructural phases aimed to provide properties (e.g., elastic modulus, hardness, strength) tailored for specific applications; it is expected that a specific property of a composite should be higher than the simple addition of that property of the constituent phases. Depending on whether metals, ceramics, or polymers comprise more than 50% by volume of a composite, it can be further classified as a metal matrix composite (MMC), a ceramic matrix composite (CMC), or a polymer matrix composite (PMC) respectively. From the microstructural point of view, a composite contains a matrix (metal, ceramic, polymer) and a reinforcement phase. The crystalline matrix phase can have an equiaxed or elongated grain structure; the reinforcement phase can have different shapes, for example particulates, whiskers, and fibers. The reinforcement shapes can be distinguished in terms of aspect ratio: particulates can be spheroidal; whiskers have a higher aspect ratio (>10); fibers have the largest aspect ratio. It is widely recognized now that the use of fibers or whiskers can lead to composites with anisotropic properties (different properties in different directions). As far as nomenclature is concerned, it is a common practice to designate a composite as $M\text{-}R_p$, $M\text{-}R_w$, or $M\text{-}R_f$, where M and R are the matrix and reinforcement, respectively, and the subscripts (p, w, f) essentially indicate the presence of reinforcement as particulates, whiskers, or fibers, respectively. One widely researched MMC is $Al\text{–}SiC_p$ composite; $Mg\text{–}SiC_p$ is being developed as a lightweight composite; several MMCs are used as automotive parts and structural components. Some popular examples of CMCs include $Al_2O_3\text{–}ZrO_{2\,p}$ and $Al_2O_3\text{–}SiC_w$; these CMCs are typically used as wear parts and cutting-tool inserts. Various resin-bonded PMCs are used for aerospace applications.

1.2 HISTORICAL PERSPECTIVE; DEFINITION AND CLASSIFICATION OF CERAMICS

As far as the history of ceramics is concerned, the word "ceramics" is derived from the Greek word *keramikos*, literally meaning potter's earth. Historically, the use of burnt clay, commercial pottery, and the existing ceramic industries can be dated back to 14,000 BC, 4000 BC, and 1500 BC, respectively. Early evidence of the use of clay- or pottery-based materials has been found in Harappan, Chinese, Greek, and many other civilizations. A large number of traditional ceramics were produced using conventional ceramic technology. Early forms of color decorative glazes date back to 3500 BC. The potter's wheel, invented around 2000 BC, revolutionized pottery making; porcelain emerged in China circa 600 AD. Glazed tiles were used to decorate the walls of the famous Tower of Babel and the Ishtar Gate in the ancient city of Babylon (562 BC). Figure 1.1 indicates the growth in ceramic technology from prehistoric ages to the 20th century. It is clear that, with technological development,

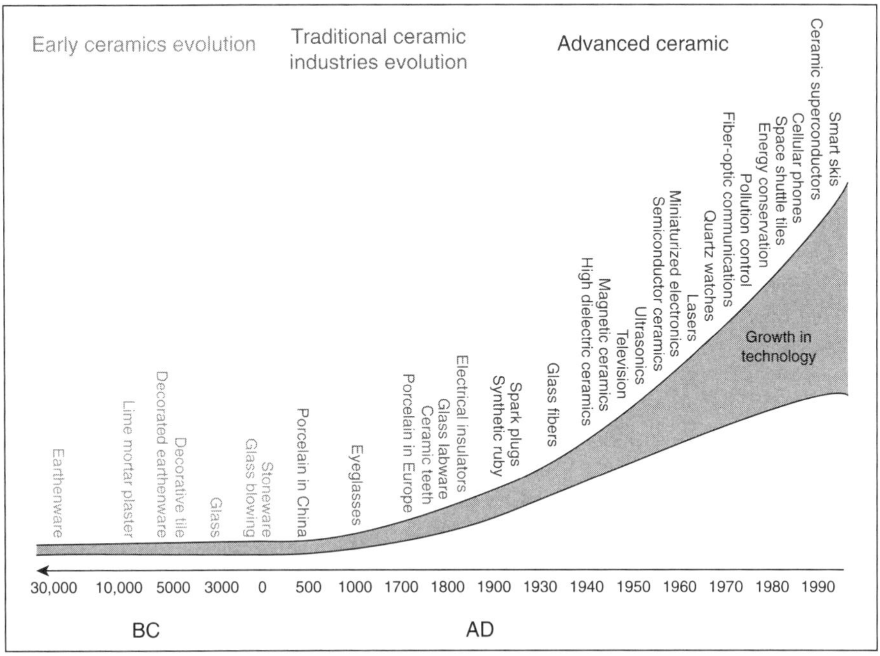

Figure 1.1 Historical evolution illustrating the growth of ceramic applications and industries.[30]

some newer applications in high-tech and important areas, for example the biomedical and electronics industries, are now possible.

A proper and exact definition of ceramics is very difficult. In general, ceramics can be defined as a class of inorganic nonmetallic materials[5] that have ionic and/or covalent bonding and that are either processed or used at high temperatures. Figures 1.1–1.4 illustrate two different aspects: (1) historical evolution of the development of ceramics right from traditional ceramics to the most advanced ceramics to composites and (2) illustration of various current uses of ceramics and their composites. For a layperson, the word "ceramic" means a coffee cup or sanitary ware—traditional ceramic products. Although the main use of ceramics in last few decades was centered on fields such as construction materials, tableware, and sanitary wares, the advancement of ceramic science since the early 1990s has enabled the application of this class of materials to evolve from more traditional fields to cutting-edge technologies, such as aerospace, nuclear, electronics, and biomedical, among others.[6] This is the reason that, in many textbooks, ceramics are classified as traditional ceramics and engineering ceramics. Traditional ceramics are largely silica or clay based and typically involve low-cost fabrication processes. A large cross section of people in the developing world is still familiar with the use of traditional ceramics. On the other hand, engineering ceramics are fabricated from high-purity ceramic powders, and their properties can be manipulated by varying process parameters and, thereby, microstructures. Also, engineering ceramics are, by far, more expensive

(a) (b)

(c)

Figure 1.2 The illustrative examples of the use of engineering ceramics: silicon nitride (Si_3N_4) ceramic cutting tool inserts and components (a), silicon nitride check valve balls ranging from around 20 mm to around 40 mm in diameter (b) and silicon nitride–based experimental automobile valve (c).[30]

Figure 1.3 The use of silicon carbide seals as structural components.[30]

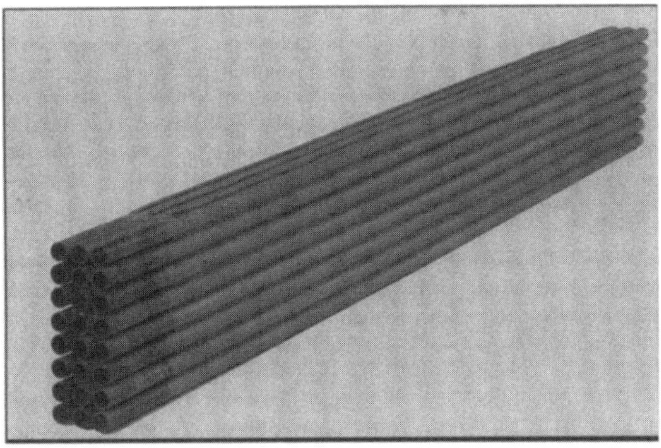

(a)

(b)

Figure 1.4 Another emerging area of oxide ceramics is shown: tubular solid oxide fuel cell module (a) and experimental planar SOFC module (b).[30]

than traditional ceramics. In this textbook, our focus is on discussing the structure, processing, properties, and applications of engineering ceramic systems, particularly on structure–property correlations. Based on their applications, engineering ceramics are usually classified into two major classes: structural ceramics and functional ceramics. While the applications of structural ceramics demands the optimization of mechanical strength, hardness, toughness, and wear resistance,[7] the performance of functional ceramics is controlled by electric, magnetic, dielectric, optical, and other properties.[6] In general, structural ceramics can be further classified into two classes: (1) oxide ceramics (Al_2O_3, ZrO_2, SiO_2, etc.) and (2) non-oxide ceramics (SiC, TiC, B_4C, TiB_2, Si_3N_4, TiN, etc.). Various chapters in this textbook focus only on several structural ceramics. Nevertheless, the crystal structure of some important functional ceramics is discussed in Chapter 2.

1.3 PROPERTIES OF STRUCTURAL CERAMICS

In general, ceramics have many useful properties, such as high hardness, stiffness, and elastic modulus, wear resistance, high strength retention at elevated temperatures, and corrosion resistance associated with chemical inertness.[7] The temporal progression of the development of advanced ceramics is presented in Figure 1.1. It has been reported that a flexural strength of more than 1 GPa can now be achieved in oxide ceramics and that a specific strength (strength-to-density ratio) of more than 2 can be obtained in some composites. Overall, a 50-fold increase in specific strength is now achievable in advanced ceramics, compared with that in primitive traditional ceramics. While various industries have still been mostly using high-speed tool steels, a 10-fold increase in cutting speed can be obtained with the use of ceramic- or cermet-based tool inserts. As far as the maximum operating temperature is concerned, Ni-based superalloys are typically used at 1000°C. In contrast, some nitride and some oxide ceramics can be used at temperatures of close to 1500°C. Although polymers have the lowest density, many of the ceramics (alumina, SiC) have half the density of steel-based materials. Therefore, high-speed turning or cutting operations are possible with ceramic- or cermet-based tool inserts. More often, density becomes a limitation or a requirement in selecting the ceramics for structural, defense, biomedical, and other applications: bone implants require density similar to that of bone; aerospace applications require minimal density with exceptional creep-resistance; and high-energy penetrators aim for high-density counterparts. In terms of elastic modulus or hardness, ceramics are much better than all the refractory metals. As an example of the hardness of commonly known ceramics, that of Al_2O_3 is around 19 GPa, which is close to 3 times the hardness value of fully hardened martensitic steel (~7 GPa). As is discussed in this book, many ceramics, such as TiB_2, can have hardness of around 28 GPa or higher. Also, the elastic modulus of Al_2O_3 is around 390 GPa, which is close to double that of steels (210 GPa). The higher elastic modulus of ceramics provides them with good resistance to contact damage. In addition, many ceramics, such as SiC and Si_3N_4, can exhibit high-temperature strength in the temperature range, where metallic alloys soften and cannot be used for structural applications. Many of these properties are realized in many of the hi-tech applications of ceramics, which include rocket nozzles, engine parts, bioceramics for medical implants, heat-resistant tiles for the space shuttle, nuclear materials, storage and renewable energy devices, and elements for integrated electronics such as microelectromechanical systems (MEMS).

Despite having many attractive properties, as just mentioned, the major limitations of ceramics for structural and some nonstructural applications is their poor fracture toughness. Over the years, it has been realized that an optimum combination of high toughness with high hardness and strength is required for the majority of the current and future applications of structural ceramics, including biomaterials (see Section Seven). To address this need, the development of ceramic composites with optimal combinations of mechanical properties is the major focus in the ceramics community.

1.4 APPLICATIONS OF STRUCTURAL CERAMICS

As mentioned earlier, ceramics are examples of high-temperature materials, which are used specifically for their high-temperature strength, hot erosion, and resistance to corrosion or oxidation at temperatures above 500°C. The need for high-temperature materials has been realized in different sectors of industry, including high-temperature machining, material production and processing, chemical engineering, high-temperature nuclear reactors, aerospace industries, power generation, and transportation, among others.

Typical examples of areas wherein engineering ceramics have found applications are illustrated in Figures 1.2–1.4. Figure 1.2 shows Si_3N_4-based materials as ball bearings, automobile valves, and cutting inserts; Figure 1.3 shows SiC used as bearing seals. In Figure 1.4, a solid oxide fuel cell (SOFC) module is shown; oxide ceramics, such as zirconia, are widely used in SOFCs. There exists a clear demand for materials that can withstand more than 1500°C; such applications include re-entry nozzles in rockets or hypersonic space vehicles. To this end, ultra-high-temperature ceramics (UHTCs) based on borides are being developed (see Section Five). Because of their high melting point, high hardness, electrical and thermal conductivity, and high wear resistance, the borides of transition metals, such as TiB_2, are used for a variety of technological applications.[8] Monolithic TiB_2, that is, without any second phase addition, has excellent hardness (≈25 GPa at room temperature), good thermal conductivity (≈64 W/m·°C), high electrical conductivity (electrical resistivity $\approx 13 \times 10^{-8}\,\Omega$ m) and considerable chemical stability.[9] Some of these attractive properties are ideally suited to be exploited for tribological applications. However, the relatively low fracture toughness (≈5 MPa $m^{1/2}$) and modest bending strength (≈500 MPa) coupled with poor sinterability of monolithic TiB_2 limits its use in many engineering applications.[10] In the materials world, TiB_2 is often used as reinforcement phase not only for ceramics, but also for metallic alloys such as stainless steel[11] and Al-alloys[12] to develop composites with improved abrasive wear resistance. The addition of TiB_2 to an Al_2O_3 or B_4C matrix increases its hardness, strength, and fracture toughness.[13] Furthermore, TiB_2 as well as TiN or TiC, is used not only to toughen Al_2O_3 and Si_3N_4 matrices, but also to obtain electroconductive materials with the incorporation of an optimum amount of an electroconductive phase.[14] These electroconductive toughened ceramics can be shaped by electrodischarge machining (EDM) to manufacture complex components, greatly increasing the number of industrial applications of these ceramic materials. The processing–property relationships of borides are discussed in one of the sections in this book, and the way sinter-aids and sintering conditions can be optimized to develop borides with high sinter density and a better combination of physical and mechanical properties is illustrated.

One application that has attracted much attention is ball bearings (see Fig. 1.2). Ceramic balls enclosed in a steel race, that is, hybrid bearings, are now used in turbopumps of the space shuttle main engine. The friction and wear properties of alumina, zirconia, and SiC in cryogenic environments are being investigated as such studies are relevant to cryotribological applications.[15–17] These ceramic balls are commercially available with diameters from 4 mm to as large as 20–30 mm and they are made from Al_2O_3, ZrO_2, SiC, Si_3N_4, or SiAlON ($Si_{6-z}Al_zO_zN_{8-z}$, with z being the substitution level). Commercial springs made of silicon nitride materials are also available. In one of the sections of this book, the microstructure and mechanical properties of such ceramics are discussed.

There is a tremendous industrial need for new tribological materials. This need is realized in metal-forming industries, bearings, gears, valve guides and tappets in engines, seals and bearings involving fluid and gas transport, often under corrosive conditions, and so on. The majority of these applications are currently served by hardened steels and WC-based hardmetals with or without surface coatings. However, new materials or improved existing materials are needed to meet the increasing demand in the tribological world. Ceramics, because of their ionic and/or covalent bonding, have a useful combination of physicomechanical properties (elastic modulus, hardness, and strength) and corrosion resistance. In many structural and tribological applications, ceramics are recognized as having great potential to replace existing materials for a series of rubbing-pairs, such as seal rings, valve seats, extrusion dies, cutting tools, bearings, and cylinder liners.[18] The materials of interest will have to combine high hardness, toughness, strength, elastic modulus, and wear resistance coupled with relatively low density, resulting in low inertia under reciprocating stresses. Furthermore, the fundamental understanding of the relationship between composition, microstructure, processing route, mechanical properties, wear behavior, and performance should be clarified in order to optimally use the engineered materials in tribological applications. The development of new tribological materials is proceeding in two main directions: the use of coatings on conventional metallic substrates and the use of monolithic ceramics and ceramic composites.

Coatings are frequently hard carbides, nitrides, or borides with recent development of diamond or diamondlike (C–H) films at the more exotic end of the hardness-versus-cost scale.[19] Coating thickness is normally between 1 and 50 μm, depending on the deposition process (physical vapor deposition [PVD], chemical vapor deposition [CVD], or electrolytic), which presents limitations in lifetime or property influence of the relatively soft substrate. Thicker coatings may be applied by thermal spraying (in the millimeter range) but are limited in chemistry, compatibility with substrate properties (thermal expansion etc.), and cohesion. An entire section of this textbook focuses on the discussion of processing and properties of coatings (Section Three).

Monolithic ceramics, especially those with improved strength and toughness, have been a focus of development in different research labs and industries since the 1970s.[20] However, monolithic ceramics are not optimal for all engineering applications. Ceramic composites such as metal matrix and PMCs are now the established approaches to designing structural materials.[21] Ceramic reinforcements are commercially available in different forms such as whiskers, platelets, particulates, and fibers. Two major classes of ceramic composites are fiber-reinforced and particle- or whisker-reinforced ceramic composites. A popular example of the first class of ceramic composites is silicon carbide fiber-reinforced glass-ceramics.[22] The alumina–silicon carbide whisker-reinforced composites are commercially fabricated for use as drilling components. Four major drawbacks normally restrict the widespread use of this material class for structural applications: high cost of ceramic fibers; the expensive composite production route; the chemical compatibility of the fiber with the matrix; and the oxidation of SiC fibers at high temperatures. To this end, particle-reinforced CMCs offer a viable and relatively cost-effective option for developing materials with improved and optimal combinations of mechanical properties (hardness, toughness, and strength).

In the world of ceramic materials, yttria-doped zirconia, in particular yttria-stabilized tetragonal zirconia polycrystalline (Y-TZP) ceramics, are regarded as a strong candidate for structural applications due to the excellent addition of strength (≈700–1200 MPa) and fracture toughness (2–10 MPa $m^{1/2}$) in addition to good chemical inertness.[23,24] The high toughness of the zirconia monoliths stems from the stress-induced transformation of the tetragonal (t) phase to the monoclinic (m) phase in the stress field of propagating cracks, a concept widely known as *transformation toughening*.[25] Basic microstructural requirements for the effective contribution from transformation toughening is the maximum retention of the tetragonal phase at the application temperature with sufficient transformability to m-ZrO_2 in the crack tip stress field. The concepts and microstructural parameters influencing transformation toughening are discussed in Section Four. Since the discovery of the concept of transformation toughening about two decades ago,[26] this approach has been successfully utilized to toughen several intermetallic,[27] glass,[28] and ceramic[29] microstructures. More recently, extensive efforts have been put into increasing the toughness of alumina by adding zirconia, a class of materials known as zirconia-toughened alumina (ZTA).[17,19]

The successful application of engineering ceramic components demands the careful selection and optimization of the initial material (i.e., powder purity, size, shape, etc.) followed by its optimal sintering (time, temperature, pressure, and environment to control grain size and densification) for achieving appropriate properties. These aspects necessitate that researchers consider the selection–processing–property–application tetrahedron, as shown in Figure 1.5.

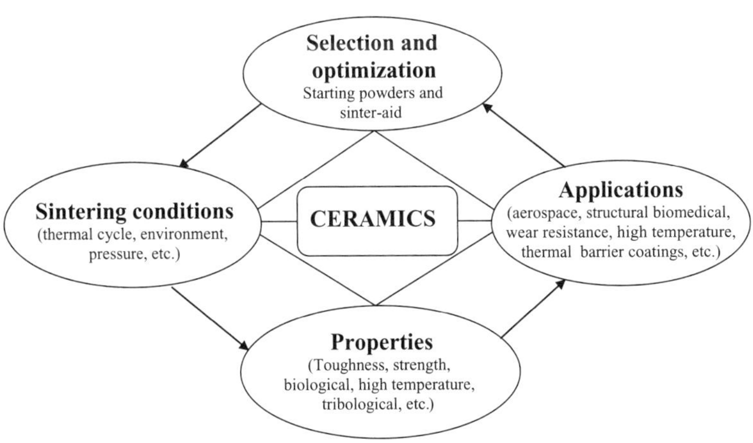

Figure 1.5 Selection–processing–property–application tetrahedron of ceramics.

REFERENCES

1 M. W. Barsoum. *Fundamentals of Ceramics*. Taylor & Francis, Boca Raton, FL, 2003.

2 C. B. Carter and M. G. Norton. *Ceramic Materials*. Springer, New York, 2007.

3 Y. M. Chiang, D. P. Birnie, and W. D. Kingery. *Physical Ceramics*. John Wiley & Sons, New York, 1997.

4 D. W. Richerson. *Modern Ceramic Engineering: Properties, Processing, and Use in Design*. CRC Press, Salt Lake City, UT, 1992.

5 W. D. Kingery, C. R. Bowen, and A. Uhlman. *Introduction to Ceramics*, 2nd ed. John Wiley & Sons, New York, 1976.

6 H. Yanagida, K. Koumoto, and M. Miyayama. *The Chemistry of Ceramics*. John Wiley & Sons, New York, 1995.

7 R. W. Davidge. *Mechanical Behavior of Ceramics*. Cambridge University Press, Cambridge, UK, 1979.

8 R. A. Cutler. *Engineering Properties of Borides, in Engineered Materials Handbook, Vol. 4, Ceramics and Glasses*. ASM International, The Materials Information Society, Materials Park, OH, 1991.

9 J. M. Sánchez, M. G. Barandika, J. G. Sevillano, and F. Castro. Consolidation, microstructure and mechanical properties of newly developed TiB_2-based materials. Scr. Metall. Mater. 26 (1992), 957–962.

10 J.-H. Park, Y.-H. Koh, H.-E. Kim, and C. S. Hwang. Densification and mechanical properties of titanium diboride with silicon nitride as a sintering aid. J. Am. Ceram. Soc. 82(11) (1999), 3037–3042.

11 S. C. Tjong and K. C. Lau. Abrasion resistance of stainless-steel composites reinforced with hard TiB_2 particles. Comp. Sci. Technol. 60 (2000), 1141–1146.

12 C. F. Feng and L. Froyen. Microstructures of the in-situ Al/TiB_2-MMCs prepared by a casting route. J. Mat. Sci. 35 (2000), 837–850.

13 G. Van De Goor, P. Sägesser, and K. Berroth. Electrically conductive ceramic composites. Solid State Ionics 101–103 (1997), 1163–1170.

14 A. Bellosi, G. De Portu, and S. Guicciardi. Preparation and properties of electroconductive Al_2O_3-based composites. J. Eur. Ceram. Soc. 10 (1992), 307–315.

15 T. K. Guha and B. Basu. Microfracture and limited tribochemical wear of silicon carbide during high speed sliding in cryogenic environment. J. Am. Ceram. Soc. 93(6) (2010), 1764–1773.

16 R. Khanna and B. Basu. Sliding wear properties of self-mated yttria-stabilised tetragonal zirconia ceramics in cryogenic environment. J. Am. Ceram. Soc. 90(8) (2007), 2525–2534.

17 R. Khanna and B. Basu. Low Friction and Severe wear of Alumina in cryogenic environment: A first report. J. Mat. Res. 21(4) (2006), 832–843.

18 K. H. Zum Gahr. Sliding wear of ceramic-ceramic, ceramic-steel and steel-steel pairs in lubricated and unlubricated contact. Wear 133 (1989), 1–22.

19 E. Vancoille. A materials oriented approach to the wear testing of titanium nitride based coatings for cutting tools. Ph. D. Thesis, Katholieke Universiteit Leuven, May, 1993.

20 J. D. Cawley and W. E. Lee. Oxide ceramics, in *Structure and Properties of Ceramics, Materials Science and Technology*, Vol. 11, R. W. Cahn, P. Haasen, and E. J. Kramer (Eds.). VCH, Weinheim, Germany, 1994, 101–114.

21 M. Rühle and A. G. Evans. High toughness ceramics and ceramic composites. Prog. Mater. Sc. 33 (1989), 85.

22 A. G. Evans. Perspective on the development of high-toughness ceramics. J. Am. Ceram. Soc. 73(2) (1990), 187–206.

23 R. H. J. Hannink, P. M. Kelly, and B. C. Muddle. Transformation toughening in zirconia-containing ceramics. J. Am. Ceram. Soc. 83(3) (2000), 461–487.

24 (a) P. F. Becher and M. V. Swain. Grain size dependent transformation behavior in polycrystalline tetragonal zirconia ceramics. J. Am. Ceram. Soc. 75 (1992), 493. (b) J. B. Wachtman,

W. R. Cannon, and M. J. Matthewson. *Mechanical Properties of Ceramics*. John Wiley & Sons, New York, 1996, 391–408.

25 D. J. Green, R. H. J. Hannink, and M. V. Swain. Microstructure—Mechanical behavior of partially stabilised zirconia (PSZ) materials, chapter 5, in *Transformation Toughening of Ceramics*, CRC Press, Boca Raton, FL, 1989, 157–197.

26 R. C. Garvie, R. H. J. Hannink, and R. T. Pascoe. Ceramic steel? Nature 258 (1975), 703.

27 D. Ostrovoy, N. Orlovskaya, V. Kovylyaev, and S. Fristov. Mechanical properties of toughened Al_2O_3-ZrO_2-TiN ceramics. J. Eur. Ceram. Soc. 18 (1998), 381.

28 T. Höche, M. Deckwerth, and C. Rüssel. Partial stabilisation of tetragonal zirconia in oxynitride glass-ceramics. J. Am. Ceram. Soc. 81(8) (1998), 2029–2036.

29 B.-T. Lee, K.-H. Lee, and K. Hiraga. Stress-induced phase transformation of ZrO_2 (3 mol % Y_2O_3)-25 vol.% Al_2O_3 composite studied by transmission electron microscopy. Scr. Mater. 38 (1998), (7)1101.

30 D. W. Richerson. *Magic of Ceramics*. Wiley–American Ceramic Society, Westerville, OH, 2000.

Chapter 2

Bonding, Structure, and Physical Properties

This chapter discusses the bonding characteristics of ceramics and mentions how to predict the physical properties, such as melting point and elastic modulus, from first-principles calculations. A large part of this chapter describes the characteristics of a number of important ceramics.

The current use of ceramics extends from pottery to refractories, abrasives, cements, ferroelectrics, glass-ceramics, magnets, and so on. Ceramics are often defined as inorganic oxides, borides, nitrides, silicides, carbides, and so on, possessing high melting point, low ductility, low density, high corrosion resistance, superior wear and abrasion resistance, and so on.

The strong bonding among various molecules overcomes the thermal effects in organizing an ordered arrangement of atoms to form crystals. To understand the properties of ceramics, it becomes essential to understand atomic structure. The development of understanding the atom as a cluster of nucleons (the nucleus) covered with an electron cloud could not explain the observed spectral lines, photoelectric emission, or even the thermal dependence of radiation based on the classical atomic model. The emission of quanta of energy as photons was established by Planck in 1900, which was complemented by Einstein's explanation of the photoelectric effect in 1905. Though the atomic model proposed by Bohr, in which electrons were supposed to orbit around the nucleus in a specified manner, could explain spectral lines, the principal quantum number as evinced in Bohr's atomic model led to the development of orbital (l), magnetic (m), and spin (s) quantum numbers to fully explain atomic structure. The restriction imposed by Pauli's exclusion principle was included to disallow any two electrons from possessing all the same quantum numbers.

Consequently, the duality of light as both particles and waves was postulated by de Broglie in 1924 to define the wavelength $\lambda = h/mv$, where h is Planck's constant (6.623×10^{-34} J s), m is the mass, and v is the velocity of the particle. The Schrödinger wave equation incorporated the restrictions imposed by de Broglie to describe wave motion as

Advanced Structural Ceramics, First Edition. Bikramjit Basu, Kantesh Balani.

$$\frac{h^2}{8\pi^2 m}\left(\frac{\partial^2\psi}{\partial x^2}+\frac{\partial^2\psi}{\partial y^2}+\frac{\partial^2\psi}{\partial z^2}\right)-P\psi=\frac{h}{2\pi i}\frac{\partial\psi}{\partial t}, \tag{2.1}$$

where ψ is the wave function describing the pattern of a wave, P is the potential energy, and the magnitude $|\psi|^2$ gives the probability of finding an electron.

The interaction of electronic charges leads to bonding, which can be (1) primary, where electrons are transferred (or shared) between atoms, and (2) secondary, where local charges create attraction with a nearby atom without actually transferring or sharing electrons between atoms. The bonding of atoms is explained in Section 2.1.

2.1 PRIMARY BONDING

Bonds in ceramics are usually ionic and covalent in nature. Therefore, metallic bonding is not discussed in this section.

2.1.1 Ionic Bonding

Ionic bonding is characterized by the *transfer* of electrons from one atom to another. This type of bonding is highly favored among the ionic species, where one atom has tendency to *donate* electrons and the other has tendency to *accept* electrons in order to attain a stable atomic configuration with a filled outer shell of electrons. Sodium has 11 electrons, with electronic configuration $1s^2\ 2s^2\ 2p^6\ 3s^1$ with 1 electron in its outermost shell. Giving out the outer-shell electron ($3s^1$) will change the configuration of Na^+ to $1s^2\ 2s^2\ 2p^6$, which will make the atom more stable as it will have its outer shell completely filled with electrons. The *donor* species (Na) is *electropositive* in nature, because once it gives out an outer-shell electron the material has lost a negative charge and hence has become an electropositive *ion* (e.g., a neutral Na atom becomes a Na^+ ion, or *cation*). Note that Na has become Na^+ after donating the electron and has acquired a positive charge. Thereby, Na will have tendency to donate its outer-shell electron in order to become more stable as Na^+.

On the contrary, Cl has 17 electrons, with electronic configuration $1s^2\ 2s^2\ 2p^6\ 3s^2\ 3p^5$ with 5 electrons in its outermost shell. It requires one more electron to complete filling its outer shell. Thus, upon accepting one electron from an electropositive material, Cl can become stable as Cl^- with configuration $1s^2\ 2s^2\ 2p^6\ 3s^2\ 3p^6$, filling its electronic orbital. Cl is an *acceptor* species (or *electronegative*) because it accepts an electron and acquires an additional negative charge, thereby becoming a Cl^- *ion* (or *anion*) from the initial Cl atom. Hence, Cl will have tendency to accept an electron in order to become more stable.

Ionizing the Na atom will require some energy (say, amount x), and adding an electron to Cl^- will release some energy (say, amount y). Additionally, the interaction between positive and negative ions will induce Coulombic attraction. Since Coulombic force increases as ions approach each other, the repulsive force because of the overlapping of electronic orbitals occurs so that only one electron stays per quantum state.

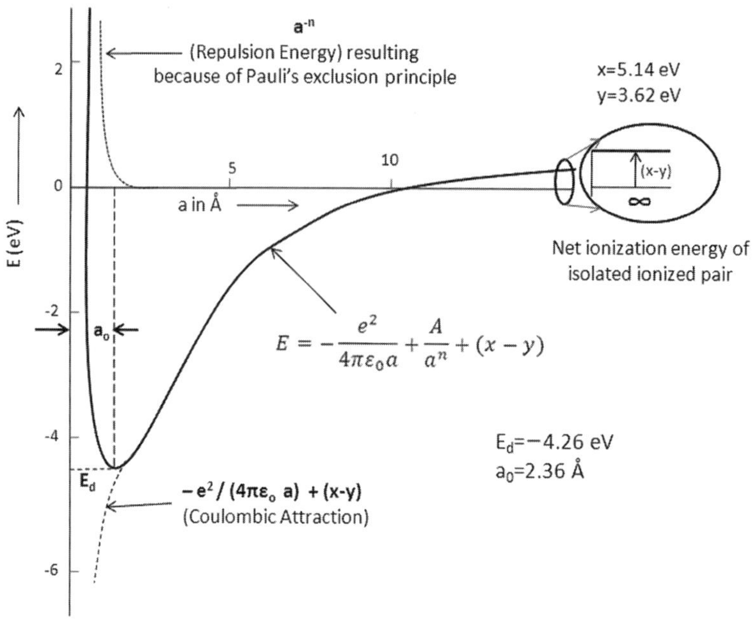

Figure 2.1 Plot of interatomic potential energy (E) vs. separation distance (a) for the attraction and repulsion among ions in NaCl crystal. Adapted from Reference 1.

Since the repulsive force is proportional to the distance between the ions as a^{-n}, the overall energy required to create the Na–Cl bond is given (refer to Fig. 2.1) as

$$E = \text{attractive energy} + \text{repulsive energy} + \text{net ionization energy},$$

$$E = -\frac{e^2}{4\pi\varepsilon_0 a} + \frac{A}{a^n} + (x - y), \tag{2.2}$$

where e is the electron charge, ε_0 is the permittivity of free space, A is an empirical constant, and the exponent n is on the order of 10.[1] The combined effect of the repulsion and Coulombic attraction results a net energy of E_d (see Fig. 2.1) for the NaCl pair.

It must also be noted that this exchange will occur when both electropositive (Na) and electronegative (Cl) atoms are present in each other's vicinity. Additionally, ionic bonds are *nondirectional* since an electropositive ion will attract any electronegative ion equally in all directions. Coulombic attraction between the atomic species leads to formation of an ionic bond (to form NaCl). The attractive force between atoms increases as the distance between them decreases (Fig. 2.1), but the repulsive forces arising among the negatively charged field of electrons and between the positive nuclei (Fig. 2.1) counteracts the bond length from shrinking to zero.

The equilibrium bond length (a_0) is given when the attractive and repulsive forces (F) balance each other to result a stable molecule (see Fig. 2.2). Correspondingly, the energy is minimum for the bond (i.e., when $F = 0$) and is related to force through

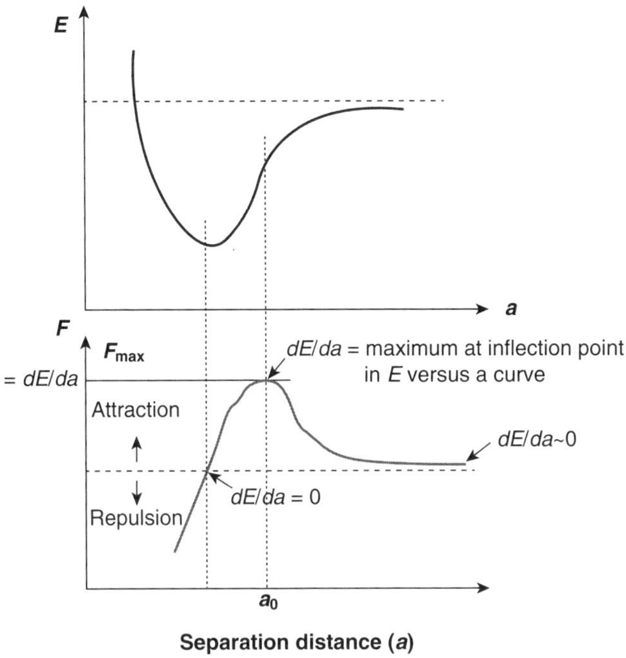

Figure 2.2 Interatomic potential energy (top), interatomic force (bottom) vs. separation, which can be used in predicting the material properties (such as melting point, coefficient of thermal expansion, and Young's modulus).

$$E = \int F \cdot da, \tag{2.3}$$

where a is the separation distance between ions. This equation is very important in materials science because for compression working (decreasing bond length) or material damage resistance (increasing bond length to cause debonding) energy must be provided. Automatically, the ionization of atoms changes the radii of the ionic species; that is, loss of electrons (in electropositive species) decreases the ionic radii, whereas gaining electrons (in electronegative species) increases the ionic radii.

Herein, the concept of coordination number arises depending on the degree of electronegativity or valence of the atomic species. Coordination number (CN) is the number of adjacent atoms or ions surrounding the specific atom or ion. CN is characterized by the radius ratio ($=r/R$, the ratio of the radius of the smaller atom or ion to the radius of the bigger atom or ion); this is essential because, depending on the size, a number exceeding CN would require overlapping of bigger atoms and would make the structure unstable. When the radius ratio r/R is in a given range, the corresponding value of CN is as follows: (1) between 0 and <0.155, CN = 2; (2) between 0.155 and <0.225, CN = 3; (3) between 0.225 and <0.414, CN = 4; (4) between 0.414 and <0.732, CN = 6; (5) between 0.732 and <1.0, CN = 8; and (6) if $r/R = 1$, then CN = 12. Correspondingly, when CN = 2, the smaller species lies between the bigger species in a line (linear geometry), which changes to smaller species sitting as follows: (1) in a triangle when CN = 3; (2) in a tetrahedron (a polygon with four triangular faces) when CN = 4; (3) in an octagon (a polygon with eight triangular faces, i.e., four atoms in one square plane and the other two located above and below the square plane, whose center is occupied by the central smaller species) when CN = 6; (4) in the body center of a cube-type lattice when CN = 8; and (5) in the center of a hexagonal/face-centered closed-type lattice when CN = 12.

It must also be stated that the nature of the curve of the plot of force versus interatomic separation decides some of the material properties. For example, the ceramics are typically characterized by a deep interatomic potential energy well and therefore, they possess high melting point, that is, the reflection of high bonding energy. In addition, the interatomic potential energy well of many ceramics has a rather sharp curvature at the equilibrium separation (a_0) and this results in ceramics having higher Young's modulus (Y), which is given as

$$Y = c\frac{dF}{da} = c\frac{d^2E}{da^2},$$

where c is a constant. Here, it can be mentioned that typically many metals are characterized by interatomic potential energy well having a shallow and a rather flat curvature at the equilibrium separation distance. This explains the relatively lower melting point and lower Young's modulus of metals, in comparison with those of ceramics.

2.1.2 Covalent Bonding

The bond between two atoms created by *sharing* of electrons is called a covalent bond. The electronic density gets concentrated between the two nuclei, giving this bond a *directionality*. The species involved are not more highly electropositive or electronegative relative to one another, and both require electrons from the other atom to make themselves stable. Hence, the self-similarity between the atoms demands sharing of electron(s) among themselves.

Considering the example of the hydrogen molecule (H_2), electrons tend to concentrate between the protons to lower their energy, increasing the probability of finding an electron between the protons. Additionally, the electrons tend to have minimum energy when they are located between the protons, and they tend to pull the two protons nearer. However, the repulsive force between the protons balances the attractive force contributed by electron–proton attraction similar to that of ionic nuclei separation in an ionic bond (Fig. 2.3).

Covalent bonding is highly prevalent in organic compounds since the four valence electrons in carbon orient themselves in a tetrahedron (diamond structure) via sp^3 hybridization. The double bonds of carbon (such as in C_2H_4) can break to create long chains of polyethylene (C_2H_4-mer unit). The energy of bonding increases when the bonding shifts from a single bond to a double or triple bond (e.g., C–C bond energies are 370, 680, and 890 kJ/mole corresponding to single, double, and triple bonds, respectively).[2] The directionality of a covalent bond, therefore, induces

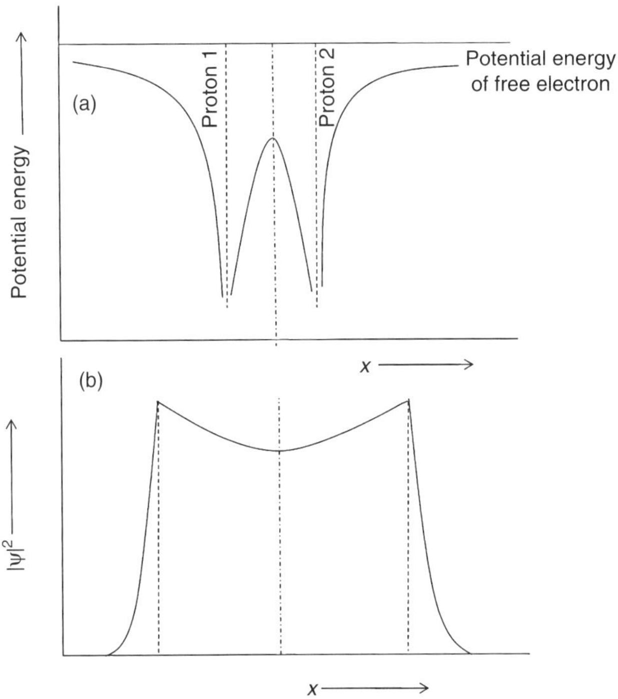

Figure 2.3 Schematic showing (a) potential energy and (b) probability of finding electron far and near proton in H_2 molecule. Adapted from Reference 1.

a characteristic "bond angle," which corresponds to a value of 109.5° for the tetrahedral configuration.

It must also be pointed out that bonding nature can be partially ionic and partially covalent [such as the $(SiO_4)^{4-}$ structure]. It must also be noticed that covalent bonding does not follow the ionic radii ratio relationship to decide the coordination number. Even when the r/R ratio is 1, the CN may just be 4 (tetrahedron structure) rather than 12.

2.1.3 Pauling's Rules

Pauling's rule predicts the probable stable crystal structure. This model assumes the hard sphere model, where ionic radius is constant for a particular valence state and a nearest neighbor coordination number (CN). Ionic radius increases as the valence decreases and the number of nearest neighbors increases. It has been stated that the cation-to-anion ratio is a determining factor as far as the likely CN is concerned. For many stable structures, minimum electrostatic energy is achieved when cation–anion attractions are maximized and like-ion electrostatic repulsion is minimized. Pauling's rules are based on the geometric stability of packing for ions of different sizes, combined with electrostatic stability arguments.

Rule 1:

Each cation will be coordinated by a polyhedron of anions, where the number of ions is determined by the relative sizes of the cation (r_c) and anion (r_a).

When anions form a regular polyhedron, a single characteristic size of cation will fill the interstices. Cations smaller than this particle size will make the whole structure unstable; that is, a particular r_c/r_a ratio will decide the largest polyhedron for which the cation can completely fill the interstice. When r_c/r_a is less than some critical value, the next lower coordination is preferred. This is true for the structure where the cation is smaller than the anions (e.g., NaCl).

For the cation polyhedron as a structured unit, use of the anion-to-cation ratio needs to be determined to predict the coordination number of cations around anions.

An exception to Pauling's rule is that ions are not really hard spheres but are somewhat deformable.

Rule 2:

The structure ensures that the basic CN polyhedrons are arranged in such a way that local charge neutrality is preserved. In other words, for A–C–A, the bond strength = (valency of ion)/CN. For example, in an octahedrally coordinated MgO, $Mg^{2+} \rightarrow$ bond strength = 2/6 = 1/3. The CN of cations around anions as well as anions around cations is important.

Rule 3:

CN polyhedra prefer linkages where they share corners rather than edges and edges rather than faces. The rationalization comes as cations always prefer to maximize their distance from other cations in order to minimize electrostatic repulsion.

Rule 4:

Rule 3 becomes important when CN is small and cation valence is high, as in $SiO_2{-}SiO_4^{4-}$, where sharing at corner is preferred.

Rule 5:

Simple structures are usually preferred over more complicated arrangements, especially when several cations of similar size and identical valence are incorporated. They occupy the same type of lattice size but get distributed at random, forming a solid solution. When cations become increasingly dissimilar, a tendency to form an ordered arrangement or superlattice may occur. Therefore, the number of different constituents in a stable structure tends to be at minimum.

Ionic crystal radii (in nanometers) for CN = 6 are as follows:

$$O^{2-} \rightarrow r_- = 0.140;$$
$$Fe^{2+} \rightarrow r_+ = 0.098;$$
$$Mg^{2+} \rightarrow r_+ = 0.072;$$

$$Ni^{2+} \rightarrow r_+ = 0.069;$$
$$Ca^{2+} \rightarrow r_+ = 0.100;$$
$$Mn^{2+} \rightarrow r_+ = 0.083.$$

2.1.4 Secondary Bonding

The primary source of strength in ceramics arises because of their ionic or covalent bonding. However, secondary bonding between atoms can also be obtained without electron transfer, such as attraction between oppositely charged species (van der Waals bonding). This type of bonding arises due to generation of dipoles resulting from the anisotropy of the crystal, whether induced or permanent. *Induced dipoles* (separation of charges) can form via changing the electric or magnetic field around the otherwise nonpolar atom. The shifting of electron density, or excitation of electron orbitals, can cause the induced dipole. On the other hand, *permanent dipoles* (such as a polar molecule of H_2O) lead to the collection of positive charges (of H atom) near the negative charges (of O atom). This *hydrogen-bridge* causes separation of charges by increasing the dipole moment, which consequently enhances the bond energy by interaction among the oppositely charged species.

The absence of secondary bonding, and the presence of primary bonding, retains the stability of a structure at much higher temperatures (and therefore results in the very high melting point of ceramics).

2.2 STRUCTURE

In summary, the structure of ionically dominated bonding (with partially covalent nature) is described by Pauling's rules:

1. Radius ratio of cation and anion decides the coordination polyhedron. Correspondingly, the cation must remain in touch with the anion for a stable structure.
2. Electrical neutrality is provided by the balance of anionic and cationic charges. Charge strength contributed by a cation equals its charge divided by its CN (e.g., for Al^{3+} with CN = 6, bond strength = 3/6 = 0.5). Hence each Al must be balanced by six charges of strength 0.5.
3. Linkage at a corner is preferred over the edge and surface. Sharing the edge and faces increases the distance between cations
4. Low coordination number and high-charge cation is favorably shared at corners.
5. Different-sized constituents do not efficiently pack as a single structure.

In ceramics, the lattice structure is dictated by the larger anions, with cations sitting at the interstices. Various ceramic structures are explained based on the close packing of ceramics.

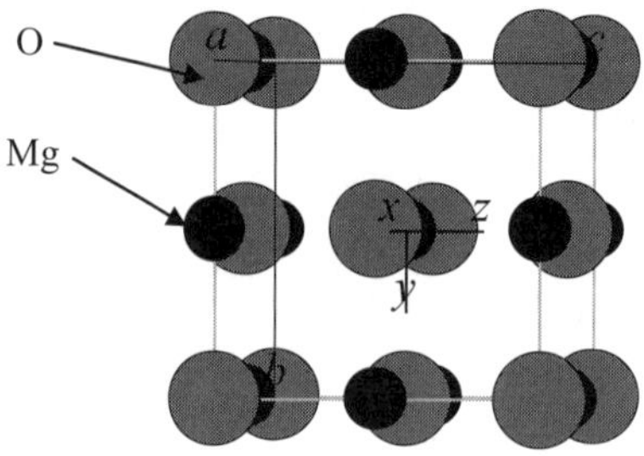

Figure 2.4 Rock salt (NaCl-type structure of MgO).

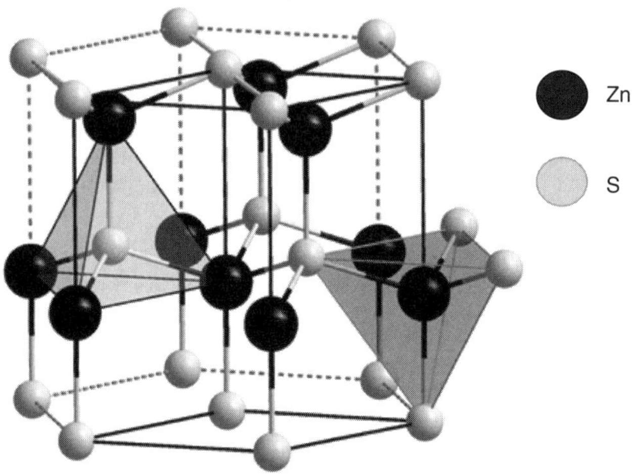

Figure 2.5 ZnS wurtzite structure.

2.2.1 NaCl-type Rock-Salt Structure

In the rock-salt-type structure, large anions occupy the simple cubic close packing and all octahedral interstitial positions are occupied by the small cations. Several oxides, such as MgO (Fig. 2.4), CaO, BaO, FeO, NiO, KF, NaCl, LiF, MgS, and CaS, follow this structure with a coordination number CN=6. Thereby the r/R ratio also falls within the range 0.414–0.732.

2.2.2 ZnS-Type Wurtzite Structure

Tetrahedral coordination of four anions distributes bond strength to 1/4 and requires coordination number of four, and the double charge (+2) renders the bond strength 1/2. Hence this requires hexagonal packing (Fig. 2.5) of large anions with half of the tetrahedral vacancies filled with cations. Examples include SiC, ZnO, and ZnS.

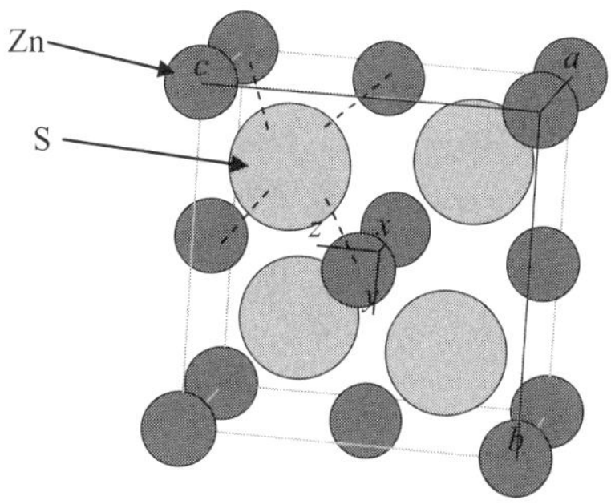

Figure 2.6 ZnS (zinc blende) structure showing Zn and S atoms.

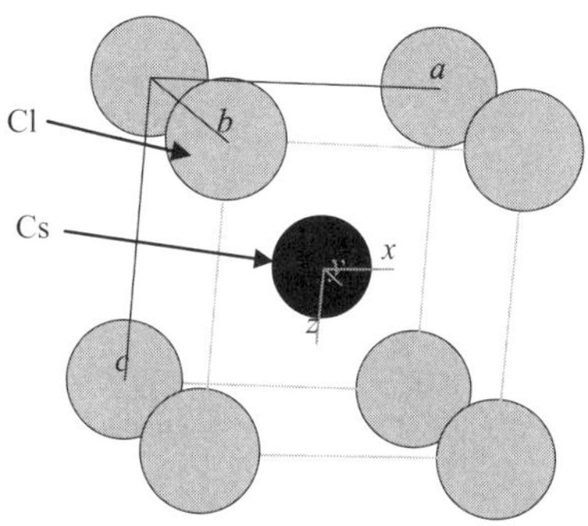

Figure 2.7 CsCl structure showing Cs^+ at interstitial location in Cl^- simple cubic lattice.

2.2.3 ZnS-Type Zinc Blende Structure

The zinc blende structure (Fig. 2.6) is based on the cubic close packing (ccp) of anions with tetrahedral coordination; BeO and SiC also display a similar structure.

2.2.4 CsCl Cesium Chloride Structure

The coordination number eight requires a simple cubic lattice with bond strength of 1/8 for Cs^+ sharing with each Cl^- ion (Fig. 2.7). Hence, the anions occupy the corners of a simple cubic lattice with cations filling all the octahedral interstices.

2.2.5 CaF_2 Fluorite Structure

The large cationic size requires a coordination number of 8, and the high valence of 4+ (of Th, Zr, Te, U, etc.) results in a bond strength of 1/2. Hence the simple cubic lattice filled with anions allows filling of cations in only half of the available octahedral sites (similar to the CsCl structure), but here only half of the octahedral sites are filled. This results in fcc packing of cations, as seen in Figure 2.8. This packing also induces a void in the center of the unit cell.

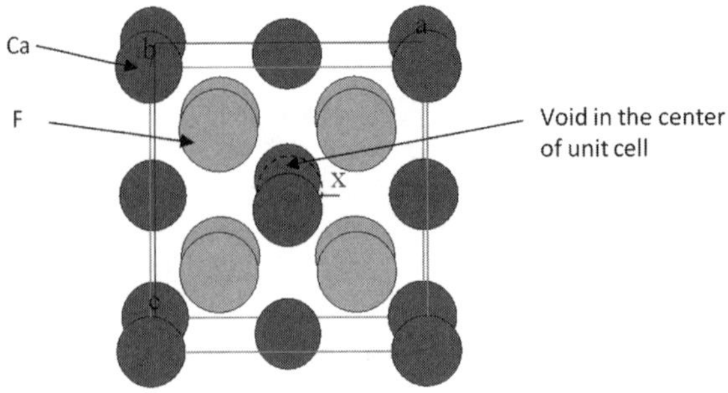

Figure 2.8 CaF_2 structure showing fcc lattice of Ca^+ with only half of the sites filled in the F^- simple cubic lattice. The generated void in the unit cell is also depicted.

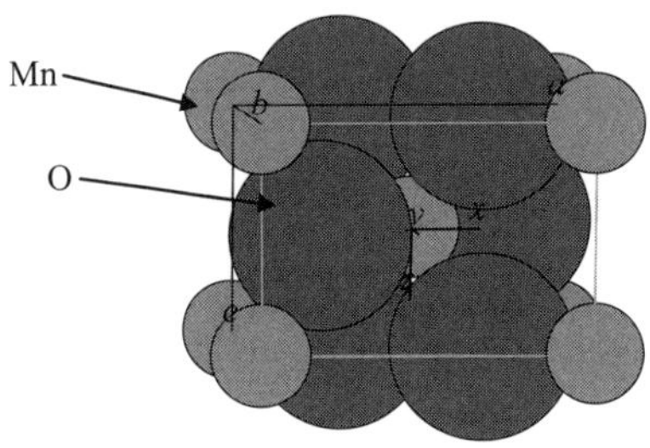

Figure 2.9 MnO_2 rutile structure showing Mn at body center octahedral site of oxygen anion lattice.

2.2.6 Antifluorite Structure

In the antifluorite structure, the fluorite structure is reversed; that is, anions are arranged in the cubic close packed array, and cations are in the tetrahedral sites. Examples include Li_2O, Na_2O, and K_2O.

2.2.7 Rutile Structure

In TiO_2, the coordination number of Ti is 6 with a valence of 4+, leading to bond strength of 2/3, which requires an oxygen ion coordination number of 3. Hence, cations occupy half of the octahedral sites as shown in Figure 2.9 for the MnO_2 structure. However, the close proximity of cations leads to distortion and results in a nearly close packed lattice. Examples include SnO_2, MnO_2, and PbO_2.

2.2.8 Al_2O_3 Corundum Structure

In the corundum structure, the Al has a valence of 3+ with coordination number 6, resulting in a bond strength of 1/2. This results in the requirement of four Al^{3+} staying around each O^{2-} by hexagonal close packing of O^{2-}, with Al^{3+} occupying two-thirds of the octahedral sites (Fig. 2.10).

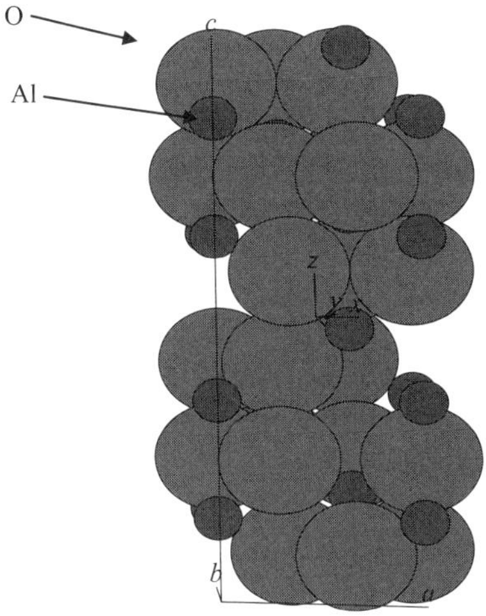

Figure 2.10 Crystal structure of Al_2O_3 ($a = b = 4.7564$ Å, $c = 12.9894$ Å, and $\alpha = \beta = 90°$, $\gamma = 1200°$).

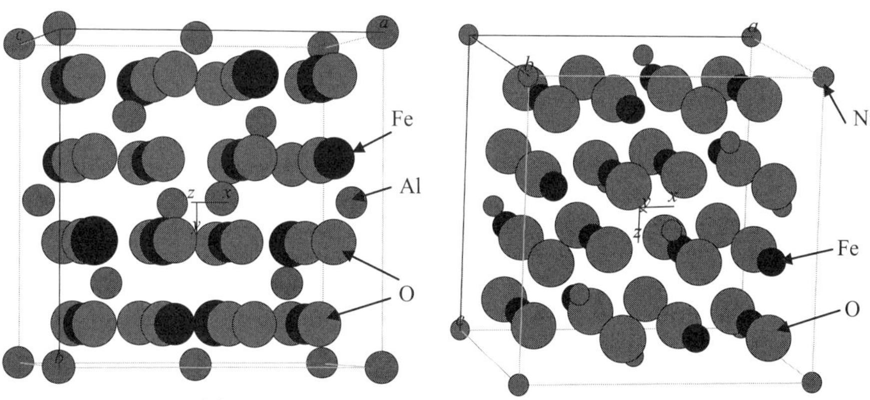

(a) (b)

Figure 2.11 (a) Normal spinel $FeAl_2O_4$ (oxygen in fcc lattice, with Fe in tetrahedral sites and Al in octahedral sites). (b) Inverse spinel $FeNiFeO_4$ showing oxygen in fcc lattice with Ni in octahedral sites, and half of Fe occupying the octahedral and the other half occupying tetrahedral interstices.

2.2.9 Spinel Structure

The spinel structure is defined by $AO{\cdot}B_2O_3$ (or AB_2O_4, where A and B are metals) such as $FeO{\cdot}Al_2O_3$, Fe_3O_4. Oxygen takes up the fcc lattice positions, which results overall in four oxygen atoms in one unit cell, resulting in four octahedral and eight tetrahedral interstices. There are two trivalent cations and one divalent cation. The manner in which the cations occupy the site results in the following variants:

1. *Normal spinel*, where the divalent cation occupies the tetrahedral site (out of the available eight, so only one-eighth site is filled) and two trivalent anions take up half of the octahedral sites (out of the available four), Figure 2.11a. This structure is followed by $ZnFe_2O_4$, $FeAl_2O_4$, and $ZnAl_2O_4$, and so on.

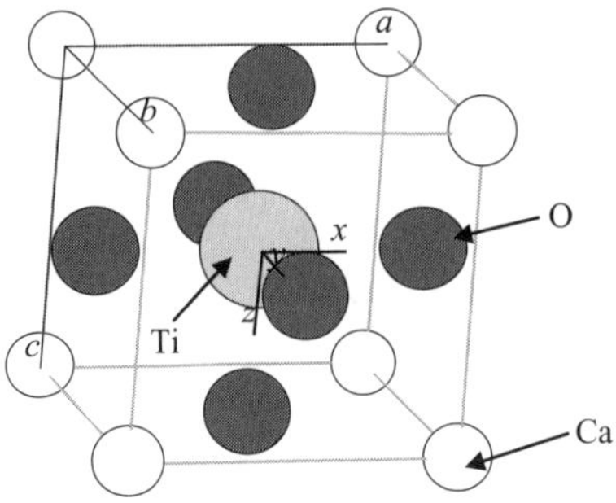

Figure 2.12 $CaTiO_3$ structure (Ca^{2+} at cube corners, Ti^{4+} at body center and O^{2-} at face center). The Ti^{4+} is exaggerated to show its presence in the center of the fcc lattice.

2. *Inverse spinel*, where one divalent and one trivalent (of the available two) anions occupy the octahedral sites, and the remaining one trivalent cation occupies the tetrahedral site, Figure 2.11b. Inverse spinel is more common and important because of the magnetic properties of Fe_3O_4, $FeTiFeO_4$, $FeNiFeO_4$, and so on.

2.2.10 Perovskite Structure

Large cations of size similar to that of anions result in a combined closed packed ABO_3 structure called the perovskite structure: one highly charged smaller cation occupies the octahedral site, and the bigger lower charged cation shares the lattice with oxygen. For example, in $CaTiO_3$, Ti^{4+} occupies the body center position coordinated with 6 O^{2-}, and Ca^{2+} is coordinated with 12 O^{2-} each of which is coordinated with 4 Ca^{2+} and 8 O^{2-} (Fig. 2.12). $YAlO_3$, $SrZrO_3$, $SrSnO_3$, and so on also share the same structure.

2.2.11 Ilmenite Structure

The ABO_3-type ilmenite structure involves alternate layering of A and B types of atoms, where half of the cation sites are occupied by A and the other half by B. This ordering with each layer of A and B makes it different from the Al_2O_3 and perovskite structures. Examples include $FeTiO_3$, $MgTiO_3$, and $NiTiO_3$.

2.2.12 Silicate Structures

The r/R ratio of 0.29 for Si/O indicates tetrahedral coordination (Fig. 2.13a). Therefore, silicates $(SiO_4)^{4-}$ involve bond strength of 1 (one) leading to four O^{2-} coordinated with each Si^{4+}, but only two Si^{4+} are coordinated with each O^{2-}. A typical crystalline cristobalite silicate structure is shown in Figure 2.13b; an amorphous silicate structure is presented in Figure 2.13d.

Silica (SiO_2) exists in three polymorphs, namely, quartz (<846 K), tridymite (846–1140 K), and cristobalite (1140–1983 K) with corresponding densities 2.65, 2.26, and 2.32, respectively. Silica structure is the connected chains of silica tetra-

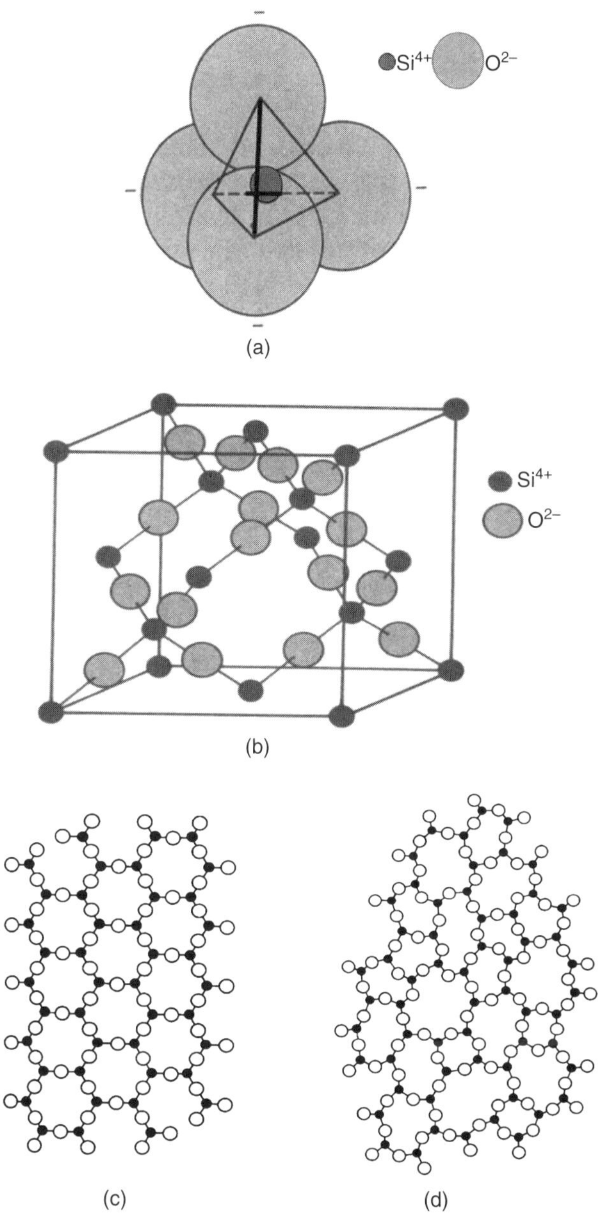

Figure 2.13 Silicate structures showing (a) tetrahedral coordination, (b) cristobalite structure, and (c) crystalline and (d) amorphous silica.

hedra. *Orthosilicates* include olivine minerals (A_2SiO_4), garnets, and aluminosilicates (such as kyanite and mullite). Oxygen ions are hexagonally close packed with Mg^{2+} in the octahedral and Si^{4+} in the tetrahedral sites, resulting in coordination of oxygen with one Si^{4+} and three Mg^{2+} or with two Si^{4+}, which can be viewed as $(SiO_4)^{4-}$ tetrahedra with Mg^{2+} in the octahedral site. *Pyrosilicates* involve $(Si_2O_7)^{6-}$

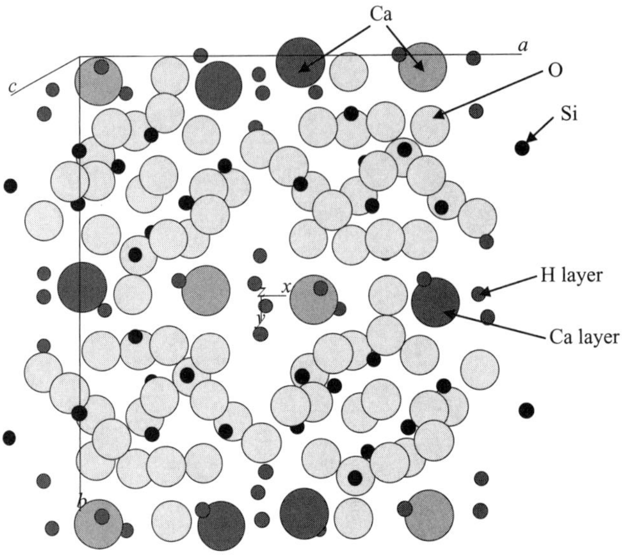

Figure 2.14 Zeolite structure showing open spaces in the unit cell.

geometry with $(SiO_4)^{4-}$ structure sharing a common oxygen at corner. *Metasilicates* can be cyclic or chain arrangements comprising $(SiO_3)_n^{2n-}$. The metasilicate chain-arrangement is built upon Si–O bonds, for example, pyroxenes (with single chains), such as $MgSiO_3$ and $MgCa(SiO_3)_2$, or amphiboles (with double chains), such as $(OH)_2Ca_2Mg_5(Si_4O_{11})_2$ and asbestos. Cyclic silicate structure includes $CaSiO_3$, $BeAl_2Si_6O_{18}$, and so on. Additionally, the three-dimensional framework of silicate family includes *feldspar*, which involves Al^{3+} replacing Si^{4+}, and balancing additional charge with a heavier atom in the interstices (e.g., orthoclase $KAlSi_3O_8$). It is to be noted that large positive ions go into interstitials, whereas smaller cations get into octahedral coordination resulting in chains or layered silicate structures. On the other hand, zeolites, another family of silicates forming a three-dimensional structure are much more open (Fig. 2.14), making them excellent molecular sieves. Moreover the alkali and alkaline earth metals present in zeolites can be exchanged in an aqueous solution removing chlorides to soften hard water.

2.3 OXIDE CERAMICS

Aluminum oxide (Al_2O_3, also known as alumina) is a white and odorless oxide ceramic with a hexagonal crystal structure as shown in Figure 2.10. In nature, Al_2O_3 is the hardest mineral after diamond, with a hardness of 18–20 GPa. Due to its high hardness and refractory nature, Al_2O_3 is widely used in thermal liners, thermal barrier installations, high-temperature insulating systems, crucibles, ceramic boards and brackets, heaters, and so on. Aluminum oxide has several polymorphs, among which α-Al_2O_3 (with space group $R\bar{3}c$) is the most thermodynamically stable form. Other

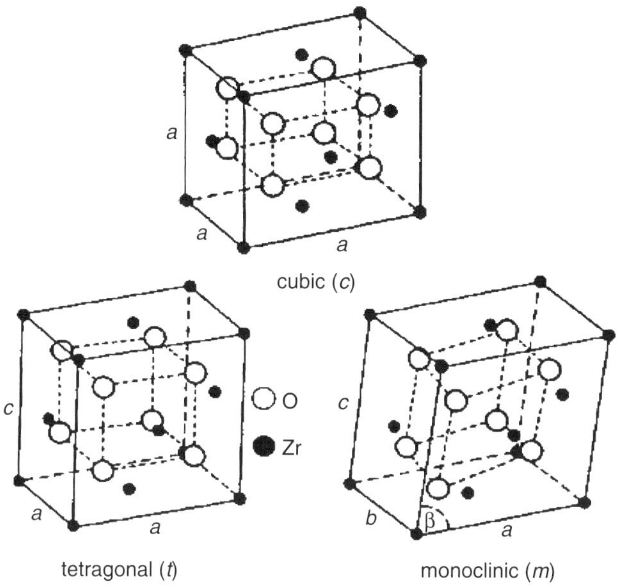

Figure 2.15 Polymorphs of zirconia, that is, cubic, tetragonal, and monoclinic phase.[8]

metastable phases can be classified as follows: (1) fcc packing cubic (γ, η), monoclinic (θ), and tetragonal or orthorhombic (δ); (2) hexagonal closed packing (hcp) rhombohedral (α), orthorhombic (κ), and hexagonal (χ). Other monoclinic phases are identified as θ′, θ″, and λ.[3]

Zirconium oxide (ZrO_2, also known as zirconia) has monoclinic crystal structure at room temperature, which transforms to tetragonal and cubic (Fig. 2.15) at higher temperatures (~1170°C and 2370°C, respectively).[4–6] Correspondingly, the volume expansion (3–5%) caused by the tetragonal-to-monoclinic transformation induces very large stresses and will cause pure ZrO_2 to crack upon cooling from high temperatures.[7] However, retaining the high-temperature tetragonal phase will allow absorption of energy that can restrict crack growth. During crack propagation, the stress-field present ahead of crack tip can induce tetragonal to monoclinic phase transformation of zirconia. Since this shear stress induced phase transformation occurs with a 4%–5% volume expansion in zirconia, it can suppress crack propagation or can even result in crack closure (see Chapter 10 for details).

Titanium oxide (TiO_2) occurs most commonly as rutile, with other structures being anatase, brookite, and other high-temperature phases. The demarcating nature of TiO_2 is its photocatalytic property, which allows enhancement of radical generation in the presence of light. Additionally, TiO_2 is widely used in marking the lanes on roads, in superhydrophobic and self-cleaning glasses, in energy generation, and so on. Additionally, the inert nature of TiO_2 makes it an excellent biocompatible material that is highly efficient for osseointegration between its layer (as a coating on a titanium implant) and bone.

Table 2.1. Physical Properties of Various Oxide Ceramics

Crystal	Crystal structure	Lattice parameters	Density (g/cm^3)	Mol. Vol. (cm^3/mol)	Mol. Wt. (g/mol)	Melting point (K)
Al_2O_3	Hexagonal	$a = b = 4.7564$ Å, $c = 12.9894$ Å	3.99	25.554	101.96	2327
ZrO_2	Monoclinic	$a = 5.142$, $b = 5.205$, and $c = 5.313$ Å, and $\beta = 99.31$	5.68	21.69	123.22	2988
TiO_2	Tetragonal	$a = b \doteq 4.5937$ Å, $c = 2.9581$ Å	4.24	18.84	79.87	

Table 2.2. Physical Properties of Various Non-Oxide Ceramics

Crystal	Crystal structure	Lattice parameters	Density (g/cm^3)	Mol. Vol. (cm^3/mol)	Mol. Wt. (g/mol)	Melting point (K)
HfB_2	Hexagonal	$a = b = 3.03$ Å, $c = 3.232$ Å,	11.1	18.03	200.11	3523
TiB_2	Hexagonal	$a = b = 3.02$ Å, $c = 3.220$ Å,	4.52	15.37	69.49	3503
Si_3N_4	Hexagonal	$a = b = 7.753$ Å, $c = 5.618$ Å	3.44	40.78	140.28	2173
B_4C	Hexagonal	$a = b = 5.633$ Å, $c = 12.164$ Å	2.52	21.93	55.26	2718
$MoSi_2$	Tetragonal	$a = b = 3.206$ Å, $c = 7.848$ Å	6.25	24.34	152.11	2303

Among various oxides a few typical highly used oxides and their physical properties are provided in Table 2.1.

2.4 NON-OXIDE CERAMICS

Non-oxide ceramics include borides, nitrides, silicides, and carbides. The advantage of oxides is that they are stable under oxidizing conditions and hence do not degrade. The advantage of carbides is their lower densities and higher melting points compared with those of oxides. Carbides form ceramics with the highest known melting points (such as HfC and TaC). However, solid-state phase transformations often are predominant and lead to spalling of carbide ceramics. Borides are also lighter (low density), but a surface B_2O_3 layer is formed that sublimes at 1500°C and limits the use of boride structures to ~1200°C. Nitrides and silicides are often highly brittle and also are heavier counterparts of their oxides. A comparison of the physical properties of a few non-oxide ceramics is provided in Table 2.2.

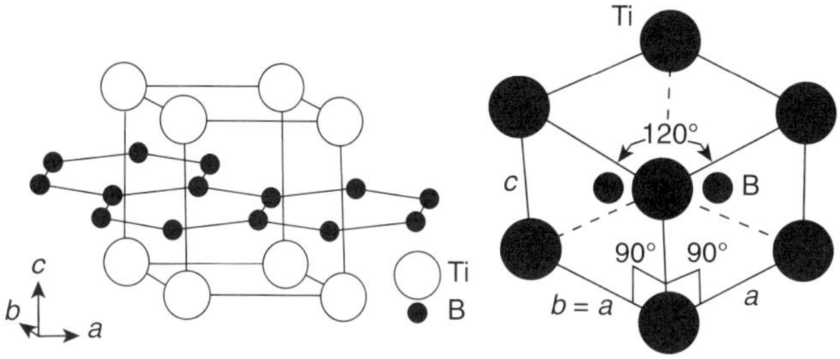

Figure 2.16 Hexagonal crystal structure of TiB_2 ceramic.

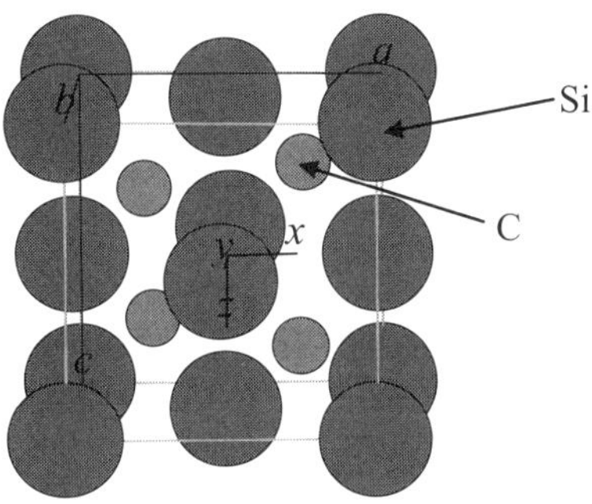

Figure 2.17 Crystal structure of SiC showing C sitting in the interstitial sites.

The excellent refractoriness of HfB_2 extends its use in wear-resistant coatings, but its application is limited to temperatures below 1500°C owing to the sublimation of the B_2O_3 layer. Hence, HfB_2 is often used as a composite reinforcement such as in SiC. Titanium diboride (TiB_2) has a hexagonal crystal structure (Fig. 2.16); it possesses excellent high-temperature wear resistance and comes under the category of ultra-high-temperature ceramics (UHTCs, with operating temperature in excess of 1600°C). TiB_2 is primarily used in ballistic armor plates, wear-resistant coatings, cutting tools, and so on.

B_4C possesses exceptional hardness, low density, high wear and tribological resistance, high thermal conductivity, and neutron-absorption capability for use as armor plating, ballistic nozzles, wear-resistance components, grinding and cutting tools, and so on.[9,10] Carbide structure involves carbon fitted into interstitials, with transition metals possessing closed pack structure. The intermediate bonding nature (between ionic and covalent) often results among transition metals and carbon, but bonding tends to be completely covalent when the other element has the same electronegativity (such as bonding in SiC), Figure 2.17.

Nitride structures are similar to those of carbides, but show less metallic bonding than carbides. Si_3N_4 has a hexagonal crystal structure (Fig. 2.18) and is a high-temperature ceramic with excellent strength, low density, and good thermal shock resistance.[11] Si_3N_4 shows excellent tribological resistance, and its superior hardness and elastic modulus extend its use to applications such as hybrid bearings, gas

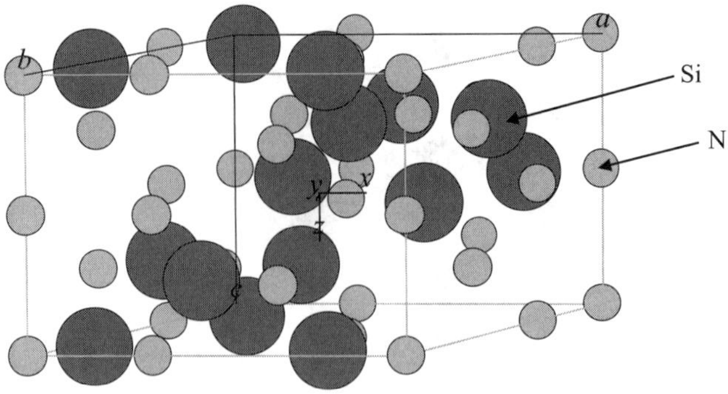

Figure 2.18 Crystal structure of Si_3N_4 intermetallic.

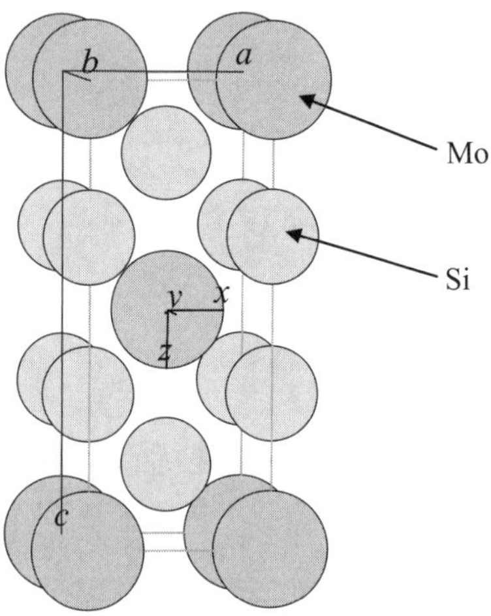

Figure 2.19 Tetragonal crystal structure of $MoSi_2$ with Mo occupying corner and body lattice position.

turbines, turbochargers, hot metal handling, and metal cutting.[11] $MoSi_2$ has excellent oxidation resistance, low density, and high electrical conductivity and therefore is used as heating elements for temperatures up to 1800°C.[12] $MoSi_2$ has a tetragonal crystal structure (Fig. 2.19) with Mo occupying the corner and body lattice positions. Physical properties of various non-oxide ceramics are provided in Table 2.2.

Additionally, graphitic structure possesses strongly covalent bonds between carbon atoms on a layered basal plane (Fig. 2.20). However, these layers are held by weak van der Waals forces and thereby graphitic structure shows highly anisotropic properties (i.e., high strength along the basal planes, and easy slipping along the basal planes).

Bonding, structure, and physical properties are quintessential in dictating the arising nature and performance of ceramic structural components. Bonding provides an insight into the energy required to separate ions and atoms, whereas structure elicits the manner in which a particular separation will occur. In addition, physical properties render a "bulk" feel of the structural ceramics. Hence a combination of

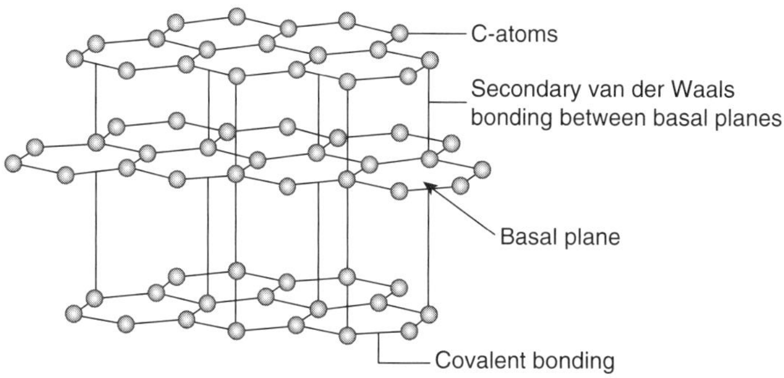

Figure 2.20 Graphitic structure of carbon.

these properties will help attain a detailed understanding of the behavior and performance of ceramics. More detailed descriptions of the structure of various engineering ceramics are provided in various textbooks.[13–15]

REFERENCES

1 W. D. Kingery, K. K. Bowen, and D. R. Uhlmann. *Introduction to Ceramics*. Wiley, New York, 2004.

2 J. F. Shackelford and M. K. Muralidhara. *Materials Science for Engineers*. Pearson Education, Upper Saddle River, NJ, USA, 2009.

3 I. Levin and D. Brandon. Metastable alumina polymorphs: Crystal structures and transition sequences. J. Am. Ceram. Soc. 81(8) (1998), 1995–2012.

4 W. Córdova-Martínez, E. D. Rosa Cruz, A. D. la Torres, P. Salas, A. Montoya, M. Avendano, R. A. Rodriguez, and O. B. Garcia. Nanocrystalline tetragonal zirconium oxide stabilization at low temperatures by using rare earth ions: Sm^{3+} and Tb^{3+}. Opt. Mate. 20(4) (2002), 263–271.

5 V. Y. Gertsman. Twin junctions in monoclinic zirconia. Interface Sci. 7(3–4) (1999), 1573–2746.

6 M. A. Choudhry and M. Z. Javed. Screw twinning in monoclinic zirconia. Indian J. Pure & Appl. Phy. 47 (2009), 506–510.

7 Q. Fang, P. S. Sidky, and M. G. Hocking. Erosion and corrosion of PSZ-zirconia and the t–m phase transformation. Wear. 233–235 (1999), 615–622.

8 D. P. Burke and W. M. Rainforth. Intermediate rhombohedral (r-ZrO_2) phase formation at the surface of sintered Y-TZP's. J. Mat. Sc. Lett. 16 (1997), 883–885.

9 L. Z. Pei and H. N. Xiao. B_4C/TiB_2 composite powders prepared by carbothermal reduction method. J. Mater. Process. Tech. 209 (2009), 2122–2127.

10 D. Jianxin and S. Junlong. Sand erosion performance of B4C based ceramic nozzles. Intl. J. Refract. Metal & Hard Matls. 26 (2008), 128–134.

11 P. Rendtel, A. Rendtel, H. Hubner, H. Klemm, and M. Herrmann. Effect of long-term oxidation on creep and failure of Si_3N_4 and Si_3N_4/SiC nanocomposites. J. Eur. Ceram. Soc. 19 (1999), 217–226.

12 J. Xu, H. Zhang, G. Jiang, B. Zhang, and W. Li. SiC whisker reinforced $MoSi_2$ composite prepared by spark plasma sintering from COSHS-ed powder. Trans. Nonferrous Met. SOC. China. 16 (2006), s504–s507.

13 C. B. Carter and M. G. Norton. *Ceramic Materials*. Springer, New York, 2007.

14 M. W. Barsoum. *Fundamentals of Ceramics*. Taylor & Francis, London, 2003.

15 Y. M. Chiang, D. P. Birnie, and W. D. Kingery. *Physical Ceramics*. John Wiley & Sons, New York, 1997.

Chapter 3

Mechanical Behavior of Ceramics

A combination of high hardness, high compressive strength, high elastic modulus, and low fracture toughness is what generally defines ceramics. Hence, these mechanical properties are among the critical requisites (with the most critical being fracture toughness) to evaluate the performance of ceramics in terms of their applications. After briefly mentioning some early theories proposed for brittleness, Griffith's theory of brittle fracture as well as indentation-induced cracking in brittle solids is discussed. It is shown how the physics of fracture behavior can be correlated with critical crack length and external stress application. From the concept of stress concentration at the crack tip, the fracture toughness is defined. After discussing theories of fracture, the variability in strength properties is discussed. Finally, various toughness measurement techniques are discussed. At the close, the physics of various toughening mechanisms is briefly mentioned.

3.1 THEORY OF BRITTLE FRACTURE

Ceramic materials are popularly known for their brittle behavior. It is probably true that ceramics are possibly best known for their brittleness rather than for some of their outstanding properties, which include the strength retention of some ceramics (SiC, Si_3N_4) at high temperature and the extremely good biocompatibility or bioactivity of others (hydroxyapatite). In this section, some of the important theories explaining the brittle fracture of ceramic and glass materials are reviewed.

3.1.1 Theoretical Cohesive Strength

Based on first-principles calculations, the theoretical cohesive strength can be determined from consideration of maxima in the total interatomic force under stress[1] (see Fig. 2.2 of Chapter 2). The initial theory of brittle fracture is therefore based on interatomic bond breakage or rupture theory, producing two additional surfaces.

Advanced Structural Ceramics, First Edition. Bikramjit Basu, Kantesh Balani.

Also, a linear elastic stress–strain relationship is assumed to be followed in bond rupture leading to failure. With such assumptions, one can arrive at the following expression for theoretical cohesive strength, σ_{th}:

$$\sigma_{th} = \left(\frac{E\gamma}{a_0}\right)^{1/2},$$

where E is the elastic modulus, γ is the surface energy, and a_0 is the equilibrium interatomic distance under unstrained conditions.

In the case of classical brittle solids, such as glass, the value of cohesive strength (σ_{th}) can be determined as 10–15 GPa, assuming $\gamma = 1\ \text{J/m}^2$, $a_0 = 3 \times 10^{-10}\,\text{m}$, and $E = 10\,\text{GPa}$.

However, a theoretical estimate shows a much larger value for strength than experimentally measured values (<1 GPa for glasses).[1] Almost an order of magnitude less strength than theoretical predictions can be attributed to the presence of a large number of cracks and flaws of varying sizes, which propagate with varying severity leading to fracture. As is explained in this chapter, the cracks act as stress concentrators or stress raisers in a brittle solid, thereby reducing the maximum load-bearing capability.

3.1.2 Inglis Theory

Inglis considered this aspect and proposed a theory based on stress concentration at a crack tip.[2] Considering an elliptical hole or cavity with major and minor axes a and b, respectively, in a solid continuum under externally applied tensile stress σ, the stress at both edges of the cavity will be higher than nominal stress and can be expressed as:

$$\sigma_{\max} = \sigma(1 + 2a/b). \tag{3.1}$$

At the tip or edge of the cavity, a tensile stress field is present; on the upper, relatively flat surface of the crack, compressive stress is realized.

For $a \gg b$, the hole or cavity can be realistically considered as a crack of length $2c$ (i.e., a semi-crack of length c) in the continuum (see Fig. 3.1), and the radius of curvature at the crack tip is

$$\rho = \frac{b^2}{a} = \frac{b^2}{c}.$$

Accordingly, Equation 3.1 can be rewritten as

$$\sigma_{\max} = \sigma\left(1 + 2\sqrt{\frac{c}{\rho}}\right). \tag{3.2}$$

From geometrical considerations, it is evident that $c \ggg \rho$ and therefore the maximum stress at the crack tip edge is

$$\sigma_{\max} = 2\sigma\sqrt{\frac{c}{\rho}}. \tag{3.3}$$

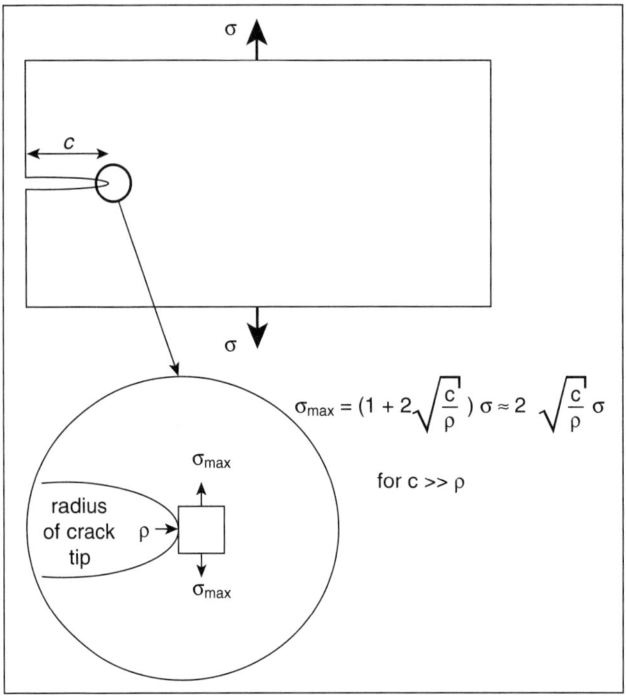

Figure 3.1 Schematic illustration showing the stress concentration at the crack tip edge.

From this expression, it should be clear that the maximum stress at the crack tip is $\sigma_{max} \gg \sigma$ and this will be reduced as one goes away from the crack tip edge. Therefore, much larger stress than externally applied stress will be realized at the crack tip. According to Inglis, fracture will take place when the stress at the crack tip is just sufficient to break interatomic bonds ahead of the crack tip. Combining Equation 3.2 with 3.3, one gets

$$\sigma_{\max} = \sigma_{th},$$
$$2\sigma\sqrt{\frac{c}{\rho}} = \left(\frac{E\gamma}{a_0}\right)^{1/2}, \tag{3.4}$$
$$\sigma_c = \left(\frac{E\gamma\rho}{4a_0 c}\right)^{1/2}.$$

Subsequently, Orowan approximated the radius of curvature of the crack tip as approximately the same magnitude as the interatomic distance (a_0) (Ref. 3), that is, $\rho = a_0$. This leads us to the following expression for the critical fracture stress:

$$\sigma_{inglis} = \left(\frac{E\gamma}{4c}\right)^{1/2}. \tag{3.5}$$

3.1.3 Griffith's Theory

By far, the most widely accepted theory of brittle fracture is Griffith's theory,[4] which is based on the total change in potential energy of a brittle solid during crack propagation under external tensile loading. Considering a rectangular plate with a through-thickness central hole loaded in tension, the total energy change for the system can be expressed as

$$\Delta u = -\Delta u_{el} + \Delta u_s, \tag{3.6}$$

where Δu_{el} is the elastic strain energy released around the elliptical hole and Δu_s is the change in surface energy as the cavity extends perpendicularly to the tensile stress direction. Assuming that the plate has thickness t that is much smaller than its width w, $t \gg w$, one can further find the expressions for Δu_{el} and Δu_s by considering that a sharp elliptical hole of major axis length c is essentially equivalent to a "real" crack with crack length c in a brittle ceramic. Griffith further proposed that elastic strain energy will be released over an elliptical volume around the crack with the major axis being twice the crack length and minor axis being the crack length (see Fig. 3.2a).

Considering Hooke's law of elasticity in deriving the change in elastic strain energy (a negative quantity) and the additional positive contribution from change in surface energy due to creation of two crack surfaces, the total change in potential energy of the cracked body can be expressed as:

$$\Delta u = \frac{-a^2}{2E}\pi \times 2c \times c \times t + 2c \times t \times 2\gamma.$$

Therefore, total change in energy per unit thickness can be expressed as:

$$\Delta u = \frac{-\sigma^2}{2E}\pi \times 2c^2 + 4c\gamma.$$

The variation of different energy terms with crack length is schematically shown in Figure 3.2b. At the maximum of the Δu–c curve, the first derivative of Δu with respect to c will be zero and the critical fracture stress or corresponding critical crack length can be derived as

$$\sigma_c = \sqrt{\frac{2\gamma E}{\pi c^*}}$$

or

$$c^* = \frac{2\gamma E}{\pi \sigma^2}. \tag{3.7}$$

On the basis of this theory, it can be said that fracture will occur if one of the following conditions is satisfied:

(a) External stress $\sigma \geq \sigma_c$

(b) Intrinsic flaw size $c \geq c^*$

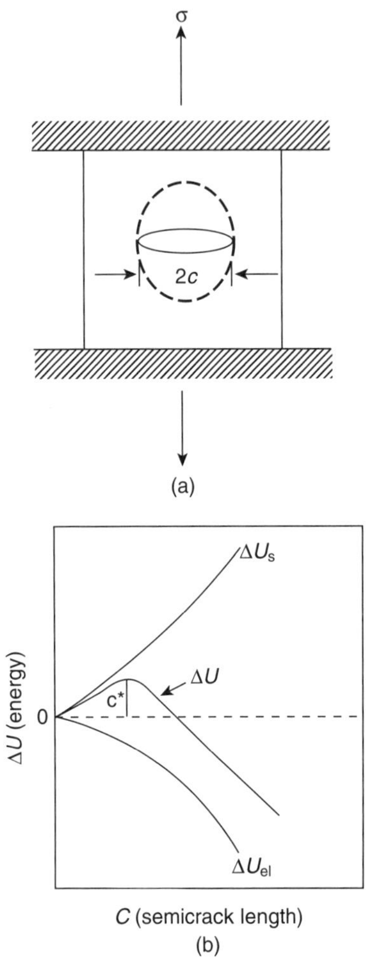

Figure 3.2 (a) Schematic illustrating a rectangular plate containing a through-thickness elliptically shaped cavity and (b) energetics involved in Griffith's theory of brittle fracture.

It can be further said that cracks of sizes less than c^* will not grow at a given external stress, as any infinitesimal growth of the crack will lead to an overall increase in energy of the system. When $c \geq c^*$, any infinitesimally small increase in crack length can lead to an overall decrease in energy along the downhill of the Δu–c curve. With an increase in external stress, the Δu–c curve will be shifted toward the left and, therefore, c^* will decrease accordingly. This implies that the critical cracks of finer sizes will be able to grow at higher stress, leading to early fracture of brittle solids.

At this juncture, one point needs to be mentioned. In glasses and some ceramics (e.g., Si_3N_4), the cracks with sizes less than the critical crack size ($c < c^*$) can grow unstably at a given stress level in a moist or humid environment, leading to fracture.[5,6] This is known as subcritical crack growth and is attributed to environmental interaction, leading to chemical attack at the crack tip, leading to easy bond breakage.

3.1.4 Irwin's Theory

Irwin later proposed that the fracture of brittle solids will depend on two terms, strain energy release rate (G) and crack resistance force (R), which are defined as follows:

Strain energy release rate, $G = \frac{\partial \Delta U_{el}}{\partial A} = \frac{\pi \sigma^2 c}{E}$,

where A is the crack surface area ($2c \times t$).

Crack resistance force, $R = \left(\frac{\partial \Delta U_s}{\partial A} \right) = 2\gamma$.

According to Irwin, fracture will occur when $G \leq R$.

At criticality,

$$G = R; \sigma_c = \sqrt{\frac{2E\gamma}{\pi c}}. \tag{3.8}$$

3.1.5 Concept of Fracture Toughness

From the preceding discussion, it is very clear that the fracture of brittle solids does not depend only on one parameter, but depends on a combination of σ_c and c^*. It is therefore evident that a combination of crack size and stress will determine the fracture of brittle materials. On the basis of this observation, the stress intensity factor is defined as $K = Y\sigma\sqrt{\pi c}$, where Y is a factor dependent on the location and orientation of the crack and on the loading condition.

In classical fracture mechanics theory, different modes of loading of the crack faces (see Fig. 3.3) are recognized—tensile or crack opening mode (mode I), shear mode (mode II), and tearing mode (mode III). Corresponding to the three different modes, the stress intensity factor can be defined as K_I, K_{II}, and K_{III}. Among the three modes, tensile mode is the most dangerous, and most failures of brittle solids are largely due to mode I failure. For this reason, we will deal with K_I only. At the critical condition, the critical mode I crack tip stress intensity factor is defined as

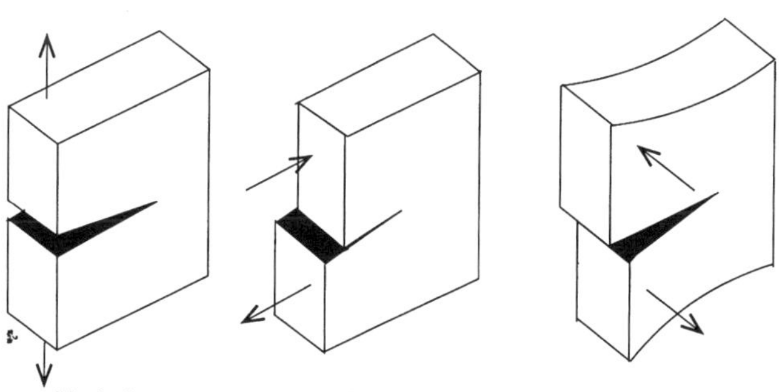

Figure 3.3 Three modes of loading of crack faces. Mode I (tensile or crack opening under mode) is the most widely observed fracture mode, whereas Mode II (shear mode) and Mode III (tear mode) are not as commonly observed.

$$K_{Ic} = Y\sigma_c\sqrt{\pi a_c}. \tag{3.9}$$

The value of Y varies around 1 to 1.1. The K_{Ic} defined in Equation 3.9 is also known as a parameter to describe fracture toughness.

3.2 CRACKING IN BRITTLE MATERIALS

As mentioned earlier and also discussed later, the presence of pre-existing flaws and their growth under the application of external stress limit the strength of brittle solids. It is therefore important to know the nature of cracking and how cracks develop during external loading.

The nature of cracking depends on the type of loading. In the case of blunt or distributed loading, for example under spherical indents, cone cracks develop in ceramics as shown in Figure 3.4. During continuous loading, cracks initially form at the indent edges and thereafter propagate at an angle downward. The entire crack

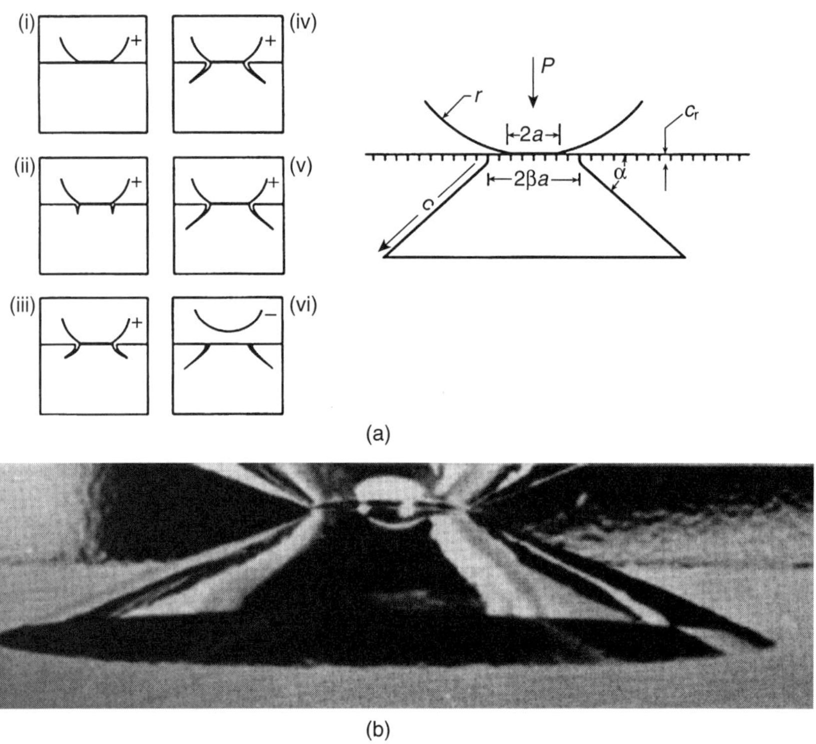

Figure 3.4 Schematic showing the development of a cone crack during loading–unloading of a spherical blunt indenter against a flat surface of a brittle material (left in [a]) and the geometrical parameter associated with cone crack configuration (right in [a]). Cone crack formation in soda lime glass is also shown (b).[6]

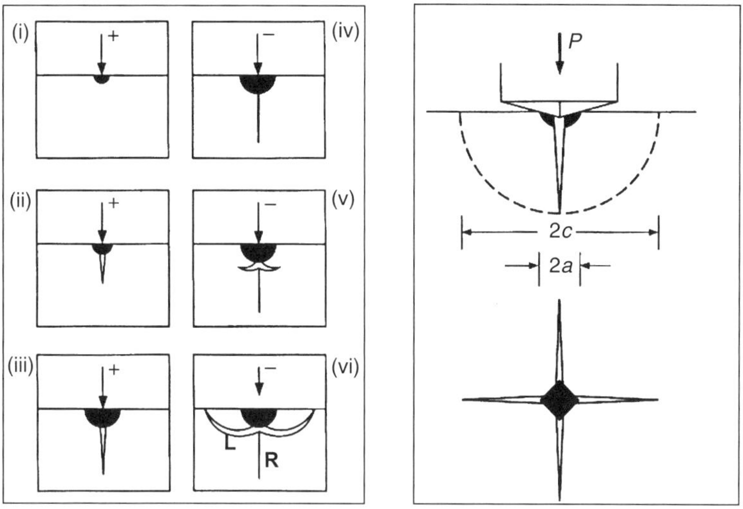

Figure 3.5 Schematic showing the development of radial–median (marked as "R") and lateral crack (marked as "L") during loading and unloading of a sharp indenter against a flat surface of a brittle material, respectively (left) and the geometrical parameter illustrating the dimension of highly deformed zone and crack length (right).[6]

pattern develops a conical geometry during unloading, as shown in the case of transparent soda lime glass (see Fig. 3.4). A different cracking pattern in brittle solids, such as ceramics and glasses, is observed under concentrated loading via a sharp indenter (e.g., Vickers indentation). Two distinct crack patterns develop: radial–median cracks and lateral cracks. During initial loading, the radial–median cracks develop from the highly deformed zone beneath the indentations. These cracks grow with time during the entire loading process. Upon unloading, a lateral crack pattern develops, with cracks starting to propagate along the transverse direction. With complete unloading, the fully developed lateral cracks meet the free surface of the brittle solid (see Fig. 3.5). During an abrasion process, characterized by a harder solid sliding on a relatively softer surface, the mechanical interaction between two solids is equivalent to multiple overlapping indentation processes under concentrated loading. Therefore, multiple lateral cracks generated from such overlapping indent-like regions will intersect each other as well as intersecting the free surfaces. Such physical phenomena will eventually lead to material removal, that is, wear, of the softer material's surface. In 2009, Tiwari proposed how the tribomechanical wear loss of ceramics[7] can be determined.

In view of the preceding discussion, it should be clear that crack lengths would determine the extent of wear damage in ceramics. Based on fundamental fracture mechanics theory,[6] Lawn demonstrated the relationship between indentation load (P) and crack length as

$$P \propto c_l^{3/2} \quad \text{or} \quad P \propto c_c^{3/2}, \tag{3.10}$$

where c_l and c_c are the lateral and cone crack length, respectively. Such a linear relationship was established after extensive experimental measurements in transparent soda lime glass in an inert environment. Such a relationship also implies that the severity of mechanical wear damage will increase with increase in indent load. A detailed fracture-mechanics-based treatment of cracking in brittle solids can be found elsewhere.[6]

3.3 STRENGTH VARIABILITY OF CERAMICS

In this section, one of the fundamental issues related to ceramics, that is, strength reliability, is discussed, with particular reference to weakest link fracture statistics. Strength reliability, one of the critical factors restricting wider use of brittle materials in various structural applications, is commonly characterized by a Weibull strength distribution function. Various strength parameters are correlated with the physics of fracture processes in brittle solids. The use of various distribution functions, that is, a gamma or lognormal distribution, is shown to characterize strength properties of some brittle solids.

Brittle materials, such as ceramics, have many useful properties: high hardness, stiffness and elastic modulus; wear resistance; high strength retention at elevated temperatures; corrosion resistance associated with chemical inertness, and so on.[1] It can be reiterated here that the advancement of ceramic science in the last few decades has enabled the use of this class of materials to evolve from more traditional applications (sanitary wares, pottery, etc.) to cutting-edge technologies, including rocket engine nozzles, engine parts, implant materials for biomedical applications, heat-resistant tiles for the space shuttle, nuclear materials, storage and renewable energy devices, fiber-optics for high-speed communications, and elements for integrated electronics such as microelectromechanical systems (MEMS).

In many engineering applications requiring load-bearing capability, that is, structural applications, it has been realized over the years that an optimum combination of high toughness with high hardness and strength reliability is required.[2] Despite having much better hardness compared with conventional metallic materials, one of the major limitations of ceramics for structural and specific nonstructural applications is their poor toughness and low strength reliability.[3] The poor reliability in strength or rather large variability in strength properties of ceramics is largely due to the variability in distribution of crack size, shape, and orientation with respect to the tensile loading axis.[4] Consequently, the strength of identical ceramic specimens under identical loading conditions is different for a given ceramic material.

3.4 PHYSICS OF THE FRACTURE OF BRITTLE SOLIDS

The variability in strength of ceramics is primarily due to the extreme sensitivity to the presence of cracks of different sizes. It can be noted that the yield strength and the fracture or failure strength of polycrystalline metals are deterministic and volume independent, when the characteristic microstructural feature (grain size) remains the same for the tested metallic samples. However, the fracture strength of a brittle material is, in particular, determined by the critical crack length according to Griffith's theory[8]:

$$\sigma_f = \frac{K_{IC}}{\sqrt{\pi c}}, \tag{3.11}$$

where σ_f is the failure or fracture strength, $K_{IC,}$ is the critical stress intensity factor (a measure of fracture toughness) under mode I (tensile) loading, and a is half of the critical or largest crack size. From this expression, it is clear that the longer the critical crack length, the lower the strength values will be. The multitude of experimental observations with ceramic samples can be summarized as follows:

(a) Brittle materials contain defects or cracks, and these cracks restrict them from achieving their potential strength.

(b) The dispersion of test results is due to dispersion in flaw sizes.

(c) The mean strength as determined from a multiplicity of similar tests depends on the size of the test specimens and the nature of loading

(d) Strength is not the same for nominally identical specimens under presumably similar experimental conditions.

(e) The larger the volume of the materials tested, the lower the strength value. This is because the larger the volume tested, the greater the probability of finding a critical crack size.

(f) Mean strength decreases as specimen volume increases.

These observations imply that (1) there should be a definite relationship between the failure probability of brittle materials and the stress level to which it is subjected and (2) brittle materials do not have a deterministic strength value that is volume dependent.

For a given ceramic material, the distribution of crack size, shape, and orientation differs from sample to sample. It is experimentally reported that the strength of ceramics varies unpredictably, even if identical specimens are tested under identical loading conditions.[4] In particular, the mean strength determined from a multiplicity of similar tests depends on the volume of material stressed, the shape of the test specimen, and the nature of loading. It is recognized that strength needs to be analyzed using different probabilistic approaches, largely because the probability of failure or fracture of a given ceramic sample critically depends on the presence of a potentially dangerous crack of size greater than a characteristic critical crack size.[4] Clearly, the probability of finding the critical crack size is higher in larger volume test specimens; consequently brittle materials do not have any deterministic strength property. Since brittle materials exhibit volume-dependent strength behavior, the mean strength decreases as the specimen volume increases. From the initial experimental observations, it was evident that a definite relationship should exist between the probability that a specimen will fracture and the stress to which it is subjected.

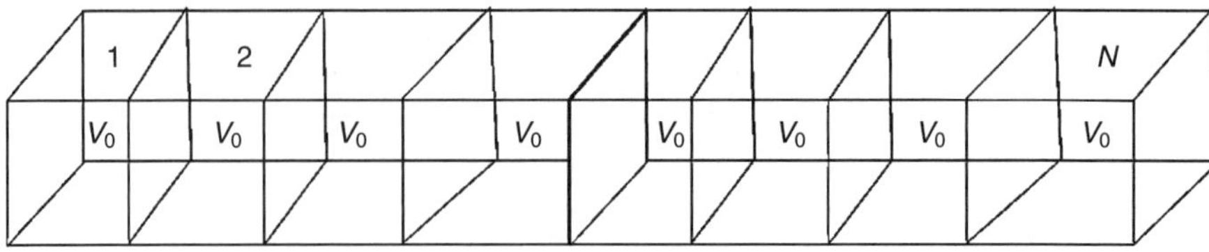

Figure 3.6 Schematic illustrating that a long ceramic test sample of rectangular cross section can be considered to contain a number N of unit cubes each having equal volume V_0. Each of the small volume elements can contain cracks of different sizes and this concept is used in developing the "weakest link fracture theory."

3.4.1 Weakest Link Fracture Statistics

Weakest link fracture statistics is based on the fact that if the weakest part of a ceramic fails due to unstable growth of a critical crack, then the entire ceramic will fail. This is the equivalent of the failure of a long chain due to failure of its weakest part at any part along the length of the chain. The weakest link fracture theory is based on the probability theory that the probability of occurrence of two events is the product of the probability that each event takes place independently. Referring to Figure 3.6, if we assume that a test sample of a brittle material with total volume V contains a number N of small unit volumes, each having equal volume V_0, then $V = NV_0$. When the entire test sample is loaded in tension, that is, mode I at a stress level σ, then each of the volume elements will experience identical stress level σ. Under such a condition, if any among the N volume elements were to fail, that would lead to the failure of the entire brittle material. From the physics of fracture, only that volume element will fail that contains the largest or critical crack oriented favorably toward the tensile stress direction in such a way that such cracks in that given volume element can grow unstably, leading to failure of the entire test sample.

If the survival probability at a given stress level (σ) of each volume element V_0 is S_0 (σ), then the failure probability is

$$F_0(\sigma) = 1 - S_0(\sigma).$$

On the basis of probability theory, the survival property of two volume elements together would be $S_2(\sigma) = S_0(\sigma) \cdot S_0(\sigma)$.

Extending this logic, the survival probability of N volume elements together would be

$$S(\sigma) = S_0[(\sigma)]^n = [S_0(\sigma)]^{V/V_0}.$$

In terms of total failure probability of N volume elements, the preceding expression can be written as

$$1 - F(\sigma) = [S_0(\sigma)]^{V/V_0}$$

$$\Rightarrow \ln[1 - F(\sigma)] = \frac{V}{V_0} \ln[S_0(\sigma)] = \frac{V}{V_0} g(\sigma),$$

where $g(\sigma) = -\ln[1 - F_0(\sigma)]$, that is, generic strength distribution law without assuming any strength distribution:

$$\therefore \quad F(\sigma) = 1 - \exp\left[-\frac{V}{V_0} g(\sigma)\right]. \tag{3.12}$$

The definition of $g(\sigma)$ demands that $g(\sigma) > 0$ and $g(\sigma)$ is a monotonic function, that is, $dg(\sigma)/d\sigma > 0$.

In 1951, Weibull[8] proposed a two-parameter distribution function to characterize the strength of brittle materials:

$$g(\sigma) = \left(\frac{\sigma}{\sigma_0}\right)^m,$$

where m is the Weibull modulus and σ_0 is the reference strength for a given reference volume V_0. The characteristic strength distribution parameter, m, indicates the nature, severity, and dispersion of flaws.[9] More clearly, a low value of m indicates a nonuniform distribution of highly variable crack lengths (broad strength distribution), while a high value of m implies a uniform distribution of highly homogeneous flaws with narrower strength distribution. Typically, for structural ceramics, m varies between 3 and 12, depending on the processing conditions.[1] The Weibull distribution function is widely used to model or characterize the fracture strength of various brittle materials, such as Al_2O_3 Si_3N_4.[2,10,11]

Based on the preceding discussion, the generalized strength distribution law can be given by the following expression:

$$F(\sigma) = 1 - \exp\left[-\frac{V}{V_0}\left(\frac{\sigma}{\sigma_0}\right)^m\right], \tag{3.13}$$

where $F(\sigma)$ is the probability of failure at a given stress level σ, V is the volume of the material tested, and V_0 is the reference volume.

In the following, some special cases are described.

In the case where the strength is controlled by surface flaws, then the total survival probability is

$$S(\sigma) = [S_0(\sigma)]^{A/A_0}.$$

Therefore, the total failure probability would be

$$F(\sigma) = 1 - \exp\left[-\frac{A}{A_0}\left(\frac{\sigma}{\sigma_0}\right)^m\right], \tag{3.14}$$

where A is the total surface area and A_0 is the reference surface area.

To characterize the strength properties of ceramics, some statistical analysis is required. Based on the weakest link fracture statistics, the mean strength can be expressed as

$$\bar{\sigma} = \int_o^\infty \sigma F(\sigma) d\sigma, \tag{3.15}$$

$$\frac{\bar{\sigma}}{\sigma_0} = \left(\frac{L}{L_0}\right)^{-1/m} \Gamma(1+1/m). \tag{3.16}$$

where the gamma function is defined as

$$\Gamma(\gamma) = \int_o^{\infty} x^{\gamma-1} e^{-x} dx.$$

From the above, it can be said that when $m = \infty$, the mean strength will have a unique value, that is, a definite strength value can be predicted as the value of m becomes large and tends to infinity.

Also, the median strength ($\sigma_{0.5}$) can be expressed as the strength level corresponding to the total failure probability of 0.5, that is,

$$F(\sigma_{0.5}) = 0.5$$

$$\Rightarrow 0.5 = 1 - \exp\left[-\frac{L}{L_0}\left(\frac{\sigma_{0.5}}{\sigma_0}\right)\right]^m$$

$$\Rightarrow \frac{\sigma_{0.5}}{\sigma_0} = \left(\frac{L}{L_0}\right)^{-1/m} (\ln 2)^{1/m}$$

The standard deviation (S) can be described as values occurring in 66% of the total population,

$$\frac{S}{\sigma_0} = \left(\frac{L}{L_0}\right)^{-1/m} \left\{\Gamma(1+2/m) - \Gamma^2(1+1/m)\right\}. \tag{3.17}$$

The coefficient of variation (COV) μ of the preceding strength distribution function is

$$\mu \equiv \frac{S}{\sigma} = \frac{\Gamma(1+2/m) - \Gamma^2(1+1/m)}{\Gamma(1+1/m)}. \tag{3.18}$$

In all the above expressions, L and L_0 are the sample span length under flexure mode and the reference length, respectively. In order to obtain the value of Weibull modulus, m from experiments, the following steps can be followed:

(a) Measure the failure strength (σ_i) of N number of samples ($N \leq 50$).

(b) Arrange the failure strength values in ascending order and rank each of the strength values as $N_f(\sigma)$

(c) For a large population of strength values, the failure probability of the sample at any given strength value can be estimated as

$$F(\sigma) = \frac{N_f(\sigma) - 1/2}{N} \tag{3.19}$$

From Eq. 3.13, one can rewrite as

$$\ln[-\ln(1-F(\sigma)] = \ln\frac{V}{V_0} + m\ln\left(\frac{\sigma}{\sigma_0}\right). \tag{3.20}$$

One can use Equation 3.20 and plot $\ln[-\ln(1 - F(\sigma))]$ versus $\ln(\sigma)$ to get the value of the Weibull modulus, m (slope).

Although Weibull's theory has been used for several decades now to characterize the strength of brittle materials, some recent studies have indicated that other distribution functions can also be considered. For example, Lu et al.[12] analyzed the fracture statistics of brittle materials using the Weibull and normal distributions. They have considered the strength data of three different ceramic materials, that is, silicon nitride (Si_3N_4), silicon carbide (SiC), and zinc oxide (ZnO). They used three-parameter Weibull, two-parameter Weibull, and normal distributions to analyze these data. It was observed that, based on the Akaike information criterion (AIC), the two-parameter Weibull or normal distribution fit better than the three-parameter Weibull distribution. Although the two-parameter Weibull distribution has been widely used in practice to model strength data, Lu et al.[12] questioned the uncritical use of Weibull distributions in general.

In a 2009 work, several statistical distribution functions were considered with an aim to critically analyze the strength data of brittle materials, such as ceramics.[13] Other than Weibull and normal, several two-parameter distributions, such as the gamma, log-normal, and generalized exponential distributions, were used. The experimentally measured strength data obtained with hot-pressed dense ceramics, such as monolithic ZrO_2 and ZrO_2–TiB_2 composites, as well as literature strength data of Si_3N_4 ceramic and glass were used to validate the statistical analysis. It was observed that the fitted Weibull and normal distributions behaved quite similarly, whereas the fitted gamma, log-normal, and generalized exponential distributions were of similar nature. Based on the limited set of strength data and using several statistical criteria, such as minimum chi-square, minimum Kolmogorov distance, and maximum log-likelihood value, the gamma or log-normal distribution function was reported to be a more appropriate statistical distribution function to characterize ceramic strength properties.

The major outcome was that the uncritical use of the Weibull distribution can be avoided and, therefore, the use of Weibull modulus as a strength reliability parameter should be made after detailed analysis of the strength reliability parameter.[13] Similar to the strength data, the grain size parameters, such as mean grain size and grain size distribution width, are equally important factors in determining critical material properties. In another study,[14] several statistical distribution functions, such as normal, lognormal, and Gumbel (extreme value of type 1), were used to evaluate the appropriate distribution function for microstructural description of sintered ceramics, such as ZrO_2. It was concluded from that study that the Gumbel distribution could describe much better (statistically) the grain size distribution. However, in many studies, the uncritical use of the Gaussian or normal distribution was made to find grain size distribution parameters for several metals and ceramic materials.

The preceding discussions evidently place the importance of detailed statistical analysis in evaluating the properties of materials on a larger scale, that is, in the field of material science.

3.5 BASIC MECHANICAL PROPERTIES

3.5.1 Vickers Hardness

Conventionally, hardness is defined as the resistance to permanent deformation. This property is measured by making indentations on flat polished surfaces. For most ceramics, hardness is measured using the Vickers indentation technique and the following expression is used:

$$H_v = 1.854\left(\frac{P}{d^2}\right), \tag{3.21}$$

where H_v is Vickers hardness, P is the applied load, and d is the average length of two diagonals.

The following three aspects need to be considered to obtain reliable and true values of hardness for engineering ceramics:

(a) The indent load should be applied in such a way that it does not cause cracking from indent corners or edges and so that a stable, well-developed indentation develops without any spalling or damage around the indentation.

(b) It is suggested that hardness of a new ceramic composition or of a ceramic processed using a new synthesis (sintering) route be measured using various indent loads. This can reveal any "indentation size effect" and a conservative estimate of "true hardness" can be obtained.

(c) To obtain a reliable measure of hardness, it is recommended to use electron microscopy to measure indent diagonals (length scale on the order of micrometers) as any small errors in measuring the diagonal length will appear as considerable errors in hardness, which has magnitude in gigapascals. This should be clear from the expression for Vickers hardness.

3.5.2 Instrumented Indentation Measurements

In conventional hardness measurements, the real-time load versus indentation depth is not recorded. In a relatively new method, hardness values were calculated from the depth of penetration using instrumented indentation (Fig. 3.7). According to the model of Oliver and Pharr (O-P),[15] which is the most commonly used method for evaluating indentation response of a material by instrumented indentation, the hardness (H) is expressed as

$$H = \frac{P_{\max}}{A_{cr}}, \tag{3.22}$$

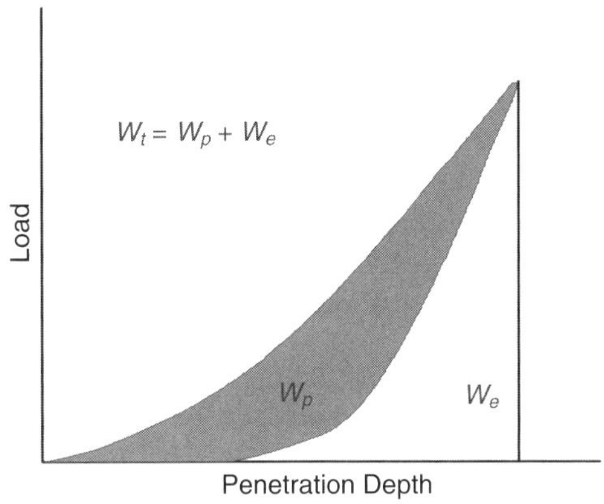

Figure 3.7 Typical plot of load versus penetration depth, as can be recorded using the instrumented indentation of a ceramic surface. The calculation of plastic work (W_p) and total work (W_e) can be made from the plot of load versus penetration depth.

where $P_{\max}$ is the maximum applied load and A_{cr} is the real contact area between the indenter and the material. Following the Oliver–Pharr model, the polynomial form of A_{cr} can be expressed as[15]

$$A_{cr} = 25.504h_c + C_1h_c + C_2h_c^{1/2} + C_3h_c^{1/4} + \cdots + C_8h_c^{1/128}, \tag{3.23}$$

where $C_1, \ldots, C_8$ are constants and can be determined by standard calibration methods, and h_c is the penetration depth, determined from the following expression[16]:

$$h_c = h_{\max} - \varepsilon(P_{\max}/S), \tag{3.24}$$

where $\varepsilon \approx 0.75$ for a Vickers indenter.[17]

The contact stiffness, S, can be calculated from the slope of the first third of the linear response, recorded during the unloading cycle, of the plot of load versus depth of penetration using the following expression[18]:

$$S = \left(\frac{dp}{dh}\right)_{h=h_{\max}} = \beta C_A E^* \sqrt{A_{cr}}, \tag{3.25}$$

where $\beta = 1.034$ and $C_A = 2/\sqrt{\pi}$ for a Vickers indenter[15,16,18] and E^* is the effective Young's modulus of the composite system comprising the indenter and the sample. Following the O-P model, E^* can be expressed as[15]

$$\frac{1}{E^*} = \frac{(1-\nu_i^2)}{E_i} + \frac{(1-\nu_s^2)}{E_s}, \tag{3.26}$$

where E and ν are Young's modulus and Poisson's ratio, respectively, and the subscripts i and s denote the indenter and sample, respectively. In general, for a Vickers diamond indenter, the values of E_i and ν_i were taken to be 1140 GPa and 0.07, respectively.[15,16,18]

Therefore, putting the values of S, β, C_A, and A_{cr} into Equation 3.25, one can easily calculate the value of E^*. The elastic modulus of the sample, E_s, can then be obtained from Equation 3.26. Thus, it may be understood that, except for a fully and perfectly elastic material, the plot of load versus depth of penetration will consist of two separate parts. The area under the unloading curve (i.e., unshaded area in Fig. 3.7) is the amount of reverse deformation energy released (W_e) when the test load is withdrawn. However, the area encompassed by the loading–unloading curve (i.e., shaded area in Fig.3.7) is the amount of plastic deformation work (W_p) performed during the indentation test. Thus, for an elastic material, $W_e \gg W_p$. However, for a plastic material, W_p should be much higher than the reverse deformation energy. The sum of these two is called the total mechanical work of indentation, W_t, which can be expressed as

$$W_t = W_e + W_p. \tag{3.27}$$

The above-mentioned analysis of the instrumented indentation has been successfully used to characterize the mechanical properties of many ceramic systems, including TiB_2-based hard ceramics[19] as well as hydroxyapatite-based bioceramic composites.[20]

3.5.3 Compressive Strength

Although ceramics are extremely weak in tension, they have superior compression properties. The difference can be ascribed to the difference in microstructural resistance to crack growth and the nature of crack propagation under two different loadings. A typical plot of stress–strain response under tension and compression is given in Figure 3.8. While ceramics behave like a perfectly linear elastic material up to fracture, the same class of material behaves nonlinearly after reaching a peak load (much higher than that in tension) in compression. As opposed to tensile crack growth, the cracks extend vertically along the compression loading direction. The serration in compression stress–strain response is essentially due to spalling of a small test volume, as the growing cracks either coalesce with each other or meet a

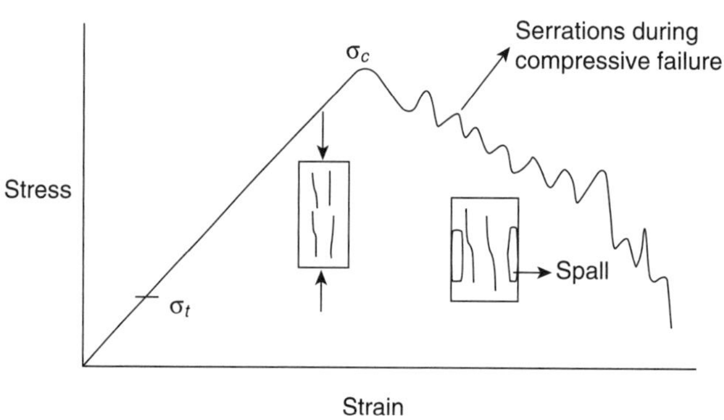

Figure 3.8 Stress–strain behavior of a brittle ceramic during compression, with σ_c indicating the compressive strength. For comparison, the tensile stress–strain plot is superimposed to illustrate around eight times higher compressive strength than tensile strength (σ_t) of a brittle ceramic. The compression failure mechanisms are also shown.

free or unconstrained surface of the material. Clearly, a delayed fracture behavior is realized in compression and, typically, compressive strength is around eight times higher than tensile strength.

Compressive strength can be measured using a universal testing machine (UTM). For this purpose, a cylindrical test sample with height-to-diameter ratio of 1.0 or larger is normally used. The samples are placed in between two parallel plates of the machine and force is applied on the appropriately aligned samples with a constant crosshead speed (typically around 0.05 mm/s). During the entire compression test, the load–displacement response can be recorded using a computer attached to the UTM. The compressive strength (σ_{cs}) of the samples can be calculated from the fracture load and the dimension of the samples using the simple formula

$$\sigma_{cs} = \frac{P}{A}, \tag{3.28}$$

where P is the maximum load (fracture load) and A is the cross-sectional area.

3.5.4 Flexural Strength

In view of the difficulty of making tensile tests on ceramics in a dogbone geometry, the strength of ceramics is measured under flexure mode either by a three-point or by a four-point bend configuration. For this purpose, samples with a bar geometry and either rectangular or circular cross section are placed in a bend fixture. Either a concentrated load is applied in the three-point configuration or a distributed load is applied at two different places in the four-point configuration (see Fig. 3.9). The flexural strength thereafter is calculated on the basis of measured fracture load and the dimension of the test sample. For three-point loading, the fracture strength can be calculated from the following expression:

$$\sigma_f = \frac{3PL}{2bd^2}, \tag{3.29}$$

where σ_f is the flexural strength of the material, P is the fracture load, L is the span length, b is the width of the sample, and d is the thickness of the specimen. Similarly, the flexural strength for the four-point bend configuration can be obtained from the expression

$$\sigma_f = \frac{3PL}{4bd^2}. \tag{3.30}$$

During flexural testing, the loading surface is placed in compression, while the opposite surface is placed under tension. Also, the stress value linearly decreases along the thickness (z direction) of the sample. In the case of four-point loading, maximum tensile stress is distributed over a larger area of the sample as opposed to the three-point flexural mode and, hence, a lower and conservative estimate of the strength is obtained in four-point flexural tests.

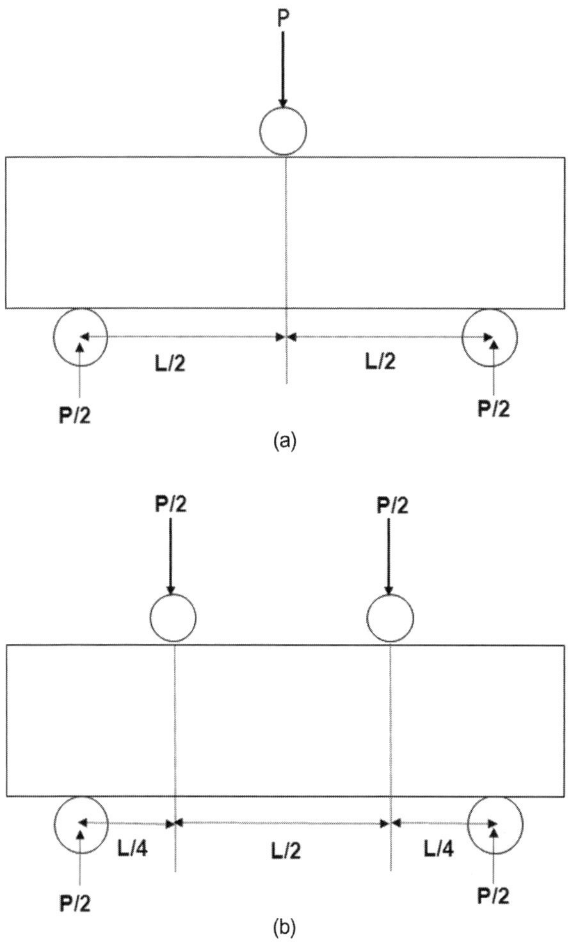

Figure 3.9 Schematic illustration of three-point (a) and four-point (b) flexural strength measurement of ceramic bar of rectangular cross section.

3.5.5 Elastic Modulus

The elastic modulus can be obtained using a dynamic elastic property analyzer, which measures Young's modulus of the sample by the impulse excitation of vibration method. The rectangular samples are impacted upon by a steel rod near the high-frequency sensor of the instrument. The values of Young's modulus are then calculated by commercial software using the following formula[21]:

$$E = 0.9465\left(\frac{mf_f^2}{b}\right)\left(\frac{L^3}{t^3}\right)T, \tag{3.31}$$

where

$$T = 1 + 6.858\left(\frac{t}{L}\right).$$

E is Young's modulus, m is mass, f_f is the natural frequency in the flexure mode, b is width, t is thickness, and L is length. This technique has been widely used in measuring elastic modulus and damping properties of various structural ceramics.[22,23]

The elastic modulus, E, and Poisson's ratio, ν, can also be evaluated by an ultrasonic method using lithium niobate crystals for reusing transmitting and receiving signals, which are typically generated at a resonant frequency of 10 MHz. The velocities of the longitudinal and shear waves can be calculated from the thickness of the specimens and the travel time of the waves across the thickness or height of the specimens. The following relationships can be used to determine the modulus properties:

$$E = \frac{(1+\nu)(1-2\nu)}{1-\nu}(\rho C_L), \tag{3.32}$$

$$\nu = \frac{1/2(C_1/C_s)^2 - 1}{(C_L/C_s)^2 - 1}, \tag{3.33}$$

where ν = Poisson's ratio, ρ = density of the specimen (g/cm^3), C_L = velocity of longitudinal wave (m/s), C_s = velocity of shear wave (m/s), and E = elastic modulus (GPa).

3.5.6 Fracture Toughness

Concerning the measurement of toughness, it should be noted that the toughness of brittle materials is dependent on the testing technique, which are widely classified into long crack and short crack methods. Long crack methods include the single edge notched beam (SENB) and single edge V-notched beam (SEVNB) techniques. Short crack techniques involve measurement of the crack lengths (radial–median) around hardness indentations, from which the toughness data can be approximated using various reported models. A significant volume of literature on the mechanistic and phenomenological description of the indentation cracking of brittle materials, like ceramics and glass-based materials, is available,[24–29] and much of the discussion in this section is based on such literature reports as well as other reports, as mentioned hereunder. While the study on indentation induced damage behavior has relevance to the wear resistance property, the indentation technique is often used to evaluate the toughness or crack growth resistance properties of ceramics. It can be noted here that the absolute toughness values of brittle materials cannot be measured by indentation techniques, for which one has to adopt the long crack fracture toughness measurement techniques, for example SENB, SEVNB, chevron notch beam (CNB). However, to compare the toughness properties of newly developed composites, the indentation technique is often used. Additionally, it is now widely recognized that a careful use can provide reproducible results for indentation toughness measurements. It should be mentioned here that the indentation method is now routinely used for convenience to compute the fracture toughness of small and relatively brittle specimens, which are otherwise hard to machine into standard test samples (e.g., SENB, SEVNB).

Broadly, there are two established ways to evaluate the fracture toughness of brittle materials. One is to use toughness evaluation with large cracks and the other one is to evaluate by short crack methods.

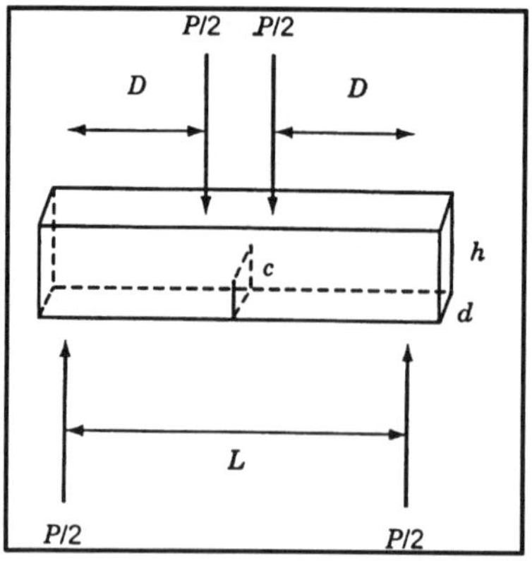

Figure 3.10 Typical geometry and loading configuration involved in SENB testing to evaluate fracture toughness.[6]

3.5.6.1 Long Crack Methods

The main difficulty in performing classical fracture toughness testing on ceramics by long crack methods is to prepare a sharp crack in front of the notch.[6] In general, two controlled methods are used for precracking the ceramic material:

(a) ***SENB Method.*** In this case cyclic compressive loads are applied to a SENB specimen with a cut notch. This results in damage accumulation leading to crack growth in the zone ahead of the notch, when the sample is loaded in flexure mode.

The testing geometry as well as various parameters is shown in Figure 3.10. The evaluation of the mode I critical stress intensity factor (K_{Ic}) by SENB follows the classical expression:

$$K_{Ic} = Y(3Pc/h^2d)c^{1/2}, \tag{3.34}$$

where $Y = 1.99 - 2.47(c/h) + 12.97(c/h)^2 - 23.17(c/h)^3 + 24.8(c/h)^4$.

(b) ***SEVNB Method.*** This is a refined version of the SENB method in which a sharp "V" notch is produced by polishing the notch root with a razor blade impregnated with diamond paste. In the SEVNB method, rectangular samples, prepared by the previously described method are used to determine the fracture toughness of the materials. A V-shaped notch was introduced along the height of the specimen using diamond saw and razor blade with diamond abrasive paste on a commercial V-notch preparation machine (made by Scientific Testing Devices, Frankfurt, Germany). The notch radius can be less than 10 μm. The samples are then fractured using a four-point bending setup on a UTM. The shape of a typical notch is shown in Figure 3.11. The fracture load is recorded and fracture toughness can be calculated from the following equation[30,31]:

$$K_{Ic} = \frac{P_f(L_o - L_i)}{BW^{3/2}} \frac{3\alpha^{1/2}}{2(1-\alpha)^{3/2}} f(\alpha), \tag{3.35}$$

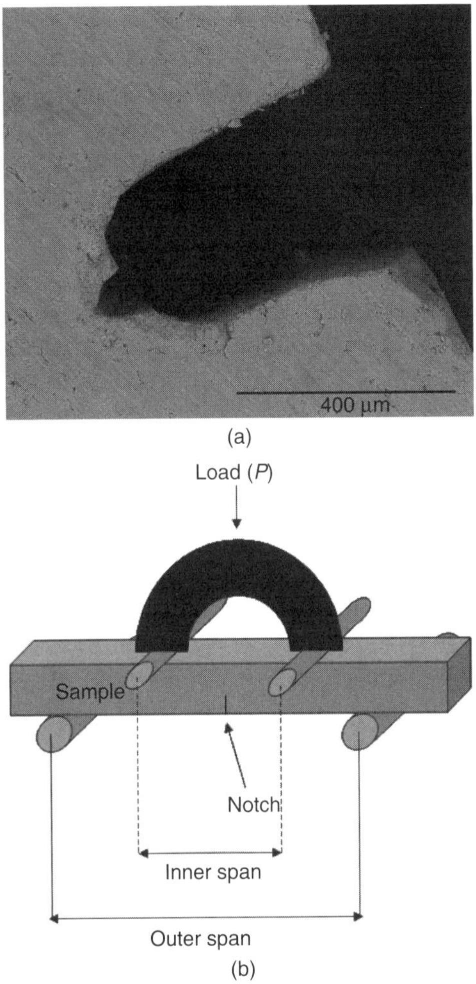

Figure 3.11 (a) A typical V-notch created on a ceramic sample (notch radius is <10 µm). (b) Four-point flexural configuration for SEVNB testing of fracture toughness measurement.

where P_f, L_0, L_i, B, and W are fracture load, outer span, inner span, specimen width, and specimen thickness, respectively, $\alpha = a/W$ with a the precrack size, and $f(\alpha)$ is

$$f(\alpha) = 1.9887 - 1.326\alpha - \frac{\alpha(1-\alpha)(3.49 - 0.68\alpha + 1.35\alpha^2)}{(1+\alpha)^2}. \tag{3.36}$$

3.5.6.2 Fracture Toughness Evaluation Using Indentation Cracking

Small surface cracks of controlled size and sharpness can be induced in brittle materials such as ceramics by hardness indenters (Fig. 3.12). This is particularly relevant for the tribological applications of ceramics, in which case it is the short

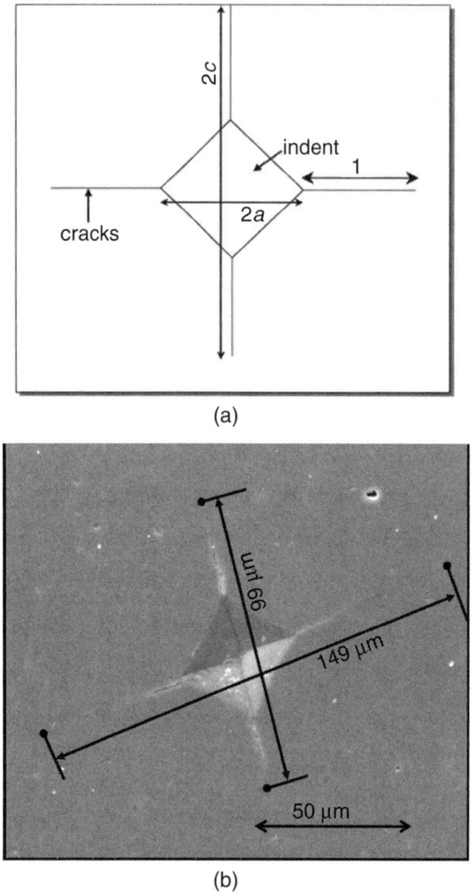

Figure 3.12 (a) Schematic of indentation-induced cracking and (b) scanning electron microscopy (SEM) observation of Vickers indentation as well as associated cracking on a ceramic surface.

crack fracture toughness that will have a larger influence on wear behavior.[32] This is particularly true as microcracks and subsequent spalling are observed on worn surfaces of many ceramics.[32] It is also recommended that the toughness of ceramics with new compositions be measured at different loads to check whether toughness increases with crack length, leading to "R"-curve behavior.

(a) ***Indentation Microfracture (IM) Method.*** The median cracks (indicated by "mc" in Fig. 3.13) that emanate from the corners of a Vickers indentation are arrested when the residual stress driving force (K_{res}) at the crack tip is in equilibrium with the fracture toughness[6]:

$$K_{Ic} = A(E/H)^n * (P/c^{3/2}) = \chi(P/c^{3/2}) = K_{res}, \tag{3.37}$$

where K_{Ic} is the indentation fracture toughness (Pa m$^{1/2}$), E is the elastic modulus (GPa), H is the Vickers hardness (GPa), P is the indentation load (N), χ is the residual stress factor, and c is the crack length (m) measured from the center of the indent impression. Furthermore, fracture toughness

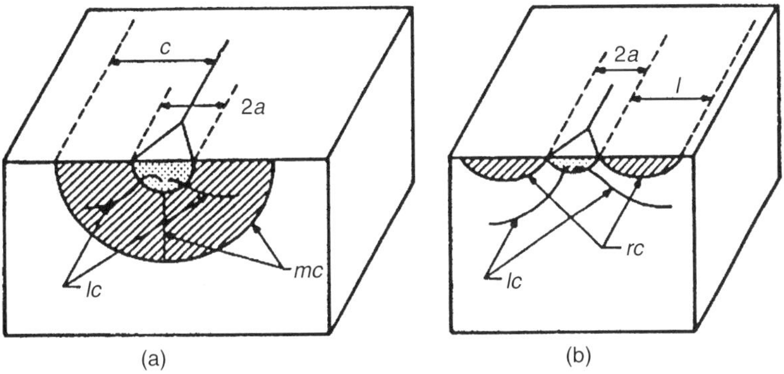

Figure 3.13 Cross-sectional views of material surface around Vickers indentations showing the formation of lateral cracks (lc) and median cracks (mc) in high-load regime (a) and radial cracks (rc) and lateral cracks in low-load regime (b). (Reprinted with kind permission from Springer Science + Business Media: Journal of Materials Science, Evaluation of K_{Ic} of brittle solids by the indentation method with low crack-to-indent ratios, vol. 1, 1982, pp. 13–16, K. Niihara et al.)

(K_{Ic}) values at different indent loads can be calculated from the measurement of radial cracks (indicated by "rc" in Fig. 3.13) around the Vickers indentations, according to the formula proposed by Anstis et al.[33]:

$$K_{Ic} = 0.016(E/H)^{1/2} P/c^{3/2}. \tag{3.38}$$

The expression for toughness estimation as proposed by Anstis et al.[33] was modified by Kaliszewski and coworkers[25] to account for the effect of the compressive stresses due to the surrounding transformation zone:

$$K_{Ic} = 0.019(E/H)^{1/2} P/c^{3/2}. \tag{3.39}$$

Evans and Wilshaw[34] confirmed that for a number of brittle materials Palmqvist cracks were formed in the low-load regime. The dimensions of the Palmqvist and median cracks are related by

$$l/a = c/a - 1. \tag{3.40}$$

For Palmqvist cracks ($0.25 < l/a < 2.5$), the fracture toughness (K_{Ic}) can be obtained from the expression given by Niihara et al.[35]:

$$(K_{Ic}\varphi/Ha^{1/2})*(H/E\varphi)^{2/5} = 0.035(l/a)^{-1/2}, \tag{3.41}$$

where $\varphi = 3$ and l = Palmqvist crack length for median cracks.

For median cracks ($c/a \geq 2.5$), the corresponding expression is

$$(K_{Ic}\varphi/Ha^{1/2})*(H/E\varphi)^{2/5} = 0.129(l/a)^{-3/2}. \tag{3.42}$$

Shetty at el.[36] modified the equation of Niihara et al.[35] and proposed the following expression for toughness determination:

$$K_{Ic} = 0.025(E/H)^{0.4}(HW)^{1/2}, \tag{3.43}$$

where $W = P/4a$ (P is the indentation load, $2a$ is the Vickers diagonal).

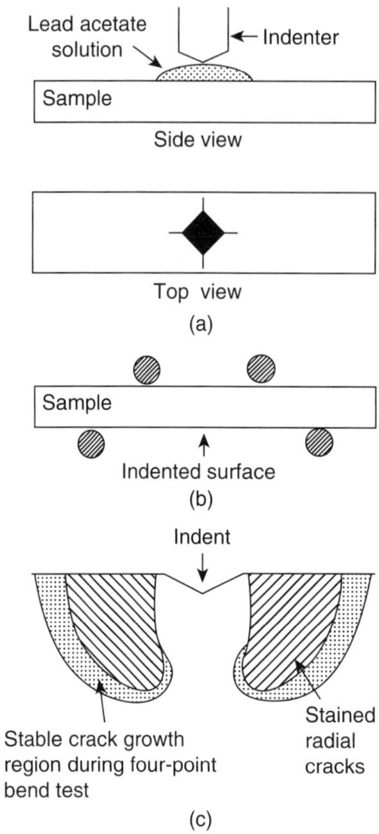

Figure 3.14 The indentation strength in bending method of evaluating the fracture toughness of brittle materials: (a) indenting the tensile surface, (b) measurement of failure stress in four-point flexure configuration, and (c) the fracture surface revealing crack propagation on the indented surface.

(b) ***Indentation Strength Bending (ISB) Method.*** In this method, a Vickers indentation is placed in the center of the tensile surface of a beam specimen, which is subsequently tested to failure in bending (Fig. 3.14).[27] The radial crack dimensions are measured optically. A drop of silicon oil was placed on the indent to minimize environmentally assisted subcritical growth. The specimen surfaces with microcracks are loaded on the tensile surface. The fracture toughness is estimated from the strength and indent load using the method of Chantikul et al.[37]:

$$K_{\mathrm{Ic}} = \eta (E/H)^{1/8} (\sigma P^{1/3})^{3/4}, \tag{3.44}$$

where $\eta = 0.59$ (a constant) and σ is the failure strength.

An estimate of toughness can also be made using the dummy indentation technique of Cook and Lawn.[38] The extended crack ($c_{m,}$) emanating from the indentation is measured and K_{Ic} is determined from the following expression:

$$K_{\mathrm{Ic}} = A\sigma c_m^{1/2} - B, \tag{3.45}$$

where $A = 2.02$ and $B = 0.68\,\mathrm{MPa\,m^{1/2}}$ are constants.

In a more refined method, the indented specimens are annealed at a temperature of lower than sintering temperature to remove both the indentation-induced residual tensile stresses and the transformation-induced compressive stresses. The crack lengths are measured before and after annealing, and the strength is determined in a flexural test. The fracture toughness is calculated according to the following expression:

$$K_{Ic} = \psi \sigma c^{1/2}, \tag{3.46}$$

where ψ is the crack geometry factor and $2c$ is the length of radial cracks. Fractographic analysis of fine-grained yttria-stabilized tetragonal zirconia polycrystalline (Y-TZP) ceramics revealed that crack profiles were semielliptical and were identical to those observed by Kaliszewski et al.[25] The shape factor $\psi = 1.025$ for a semielliptical crack can be used.

3.6 TOUGHENING MECHANISMS

Prior to the discussion of various possible toughening mechanisms in ceramics, it is important to revisit the issue of brittleness in ceramics. It has been widely recognized that the brittleness of ceramics is due to a multitude of factors.[6]

In the case of ceramics with predominantly ionic binding, the dislocations cannot glide on all possible slip planes as displacements of ions, as schematically shown in Figure 3.15a, can lead to violation of the local electrical charge neutrality condition. Therefore, dislocations can only be allowed to glide on specific planes oriented at a 45° angle, which can restore the condition of electroneutrality.

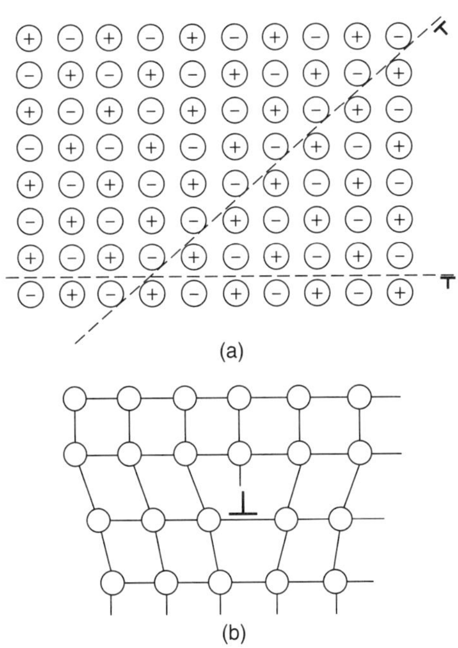

Figure 3.15 Schematic illustration showing (a) the possibility of dislocation glide at specific planes in an ionic ceramic and (b) the difficulty in dislocation glide due to rigid bond network in a covalent ceramic.

For ceramics with predominantly covalent bonds, dislocation movement is quite difficult due to directional properties and the inherently rigid bond network, as such movement requires bonds to be broken and remade as well as bond angles to be distorted (see Fig. 3.15b).

Many ceramics do not have five independent active slip systems and, therefore, any homogeneous deformation without localized fracture is not possible.

The dislocation core width of ceramics is narrower than that in metals, leading to the requirement of high Pierls–Nabarro stress for dislocation glide.[1]

All of the preceding factors, often in combination, make it difficult for a ceramic grain in a constrained microstructure to accommodate shape change; otherwise it leads to strain incompatibilities at the grain boundary, leading to cracking. Once cracking starts, it is rather easy for cracks to grow in ceramics. In the case of metals, yielding occurs or localized plasticity due to dislocation movement in the crack tip stress field takes place. This absorbs a fraction of available energy at the crack tip, thereby reducing the total driving force for further crack propagation in metals. In view of the lack of dislocation glide in ceramics, such a phenomenon is ruled out; therefore, cracks, once they attain critical size, often can grow in an unstable manner, leading to fracture or failure of the ceramic component.

In view of the lack of any room-temperature ductility, there has been a continuous effort in the ceramics community to discover newer microstructural designs to enhance crack growth resistance. From a fundamental point of view, if any interaction between a growing crack and the microstructure can absorb a fraction of the energy available at the crack tip stress field, then the driving force for crack propagation will be reduced. In other words, the crack opening displacement will be consequently reduced and, as a result, the crack tip will be blunted. Such mechanisms to improve crack growth resistance, that is, toughening mechanisms, can be broadly classified into two generic types (see Fig. 3.16):

1. ***Process Zone Mechanisms.*** Enhanced crack growth resistance is realized due to phase-transformation-induced volume expansion or microcracking or crack deflection in the process zone around the crack tip. Since all these mechanisms will be discussed in more detail in later chapters of this book, only fundamental concepts will be summarized here. In the case of toughening induced by phase transformation (popularly known as transformation toughening), for example the transition from tetragonal to monoclinic zirconia, in the crack tip stress field involves volume expansion. In a constrained microstructure, this will result in compressive stress on the crack faces, leading to closure of the crack tip. Crack deflection is mostly observed in composites containing particulates, whiskers, or fibers as the second phase. Essentially, the crack path tortuosity is increased as the crack bypasses around the hard or rigid second phase, and the model of Faber and Evans[39,40] predicts that the maximum increment in crack deflection induced toughening can be realized in composites with around 30 vol% particulate reinforcement. The particle size and shape as well as the distribution of second-phase particles also has an influence on achievable toughness increment.

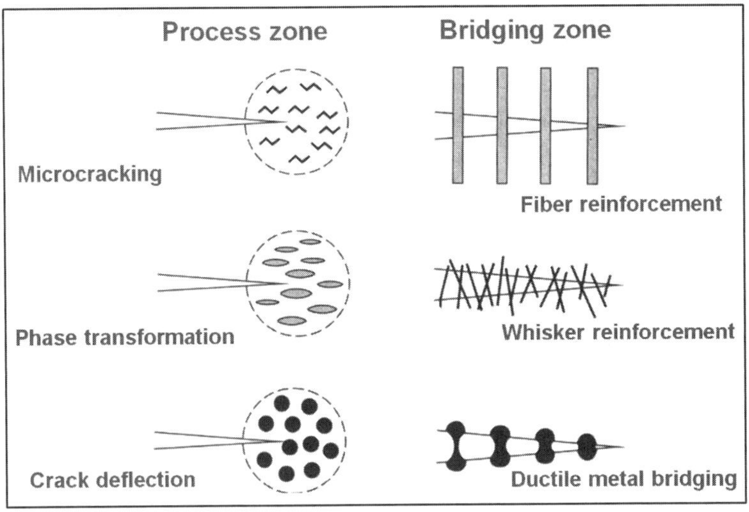

Figure 3.16 Summary of various toughening mechanisms in ceramic-based materials (adapted from Ref. 6).

2. ***Bridging Zone Mechanisms.*** This type of toughening is realized in fiber-reinforced or whisker-reinforced composites. For example, both SiC matrix and SiC fibers are inherently brittle. However, the toughness of SiC/SiC_f composites can be very high and the underlying mechanisms involve crack bridging due to fiber–crack interaction. Such interaction in composites under externally applied tensile stress essentially involves matrix–fiber debonding or matrix microcracking in the crack wake of the matrix crack or mode I crack, crack deflection at the interface, fiber pullout, and, finally, crack bridging. The combination of these mechanisms essentially results in the nonlinear stress–strain tensile response of the composite, leading to damage-tolerant behavior, as opposed to the linear elastic fracture of ceramic monoliths (see Fig. 3.17).[41] The extent of fiber pullout as well as properties of the matrix–fiber interface strongly determines the toughness increment. Fiber toughening is also accompanied by modulus enhancement and this is governed by the rule of mixtures. The fiber-reinforced composites, however, have their own limitations, which include expensive production cycle, anisotropic properties with higher property along the fiber direction, and so on.

Among various reinforcements, fibers have the largest aspect ratio, followed by whiskers and then particulates. In whisker-reinforced composites with random whisker alignment, crack bridging also takes place but to a lower extent than fiber toughening. Such bridging also involves whisker pullout and the toughness increment depends on the volume fraction of whiskers. A third type of toughening generally occurs in an important class of materials known as cermets, which are characterized by dispersion of a ceramic phase in a metallic matrix, for example WC–Co. In the crack tip stress field, the ductile flow of metallic particles leading to bridging of crack faces reduces

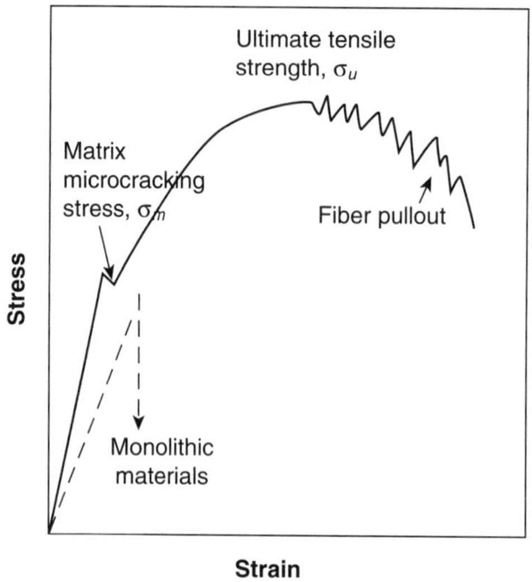

Figure 3.17 Schematic illustration of tensile stress–strain response of ceramic fiber-reinforced ceramic composites vis-à-vis response of monolithic ceramic.

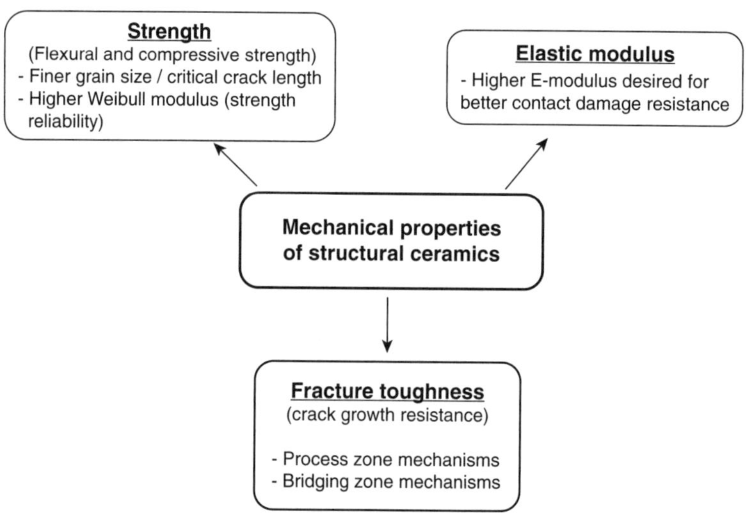

Figure 3.18 Schematic of various properties relevant to the mechanical behavior of structural ceramics.

the available driving force for crack propagation. The toughness of WC–Co cermet therefore is reduced with lowering of Co content. From the material-development point of view, a critical amount of Co is required to facilitate liquid phase sintering and the balance of hardness and toughness requires tailoring of Co content.

As a concluding note, various mechanical properties relevant to structural applications of ceramics are summarized in Figure 3.18, and more descriptions of the mechanical behavior of ceramics are available in various textbooks.[42–49] While a high elastic modulus would ensure better contact damage resistance, that is, wear resistance, in tribological applications, a finer microstructural scale (grain size) or critical crack length can result in better strength properties. As discussed in this chapter, higher Weibull modulus value (m) is desired for better strength reliability. An important issue in the development of structural ceramics is the enhancement of fracture toughness properties. Microstructural tailoring or reinforcement with a stronger ceramic phase (whiskers, fibers) can be adopted to enhance fracture toughness via process zone or bridging zone mechanisms in structural ceramics.

REFERENCES

1 G. E. Dieter and D. Bacon. *Mechanical Metallurgy*. McGraw Hill, New York, 1988.

2 C. E. Inglis. Stresses in a plate due to the presence of cracks and sharp corners. Trans. Inst. Nav. Archit. 55 (1913), 219–241.

3 E. Orowan. *Fatigue and Fracture of Metals, Symposium at Massachusetts Institute of Technology*. John Wiley & Sons, NewYork, 1952.

4 A. A. Griffith. The phenomena of rupture and flow in solids, Philos. Trans. R Soc London 221A (1920), 163–198.

5 S. M. Wiederhorn. Moisture assisted crack growth in ceramics. Int. J. Fracture Mech. 4(2) (1968), 171–177.

6 B. R. Lawn. *Fracture of Brittle Solids*. Cambridge University Press, Cambridge, 1993.

7 A. Tiwari, B. Basu, and R. Bordia. Model for fretting wear of brittle ceramic. Acta Mater. 57 (2009), 2080–2087.

8 W. Weibull. A statistical distribution function of wide applicability. J. Appl. Mech. 18 (1951), 293–305.

9 R. W. Davidge. *Mechanical Behavior of Ceramics*. Cambridge University Press, Cambridge, 1979.

10 Z. P. Bazant. Probability distribution of energetic-statistical size effect in quasibrittle fracture. Probabilistic Eng. Mech. 19 (2004), 307–391.

11 R. Danzer, T. Lube, and P. Supancic. Monte Carlo simulations of strength distributions of brittle materials—Type of distribution, specimen and sample size. Z. Metallkd. 92 (2001), 773–783.

12 C. Lu, R. Danzer, and D. Fishcer. Fracture statistics of brittle materials: Weibull or normal distribution. Phys. Rev. E 65 (2002), 067102.

13 B. Basu, D. Tiwari, D. Kundu, and R. Prasad. Is Weibull distribution the most appropriate statistical strength distribution for brittle materials? Ceram. Int. 35 (2009), 237.

14 D. Dierickx, B. Basu, J. Vleugels, and O. Van Der Biest. Statistical extreme value modelling of particle size distributions: Experimental grain size distribution type estimation and parameterisation of sintered zirconia. Mater. Characterisation 45 (2000), 61–70.

15 W. C. Oliver and G. M. Pharr. An improved technique for determining hardness and elastic modulus using load and displacement sensing indentation experiments. J. Mater. Res. 7 (1992), 1564–1583.

16 Z. Ling and J. Hou. A nanoindentation analysis of the effects of microstructure on elastic properties of Al_2O_3/SiC composites. Compo. Sc. Tech. 67 (2007), 3121–3129.

17 M. Kanari, K. Tanaka, S. Baba, and M. Eto. Nanoindentation behavior of a two-dimensional carbon-carbon composite for nuclear applications. Carbon 35 (1997), 1429–1437.

18 S. Guicciardi, A. Balbo, D. Sciti, C. Melandri, and G. J. Pezzotti. Nanoindentation characterization of SiC-based ceramics. J. Eur. Ceram. Soc. 27(2–3) (2007), 1399–1404.

19 A. Mukhopadhyay, G. B. Raju, B. Basu, and A. K. Suri. Correlation between phase evolution, mechanical properties and instrumented indentation response of TiB_2-based ceramics. J. Eur. Cer. Soc. 29 (2009), 505–516.

20 S. Nath, A. Dey, A. K. Mukhopadhyay, and B. Basu. Nanoindentation response of novel hydroxyapatite-mullite composites. Mat. Sci. Eng. A 513–514 (2009), 197–201.

21 Standard Test Method for Dynamic Young's Modulus, Shear Modulus and Poisson's Ratio for Advanced Ceramics by Impulse Excitation of Vibration ASTM Designation. June, 1995, C1259-95, 375.

22 G. Roebben, B. Basu, J. Vleugels, J. Van Humbeeck, and O. Van Der Biest. The innovative impulse excitation technique for high temperature mechanical spectroscopy. J. Alloys Com. 310(1–2) (2000), 284–287.

23 B. BASU, L. DONZEL, J. VAN HUMBEECK, J. VLEUGELS, R. SCHALLER, and O. VAN DER. Biest; Thermal expansion and damping characteristics of Y-TZP. Scr. Mater. 40(7) (1999), 759–765.
24 S. PALMQVIST. Occurrence of crack formation during Vickers indentation as a measure of the toughness of hard materials. Arch Eisenhuettenwes 33 (1962), 629–633.
25 M. S. KALISZEWSKI, G. BEHRENS, A. H. HEUER, M. C. SHAW, D. B. MARSHALL, G. W. DRANSMANN, and R. W. STEINBRECH. Indentation studies on Y_2O_3-stabilized ZrO_2: I, development of indentation-induced cracks. J. Am. Ceram. Soc. 77 (1994), 1185–1193.
26 G. D. QUINN and R. C. BRADT. On the Vickers Indentation Fracture Toughness Test. J. Am. Ceram. Soc. 90(3) (2007), 673–680.
27 B. R. LAWN. Indentation of ceramics with spheres: a century after Hertz. J. Am. Ceram. Soc. 81(8) (1998), 1977–1994.
28 B. R. LAWN, N. P. PADTURE, H. CAIT, and F. GUIBERTEAU. Making ceramics "ductile." Science 263 (1994), 1114–1116.
29 J. B. QUINN and G. D. QUINN. Indentation brittleness of ceramics: a fresh approach. J. Mater. Sci. 32 (1997), 4331–4346
30 H. TADA. *The Stress Analysis of Cracks Handbook*, 2nd edition. Paris Productions Inc., St. Louis, MO, 1985.
31 Standard Test Methods for the Determination of Fracture Toughness of Advanced Ceramics at Ambient Temperatures, ASTM Provisional Standard Designation No. PS070-97. American Society for Testing and Materials, West Conshohocken, PA.
32 B. BASU, J. VLEUGELS, and O. VAN DER BIEST. Microstructure-toughness-wear relationship of tetragonal zirconia ceramics. J. Eur. Cer. Soc. 24(7) (2004), 2013–2040.
33 G. R. ANSTIS, P. CHANTIKUL, B. R. LAWN, and D. B. MARSHALL. A critical evaluation of indentation techniques for measuring fracture toughness: I, direct crack measurements. J. Am. Cer. Soc. 64 (1981), 533–538.
34 A. G. EVANS and T. R. WILSHAW. Quasi-static solid particle damage in brittle solids—I. Observations analysis and implications. Acta Metallurgica 24(10) (1976), 939–956.
35 K. NIIHARA, R. MORENA, and D. P. H. HASSELMAN. Evaluation of K_{Ic} of brittle solids by the indentation method with low crack-to-indent ratios. J. Mater. Sci. Lett. 1 (1982), 13–16.
36 D. K. SHETTY, I. G. WRIGHT, P. N. MINCER, and A. H. HEUER. Indentation fracture toughness of WC–Co ceramics. J.Mater. Sc. 20 (1985), 1873–1882.
37 P. CHANTIKUL, G. R. ANSTIS, B. R. LAWN, and D. B. MARSHALL. A critical evaluation of indentation techniques for measuring fracture toughness: II, Strength method. J. Am. Ceram. Soc. 64(9) (1981), 539–543.
38 R. F. COOK and B. R. LAWN. A modified indentation toughness technique. J. Am. Cer. Soc. 66(11) (1983), C200–C201.
39 K. T. FABER and A. G. EVANS. Crack deflection processes—I. Theory. Acta Metallurgica 31(4) (1983), 565–576.
40 K. T. FABER and A. G. EVANS. Crack deflection processes—II. Experiment. Acta Metallurgica 31(4) (1983), 577–584.
41 A. G. EVANS and R. M. MCMEEKING. On the toughening of ceramics by strong reinforcements. Acta Metallurgica 34(12) (1986), 2435–2441.
42 C. B. CARTER and M. G. NORTON. *Ceramic Materials*. Springer, New York, 2007.
43 M. W. BARSOUM. *Fundamentals of Ceramics*. Taylor & Francis, London, 2003.
44 W. D. KINGERY, H. K. BOWEN, and D. R. UHLMANN. *Introduction to Ceramics*, 2nd ed. John Wiley and Sons, New York, 1976.
45 D. J. GREEN. *An introduction to the Mechanical Properties of Ceramics*. Cambridge University, Cambridge, 1998.
46 J. B. WACHTMAN, W. R. CANNON, and M. JOHN MATTHEWSON. *Mechanical Properties of Ceramics*. John Wiley & Sons, NewYork, 2009.
47 K. K. CHAWLA. *Ceramic Matrix Composites*. Chapman and Hall, New York, 2003.
48 I. J. MCCOLM. *Ceramic Hardness*. Plenum Publishing Corporation, New York, 1990.
49 D. J. GREEN, R. H. J. HANNINK, and M. V. SWAIN. *Transformation Toughening of Ceramics*. CRC Press, Boca Raton, FL, 1989.

Section Two

Processing of Ceramics

Chapter 4

Synthesis of High-Purity Ceramic Powders

Conventional processing routes of synthesizing materials (specifically polymers and metals) do not work for ceramics. It is purely because of their high melting points and extreme brittleness. Melting would require a container where the molten ceramics need to be held and poured. The irony is that ceramics themselves act as crucibles for pouring liquid metal. So they cannot be used to (1) melt a self-similar ceramic, (2) be able to contain a molten and superheated ceramic for consequent pouring, and (3) solidify ceramics in a cast to result in a final component. In addition, their extremely poor ductility and malleability restricts their formability via processes such as rolling, low-temperature extrusion, and bending. Inherently, ceramics processing requires sintering to achieve full (or controlled) densification without reaching the melting point of the ceramics. In addition, a viable combination of heat and pressure can be adopted for successful compaction of ceramics. Synthesis of high-purity ceramics requires key elements described in Chapter 4. The sintering mechanisms involved are presented in Chapter 5, followed by sintering of ceramics via conventional and advanced routes in Chapter 6.

It is commonly accepted that the starting powder is critically important for the fabrication of advanced ceramics with improved mechanical properties and reliability as well as reproducible behavior. In particular, strength and fracture toughness are observed to be strongly dependent on the particle size, the chemistry of the starting powder, and the sintering parameters. Herewith, the synthesis of various ceramic powders is described.

4.1 SYNTHESIS OF ZrO_2 POWDERS

Over the last few years, several new technologies have been developed for the production of submicron, ultrapure powders with a narrow size distribution.[1,2] The factors controlling the quality of zirconia starting powders include an ultrafine particle size, a narrow size distribution, the amount and distribution of stabilizing oxides

Advanced Structural Ceramics, First Edition. Bikramjit Basu, Kantesh Balani.

(yttria [Y_2O_3], calcia, etc.), a uniform particle shape, and a low impurity content. The different powder processing routes to obtain zirconia starting powders include hydroxide coprecipitation or alkoxide hydrolysis, gel precipitation, microemulsion techniques, sol-gel synthesis, gas phase reaction, and hydrothermal synthesis.[3–6] Only the first and last of these methods are exploited in the commercial production of zirconia powders for the fabrication of tetragonal zirconia ceramics. In the coprecipitation route,[6] the yttria-doped powders are obtained by ammonia leaching of $ZrOCl_2$ and $YCl_3/Y(NO_3)_3$ solution. In the obtained powders, the yttria dopant is homogeneously distributed in the zirconia particles. In the plasma coating route, yttria is co-milled with ZrO_2 base powders produced by plasma decomposition of $ZrCl_4$. More recently, Belgian researchers have developed a new route to synthesize Y_2O_3-doped powders using a "suspension drying" method.[7,8] In this route, a stock solution of Y_2O_3 dissolved in HNO_3 (65 vol%) is mixed with commercial monoclinic zirconia powders and is subsequently milled with Y-TZP milling balls and calcined in air at 800°C for 1 hour to obtain submicron-sized yttria-coated zirconia (Y-ZrO_2) powders. The impurity content (wt%) of the high-purity commercial Y-ZrO_2 starting powders includes Al_2O_3 (0.005), SiO_2 (<0.002), Fe_2O_3 (0.004), and Na_2O (<0.01). The prominent commercial producers and suppliers of high-purity zirconia powders are Japanese companies such as Tosoh, Nikkato, Toyo Soda, Daiichi, and the U.K.-based Tioxide Specialities.

4.2 SYNTHESIS OF TiB_2 POWDERS

Titanium diboride (TiB_2) powder can be prepared by a variety of high-temperature methods, such as the direct reactions of titanium or its oxides or hydrides with elemental boron over 1000°C; carbothermal reduction of titanium oxide and boron oxide; or hydrogen reduction of boron halides in the presence of the metal or its halides. Among various synthesis routes, electrochemical synthesis and solid-state reactions have been developed for preparing finer titanium diboride in large quantities. An example of a solid-state reaction is the borothermic reduction, which can be illustrated by the following reaction:

$$2TiO_2 + B_4C + 3C \rightarrow 2TiB_2 + 4CO. \qquad (4.1)$$

Typical oxygen and carbon content in the as-synthesized TiB_2 is around 0.5 and 0.6 wt%, respectively. Synthesized TiB_2 powder is observed to have finer sizes with D_{50} around 1.1 µm (see Fig. 4.1). It is worthwhile to mention here that it has now become possible to produce large quantites (kilogram scale) of phase-pure TiB_2 powders on both laboratory scale and commercial scale using borothermic reduction process. The aforementioned powder synthesis routes, however, cannot produce nanosized powders. Bates and coworkers[9] prepared nanocrystalline TiB_2 of 5- to 100-nm size using a solution phase reaction of $NaBH_4$ and $TiCl_4$, followed by annealing the obtained amorphous precursor at 900–1100°C. Axelbaum et al.[10] developed a gas phase combustion process, which directly yielded nonagglomerated, low-oxygen-content TiB_2 nanoparticles by the reaction of sodium vapor with $TiCl_4$ and BCl_3.

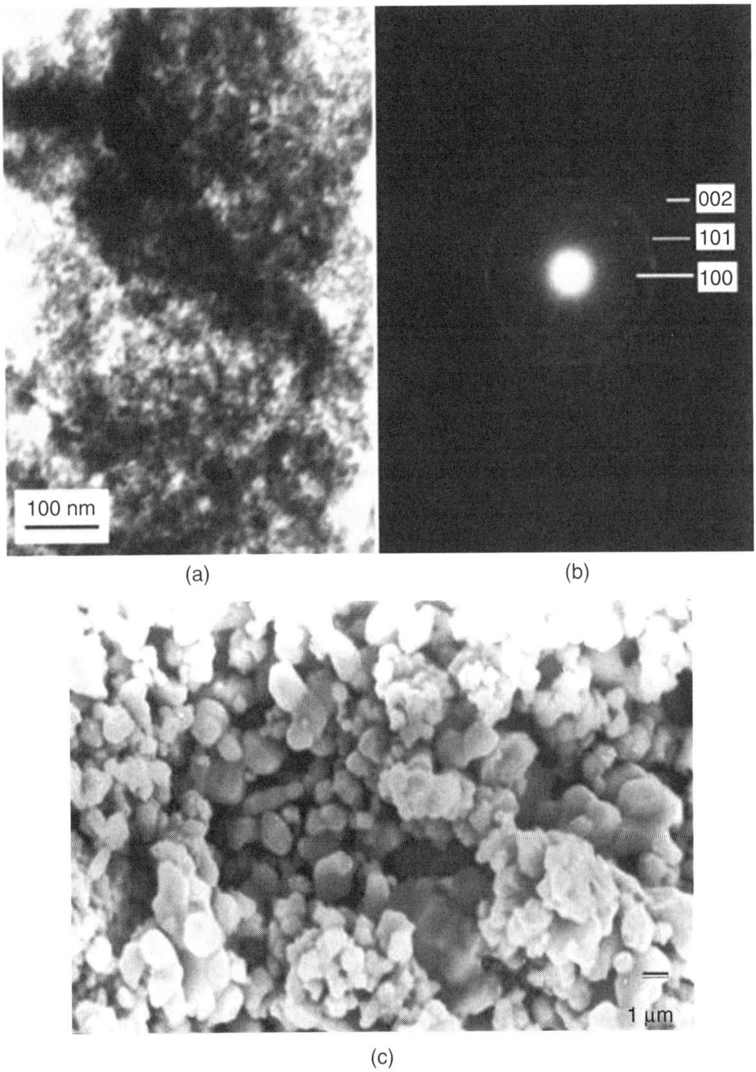

Figure 4.1 (a) Transmission electron microscopy (TEM) image and (b) selected area diffraction pattern (SADP) of nanocrystalline TiB_2 powders, synthesized by reaction between $TiCl_4$ and $NaBH_4$ for 12 hours at 600°C.[11] (c) SEM image of micron-sized TiB_2 powders, produced by borothermic reaction among TiO_2, B_4C, and C.

However, the products were reported to be contaminated with metallic titanium and titanium oxide.

Another potential route to produce submicrometer-sized TiB_2 powder is mechanical alloying of the mixture of elemental Ti and B powders. No amorphization is reported to occur during the process, because of the negative heat of formation of TiB_2. The elemental powders were found to react to form stable TiB_2. It was noted that the size of the transition metal and the heat of formation of borides greatly affected the mechanical alloying time, while producing finer sized TiB_2. Ultrafine (nanometric) TiB_2 powder could also be produced through a self-propagating high-temperature synthesis (SHS) process involving addition of varying amounts of NaCl. As the amount of NaCl (diluents) increased, the particle size of TiB_2 was found to decrease, reaching a size of 26 nm in the case of 20 wt% NaCl addition. The ignition temperature for the stoichiometric mixture of TiO_2, H_3BO_3, and Mg was found to be as low as 685°C.[11]

In an effort to synthesize nanocrystalline titanium diboride, Gu and coworkers[12] developed a novel route using solvothermal reaction of metallic sodium with amorphous boron powder and $TiCl_4$ at 400°C. Such a synthesis route can be represented by the following reaction:

$$TiCl_4 + 2B + 4Na \xrightarrow{\text{Benzene at } 400^\circ\text{C}} TiB_2 + 4NaCl. \quad (4.2)$$

It has been reported that active (nascent) titanium, generated by the reduction of $TiCl_4$ using metallic sodium, helps in the formation of nanocrystalline TiB_2. The benzene serves as a reaction medium to control the rate of reaction and particle size. In another work, Chen et al.[13] also prepared nanocrystalline TiB_2 by the reaction of $TiCl_4$ with $NaBH_4$ in the temperature range of 500–700°C for 12 hours in an autoclave. As shown in Figure 4.1, the nanocrystalline TiB_2 exhibits a size range of 10–20 nm.

From the preceding discussion, it should be evident that a number of laboratory-scale synthesis routes have been successfully developed to synthesize micron- or submicron-sized TiB_2 powders. Such success is, however, limited as far as industrial-scale production of large quantities of TiB_2 is concerned. From the point of view of sintering and consolidation, an emphasis should be placed on obtaining finer TiB_2 powders with a narrow size distribution as well as with limited agglomeration. From classical Herring's scaling law, it can be theoretically predicted that a decrease in particle size by one order of magnitude can cause three to four orders of magnitude reduction in sintering time, depending on the dominant densification mechanism (such as lattice diffusion, grain boundary diffusion). The presence of agglomerates in the starting powders causes a decrease in sinterability and the formation of microcracks and macrocracks in the sintered ceramics. The problem of agglomeration, therefore, has been recognized as an important issue in the synthesis of nanosized ceramic powders. Besides the size of the initial TiB_2 powder, its purity, in terms of oxygen content, largely influences sinterability. For example, TiB_2 powders containing ≥1 wt% oxygen could be densified only up to 90% of the theoretical density, even at higher sintering temperature.

4.3 SYNTHESIS OF HYDROXYAPATITE POWDERS

Hydroxyapatite (HAp) is reported to be synthesized widely using a suspension–precipitation route.[14] The precursor chemicals commonly used are calcium oxide (CaO) and phosphoric acid ($H_3(PO)_4$). At the initial stage, CaO is dispersed in distilled water with a concentration of 18.6 g/L. The dispersed medium should be kept on a hot plate and should be stirred using a magnetic stirrer. Following this, an equivalent amount (keeping the Ca/P ratio the same as HAp) of H_3PO_4 solution (0.17 M) is added dropwise to the dispersed CaO medium. The total solution needs to be stirred at around 80°C for 3–4 hours to allow the reaction to take place toward completion. Thereafter, concentrated NH_4OH (liquor ammonia) is added dropwise until the pH of the solution increases to 10. Subsequently, the solution is kept at room temperature for one day to facilitate the precipitation of the reaction product, which is collected with the help of a filter paper. After drying at 100°C for one day, the lump needs to be crushed by using an agate mortar to make the material into powder form. This method is capable of producing phase-pure HAp powder. The calcined powders have mean particle size d_{50} of 1.3 µm (determined with a laser particle size analyzer). Inductive coupled plasma–atomic emission spectroscopy (ICP-AES) analysis using a complexometry technique indicated that the aforementioned synthesis route can produce HAp powders with a Ca/P ratio of 1.64, after calcination at 800°C. A schematic illustration of the HAp powder synthesis, as described previously, is shown in Figure 4.2a, and a scanning electron microscopy (SEM) image of the as-synthesized powder is shown in Figure 4.2b.

4.4 SYNTHESIS OF HIGH-PURITY TUNGSTEN CARBIDE POWDERS

Tungsten carbide (WC), an important member of the hard carbides family and in particular WC–Co cermet is widely used due to its relatively high hardness and high elastic modulus, fracture toughness, and wettability to cobalt.[15–17] Due to decomposition before melting, WC carbide cannot be produced by melting. Conventionally WC powders are produced by direct reaction of pure W with carbon.[18] In this method, the particle size of WC powders can vary in the range of 0.15–12 µm, depending on the initial pure W powder size and carburization temperature. Coarse WC powder (size >2 µm) can be produced using coarse W powder and high carburization temperature. Similarly, submicron W powder is required to obtain a WC cuboid size of less than one micron at 1350–1400°C carburization temperature. A schematic diagram illustrating the relationship between carburization temperature and WC size is shown in Figure 4.3. The W powder used in cemented carbide technology is usually produced by reduction of tungsten trioxide WO_3, tungsten hydratic acid (H_2WO_4), blue tungsten oxide (W_4O_{11}), or ammonium paratungstenate ($5NH_3{\cdot}12WO_3{\cdot}5H_2O$). To provide an illustrative example, SEM images of typical W and WC powders are shown in Figure 4.3.

To obtain WC–Co materials, tungsten carbide is mixed with cobalt and subsequently sintered to full density using conventional powder metallurgical techniques. Several methods have been developed to synthesize finer-sized WC cuboids and sintered WC products. The synthesis of WC by gas–solid reaction is the method used to obtain finer WC. Other methods include high-frequency induction heated combustion synthesis (HFIHCS), spark plasma sintering (SPS), field-activated and

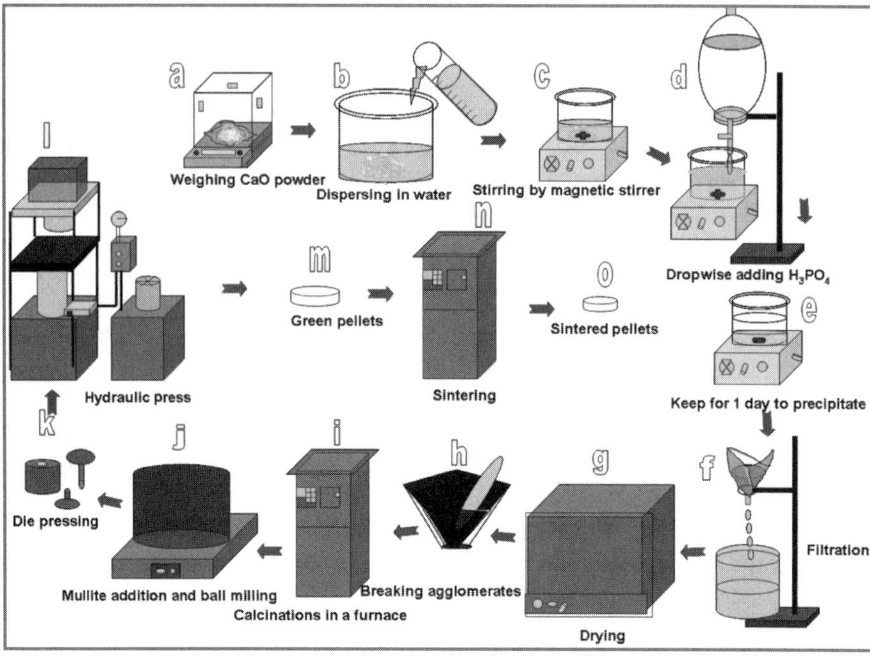

(a)

(b)

Figure 4.2 (a) Illustration of the synthesis of HAp powders and subsequent cold pressing to obtain green pellets (see color insert) and (b) SEM image of as-synthesized HAp powders, after 16 hours of ball milling.

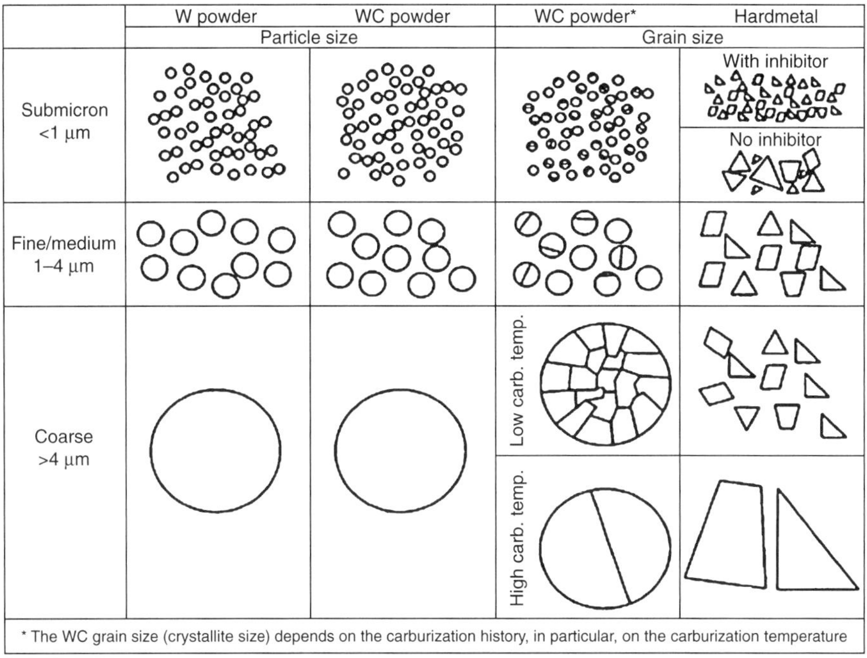

Figure 4.3 WC size as a function of carburization temperature and initial W powder size.[17]

pressure-assisted combustion synthesis (FAPACS), plasma enhanced chemical vapor deposition, ion arc exchange.

Eskandarany et al.[19] prepared pure WC powders (~7-nm size) via mechanical solid-state reduction of WO_3 and Mg, followed by solid-state reaction of W and C using high-energy ball milling (HEBM). These powders were consolidated at 1690°C via PAS. It was reported that the as-consolidated WC had a grain size of around 25 nm, which resulted in extremely high hardness of around 23 GPa. Eskandarany et al.[19] also fabricated bulk nanocrystalline WC (grain size ~95 nm) via hot pressing (1690°C, 1.5 GPa) nanocrystalline WC powders (~7 nm), produced by HEBM of elemental W and C at room temperature. The as consolidated WC nanoceramics exhibited a hardness of ~21 GPa. An SEM image of pure W and WC is shown in Figure 4.4.

As a concluding note, the following points need to be emphasized (see Fig. 4.5) as far as ceramic processing is concerned; more details on ceramic powder characteristics can be found elsewhere.[20–23] First, if the purity of starting powders is not controlled, then the impurities can react to form undesired phases during sintering. This will eventually lead to degradation in consequent fabrication and postprocessing stages. Second, the sizes and size distribution need to be fine and uniform, respectively, in order to attain homogeneous consolidation and densification. In the case of a bimodal distribution with a large fraction of coarser particles, the sinterability of coarse-particle regions will be inferior to that of finer-particle regions. Such differences in densification lead to residual porosity, which can cause eventual

Figure 4.4 SEM image of W powder (a) and WC powder (b).[24]

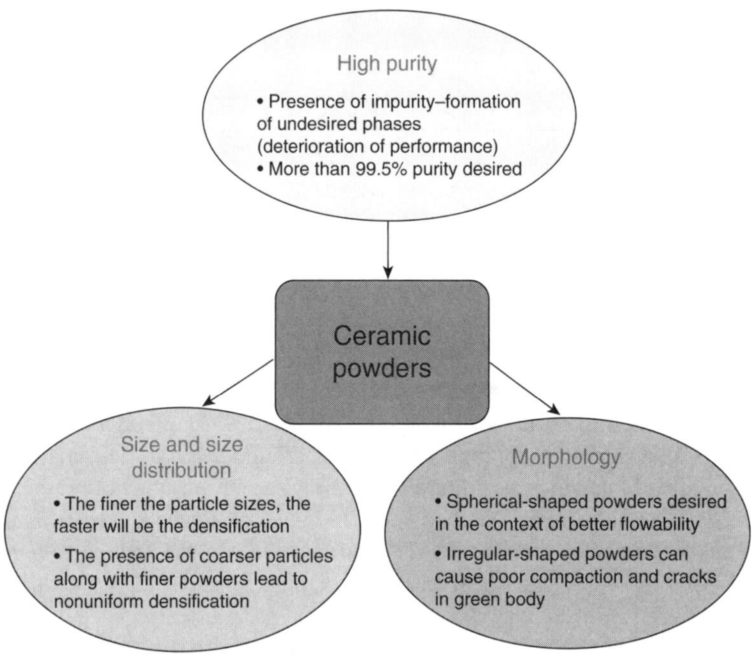

Figure 4.5 Schematic illustration showing the properties required for ceramic powders from the point of view of enhanced sinterability and better properties.

cracking during mechanical loading. As far as the powder morphology is concerned, the proper filling of a mold cavity requires good flow characteristics (flowability), which can be attained from the use of spherically shaped powder particles. Irregularly shaped particles cause inferior compaction properties and result in eventual cracking during sintering.

REFERENCES

1 A. J. A. Winnubst and M. M. Boutz. Sintering and densification; new techniques sinter forging. Key Eng. Mater. 301 (1998), 153–154.

2 M. J. Mayo. Processing of nanocrystalline ceramics from ultrafine particles. Int. Mater. Rev. 41(3) (1996), 85–115.

3 M. M. R. Boutz, R. J. M. Olde Scholtenhuis, A. J. A. Winnubst, and A. J. Burggraaf. British ceramic proceedings, in *Nanoceramics*, Vol. 51, R. Freer (Ed.). The Institute of Materials, London, 1993, 75–86.

4 J. Tartaj, J. F. Fernandez, C. Moure, and P. Duran. Effects of seeding on the crystallisation kinetics of air-calcined yttria-doped hydrous zirconia. J. Eur. Ceram. Soc. 18 (1998), 229–235.

5 D. Burgard, C. Kropf, R. Nass, and H. Schmidt. Better ceramics through chemistry. MRS 346 (1994), 101–107.

6 G. Dell'Agli and G. Mascolo. Hydrothermal synthesis of ZrO_2–Y_2O_3 solid solutions at low temperature. J. Eur. Ceram. Soc. 20 (2000), 139–145.

7 R. Singh, C. Gill, S. Lawson, and G. P. Dransfield. Sintering microstructure and mechanical properties of commercial Y-TZPs. J. Mat. Sc. 31 (1996), 6055.

8 W. Burger, H. G. Richter, C. Piconi, R. Vatteroni, A. Cittadini, and M. Boccalari. New Y-TZP powder for medical grade. J. Mat. Sc. Mater. Med. 8 (1997), 113.

9 S. E. Bates, W. E. Buhro, C. A. Frey, S. M. L. Sastry, and K. F. Kelton. Synthesis of titanium boride $(TiB)_2$ nanocrystallites by solution-phase processing. J. Mater. Res. 10 (1995), 2599.

10 R. L. Axelbaum, D. P. DuFaux, C. A. Frey, K. F. Kelton, S. A. Lawton, L. J. Rosen, and S. M. L. Sastry. Gas-phase combustion synthesis of titanium Boride [TiB2] nanocrystallites. J. Mater. Res. 11 (1996), 948–954.

11 A. K. Khanra, L. C. Pathak, S. K. Mishra, and M. M. Godkhindi. Effect of NaCl on the synthesis of TiB_2 powder by a self-propagating high-temperature synthesis technique. Materials Letters 58 (2004), 733–738.

12 Y. Gu, Y. Qian, L. Chen, and F. Zhou. A mild solvothermal route to nanocrystalline titanium boride. J. Alloys Comp. 352 (2003), 325–327.

13 L. Chen, Y. Gu, Y. Qian, L. Shi, Z. Yang, and J. Ma. A facile one step route to nanocrystalline TiB_2 powders. Mater. Res. Bull. 39 (2004), 609–613.

14 M. H. Santos, M. Oliveira, L. P. F. Souza, H. S. Mansur, and W. L. Vasconcelos. Synthesis control and characterization of hydroxyapatite prepared by wet precipitation process. Mater. Res. 7 (2004), 625–630.

15 H. E. Exner. Physical and chemical nature of cemented carbides. Int. Mat. Rev. 4 (1979), 149–173.

16 H. J. Scussel. *Friction and Wear of Cemented Carbides*, Vol. 18. ASM Handbook, ASM Int., Metals Park, OH, 1992, 795.

17 E. Lassner and W. D. Schubert. *Tungsten—Properties, Chemistry, Technology of the Element, Alloys and Chemical Compounds*. Kluwer Academic/Plenum Publishers, New York, 1999.

18 B. Aronsson. Influence of processing on properties of cemented carbide. Int. Powder Met. 30(3) (1987), 175.

19 M. S. E. Eskandarany, A. A. Mahday, H. A. Ahmed, and A. H. Amer. Synthesis and characterizations of ball-milled nanocrystalline WC and nanocomposite WC-Co powders and subsequent consolidations. J. Alloys Comp. 312 (2000), 315–325.

20 J. S. Reed. *Introduction to the Principles of Ceramic Processing*. John Wiley & Sons, New York, 1988.

21 C. B. Carter and M. G. Norton. *Ceramic Materials*. Springer, New York, 2007.

22 M. W. Barsoum. *Fundamentals of Ceramics*. Taylor & Francis, London, 2003.

23 L. S.-J. Kang. *Sintering*. Elsevier, Burlington, VT, 2005.

24 M. Tata, D. Miroud, S. Lebaili, and T. Cutard. The study of properties of WC-based and W-based composites fabricated by infiltration with liquid Cu-Mn binder. Asian J. Sci. Res. 2 (2009), 76–86.

Chapter 5

Sintering of Ceramics

In this chapter, the thermodynamic and kinetics aspects of sintering of ceramics are discussed. The mechanisms of solid-state and liquid-phase sintering (LPS) are reviewed with an emphasis on illustrating how various parameters can be optimized to achieve densification without grain growth. One of the advanced sintering techniques, for example microwave sintering (MW), is also briefly discussed.

5.1 INTRODUCTION

Sintering refers to the process of firing and consolidation of powders at $T > 0.5T_m$, where diffusional mass transport leads to the formation of a dense body. It is a technique based on powder metallurgy and produces high-density materials and components from metal or ceramic powders by applying thermal energy and/or mechanical pressure. Typically, sintering, a variant of powder metallurgy process, involves preparation of a powder blend (powder + binder or sinter additive) followed by compaction and consolidation at high temperature (see Fig. 5.1). Sintering involves a process of transformation from a porous state to a state of dense material and it must involve the process of neck formation and growth. As shown in Figure 5.2, the sintering process essentially involves the dynamic and continuous change in pore size and shape. As is explained later, the removal of porosity can take place by mass transport from powder particles to porous region either via lattice diffusion (i.e., from bulk of powder) or via grain boundary (GB) (particle boundary) diffusion. It is now widely recognized that sintering is the only processing route for the refractory metals and ceramics. This is because of the following factors: (1) The conventional melting–solidification route, as is widely used for metals, cannot be applied in the case of ceramics. This is because the melting points of most of these materials lie in the range of 2000–3500°C, thereby necessitating the use of furnaces capable of operating at such extreme conditions. This is not a cost-effective engineering solution. (2) Ceramics are essentially brittle and lack any room-temperature ductility. Therefore, conventional metal-forming processes, for example forging and rolling, cannot be adopted to produce products with desired shapes in the case of refractory

Advanced Structural Ceramics, First Edition. Bikramjit Basu, Kantesh Balani.

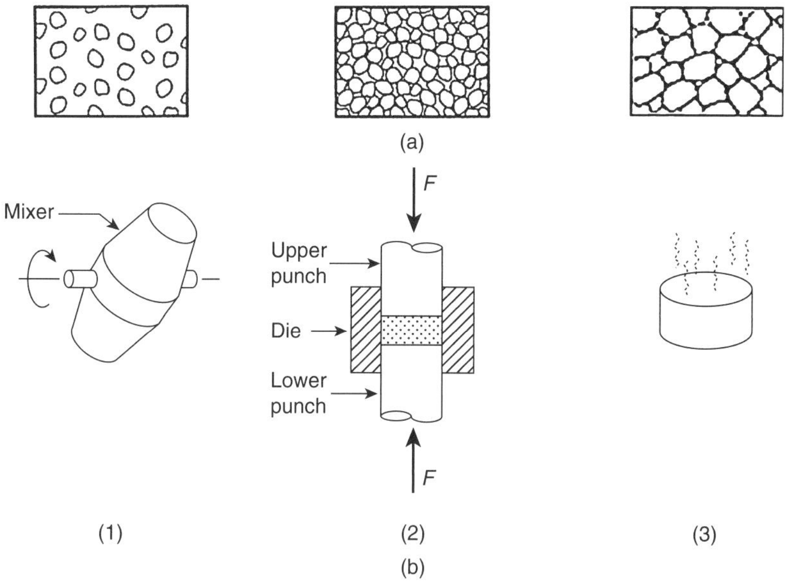

Figure 5.1 Stages involved in the conventional sintering route: (a) blending, (b) compaction, and (c) sintering.

metals and ceramics. (3) Low thermal conductivities of ceramics (2–50 W/m·K), in contrast to the conductivity of metals cause large temperature gradients, resulting in thermal stress and shock during melting and solidification of ceramics.

Sintering can be accomplished by using pressure (e.g., hot pressing, hot isostatic pressing [HIP or HIPing]) or without pressure (e.g., pressureless sintering in air). At this juncture, it can be mentioned that this sintering route is also adopted to produce porous ceramics for various engineering applications, including biomedical applications. To fabricate ceramics with a desired porosity, two strategies can be adopted: (1) Powders can be mixed with volatile material (naphthalene, poly(methyl methacrylate) [PMMA], polyvinyl alcohol [PVA], sugar, etc.) and subsequently can be sintered. The idea is that the evaporation of the volatile material will leave pores. With the selection of the amount and size of volatile material, a porous ceramic with the desired microstructure can be obtained. Adopting this strategy, one can control volume fraction and size of porosity. (2) Sintering at a temperature much lower than that used to obtain a dense ceramic can also produce a porous material. The idea here is not to allow the complete removal of open porosity. This is an uncontrolled way to produce porous ceramics. As far as the distinction between closed and open pores is concerned, it is widely observed that when sintered density is above 95% of the theoretical density, then closed porosity is left in the sintered material. Below 95% theoretical density, pores are open to the free surface and often are interconnected in the bulk of the sintered ceramics.

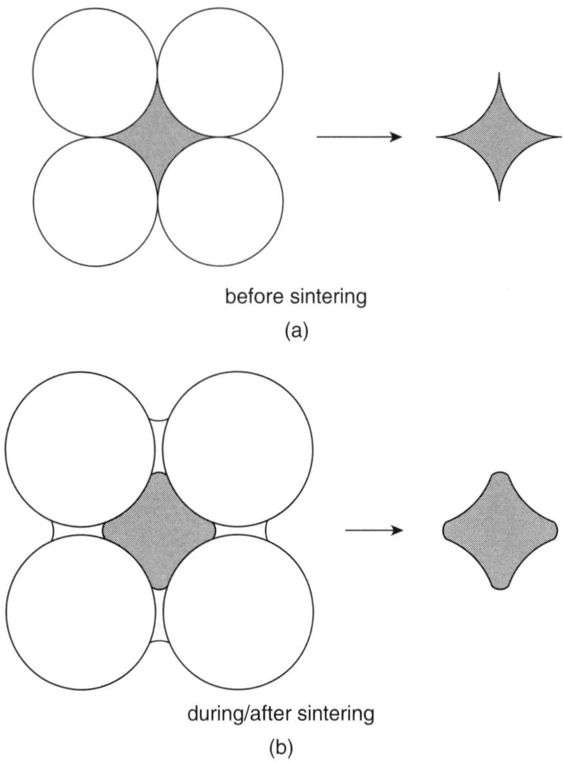

Figure 5.2 Illustration of the pore size and shape change during pressureless sintering: (a) initial stage and (b) intermediate stage.

5.2 CLASSIFICATION

Sintering can be broadly classified into two categories:

(**a**) Liquid phase sintering (LPS)

(**b**) Solid-state sintering (SSS)

In the case of LPS, a liquid phase forms either prior to reaching the sintering temperature or at the sintering temperature, and the liquid phase enhances the mass transport required for sintering. LPS is effective in the case of refractory metals (W, Ta, etc.) or non-oxide ceramics (SiC, Si_3N_4, etc.). In the case of solid-state sintering, the liquid phase does not form and the entire solid–solid (S/S) and solid–vapor (S/V) interfacial areas are replaced by only S/S interfacial areas. The solid-state sintering process takes place purely by diffusional process induced neck growth. This is in contrast to LPS, which involves melting of the liquid phase followed by rapid diffusional mass transport. A third type, although not listed in the conventional classification, is reactive sintering. This involves the *in situ* sintering reaction between starting powders and enhanced mass transport in the presence of the product phase, which is necessarily a liquid phase at sintering temperature. Later in this chapter, reactive sintering is discussed. Often reactive sintering is considered as synonymous to LPS. However, the fundamental difference is that LPS typically involves formation of liquid resulting from an oxide layer on the powder surface or its reactivity with dopants, sinter-aid, or sintering atmosphere.

5.3 THERMODYNAMIC DRIVING FORCE

In simplistic terms, sintering is described as a process of transformation of a porous powder compact to a dense solid. From a thermodynamic point of view, all the S/V interfaces in the porous compact are essentially replaced by S/S interfaces. Total driving force for sintering is therefore the reduction in surface energy of the system and any irreversible process with reduction in total energy of the system is thermodynamically preferred. Thermodynamically, this can be expressed as

$$d(\gamma A) = Ad\gamma + \gamma dA < 0. \tag{5.1}$$

Typically S/S or S/V surface area is replaced by only S/S during the initial or second stage of sintering and, therefore, the $(Ad\gamma)$ term dominates. During the final stage of sintering, grain coarsening occurs (if sintering time or holding time is not optimized) and, therefore, the (γdA) term dominates at the last stage of sintering. Ideally, sintering of the powder compact should lead only to densification into the most close-packed arrangement of a hexagonal array of grains and should ideally evolve from a state of largely loose-packed spherical powder particles (see Fig. 5.3). It can be reiterated here that the sintering process is associated with dynamic changes in pore size and shape. For example, a cylindrical interconnected pore channel is developed during the intermediate stage of sintering. Once the shrinkage of these pores takes place to a significant extent (resulting in an increase in relative density by ~30%),

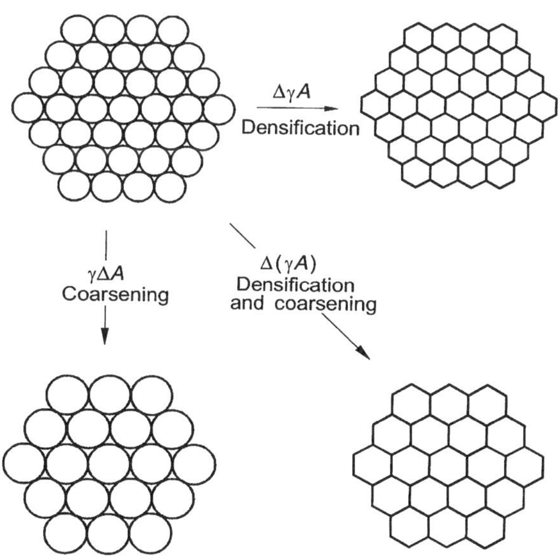

Figure 5.3 Basic phenomena occurring during sintering under the driving force for sintering (adapted from Reference 1).

the final stage of sintering starts. The isolated pores can now be removed by lattice diffusion alone. The transformation of a porous compact to a dense solid as well as the change in pore size and shape during sintering can be analyzed from first-principles calculations, which are based on the fact that the pressure applied to the solid by the curved surface increases the chemical potential of the constituents and the pressure of the vapor phase in equilibrium with them. The increase in chemical potential can be derived by considering the transfer of 1 mole of material from the flat surface through the vapor or liquid to the spherical surface. Considering this condition and from the first-principle calculations, the mass transport leading to removal of interparticle porous region involved in sintering can be analyzed; such an analysis, as described next, leads to the derivation of the Thompson–Freundlich equation (TFE). This also provides an estimate of the equilibrium vapor pressure in a powder compact. From fundamental thermodynamic concepts, the work done in expanding a sphere of radius r against external constraint is equal to the increase in surface energy; this can be written as $-\Delta p{\cdot}dV = \gamma dA$. This scenario is realized for an individual powder particle in a powder compact.

For a spherical particle $dA = 8\pi r dr$ and $dV = 4\pi r^2 dr$ and therefore,

$$\Delta p = \gamma\left(\frac{2}{r}\right), \tag{5.2a}$$

where γ is the surface energy, $r > 0$ for convex surfaces, $r = \infty$ for flat surfaces, and $r < 0$ for concave surfaces.

For irregular particles of sizes r_1 and r_2, Equation 5.2a can be written as

$$\Delta p = \gamma\left(\frac{1}{r_1} + \frac{1}{r_2}\right). \tag{5.2b}$$

Now, for curved surfaces in equilibrium, the difference in chemical potential can be realized by the work done in getting 1 mole of material from the flat interface to the curved interface.

At constant temperature and pressure, this can expressed as

$$\begin{aligned}\Delta\mu &= RT\ln C\text{-}RT\ln C_o = RT\ln(C/C_o) = RT\ln(p/p_o)\\ &\equiv \text{work done in increasing the surface area,}\end{aligned} \tag{5.3}$$

where C and p are the concentration and vapor pressure, respectively, at the curved interface, and C_0 and p_0 are the concentration and vapor pressure, respectively, at the flat interface.

From thermodynamic point of view, the chemical potential for the ith species can be expressed as

$$\mu_i = \left(\frac{\partial G}{\partial n_i}\right)_{T,P,n_1,n_2,\ldots n_{i-1},n_{i+1}\ldots}, \tag{5.4}$$

where the free energy change (dG) of the system can be expressed as

$$dG = VdP - SdT + \Delta\mu. \tag{5.5}$$

From Equation 5.4, the chemical potential can be described as the change in free energy of the system due to addition of 1 mole of the *i*th species, when temperature (T), pressure (P), and the amount of all other components remain constant. For 1 mole of material transfer, $dV = V_m$ = molar volume = $4\pi r^2 dr$. Therefore, $dr = V_m/4\pi r^2$ and $dA = 8\pi r V_m/4\pi r^2 = 2V_m/r$.

With temperature, pressure, and overall composition remaining constant, the work done is equivalent to the change in chemical potential; therefore, from the preceding set of equations, the following emerges,

$$\mathrm{RT}\ln(C/C_o) = \gamma dA = \gamma(2V_m/r); \tag{5.6}$$

that is,

$$C = C_o \exp[(\gamma V_m/\mathrm{RT})(2/r)]. \tag{5.7}$$

Similarly, one can also write the following expression:

$$p = p_o \exp[(\gamma V_m/\mathrm{RT})*(2/r)]. \tag{5.8}$$

Since, $e^x = 1 + x$ (approximately for small x),

$$C = C_o[1+(\gamma V_m/\mathrm{RT})*(2/r)] \tag{5.9a}$$

or

$$p = p_o[1+(\gamma V_m/\mathrm{RT})*(2/r)]\ (\mathrm{TFE}) \tag{5.9b}$$

From TFE, the equilibrium vapor pressure over a surface of positive curvature exceeds that over a flat surface and that over a surface of negative curvature. One result of the pressure difference across a curved surface is a change in solubility or vapor pressure compared with a planar surface. From TFE, the change in solubility or vapor pressure is approximately inversely proportional to particle size. Therefore, it is common in ceramics processing to synthesize powders of submicron size, which give rise to capillary forces of appreciable magnitude (2–5 MPa).

The neck region can be described by a saddle curvature with neck dimension x and ρ (small and negative), as shown in Figure 5.4. The pressure that is applied to

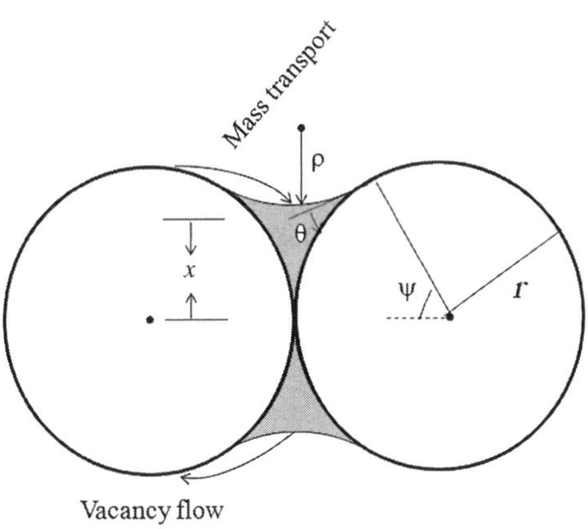

Figure 5.4 Schematic of the contact between two spherical particles during solid-state sintering; x, ρ, and r refer to neck dimension, neck radius, and particle size, respectively.

the GB plane due to the characteristic neck curvature is compressive and can be expressed by the following empirical relationship:

$$\Delta P = \gamma(1/\rho + 1/x) \approx \gamma\left(\frac{1}{\rho}\right). \quad (5.9a)$$

The approximation in Equation 5.9a is due to the fact that $x \gg \rho$ and, therefore, the other term can be neglected. For pressure-assisted sintering, Equation 5.9a can be modified with the assumption that $\rho \sim r$ (particle size),

$$\Delta P = \gamma/r + P_{\text{ext}}\ f(r, \text{geometry}), \quad (5.9b)$$

where P_{ext} is the externally applied pressure, and $f(r$, geometry) is a function illustrating the change in density and geometry of pores during sintering. The neck-curvature-induced pressure with or without external pressure will influence the mass transport and vacancy transport between the GB plane and neck region during sintering processes. One can realize this of one considers the following expression, correlating the change in vacancy concentration with stress (σ):

$$C_{\text{v}}(\sigma = \sigma) = C_{\text{v}}(\sigma = 0)\exp[\sigma\Omega/\text{KT}], \quad (5.10a)$$

where Ω = vacancy volume. Considering the small value of the ($\sigma\Omega$/KT) term, one can write

$$C_{\text{v}}(\sigma = \sigma) = C_{\text{v}}(\sigma = 0)[1 + \sigma\Omega/\text{KT}]. \quad (5.10b)$$

Due to the pressure difference, C_{v} is higher at the neck region and vacancy diffusion occurs down the vacancy concentration gradient. From the preceding mathematical background, it should now be clear that vacancies flow from neck region to GB plane and this is equivalent to material transport from GB plane to neck region. In addition, one should note that the pressure, as estimated from Equation 5.9a, is negative and compressive stress on the GB plane will lead to the interpenetration of spherical particles, leading to overall shrinkage of a powder compact. From Figure 5.4, it can be said that if P_{ext} is applied, ΔP increases. Therefore, vacancy concentration increases, leading to more mass transport. This is the basis for better densification in the case of hot pressing or other pressure-assisted sintering processes.

5.4 SOLID-STATE SINTERING

In this section, the mechanisms of solid-state sintering are discussed. An emphasis is placed on illustrating how neck growth will vary with sintering time. The change in relative density with sintering time is schematically shown in Figure 5.5. Three distinguishable stages can be identified: (1) initial stage, characterized by neck formation at the interparticle region; (2) intermediate stage, characterized by neck growth leading to interconnected pore channel formation and dynamic removal of pores, leading to breakdown of continuous pore channel; and (3) final stage, characterized by achieving nearly theoretical density with isolated pores. In the initial stage of sintering, vapor phase transport of atoms by surface diffusion takes place.

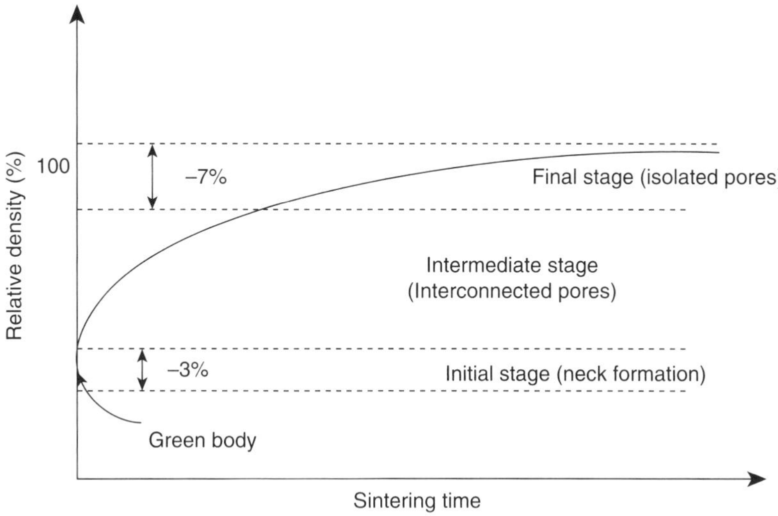

Figure 5.5 Schematic illustration of three stages in solid-state sintering of a powder compact (adapted from Reference [1]).

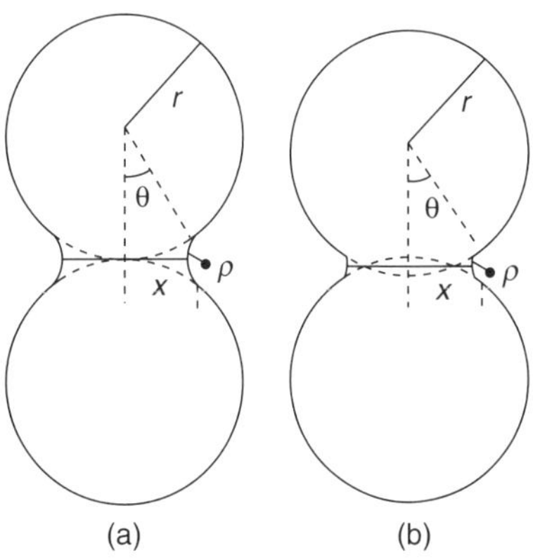

Figure 5.6 Two-particle model for initial-stage sintering (a) without shrinkage and (b) with shrinkage; x, ρ, and r are neck dimension, neck radius of curvature, and particle size, respectively (adapted from Reference 1).

If held for longer time at the initial stage, coarsening instead of shrinkage takes place for the powder compact (see Fig. 5.6).

During continuous heating to sintering temperature, effective neck growth takes place during intermediate-stage sintering by GB diffusion as well as by lattice diffusion during the final stage of sintering. It is clear from Figure 5.5 that maximum densification takes place during the intermediate stage. Therefore, much attention is paid to developing an understanding of how neck sizes increase with time in the intermediate stage of sintering. Following the classical two-particle sintering model, neck growth in the case of solid-state sintering can be explained by the following relationships:

$$\left(\frac{x}{r}\right)^6 = \left[\frac{192\delta\gamma D_b\Omega}{KT}\right]\frac{t}{r^4}, \tag{5.11}$$

$$\left(\frac{x}{r}\right)^5 = \left[\frac{80\delta\gamma D_l\Omega}{KT}\right]\frac{t}{r^3}, \tag{5.12}$$

where D_l is the lattice self-diffusion coefficient for the rate-limiting diffusing species, D_b is the boundary self-diffusion coefficient for the rate-limiting diffusing species, δ is boundary thickness, K is the Boltzmann constant, Ω is the vacancy volume, x is neck dimension, t is the sintering time, T is the sintering temperature, γ is the surface energy, and r is the particle size of the powders. While Equation 5.11 was derived for the intermediate stage of sintering, in which neck growth is controlled by GB diffusion, Equation 5.12 is related to the lattice-diffusion-controlled neck growth process, occurring at the final stage of sintering. From Equations 5.11 and 5.12, it is evident that the neck growth rate increases significantly with reduction in particle size. The dependency of sintering kinetics on powder particle size can be described on the basis of Equations 5.11 and 5.12.

Based on Equations 5.11 and 5.12, the neck growth at a given sintering temperature can be expressed by the following generic equation: $(x/r)^n = \text{constant}*(t/r^p)$, where $n = 2, \ldots, 6$ and $p = 2, \ldots, 4$.

Also, if the relative neck size with respect to particle size remains similar for a particular densification mechanism at a given sintering temperature, then one can write

$$\frac{t(r_2)}{t(r_1)} = \left[\frac{(r_2)}{(r_1)}\right]^p, \tag{5.13}$$

where $p = 2, \ldots, 4$, $t(r)$ is the time taken to sinter a compact of uniform particle size r, and r_1, r_2 are two different particle sizes.

Equation 5.13 is known as Herring's scaling law and this law essentially states that finer powders will sinter at a much faster rate, that is, in a shorter sintering time, compared with coarser powders.[2] For example, 10-nm-sized powders will be densified at the sintering temperature in six orders of magnitude less time than 10-μm-sized powders of the same composition at the same sintering temperature. This establishes the basis for wider research interests in synthesizing finer (nanosized) ceramic powders.[3]

5.5 COMPETITION BETWEEN DENSIFICATION AND GRAIN GROWTH

From Figure 5.5, it can be clearly seen that the densification becomes slower and typically around 5–7% increase in relative density can be expected in the final stage of sintering. As far as the activation energy of diffusion is concerned, lattice diffusion can favorably take place at the final stage of sintering (see Fig. 5.7). The closed pores or even pores detached from the GB are experimentally observed in the final stage of sintering. From a kinetics point of view, both the grain growth and pore

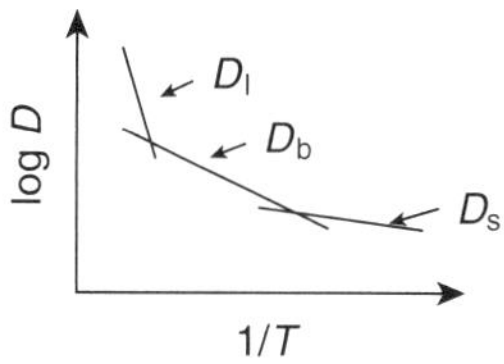

Figure 5.7 Variation of diffusion coefficients with temperature. D_l, lattice (or bulk) diffusion coefficient; D_b, boundary (or interface) diffusion coefficient; D_s, surface diffusion coefficient.

growth take place simultaneously in the final stage. Therefore, the optimization of sintering temperature and time is important to avoid both processes, while ensuring elimination of closed porosity.

The grain growth process can be analyzed by obtaining the mathematical expression for GB mobility as a function of related parameters. The velocity at which each GB moves during the grain growth process can be described as

$$V = \text{mobility term} \times \text{driving force}.$$

The driving force for grain growth stems from the reduction in overall surface energy as well as from the curvature effect of each boundary. In a dense polycrystalline solid, the curvature effect is six times than that of spherical grains of equivalent microstructure.

For spherical grains, the driving force $F = \gamma^* S_v = \gamma^*(3/g)$, where g is grain size and S_v is the ratio of surface area to volume.

Considering that each boundary is shared by two neighboring grains, the GB mobility can be expressed as

$$V = M_b{}^*2^*F$$

or

$$V = M_b{}^*2^*\gamma^*(3/g). \tag{5.14}$$

Taking into account the curvature effect for a polycrystalline solid,

$$V = dg/dt = M_b{}^*\gamma/g. \tag{5.15}$$

Integrating Equation 5.15 and assuming an initial grain size g_0 at $t = 0$, one arrives at

$$g^2 - g_0^2 = 2M_b\gamma t, \tag{5.16}$$

where t is the overall sintering time at the final stage of sintering or the holding time at sintering temperature.

To determine grain growth during sintering, it is therefore important to know the values of "grain boundary mobility" (M_b). There exist three different cases, which can be used to determine the M_b term.

Case I: M_b depends upon the intrinsic diffusivity of host (matrix) atoms, leading to atomic jumps across moving GB:

$$M_b = D_b\Omega/\delta\text{KT}, \tag{5.17}$$

where

D_b = bulk diffusivity of host atom,

δ = boundary layer thickness, ~0.5 nm, and

Ω = vacancy volume.

In finding the D_b of ionic ceramics, the diffusivity of slower of cations or anions along its fastest path should be considered.

Case II: In the presence of a liquid film at GB, the following expression can be used:

$$M_b = D_l C_l \Omega / \delta \mathrm{KT}, \tag{5.18}$$

where

D_l = diffusivity of atoms through the liquid film,

C_l = solubility of host solid in the liquid film, and

δ = liquid film thickness.

Case III: The case of solute drag, that is, the redistribution of solute near the GB region, takes place due to the coupled effect of inherent interaction of segregated solute with GB plane and mobility of GB. The redistribution of solute during grain growth results in a drag force on GB and this can be expressed as

$$M_b = D_s \Omega / [(2\delta)\mathrm{KT} C_s \exp(U/\mathrm{RT})], \tag{5.19}$$

where

D_s = diffusivity of solute;

δ = region of solute segregation;

C_s = overall solute concentration at bulk of grains, away from GB;

U = interaction energy between solute and GB plane; and

$C_s \exp(U/RT)$ = segregated solute concentration at GB region.

Therefore, the net mobility of GB can be written as

$$M_{\mathrm{net}} = [1/M_{\mathrm{b(intrinsic)}} + 1/M_{\mathrm{b(solute\ drag)}}]^{-1}.$$

Examples of Case III include MgO doping in Al_2O_3 or Nb doping in $BaTiO_3$. During sintering, such dopants tend to segregate near GB region (region of higher disorder) and influence GB mobility.

Case IV: The pores attached to GB can exert pore drag, and pore mobility is essentially due to diffusion of materials to pores via lattice–GB diffusion. The pore mobility term can be expressed as

$$M_p = A \exp(-Q/\mathrm{RT})/r^n, \tag{5.20}$$

where $n = 3$ for lattice diffusion and $n = 4$ for GB diffusion, Q = activation energy for rate-limiting process (lattice–GB diffusion), r = pore radius, and A = constant.

Under the preceding condition, the net mobility term can be expressed as

$$M_{\text{net}} = M_b M_p / (NM_b + M_p), \tag{5.21}$$

where M_p = pore mobility, M_b = intrinsic mobility in the absence of pores, and N = number of pores per unit area of GB region.

Now, let us discuss two extreme scenarios that can occur during the final stage of sintering.

When pore density is large or boundary mobility (M_b) is large, then $NM_b \gg M_p$. The net boundary mobility can therefore be approximated as $M_{\text{net}} = M_p/N$. When the number of pores per unit area is small, then $NM_b \ll M_p$ and, therefore, $M_{\text{net}} = M_b$ (boundary migration controlled). Based on the preceding analysis, and on experimental verifications, two important relationships—one between grain size and pore size and the other between grain size and density—are plotted in Figures 5.8 and 5.9. In both diagrams, the region of pore separation has been clearly identified. An important observation is that a small amount of doping, for example MgO doping in Al_2O_3, can shift the pore separation region diagonally upward, so that the entry into this region can largely be avoided during sintering processes. In Figure 5.8, besides the pore separation region, two more regions are boundary control and pore control. With the addition of dopant, the boundary between these two regions also shifts to the right. This means that, at a given grain size, the transition to the pore control region will occur at a larger pore size in doped ceramic, compared with

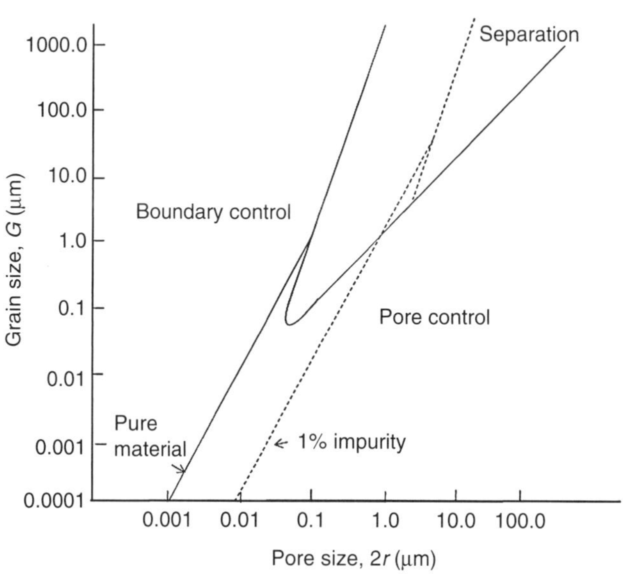

Figure 5.8 Pore separation during sintering (adapted from Reference 2).

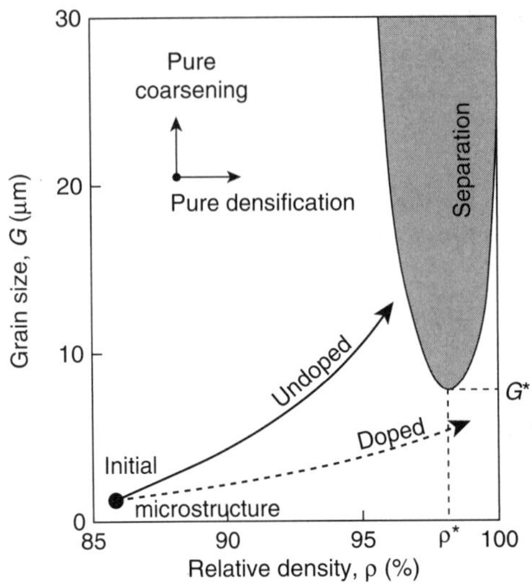

Figure 5.9 Schematic of microstructure development in terms of a plot of grain size versus density (adapted from Reference 1).

undoped ceramic. In Figure 5.9, it can be seen that, at a given densification level, grain growth can be restricted in a doped ceramic. From Figure 5.9, a critical grain size for pore separation at a given densification level (P^*, G^*) can be identified to avoid pore separation. This is primarily due to the fact that the dopants, while being attached to the grain boundaries, will reduce the driving force for GB mobility. From the preceding discussion, it should be clear that the following two ways prevent pore separation during the final stage of sintering and, thereby, allow the attainment of full densification without grain growth:

(a) use of appropriate type of dopant in the required amount and

(b) tailoring sintering temperature and time, so that grain growth can be limited.

5.6 LIQUID-PHASE SINTERING

As mentioned earlier, LPS involves the formation of a liquid phase, constituting primarily the component with lower melting point during sintering. The molten component surrounds the particles that have not melted and, due to enhanced diffusion via the liquid phase, high compact density can be quickly attained in difficult-to-sinter materials. Important variables that are taken into consideration are the nature of the alloy, molten component–particle wetting, and capillary action of the liquid. The formation of sintering liquid occurs via reaction of the ceramic matrix with the sinter-aid. For example, the reaction of Y_2O_3, SiO_2, and Al_2O_3 with Si_3N_4 leads to Y-SiAlON at GB. In some ceramic systems, the liquid phase is often an amorphous or a glass phase. The presence of a glass phase induces brittleness in LPSed materials. For example, in Si_3N_4-based materials, the intergranular glass phase can be crystallized by providing post-sintering heat treatment, that is, heating to a temperature lower than the original sintering temperature. Compared with solid-state sintering, the liquid phase enhances mass transport, as the liquid phase penetrates the interparticle boundary. However, if LPS is not carefully optimized, the liquid phase also enhances grain growth (coarsening), leading to degradation in material properties.

The essential requirements of LPS are as follows:

(a) good wettability, that is, an appreciable amount of liquid formation with good combination of contact angle and wetting angle, so that liquid wets GB region and neck region;

(b) solubility of principal ceramic component in the wetting liquid (appropriate amount of additive needed); and

(c) low-viscosity liquid so as to allow rapid diffusion kinetics.

A typical illustration showing the principles of LPS is shown in Figure 5.10. The contact angle at the S/L/V interface and wetting angle at the S/L interface are the key factors in determining the wettability of solid particles by liquid. The presence of liquid film essentially applies a compressive force, $\Delta P = -2\gamma_{LV}/\rho$, where ρ is the radius of the liquid meniscus as indicated in Figure 5.10. The neck growth during LPS can be described by the following expression:

$$(x/r)^6 = 768\,\delta\,D_{liq}C_o\,\gamma_{LV}\,\Omega/KT^*(t/r^4), \tag{5.22}$$

where r = particle size, x = neck size, and δ = distance between interfaces of two grains.

In Equation 5.22, the interfacial energy term (γ) is dependent on the particle size (r) and is directly dependent on liquid diffusivity (D_{liq}). The higher the D_{liq} or the higher the solubility of particles in the liquid phase, the faster the neck growth will be.

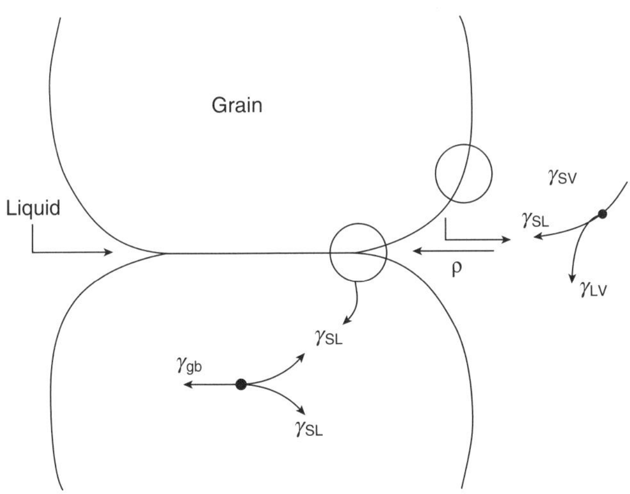

Figure 5.10 Different interfacial energies involved in liquid-phase sintering. The subscripts S, L, and V indicate solid, liquid, and vapor, respectively (adapted from Reference 2).

Three overlapping stages are used to describe the phenomenology of LPS:

(a) ***Rearrangement of Particles.*** This stage occurs within ~10 minutes of the formation of low-viscosity sintering liquid and is driven by the attractive capillary pressure, formed due to lower contact angle (<60°) at the S/L interface.

(b) ***Dissolution–Reprecipitation.*** This stage is faster for finer spherical particles and with low dihedral angle at the S/L interface (<50°). This can be limited either by reaction at the S/L interface or by mass transport (diffusion) via the liquid phase.

(c) ***Solid-State Sintering and Microstructural Coarsening.*** This stage occurs when S/S contact is established at the last stage of LPS.

5.7 IMPORTANT FACTORS INFLUENCING THE SINTERING PROCESS

The sintering process depends on various parameters, which can be classified into intrinsic or extrinsic variables, such as

(a) ***Powder Characteristics*****:** The powder size, size distribution, shape, and agglomeration all are important powder-related variables. Depending on the particle size, the degree of internal space can be controlled, whereas the distribution of powder size can result in filling of the internal spaces by introducing smaller particles that fit in the spaces between bigger particles, thus reducing the overall porosity and resulting in higher densification. This requires the presence of "bimodal" (i.e., two types of powders with the first one creating spaces, and the second one filling the spaces) distribution of particles. The shape of the powder particles also holds a key since spherical particles possess the least surface area and thus require higher energies for mass diffusion. Additionally, they render point contacts with the adjacent particles, limiting the channel (or path) of diffusion. Thus, irregularly shaped particles create higher roughness and introduce more contact paths, enhancing mass diffusion. Powder agglomeration may not be appreciated when a dense compact is required after sintering. Agglomeration of particles eliminates control from processing since transfer of heat and pressure is blocked by the spaces between spongy agglomerated particles. Additionally, agglomeration blocks the generation of uniform properties across the sintered material and results in an anisotropic structure. Hence material properties will vary in service, with certain areas becoming weak due to stress concentration and poor mechanical properties or improper sintering at localized sections.

As referred to in earlier sections, the change in the surface curvature dictates the change in free energy (or chemical potential); thus it can also alter the sintering thermodynamics and kinetics. The pressure difference (proportional to $2\gamma/r$) across a curved surface creates a different scenario of sintering than it does with a planar surface. As far as the particle size dependence is concerned, Equation 5.13 is a significant result as it establishes the particle size dependence of sintering kinetics. Overall, powder size, shape, and distribution become highly important variables because of the manner in which densification and grain growth can be affected by proper selection of these powder variables.

(b) ***Powder Chemistry*****:** Powder compositional homogeneity, impurity content, sinter-aid or sinter additive (content and type), presence of any surface scale (e.g., SiO_2 layer on SiC surface or TiO_2 on TiB_2 powder surface) are powder-chemistry-related variables. Optimal amounts of binder or sinter-additive should be added in the case of refractory metals or ceramics to ensure formation of a sufficient amount of sintering liquid to wet the matrix grains.

Powder purity is also one of the most important concerns in sintering powders, since the presence of secondary phases tends to alter the response of a material (free energy change proportional to $d\gamma$) under temperature and pressure. Additionally, the phase formation and distribution also alter the way the densification or grain growth might occur. Consequently, the final application of a component depends on the physical, mechanical, chemical, and other (such as insulating) properties, which are highly dictated by the development of phases (including the porosity content and distribution, phase size, fraction, and distribution), and the residual stresses that are induced during processing. Hence, the chemistry of the powder particles used as initial feedstock is a very important parameter as the grain growth is guided by the phase stability, thus dictating the change of overall free energy in reducing the interfacial area and/or surface energy.

(c) ***Sintering Conditions*****:** Sintering temperature, sintering time (holding time), sintering atmosphere, and heating rate are all important sintering-related parameters. Optimization of sintering conditions (temperature, time, atmosphere, heating rate) is necessary. For monolithic material, sintering temperature should be greater than half the melting temperature for solid-state sintering ($T > 0.5T_m$). For LPS, the temperature should be above the melting temperature of the binder phase. For example, the melting temperature of cobalt is 1400°C and WC–Co cermets are conventionally sintered at 1500°C. Sintering time, that is, holding time at the sintering temperature, should be optimized so that sufficient diffusion, leading to removal of closed pores, or sufficient mass transport in the liquid phase takes place without promoting grain growth.

As far as sintering atmosphere is concerned, it is generally known and followed in reality that oxide ceramics are sintered in air, while non-oxide ceramics, such as borides and carbides, are densified in vacuum. However, nitrides are recommended to be sintered in N_2 atmosphere. For metallic samples (Cu, Ni, steel), H_2 or argon/H_2 atmosphere is used. While argon provides an inert atmosphere, the oxides on metallic powders are reduced by a generic reaction, occurring during the initial or intermediate stage of sintering:

$$MO_x + xH_2 \rightarrow xH_2O + [M].$$

Sintering conditions in the case of composites critically depend on the type and amount of the second phase. For example, in the case of A–B composites with B having higher T_m than A, the densification becomes more difficult with an increase in the amount of B; neck growth for B particles would be difficult, while the matrix A can be sintered easily. For this reason, the second phase addition is restricted to around 20–30 wt% to achieve better densification. This will necessitate the optimization of the addition of B. At a sintering temperature $T > 0.5\ T_{mA}$, the matrix grains of A will densify, while the densification of regions with B particles, would be difficult, leading to desintering.

5.8 POWDER METALLURGICAL PROCESSES

In this section, we will discuss the various process steps involved in obtaining dense ceramics using the standard powder metallurgical processes, including conventional and advanced sintering processes. Following the discussion in the preceding section, it bears repeating here that in order to ensure faster densification, it is important to use agglomerate-free high-purity ceramic powders with finer particle size. To enhance sintering, often a desired amount and type of binder or sinter-aid is used. This requires efficient milling, which is otherwise also used for two reasons: (1) to reduce the starting particle size in the powder feedstock and (2) to mix a second phase of a different composition with the matrix phase.

5.8.1 Ball Milling

Ball milling utilizes mixing of elemental and prealloyed powders along with the grinding balls to cause milling of powders, as shown schematically in Figure 5.11. Usually stainless steel, WC, or other ceramic (Al_2O_3, ZrO_2) balls are used for the mixing and homogenization of ceramic powders. This process leads to particle size reduction by comminution due to repeated crushing of powder particles between balls moving at high speeds. The physical mechanisms include extensive deformation (for metallic powders) and subsequent fracturing. During the initial stage, the particles fracture due to the impact of balls as well as due to the friction between the balls and particles. In the second stage, cold-welding occurs, which causes particles to adhere to other particle surfaces. Consequently, the particles become spherical and their size reduces. As the ball milling progresses, the powders repeatedly flatten, cold weld, fracture, and reweld. The flattening of particles primarily occurs for ductile metals, which consequently work harden and become brittle. Thereby, successive impacts fragment the particles. Similarly, in the case of brittle powders (ceramic powder), the particles undergo repeated fracture. It must be noted that the most effective milling requires a critical speed, which is inversely proportional to the internal diameter of the mill. If the speed of a ball mill is lower than the critical speed, then the ball impact, which causes milling, is minimal; any speed greater than the critical speed causes the ball to revolve with the container due to

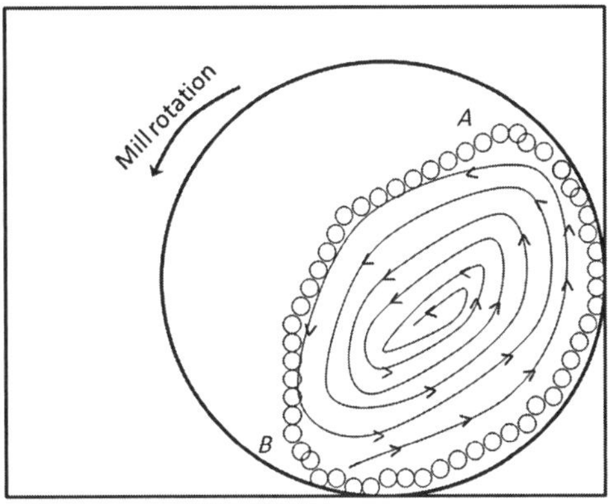

Figure 5.11 Schematic illustration showing the movement of balls and powders in a ball mill.

centrifugal force. Therefore, only the critical speed causes cascading (frictional milling) and impacting (impact milling) of the powder media. An ideal movement of balls and powder is shown in Figure 5.11. It should be also mentioned that the ball-to-powder ratio (BPR) is critical in achieving the desired results. Ideally, the value of BPR is 4:1 for effective milling. Increasing the ball diameter or the density of balls also leads to increase in the BPR. The BPR decides the ball milling time, which needs to be optimized to save power and cost.

As mentioned earlier, effective ball milling can refine the powder particle size. The milling time needs to be carefully optimized to save power and optimize total production time. It has been reported that a steady-state particle size reinforcement is attained at longer milling durations (24 or 48 hours), depending on material type. Any additional milling time often causes agglomeration of powders and therefore is not desired. In summary, the ball milling variables are powder-to-ball ratio (1:4 preferred), type of ball (harder than powders), liquid medium (acetone, toluene), or dry medium (air, vacuum, inert), and mill speed. The contamination from milling balls as well as from milling vials is also of great concern as balls of lower hardness than powders can lead to wear of balls. Dry milling in specific cases is preferred. In the case of high-speed milling, high temperatures are generated at particle–particle interfaces, leading to localized temperature rise or oxidation in the case of metallic powders. The milling medium (dry or wet) as well as milling time should be optimized to assure good homogeneous mixing of binder or sinter-additive with the matrix phase. There is an enhanced tendency to form agglomerates, and handling in vacuum or inert gas conditions and special chemical treatments are generally required. Some typical problems experienced with handling nanopowders involve agglomerate formation, as van der Waals force of attraction, being larger with nanopowders, causes larger interagglomerate porosity, as shown in Figure 5.12. Since the intercrystallite pore sizes are smaller, they can be removed during the sintering process. However, it would be difficult for interagglomerate pores to be completely removed because of their large sizes. Any residual porosity in the sintered ceramic would act as potential sites for stress concentration, leading to cracking and, therefore, degradation in mechanical properties.

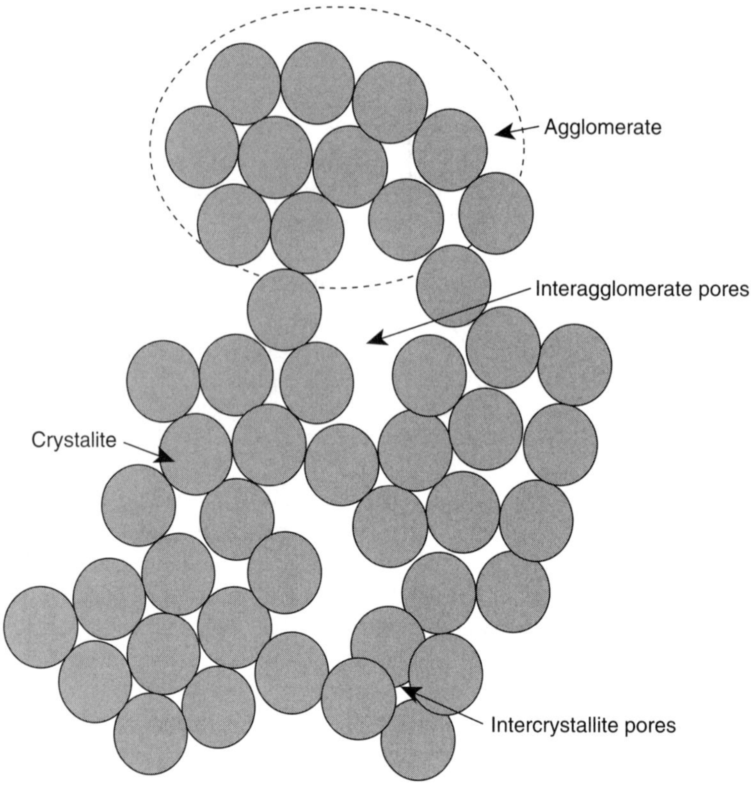

Figure 5.12 Schematic illustration showing the problems associated with nanosized powders—interparticle–intercrystallite porosity with small sizes and interagglomerate porosity with larger sizes.

5.8.2 Compaction

The second step, after ball milling, is compaction, wherein the powders are pressed into the desired shape and size using a hydraulic or mechanical press to obtain the "green compact." Compaction can be carried out at room temperature without heating the powders and is called cold pressing. In contrast, high-temperature compaction is carried out at above the recrystallization temperature to achieve high densification. The cold compaction processes are discussed next, whereas thermomechanical processes (high-temperature compaction via mechanical deformation) are detailed in a later chapter of this book.

5.8.2.1 Cold Pressing

Cold pressing involves compacting the powders in a steel die with extreme applied uniaxial pressure. Cold pressing involves the enhancement of the packing of loose powders (apparent density) by keeping them in dies or punches and applying pres-

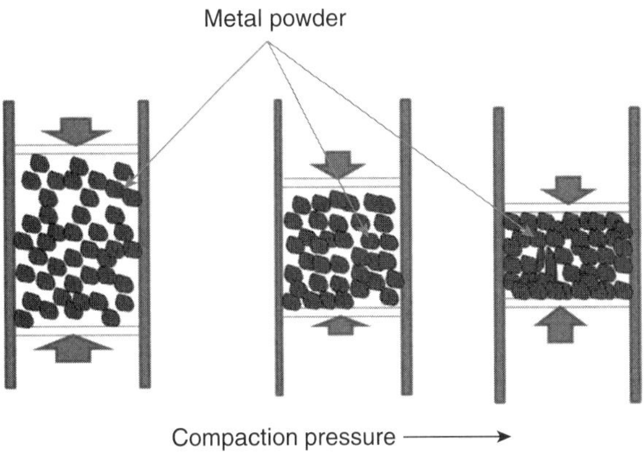

Figure 5.13 Typical compaction during cold pressing.

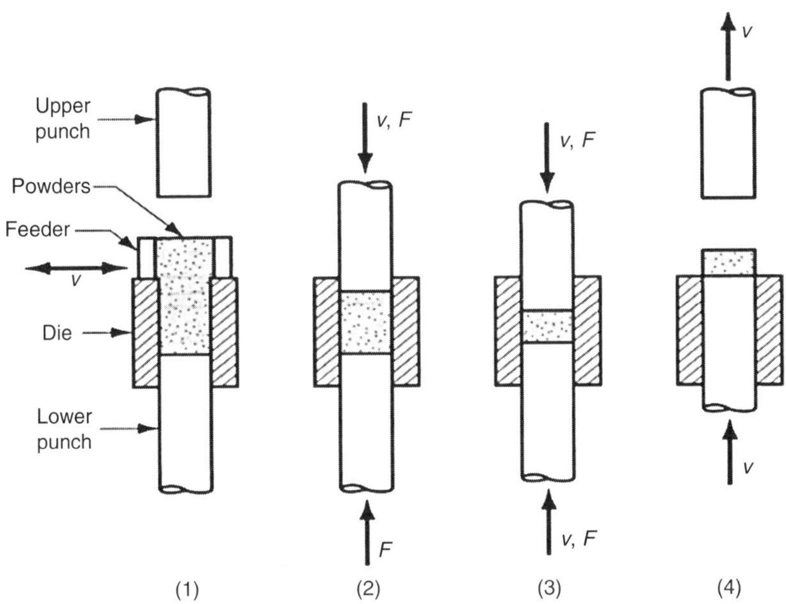

Figure 5.14 Conventional method of compaction: (a) filling die cavity with powder by automatic feeding system, (b) initial and (c) final positions of upper and lower punches, and (d) ejection of part.

sure to reach their green density. Herein, the rearrangement of powders occurs, followed by mechanical deformation, resulting in compaction, as seen in Figure 5.13. Consequently, cold pressed pellets can be further densified to their full density (or 100% theoretical density) via high-temperature sintering. A typical cold compaction cycle involves filling the die with loose powders (Fig. 5.14). This can be done via automatic feeding, where the previous compacted part is removed and the die cavity is filled by a hopper before the compaction occurs for the consequent part. Then, pressure is applied by a uniaxial punch to compact the powders to obtain a green body. Correspondingly, one of the dies can push the part out for its removal after cold compaction. These parts may be engineered to achieve controlled porosity in the material, as required for filters, for brake pads, for retaining lubrication in service, and so on.

5.8.2.2 Cold Isostatic Pressing

The basic problem with cold pressing is the variation of apparent density across its thickness. Since the loading is applied uniaxially, the compact is denser along the pressure direction than across the transverse direction. Therefore, to achieve a more uniform compacted part, cold isostatic pressing (CIP) becomes highly useful. Herein, a pressure of ~300 MPa is applied, while immersing the sealed container of powder in water or oil (Fig. 5.15). CIP is used to obtain desired complicated shaped parts without density gradient. The ceramic powder is sealed in a flexible bag (such as a metallic container) and submerged in a fluid to achieve uniform hydrostatic pressure in all directions. Consequently, the pressure is applied to obtain high green density. Herein, the density is uniform because of the hydrostatic pressure. The CIP process can be classified into two schemes: wet-bag and dry-bag processes. In a wet-bag process, the ceramic powder is filled in a rubber mold, and then this bag is immersed in the fluid for compaction. Herein, the isostatic pressure is achieved via transfer from pressurizing the fluid (Pascal's law), which then passes it on to the rubber mold to cause compaction of the ceramic powders. In the dry-bag process, only radial pressure (along the horizontal direction) is applied between the flexible mold and a rigid shell, while the bag can rest on its top or bottom surface. This becomes essential to obtain a better quality surface finish and to handle complicated parts.

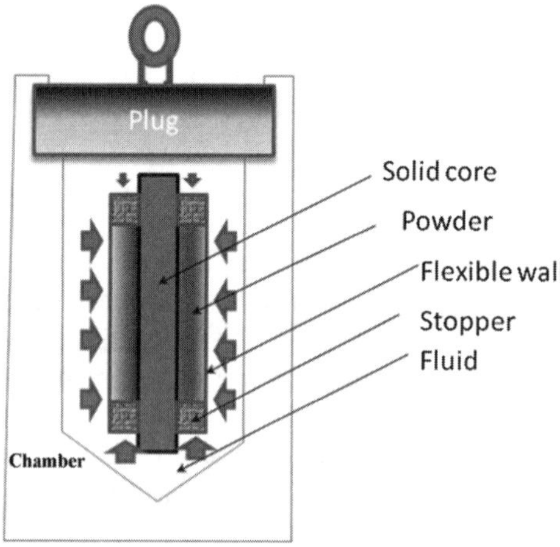

Figure 5.15 Schematic illustration of cold isostatic pressing of ceramic powders.

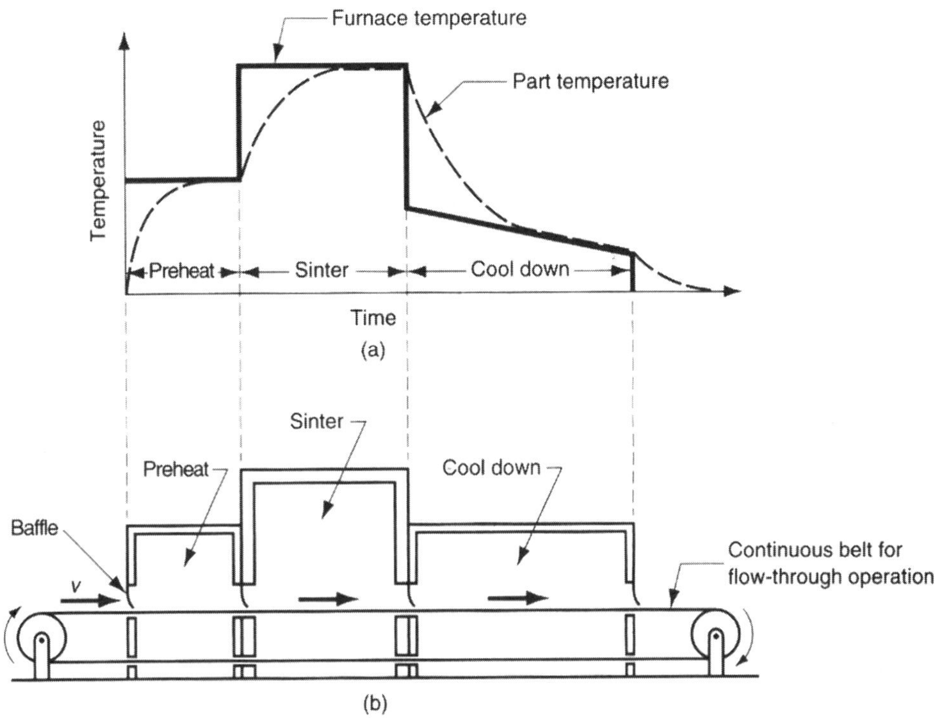

Figure 5.16 (a) Typical heat-treatment cycle in sintering. (b) Schematic of continuous sintering furnace. (Reprinted from M. P. Groover, *Fundamentals of Modern Manufacturing: Materials, Processes, and Systems*. Copyright 2010, reprinted with permission of John Wiley & Sons, Inc.)

5.8.3 Pressureless Sintering

In pressureless sintering, a powder compact is heated to sintering temperature in a furnace and no external gas pressure or mechanical pressure is provided. Depending on whether the compact is to be densified via solid-state sintering or LPS, both sintering temperature and time are to be selected for a given powder compact composition. A typical heating cycle followed in industrial-scale production of ceramic parts is provided in Figure 5.16. As observed in Figure 5.16a, a typical heat-treatment cycle includes a first stage of preheating to allow equilibrium–homogenization to be attained without inducing drastic thermal stresses. This stage is followed by sintering, where the ceramic component is held at sintering temperature for a specified dwell time. This is the most critical parameter because the selection of temperature and time of sintering, and the sintering environment the ceramic component is exposed to, decides the evolution of microstructure (phase content), porosity elimination, degree of phase homogenization, and grain growth. The sintering hold is followed by the cooling of the ceramic component. A rapid cooling might allow grain refinement, but it can induce thermal stress as well in the final component. High thermal stresses can also lead to cracking in the ceramic component; hence proper control of the cooling cycle also becomes essential. Typically, furnace cooling is adopted to minimize the thermal stresses.

In industries where the furnace cannot be blocked for sintering a single component at a time, a continuous sintering furnace can be adopted (Fig. 5.16b). Herein, the samples are fed from one side, and the conveyer belt speed is maintained in such a manner that the components have received the exact sintering cycle of heating, hold, and cooling down, when the component comes out the exit side. It can happen, in certain conditions, that pressureless sintering does not completely densify the material, in which case pressure sintering becomes essential. Hot pressing is only suited to relatively simple shapes, with components usually requiring diamond grinding to achieve the finished tolerances. HIP or HIPing results in good metallurgical bonding between particles, and good mechanical strength is achieved. Examples are superalloy components (aerospace), WC cutting tools, and powder metallurgy tool steels. The operating conditions for HIP include inert gas pressure of 100 MPa at $T > 1100°C$. Other high-temperature pressing techniques, such as hot pressing and high-temperature extrusion, are covered in the next chapter. In this chapter, other sintering techniques such as reactive and MW techniques are presented in the following subsections.

5.8.4 Reactive Sintering

Reaction or reactive sintering involves initiating a reaction combining the starting ceramic powders and resulting in densification being enhanced in the presence of a third phase. The generated secondary phases (of the initial ceramic composition) get entrapped in the matrix. Reactive sintering is to be approached as a combination of isolated processes: (1) reaction and (2) densification. Hence, a balance is required to achieve densification before the reaction interferes with the densification process (Fig. 5.17). Ideally, the molar volumes of the initial and final phases should match; otherwise the continued reaction can lead to interfacial cracking due to volume change during reactive sintering. When the reaction rate is higher, then the process will complete before the densification is achieved. Hence, sintering at lower tem-

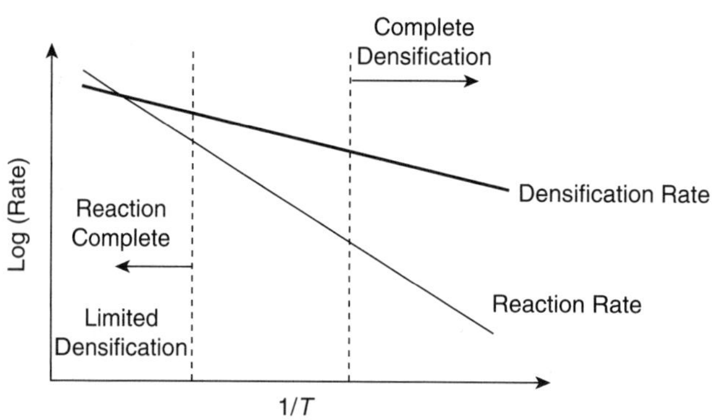

Figure 5.17 Balance between densification and sintering reaction during reactive sintering.

peratures allows achieving full densification before the reaction interferes with the densification process. When the molar volume of the reaction and product is similar, compact densification can be achieved. The annihilation of GB (grain growth) and reduction in the surface energy (densification) are the primary dictating factors in choosing the sintering temperature and time. Selection of temperature plays an essential role in creating a balance between densification and reaction kinetics as well. On the one hand, when increasing temperature starts enhancing the reaction rate, the increased densification does not have direct proportionality with increasing temperature. Increasing the temperature promotes formation of a liquid phase (assisted by the particle size as well), allowing LPS to dominate, and thereby the high reactivity of liquid assists phase transition while resulting in densification. Due to higher change in the free energy associated with ultrafine ceramic particles (<1 μm), the kinetics of reactivity increases, which in turn also supplement high reactivity during reactive sintering. Hence, the ultrafine ceramic content should also be considered in estimating the reactive sintering temperature and time. These secondary considerations make the process highly complicated, since the reaction should be over only after complete densification has occurred; thus, lower-temperature sintering becomes more appropriate when sintering parameters are unknown for reactive sintering of ceramics. Aluminum is added as a sintering aid to result in densification of tungsten, as well as a reducing agent to free tungsten trioxide.[4] Aluminum thus melts and acts as a glue holding the tungsten particles to achieve densification (Fig. 5.18). Figure 5.19 elicits the difference between the 80% dense die pressed and conventionally sintered tungsten (Fig. 5.19a), MW tungsten (Fig. 5.19b), and reactive phase sintered tungsten (Fig. 5.19c). It can be noticed that reactive sintering provides a very dense compact, wherein the grain boundaries are decorated with a low-melting-point aluminum phase. Thus reactive sintering of tungsten elicits the sintering with a low-melting-temperature sintering additive serving as reactive agent as well to result in densification.

5.8.5 Microwave Sintering

Microwaves initiate uniform heating of the entire volume of a powder compact via oscillation of free electrons and ions (frequency of ~2.5–85 GHz). It becomes easier

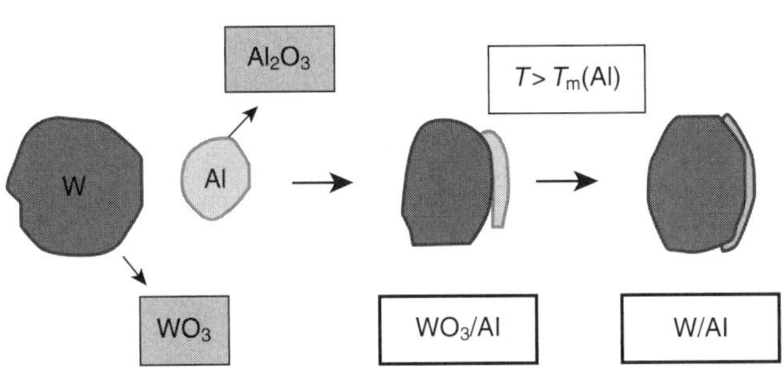

Figure 5.18 Reactive sintering of tungsten with sintering addition of aluminum.[4]

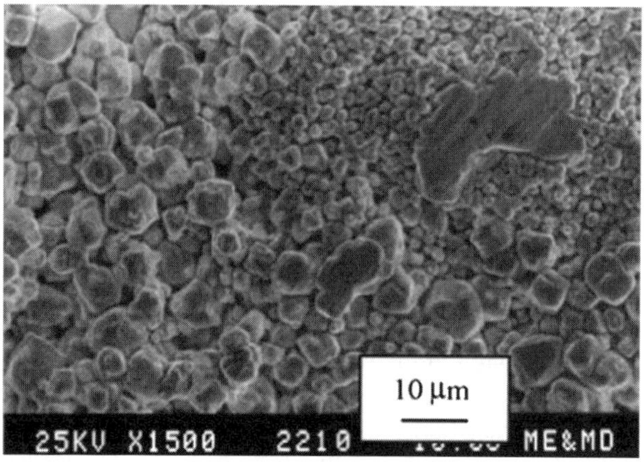

(a)

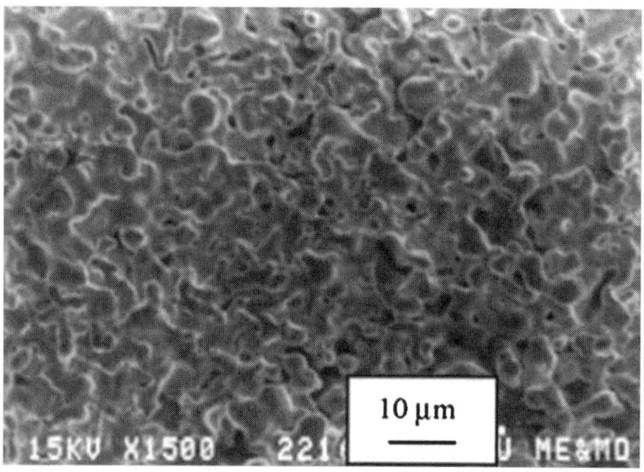

(b)

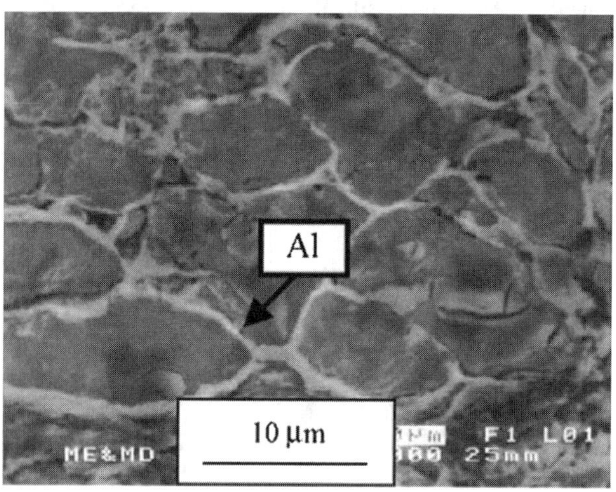

(c)

Figure 5.19 SEM image of 80% die pressed (a) conventionally sintered and (b) microwave sintered tungsten, and (c) reactive sintered tungsten showing aluminum at grain boundaries. The label A1 points toward the presence of reactive phase around the grain boundaries.[4]

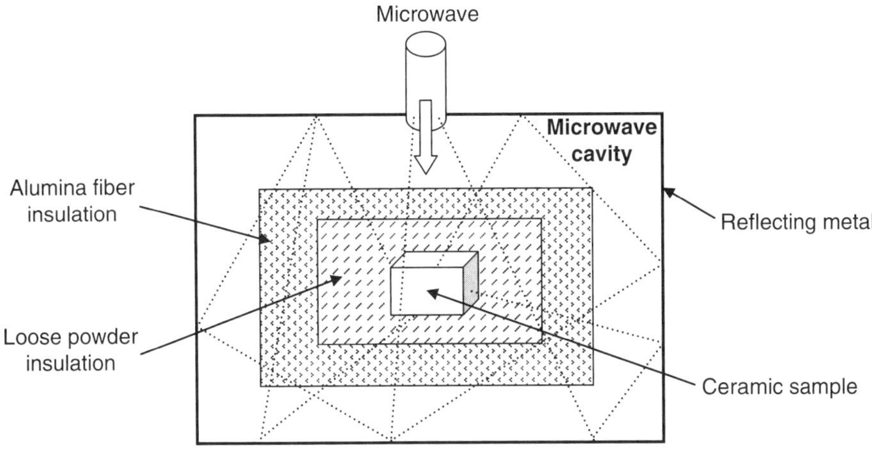

Figure 5.20 Schematic of microwave sintering setup.

to achieve volumetric heating even in complicated shapes, with heating rates on the order of 1000°C/min. The ceramic powder compact is contained in the microwave cavity for sintering and densification (Fig. 5.20). Since high-frequency microwaves cause skin effects, that is, preferentially heating the surface, a combination of low and high frequencies during MW may be required to realize uniform heating. Otherwise, strong thermal gradients can be generated within the body and cause immediate fracture. Hence, controlling the degree of microwave interaction specific to the material's absorption should be considered. In addition, the ceramic requires an envelope of nonabsorbing material to create insulation and limit thermal loss (Fig. 5.21a). An advantage associated with MW is the short heating and cooling cycle, that is, rapid volumetric heating, which restricts grain growth during sintering. In addition, uniform heating of the material does not allow generation of steep surface-to-bulk thermal gradients, which are predominant in the conventional sintering processes. Specific advantages of MW vis-à-vis conventional sintering in terms of overall process cycle time and energy consumption are shown in Figure 5.21b.

Some early work to realize the efficacy of microwave heating to achieve the densification of various ceramic systems as well as to understand the densification mechanism in an electromagnetic field can be found in References [5–12]. The fundamental mechanism of MW involves the coupling of microwaves with materials, volumetric absorption of electromagnetic energy, and subsequent transformation to heat energy. The primary material property determining the effectiveness of microwave coupling is the dielectric loss factor. Since many ceramics are insulators, the concept of hybrid sintering was later proposed to realize more effectively the MW sintering of insulating ceramics, such as zirconia.[13,14] The concept of hybrid sintering employs two facts: (1) The dielectric loss factor increases exponentially with temperature and (2) some ceramics, such as SiC, exhibit high loss factors even at room temperature. This motivated the researchers to use a SiC susceptor in the microwave cavity to enable initial rapid heating to a temperature level at which

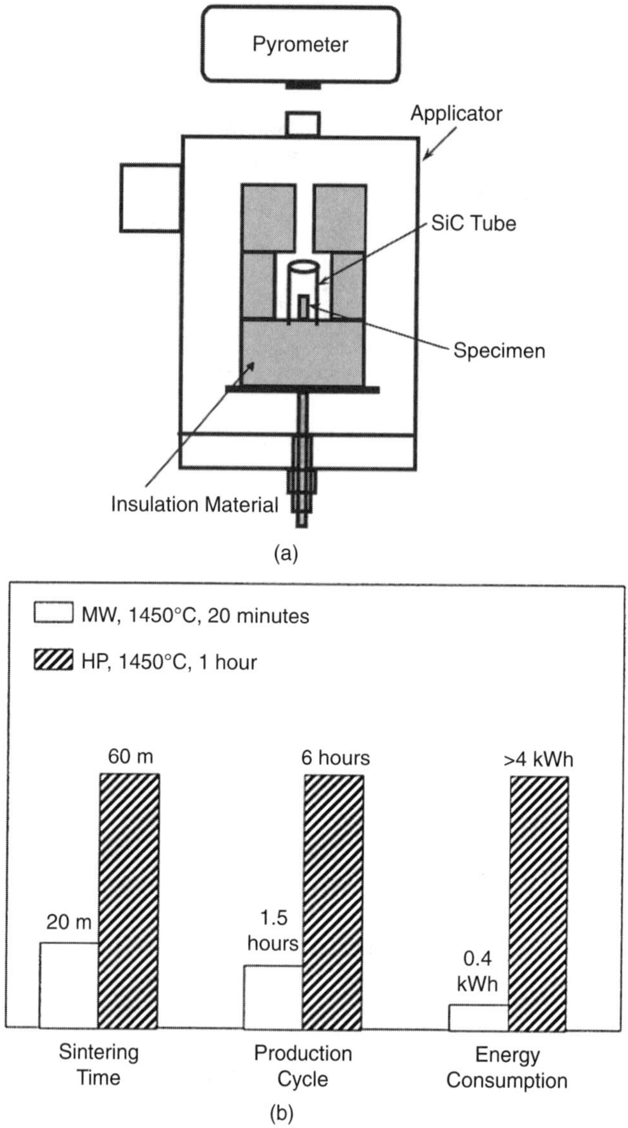

Figure 5.21 (a) Schematic illustration of hybrid sintering in a microwave cavity for microwave sintering and (b) a comparison in terms of total sintering time, process cycle, and energy consumption as recorded in hot pressing (HP) and microwave sintering (MW) of insulating zirconia ceramic.[13]

the electromagnetic energy absorption capability of even insulating ceramics is sufficient to cause volumetric heating of the ceramic, which is otherwise microwave transparent at room temperature. The term "hybrid" essentially refers to the combination of pure microwave heating and conventional heating by the susceptor. The technology of hybrid sintering solves the long-standing problem of room-temperature

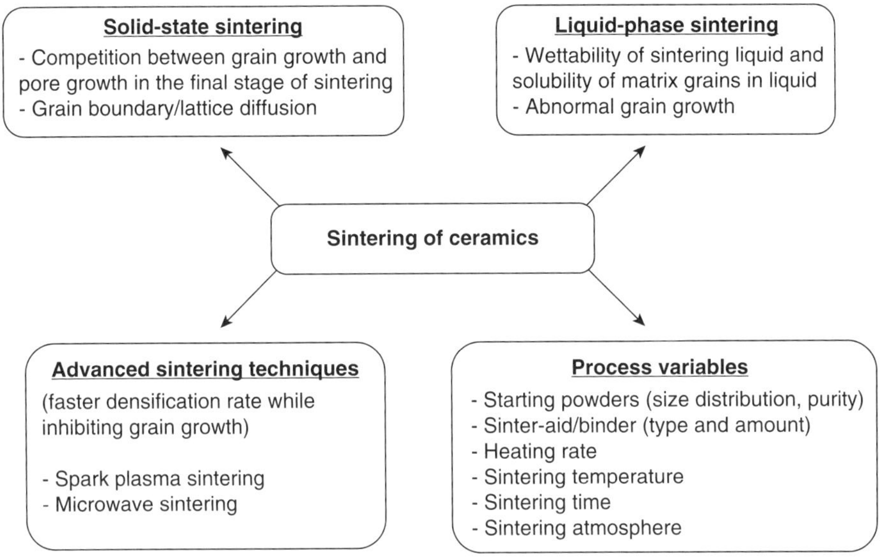

Figure 5.22 Summary of various factors controlling the sintering of ceramics as well as issues related to solid-state and liquid-phase sintering. Also, some advanced sintering techniques are mentioned.

coupling of otherwise insulating ceramics in a microwave cavity. Electrophoretic deposition followed by MW sintering is utilized to obtain materials with designed microstructures and properties, for example porous titanium coatings and functionally graded oxide-based composites.[15,16]

As a concluding note, the various issues in the consolidation of ceramics via solid-state and LPS are summarized in Figure 5.22. Also mentioned in Figure 5.22 are the various factors determining the sintering of ceramics; these parameters need to be carefully considered while optimizing the process of consolidation for a new ceramic composition. Some advanced sintering techniques, such as SPS and MW, used to densify ceramics at a much faster rate than conventional sintering are also mentioned in Figure 5.22. More details of various conventional and advanced ceramic processing techniques can be found in various ceramic textbooks.[17–22]

REFERENCES

1 S.-J. L. Kang. *Sintering: Densification, Grain Growth and Microstructure*. Elsevier, London, 2005.

2 Y.-M. Chiang, D. P. Birnie, and W. D. Kingery. *Physical Ceramics*. John Wiley & Sons, Hoboken, NJ, 1997.

3 D. Sarkar, M. C. Chu, S.-J. Cho, Y. I. Kim, and B. Basu. Synthesis and morphological analysis of titanium carbide nanopowder. J. Am. Ceram. Soc. 92(12) (2009), 2877–2882.

4 C. Selcuk and J. V. Wood. Reactive sintering of porous tungsten: A cost effective sustainable technique for the manufacturing of high current density cathodes to be used in flash lamps. J. Mater. Process. Tech. 170 (2005), 471–476.

5 R. Roy, D. Agrawal, J. Cheng, and S. Gedevanishvili. Full sintering of powdered-metal bodies in a microwave field. Nature. 399 (1999), 668.
6 D. K. Agrawal. Microwave processing of ceramics: A review. Curr. Opin. Solid State Mat. Sci. 3(5) (1998), 480–486.
7 J. Majling, P. Znasik, J. Cheng, D. Agrawal, and R. Roy. Conventional and microwave sintering of condensed silica fume. J. Mat. Res. 10(10) (1995), 2411–2414.
8 Y. Fang, D. K. Agrawal, D. M. Roy, and R. Roy. Microwave sintering of hydroxyapatite. J. Mater. Res. 9(1) (1994), 180–187.
9 Y. Fang, D. K. Agrawal, D. M. Roy, and R. Roy. Fabrication of porous hydroxyapatite ceramics by microwave processing. J. Mater. Res. 7(2) (1992), 490–494.
10 R. Roy, D. K. Agrawal, and J. Cheng. Process for Sintering Powder Metal Components, U.S. Patent #6,183,689 (issued February 6, 2001).
11 J. Cheng, R. Roy, and D. Agrawal. Microwave processing in pure H fields and pure E-Fields, U.S. Patent #6,365,885, Issued April 2, 2002. 42.
12 J. Cheng, D. Agrawal, Y. Zhang, B. Drawl, and R. Roy. Fabricating transparent ceramics by microwave sintering. Am. Cer. Soc. Bull. 79(9) (2000), 71–74.
13 C. Zhao, J. Vleugels, O. Van Der Biest, C. Groffils, and C. P. J. Luypaert. Hybrid sintering with a tubular susceptor in a cylindrical single mode microwave furnace. Acta Mater. 48(14) (2000), 3795–3801.
14 S. G. Huang, L. Li, O. Van der Biest, and J. Vleugels. Microwave sintering of CeO_2 and Y_2O_3 co-stabilised ZrO_2 from stabiliser-coated nanopowders. J. Eur. Ceram. Soc. 27 (2007), 689–693.
15 B. Neirinck, T. Mattheys, A. Braem, J. Fransaer, O. Van der Biest, and J. Vleugels. Porous titanium coatings obtained by electrophoretic deposition (EPD) of Pickering emulsions and microwave sintering. Adv. Eng. Mater. 10(3) (2008), 246–249.
16 C. Zhao, J. Vleugels, L. Vandeperre, B. Basu, and O. Van Der Biest. Y-TZP/Ce-TZP functionally graded composite. J. Mat. Sc. Lett. 17 (1998), 1453–1455.
17 C. B. Carter and M. G. Norton. *Ceramic Materials*. Springer, New York, 2007.
18 M. W. Barsoum. *Fundamentals of Ceramics*. Taylor & Francis, Boca Raton, FL, 2003.
19 W. D. Kingery, H. K. Bowen, and D. R. Uhlmann. *Introduction to Ceramics*, 2nd ed. John Wiley and Sons, New York, 1976.
20 M. N. Rahaman. *Ceramic Processing and Sintering*. CRC Press, Boca Raton, FL, 2003.
21 D. W. Richerson. *Modern Ceramic Engineering: Properties, Processing, and Use in Design*. CRC Press, Boca Raton, FL, 1992.
22 J. S. Reed. *Introduction to the Principles of Ceramic Processing*. John Wiley & Sons, New York, 1988.
23 M. P. Groover, *Fundamentals of Modern Manufacturing: Materials, Processes, and Systems*. John Wiley and Sons Inc., Hoboken, NJ, 2010.

Chapter 6

Thermomechanical Sintering Methods

Thermomechanical treatment comprises a combination of thermal and mechanical treatments to impart superior properties to a material. Often there is limited plastic deformation in ceramics; hence, high-temperature processing can induce some softening during ceramic processing. In addition, high temperature assisted with mechanical refinement can induce phase transformation, which can yield certain crystal defects in the material. Herein, two degrees of freedom, namely pressure and temperature, are played with to attain the high-density components (Fig. 6.1). Various techniques fall under the umbrella of thermomechanical processing, such as hot rolling, sinter forging, hot pressing, hot isostatic pressing (HIP or HIPing), or extrusion. These are considered one by one in the upcoming sections.

6.1 HOT PRESSING

Hot pressing is a conventional compaction process that utilizes simultaneous application of pressure while the material is heated to high temperatures (above the recrystallization temperature). A schematic of the process and actual hot-pressing unit is shown in Figure 6.2a and b, respectively. Hot pressing differs from the powder metallurgy technique in that there is simultaneous application of heat and pressure (to induce sintering and creep) as opposed to power metallurgy techniques where heat application is independent of the pressure cycle. Ceramic powders are fed into a die and are pressed to ~10–50 MPa at temperatures in the range 1000–2200°C for a certain duration (from minutes to a few hours) to achieve a dense ceramic. Ceramic composites can achieve up to 100% theoretical densities, and often, densities are more than ~95% of the theoretical density. High pressure and temperature synergistically allow rearrangement of particles, allow plastic flow because of both high temperature and applied pressure, and sinters the ceramic by elimination of porosity (with application of high pressure) in a vacuum or inert atmosphere. Typical hot-pressing times, temperatures, and pressures for various ceramics are presented in Table 6.1. The crux of material selection lies in choosing a die that can withstand

Advanced Structural Ceramics, First Edition. Bikramjit Basu, Kantesh Balani.

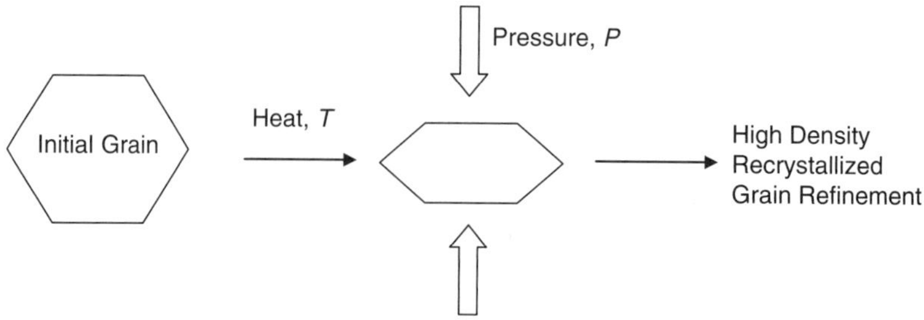

Figure 6.1 Schematic of heat and pressure involvement in the thermomechanical treatment to achieve refined grains and dense structure.

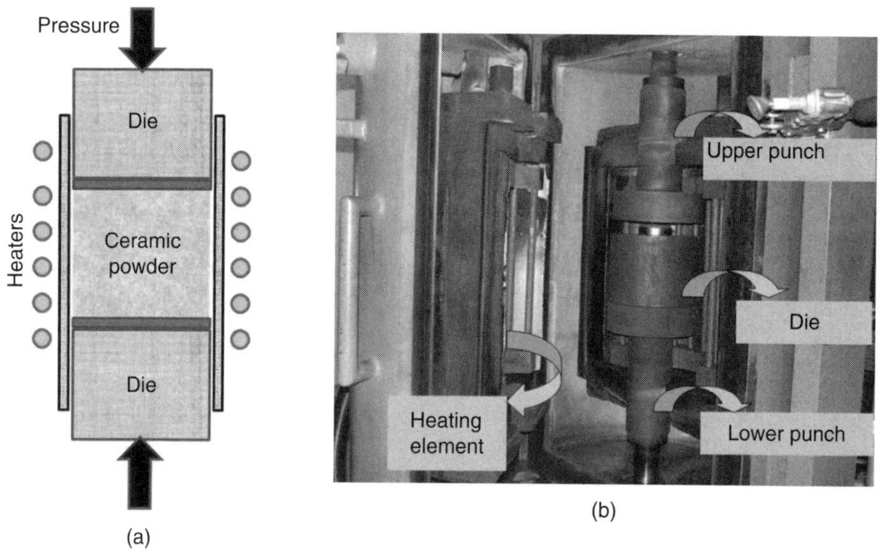

Figure 6.2 (a) Schematic and (b) industrial hot-pressing setup. See color insert.

Table 6.1. Hot-Pressing Temperatures, Pressures, and Time for Various Ceramics

Material	Temperature (°C)	Pressure (MPa)	Time	Others	Reference
ZrB_2	1000–1600	40	2–60 minutes	5×10^{-5} Torr ~98% density	1
TiB_2	1600–1900	30	30–120 minutes	>98% density	2
Si_3N_4	1750	30	24 hours	–	3
TiO_2	1100	5000-kg uniaxial	2 hours	~91% density	4
AlN	1400–1700	20	~30 minutes	~100% dense	5
B_4C	2000	30	1 hour	~99% dense Ar atmosphere	6

such high temperature and pressure without reacting with the ceramic material. Most often graphite is utilized as a die material for containing the powder, and withstanding high temperature and consequently compacting the ceramic powders at high pressures. Sintering mechanisms are standard surface diffusion, vacancy migration, and bulk diffusion.

Heating of a ceramic powder compact can be attained by one of the following routes:

1. ***Inductive Heating.*** Inductive heating is generated via application of a high-frequency electromagnetic field using an induction coil. Since the pressure and temperature can be controlled independently, the liquid phase can also be contained via utilization of low pressures. Good inductive coupling rendered by the graphite mold can allow magnetic field penetration of up to a few millimeters from where the good thermal conductivity of the graphite mold carries from the mold to the ceramic material. However, the nonuniform air-gap difference between the mold and the inductive coil can result in nonuniform heating of the mold (and thereby the ceramic). Also, the heating rate has to be controlled since an extremely slow cooling rate will result grain growth, whereas extremely fast cooling rates will generate thermal stresses and can damage the mold itself.

2. ***Indirect Resistance Heating.*** In indirect resistance heating, the mold containing the ceramic is placed in a chamber, and the chamber is resistive heated via a separate heating element present in the chamber itself. Electric current is passed through the heating element and consequently the I^2R effect allows heating of the heating element and convective heating heats up the mold. Since the ceramic in the mold is heated indirectly, that is, from the heating element then to the mold, this type of heating is called indirect resistance heating.

   Since heating is indirect, high temperatures can be attained without worrying about the mold conductivity. However, it generally takes longer since heating is indirect, and since first the chamber attains the targeted temperature and then only the mold surface attains the required temperature for pressure-assisted sintering of the ceramic.

3. ***Direct Heating.*** In direct heating, the mold is connected to the power supply and is heated directly via resistive heating when current is passed through it. The direct contact allows rapid heating of the mold and enhances power activity in achieving sintering in short durations and at lower temperatures. The advantage lies in the fact that local air-gaps between powder particles provide a high-resistance path, hence the heating is higher (due to the I^2R effect), and more heat therefore softens the material and provides uniform densification. Sintering times are reduced to only a few minutes at temperatures ~200–500°C below the conventional sintering temperatures. Hence direct heating provides a reduction in processing cost by (1) decreased processing time, (2) lowering of processing temperature, and (3) enhanced die life.

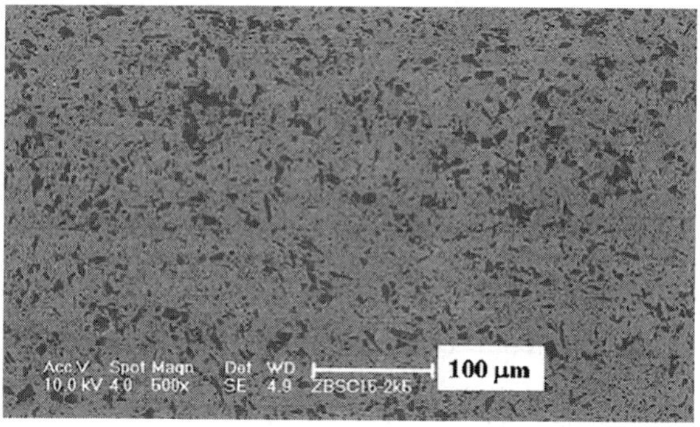

Figure 6.3 The dense microstructure with uniform distribution of SiC is observed in ZrB_2 composite reinforced with 25 vol% SiC and 1 wt% Ni produced at 40 MPa, 1600°C for 30 minutes.[1]

Rangaraj[1] showed a highly dense microstructure of hot-pressed ZrB_2 reinforced with 25 wt% SiC and 1 wt% Ni (Fig. 6.3). Density approached nearly 98% when the ZrB_2–(25 wt%)SiC was reinforced with 1 wt% Ni due to transient liquid phase densification. Sintering often leads to grain growth with dwell time, which can be expressed as

$$G = G_0 + kt^n, \tag{6.1}$$

where G is the average grain size, G_0 is the initial grain size, k is a constant, t is sintering dwell time, and n is the growth coefficient. Theoretically n is ~0.5, but impurities and pores impede the grain growth and the grain growth coefficient decreases. Effect of sintering time and temperature on the grain size of TiB_2 is presented in Figure 6.4.

6.2 EXTRUSION

Extrusion is a process where a ceramic billet is pushed through a die to get bars, billets, tubes, and so on of fixed cross section (Fig. 6.5a). Horizontal extrusion presses with ~200–12,000 tons capacity are common, with pressures going from 30 to 1000 MPa. Lubrication is generally provided by glass for high-temperature processing. The inherent advantage of extrusion is that complex cross-sectional geometry can be attained even for brittle materials since the material experiences compressive forces or shearing. This process is analogous to squeezing a toothpaste tube to get a uniform cross section of toothpaste through its length. In extrusion, ceramic is pressed against (1) a die whose one end is kept open (which is the required cross section) and (2) a ram to push it through the die opening. This setup is classified as either direct extrusion or backward (or indirect) extrusion:

1. ***Direct Extrusion.*** In this setup, the direction of material flow and ram-pushing are in the same direction, that is, the material is squeezed between the ram and die with an opening in the die side so the outcoming billet (or tube or bar) traverses in the same direction as that of the pushing ram (Fig. 6.5b). An inherent disadvantage of direct extrusion is that higher frictional

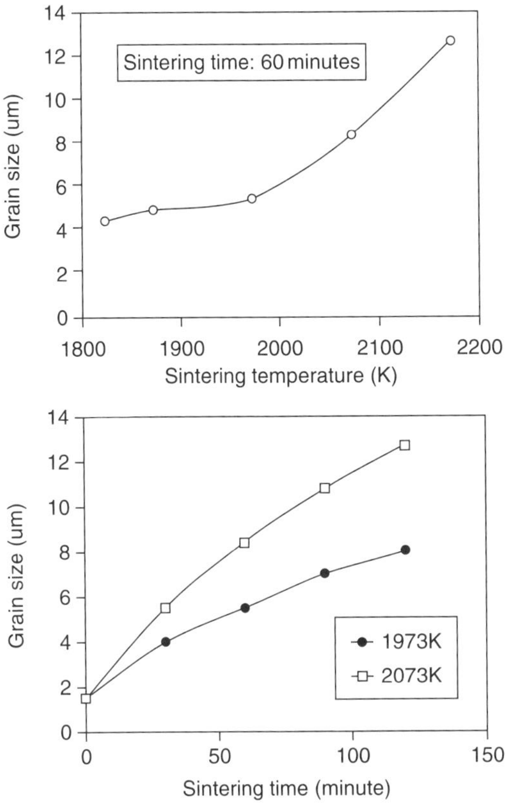

Figure 6.4 Effect of sintering time and temperature on the grain size of TiB_2.[2]

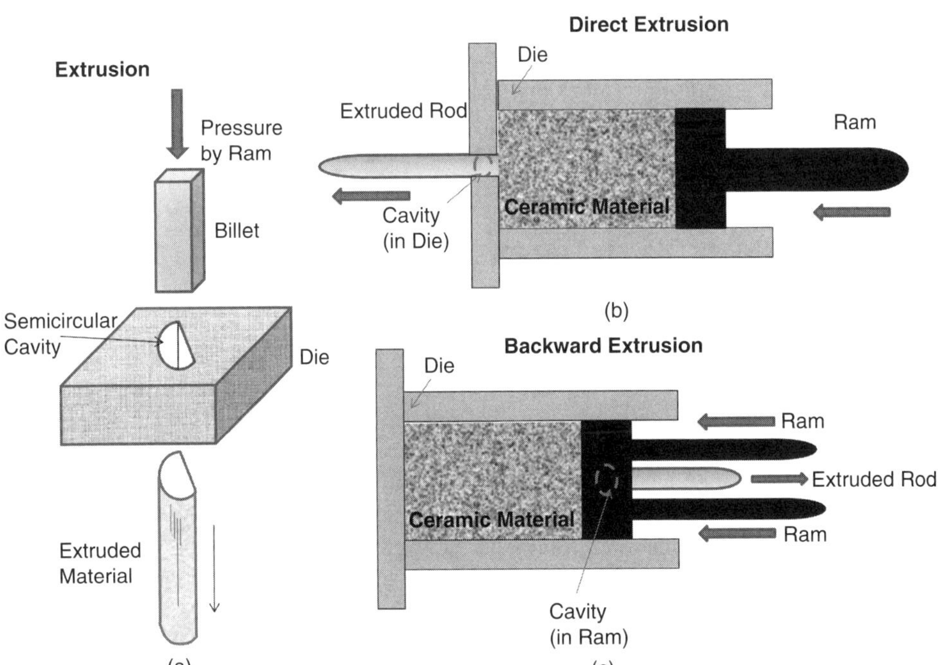

Figure 6.5 Schematic of (a) extrusion process, (b) direct extrusion, and (c) backward extrusion to produce ceramic rod.

forces incurred on the entire surface of the billet must be overcome in forcing the billet through the opening of the die on the opposite end. Hence very high pressures are required in the beginning of the extrusion process. Though the pressure requirement decreases as the material is used up, again, during the end when the ceramic feed is almost gone, the material will have to traverse radially to feed the cavity and exit. This, again, requires high energy for pushing it through the opening. Hence the butt-end of the billet unbalances the economy of processing and is therefore generally discarded.

2. ***Backward Extrusion.*** In this setup, the direction of material flow and ram-pushing are opposite to each other, that is, the material is squeezed between the ram and die with an opening in the ram side so the outcoming billet (or tube or bar) traverses in the opposite direction as that of pushing ram (Fig. 6.5c). Here the advantage lies in the fact that the material comes out as the ram is being pushed in. So only the frictional forces of the material coming out need to be overcome. This allows extrusion of bigger cross sections, enhancing extrusion speeds, and with less heating due to the reduced friction. Therefore the life of the die is enhanced as well. Also, the use of the initial billet is more uniform, and peripheral defects (arising from nonsupply due to radial filling, as in direct extrusion) is not limiting anymore. The inherent disadvantage of backward extrusion is the limitation on the length of the extruded material dictated by the maximum length of the ram push-rod. In addition, the surface impurities of the initial billet may affect the surface of the extruded material and spoil its aesthetics.

Apart from producing dense structures, extrusion can also be utilized to engineer ceramics that require continuous porosity. To prepare a porous SiC–Si_3N_4 composite, SiC is extruded with a pore-forming agent and polymer core, which consequently elongates and thins with consequent extrusion passes (Fig. 6.6).[7] Later the core polymer–filler can be oxidized or burnt to produce a continuously porous ceramic composite.

6.3 HOT ISOSTATIC PRESSING

Often, the uniaxial pressing compresses the surface from where the pressure is being applied, and the force then has to traverse along the axis of load application. This brings a drawback that top and bottom surfaces get compressed more when compared to the core and sides of the ceramic sample. Hence, the term "HIP or HIPing" evolves from the "similar" static pressure arising from a fluid (gas/liquid) around a ceramic to apply pressure from all around the sample surface (Fig. 6.7a). A typical HIPing unit is shown in Figure 6.7b. The ceramic body is submerged in the chamber at high temperatures, and a high-pressure inert gas compresses the ceramic to uniformly sinter the ceramic. Hence, a consolidated high-density ceramic can be easily obtained showing homogeneous microstructure. Reduction in porosity, increased strength and hardness, and uniform microstructure are common in HIP. HIPing utilizes a vessel that acts as a chamber and a ceramic green is placed here (Fig. 6.7c).

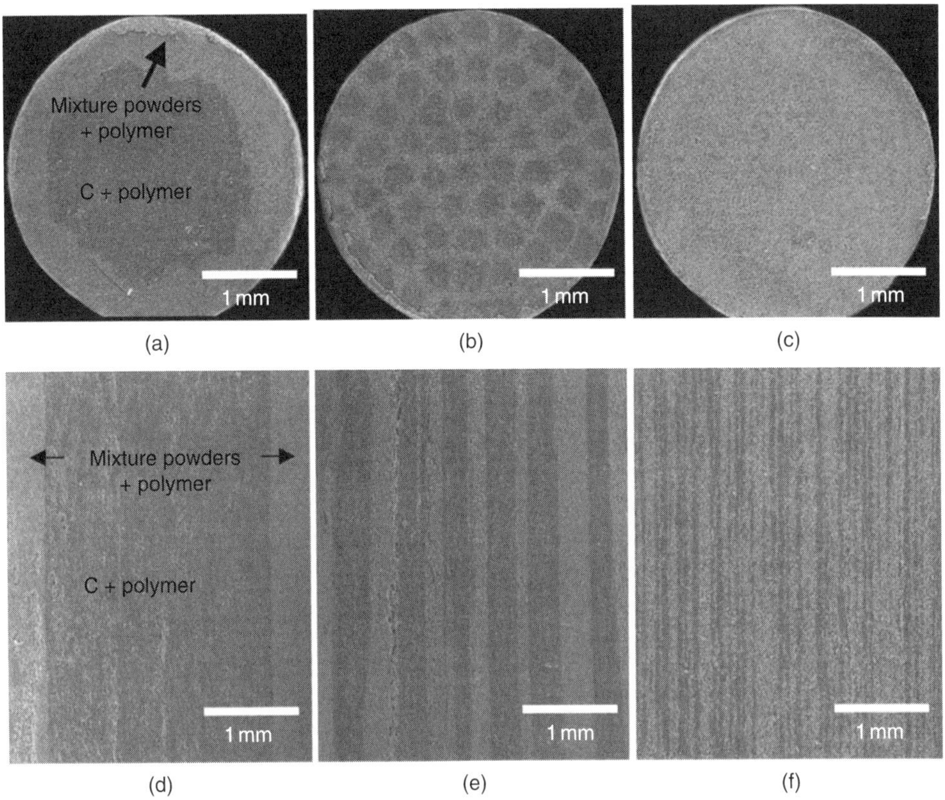

Figure 6.6 SEM micrographs of transverse (a) first-pass, (b) second-pass, and (c) third-pass extruded section, and SEM micrograph of longitudinal (d) first-pass, (e) second-pass, and (f) third-pass extruded bodies of SiC.[7] Later this filler–polymer can be removed by burning or oxidation, resulting in a continuously porous ceramic.

Consequently, the chamber is sealed and gas pressure is increased to the required value at selected sintering temperature. The material is not encapsulated in a flexible vessel, and densification can occur even when closed porosity is present in the material. A cycle involves filling the can with ceramic powder, vacuum baking to remove moisture, and consequent application of pressure at elevated temperature to achieve a fully dense part.

Hot pressing of Al_2O_3-reinforced SiC at 1850°C for 1 hour at 200 MPa has resulted in a dramatic increase in fracture toughness (from 3.1–4.3 $MPa{\cdot}m^{1/2}$ to 4.3–5.8 $MPa{\cdot}m^{1/2}$).[8] This is attributed to crack propagation along grain boundaries that receive resistance from higher density and enhanced grain boundary area. A scanning electron microscopy (SEM) fractograph of the composite is presented in Figure 6.8a.[8] In addition, owing to the high temperature of sintering, formation of eutectic liquid was inevitable and is observed to agglomerate at the triple grain junction, as shown in the transmission electron microscopy (TEM) imaging (Fig. 6.8b).[8]

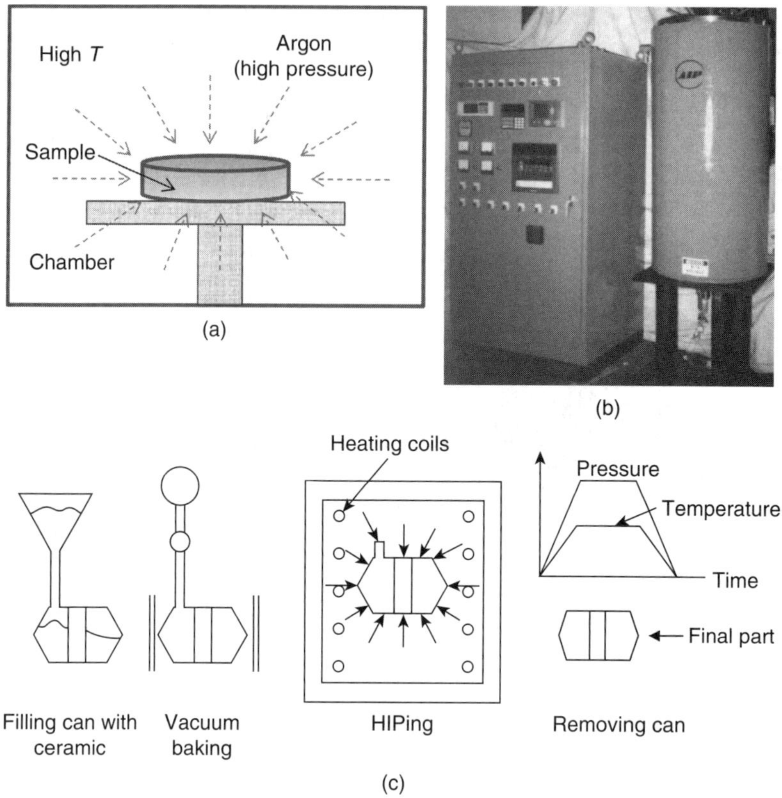

Figure 6.7 Schematic showing (a) the principle involved, (b) a commercial unit for hot isostatic pressing (HIPing), and (c) the sequence of HIPing. See color insert.

6.4 HOT ROLLING

Hot rolling is the process of passing the slab or billet through rolls (at high temperatures) for achieving texture and reduced cross section or required shape (Fig. 6.9). Two or more horizontal rolling cylinders rotate in opposite directions and allow the billet to pass through a cross section that is reduced from the original cross section of the billet. Hence the ceramic billet experiences compressive forces perpendicular to the rolling direction. Consequently, the material is pushed forward owing to compressive stresses, and frictional forces induce shear and help in refining the grains. Hot rolling is carried out at temperature above the recrystallization temperature of the material. Two rolling cylinders compress the billet and result in mechanical working, whereas recrystallization occurring at high temperatures reduces the grain size during hot working of the ceramic; it can also result in an equiaxed structure. Hot rolling allows easier deformation and thinning of ceramics due to plasticity induced at high temperatures. Hot rolling provides enhanced density, and mostly

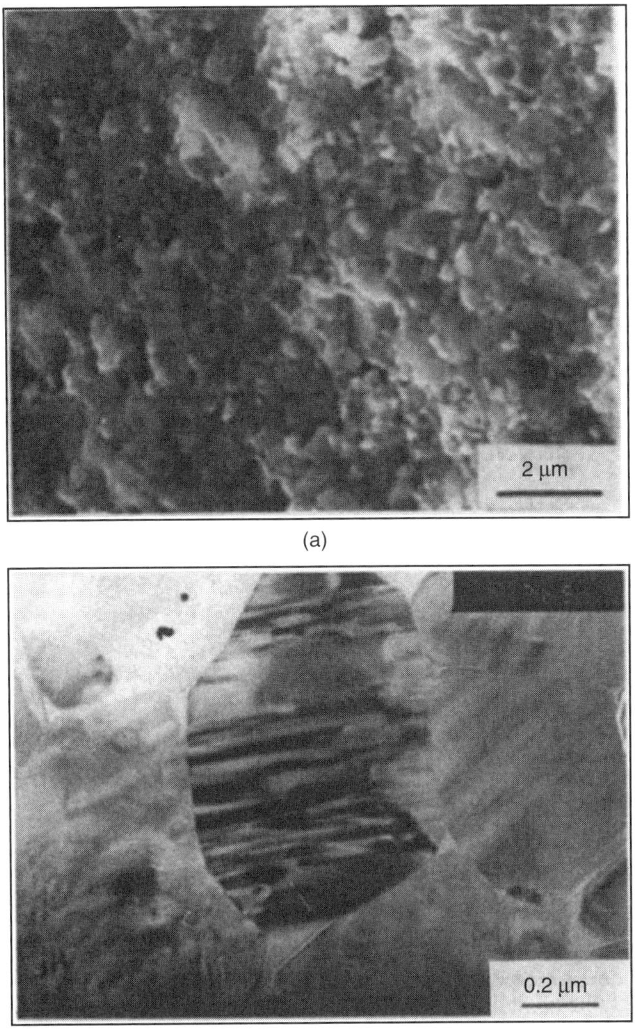

Figure 6.8 (a) SEM fractograph and (b) TEM micrograph of 2 wt% Al_2O_3-doped SiC. Fractograph shows fine-grained structure, and TEM micrograph shows formation of low-temperature eutectic at the triple grain junction.[8]

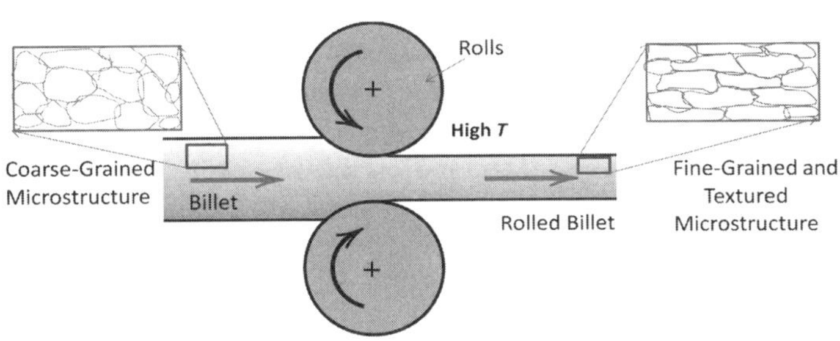

Figure 6.9 Schematic of hot-rolling process.

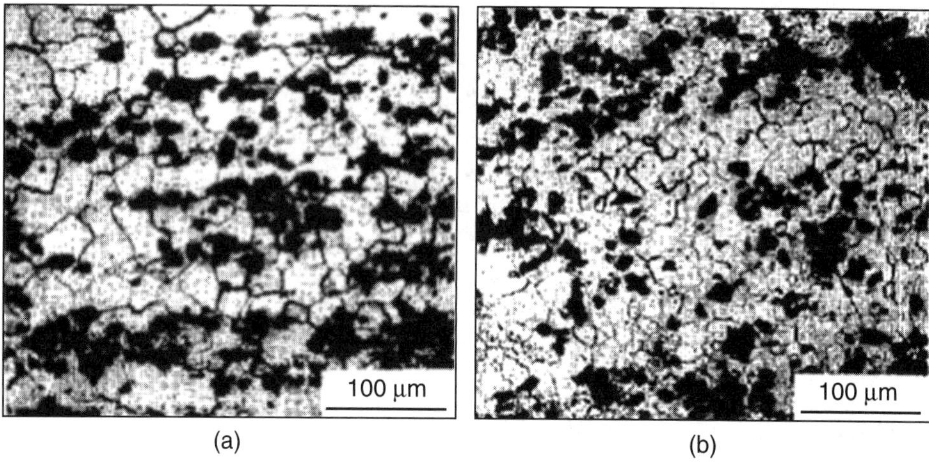

Figure 6.10 (a) Agglomeration of Al_2O_3 in Al matrix. (b) Distribution of Al_2O_3 in Al matrix after hot rolling.[9]

because of refined grain structure, high strength, high ductility, high toughness, and enhanced performance of the rolled material can be achieved.

Hot rolling of Al_2O_3-reinforced Al using a 10-cm roll at a speed of 8 m/min with 10% reduction in thickness per pass to a total of 70% reduction at a temperature of 530°C[9]. The rolled composite was heated to 590°C for 2–3 minutes between each pass as well. Agglomeration of Al_2O_3 in the Al matrix is visible (Fig. 6.10a), which distributes due to hot rolling (Fig. 6.10b).

6.5 SINTER FORGING

Sinter forging involves high-temperature pressing to achieve simultaneous deformation, consolidation, and densification of a ceramic. Temperatures are lower than those used in HIPing (by ~200–300°C), thus the initial grain size can be easily retained by sinter forging. A schematic of sinter forging is provided in Figure 6.11. High pressures induce high shear strains in deforming the material, which allows rapid reduction in pore volume and increases surface contacts to enhance atomic diffusion toward achieving higher densification. Table 6.2 shows typical temperature and pressure requirements for various ceramics. High pressure, in turn, also restricts grain growth. Lower temperatures utilized in sinter forging limit surface oxidation and, thereby, oxide inclusion, compositional transformations, residual stresses, cracking, and so on are highly restricted during ceramic processing. Higher surface curvature ($=2\gamma/r$, where γ is the surface energy, and r is the pore radius) of the pores (or smaller pores) requires high pressures for their annihilation; hence this process serves as an apt technique for achieving highly dense ceramics.

Deliberate induction of porosity can be attained in sinter-forged ceramics. It has been demonstrated by Kondo et al.[10] that the porosity increases as the soaking time is increased for the ceramic (Fig. 6.12). Increased porosity with soaking time is

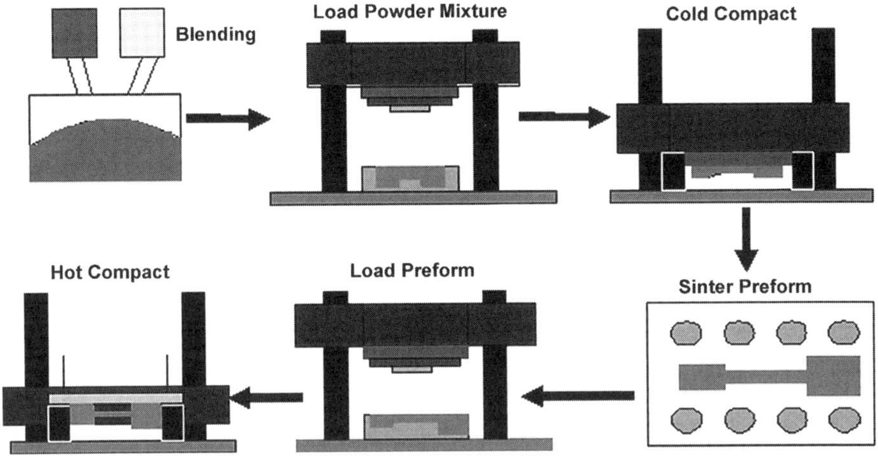

Figure 6.11 Schematic of sinter forging process.[13]

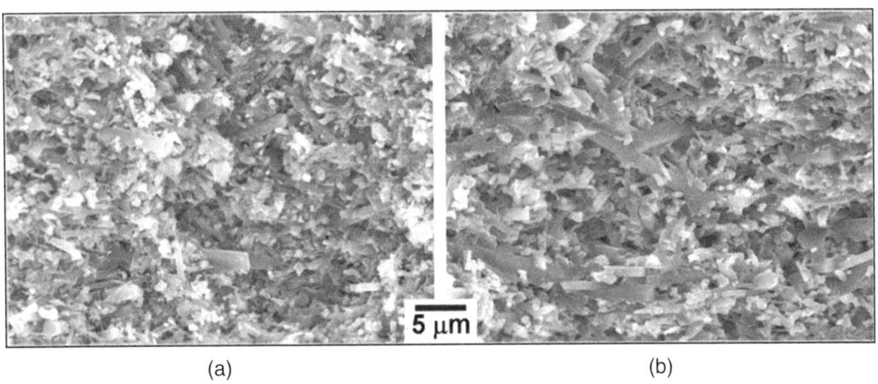

Figure 6.12 Fractured surface of sinter-forged Si_3N_4 soaked at 30 minutes each, with a total sintering time of (a) 3 hours and (b) 8 hours, respectively.[10]

Table 6.2. Pressure and Temperatures Required for Hot-Forging of a Few Ceramics

Material	Temperature (°C)	Pressure (MPa)	Time/atmosphere	Reference
Si_3N_4	1850	30	30–120 minutes N_2 atmosphere ~40% porosity	10
Al_2O_3–ZrO_2(Y_2O_3)-glass	1250	50	1 hour	11
3-YTZ	1300–1400	10, 16, 26	Varied time	12

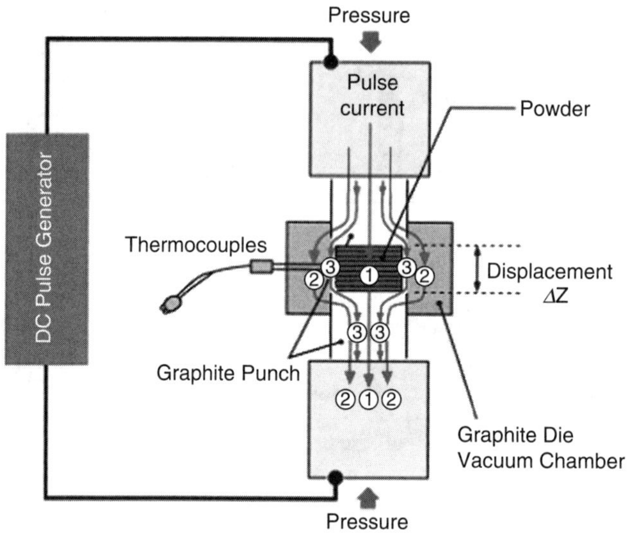

Figure 6.13 Schematic of spark plasma sintering showing the electrical path.

attributed to the difficulty in densification by the grain-coarsening during soaking (by formation of rodlike Si_3N_4).[10]

6.6 SPARK PLASMA SINTERING

Spark plasma sintering (SPS) utilizes release of electrical energy via arcing at the porous regions between powders to create local plasma and allow enhanced mass transport in the neck region to achieve full densification. SPS equipment comprises a graphite die in which powder is kept, and then current is passed through the graphite die (Fig. 6.13). Current traverses through the (1) graphite die to the bulk powder (1–1), (2) interface of graphite and ceramic powder, and (3) graphite–graphite (Fig. 6.13). Localized heating of particles occurs due to electrical discharge leading to heating and cleaning of surfaces.

Additionally, the powders compacted using the pressure exerted by the die (~15–100 MPa) assist mass flow in addition to the diffusion in the neck regions due to Joule heating resulting in rapid sintering (Fig. 6.14). Pressure application breaks down the surface films as well. The temperature gradient across the powder is strongly sensitive to both (1) power input and (2) thermal conductivity of the powder. Using finite element simulation, it has been analytically shown that a difference exists in the temperature being experienced by ceramic powders at die center and die surface, depending on thermal conductivity and total power input.[14] The higher the thermal conductivity, the less the temperature gradient across the powder compact, as shown in Figure 6.15. No significant temperature gradient is expected, however, when the thermal conductivity of the powder compact is more than 40 W/m·K.

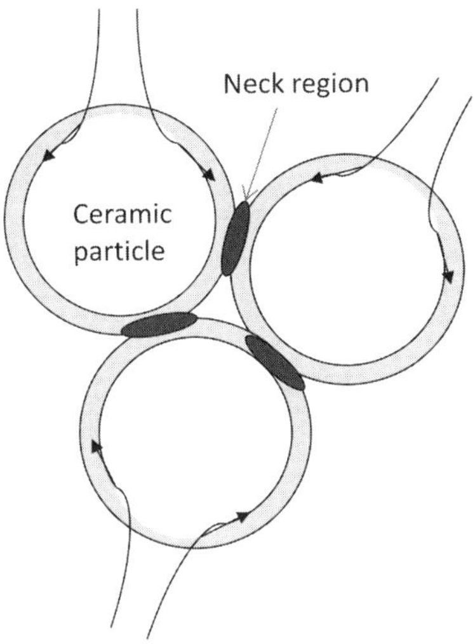

Figure 6.14 Schematic of Joule heating effect in resulting sintering of the powders during spark plasma sintering.

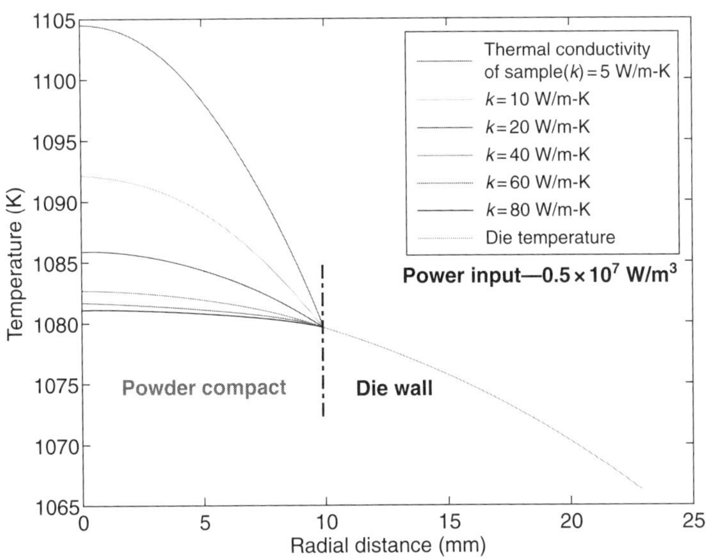

Figure 6.15 Effect of thermal conductivity in generating temperature gradient from surface to core of the ceramic powders inside a graphite die.[14] See color insert.

A few sinter-forged samples require pretty short processing times, as shown in Table 6.3. Duan et al. have consolidated nano-Al_2O_3 and nano-TiO_2 powders using SPS at 63 MPa for 3 minutes at 1150°C.[15] A typical high-density image is shown in Figure 6.16 showing densification. The use of SPS to obtain nanostructured ceramics will be described in later chapters of this book, and more details of various sintering techniques can be found in some textbooks as well as in review papers.[19–23]

Table 6.3. Sintering Time, Temperature, and Pressure Required for a Few Ceramics

Material	Temperature (°C)	Pressure (MPa)	Time (minutes)	Density (%)	Reference
AlN	1500–1600	30–40	15–30	~100	15
Al_2O_3	1150	63	3	~100	16
TiO_2	1150	63	3	~100	15
Si_3N_4	1500	20	5	>98	17
TiB_2/ZrB_2	1200–1500	40	10–15	>95	18

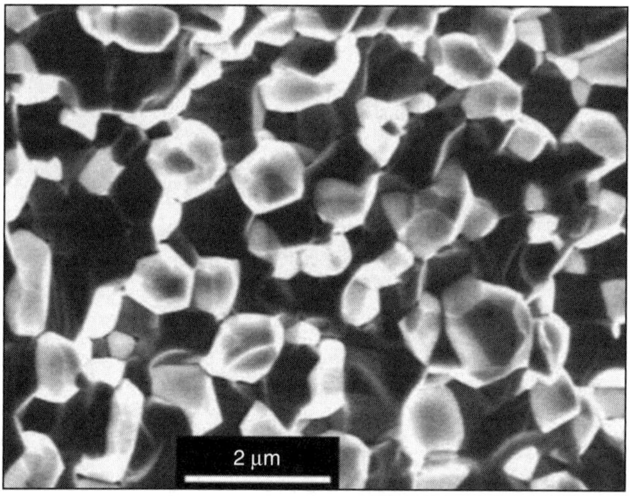

Figure 6.16 High densification of nano-Al_2O_3 and nano-TiO_2 is observed via SPS processing.[15]

Thermomechanical processing induces ease of sintering the material under reduced thermal exposure. Additionally, thermomechanical sintering methods provide an advanced control parameter to tailor densification via pressure, while retaining or reducing the grain size of structural ceramics. Thus, the role of controlling the processing parameters can yield superior mechanical properties as sintered parts with ~100% theoretical density can be easily achieved under optimized conditions.

REFERENCES

1 L. Rangaraj, C. Divakar, and V. Jayaram. Fabrication and mechanisms of densification of ZrB2-based ultra high temperature ceramics by reactive hot pressing. J. Eur. Ceram. Soc. 30(1) (2010), 129–138.

2 W. Wang, Z. Fu, H. Wang, and R. Yuan. Influence of hot pressing sintering temperature and time on microstructure and mechanical properties of TiB_2 ceramics. J. Eur. Ceram. Soc. 22 (2002), 1045–1049.

3 K. Rajan and P. Sajgalik. Local chemistry changes in Si_3N_4 based ceramics during hot-pressing and subsequent annealing. J. Eur. Ceram. Soc. 19 (1999), 2027–2032.

4 A. Weibel, R. Bouchet, R. Denoyel, and P. Knauth. Hot pressing of nanocrystalline TiO_2 (anatase) ceramics with controlled microstructure. J. Eur. Ceram. Soc. 27 (2007), 2641–2646.

5 P. Boch, J. C. Glandus, J. Jarrige, J. P. Lecompte, and J. Mexmain. Sintering, oxidation and mechanical properties of hot pressed aluminium nitride. Ceram. Intl. 8(1) (1982), 34–40.

6 Q. C. Ma, G. J. Zhang, Y. M. Kan, Y. B. Xia, and P. L. Wang. Effect of additives introduced by ball milling on sintering behavior and mechanical properties of hot-pressed B_4C ceramics. Ceram. Int. 36 (2010), 167–171.

7 A. K. Gain, J. K. Han, H. D. Jang, and B. T. Lee. Fabrication of continuously porous SiC–Si_3N_4 composite using SiC powder by extrusion process. J. Eur. Ceram. Soc. 26 (2006), 2467–2473.

8 S. Jihong, G. Jingkun, and J. Dongliang. Hot isostatic pressing of alpha-silicon carbide ceramics. Ceram. Intl. 19 (1993), 347–351.

9 J. C. Lee and K. N. Subramanian. The tensile properties of hot-rolled $(A1_2O_3)$p-Al composites. Mat. Sci. Eng. A 196 (1995), 71–78.

10 N. Kondo, Y. Inagaki, Y. Suzuki, and T. Ohji. Fabrication of porous anisotropic silicon nitride by using partial sinter-forging technique. Mat. Sci. Eng. A 335 (2002), 26–31.

11 S. Balasubramanian, H. Keshavan, and W. R. Cannon. Sinter forging of rapidly quenched eutectic Al_2O_3–$ZrO_2(Y_2O_3)$-glass powders. J. Eur. Ceram. Soc. 25 (2005), 1359–1364.

12 D. M. Owen and A. H. Chokshi. Final stage free sintering and sinter forging behavior of a yttria-stabilized tetragonal zirconia. Acta Mater. 46(2) (1998), 719–729.

13 V. Viswanathan, T. Laha, K. Balani, A. Agarwal, and S. Seal. Challenges and advances in nanocomposite processing techniques. Mat. Sci. Eng. R 54 (2006), 121–285.

14 D. Tiwari, B. Basu, and K. Biswas. Simulation of Thermal and electric field evolution during spark plasma sintering. Ceram. Int. 35 (2009), 699.

15 X. Du, M. Qin, A. Rauf, Z. Yuan, B. Yang, and X. Qu. Structure and properties of AlN ceramics prepared with spark plasma sintering of ultra-fine powders. Mat. Sci. Eng. A 496 (2008), 269–272.

16 R. G. Duan, G. D. Zhan, J. D. Kuntz, B. H. Kear, and A. K. Mukherjee. Spark plasma sintering (SPS) consolidated ceramic composites from plasma-sprayed metastable Al_2TiO_5 powder and nano-Al_2O_3, TiO_2, and MgO powders. Mat. Sci. Eng. A 373 (2004), 180–186.

17 L. Bai, M. Xiaodong, S. Weiping, and G. Changchun. Comparative study of β-Si_3N_4, powders prepared by SHS sintered by spark plasma sintering and hot pressing. J. Uni. Sci. Tech. Beijing 14(3) (2007), 271–275.

18 T. Venkateswaran, B. Basu, G. B. Raju, and D.-Y. Kim. Densification and properties of transition metal borides-based cermets via spark plasma sintering. J. Eur. Ceram. Soc. 26 (2006), 2431–2440.

19 L. S.-J. Kang. *Sintering*. Elsevier, Burlington, VT, 2005.

20 W. D. Kingery, H. K. Bowen, and D. R. Uhlmann. *Introduction to Ceramics*, 2nd ed. John Wiley and Sons, New York, 1976.

21 Y. M. Chiang, D. P. Birnie, and W. D. Kingery. *Physical Ceramics*. John Wiley & Sons, New York, 1997.

22 M. N. Rahaman. *Ceramic Processing and Sintering*. CRC Press, Boca Raton, FL, 2003.

23 A. Mukhopadhyay and B. Basu. Consolidation-microstructure-property relationships in bulk nanoceramics and ceramic nanocomposites: A review. Int. Mater. Rev. 52(5) (2007), 257–288.

Section Three

Surface Coatings

Chapter 7

Environment and Engineering of Ceramic Materials

This chapter serves two purposes: first, to discuss various properties required to ensure longer and reliable application of structural ceramics in various environments; second, to give a brief summary of some of the important oxide and non-oxide ceramics, with a particular focus on the combination of properties achievable with them.

For most practical applications, macroscopically homogeneous composites are indistinguishable to the naked eye. However, composites mainly have two components: matrix (ceramic, metal, or polymer) and reinforcement (fiber, particulate, platelet, whisker, etc.). Each category can be broadly categorized as polymer matrix composites (PMCs), metal matrix composites (MMCs), and ceramic matrix composites (CMCs). Each class of composites is recognized by the major constituent material present in the fabricated component.[1,2]

PMCs can be tailored to specific applications utilizing their unmatched properties of being lighter, stronger, and stiffer than unreinforced polymers or conventional metals. Owing to their lower fabrication temperatures, PMCs are easy to fabricate compared with MMCs and CMCs.[1,2] The specific strength and lightness of PMCs give them wide applicability in automotive applications. Liquid crystal polymers (LCPs), carbon fibers, Kevlar fibers, Aramid fibers, and ultra-high-molecular-weight polyethylene (UHMWPE) fibers are widely used in military and aerospace applications as fiber reinforcements in polymer matrix.

MMCs are used for high-temperature strength in parts such as cylinder sleeves, carbide drills, tank armor, rotors, jet landing gear, and disk brake calipers.[2] MMCs are superior to PMCs in showing resistance to fire, resistance to absorbing moisture, operation over a wide temperature range, resistance to radiation, absence of outgassing, and superior electrical and thermal conductivity. However, MMCs tend to be more expensive, heavy, and require special methods of fabrication.

CMCs display enhanced wear resistance, hardness, corrosion resistance, and temperature resistance. Applications of CMCs that have been on the market for a number of years include cutting tools and wear parts, space shuttle engine thrust cells, exhaust nozzle flaps, seals, propulsion thrusters and hot gas valve components,

---

Advanced Structural Ceramics, First Edition. Bikramjit Basu, Kantesh Balani.

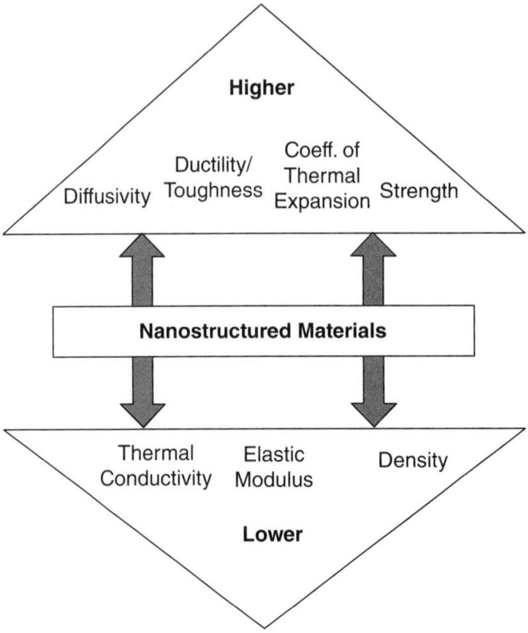

Figure 7.1 Advantages of nanostructured composites.

turbine engine combustors, and xenon ion propulsion units.[2] Yet the very nature of ceramics—their brittleness—halts the potentially exhaustive use of CMCs.

The class of "nanocomposites" comprises those composites—that is, materials that are useful combinations of two or more physically and/or chemically distinct materials —in which the size of one of the phases is *less than a hundred nanometers*.[3] Constituent materials maintain their identities even after the composite is completely formed. The advantage of using nanocomposites lies in achieving the best qualities of the constituted material (Fig. 7.1) and often some that are even superior to those of the constituent materials. Some improvements gained in using nanocomposite materials include, but are not limited to, strength, fracture toughness, corrosion resistance, thermal conductivity, weight reduction, stiffness, and wear resistance.[1,2,4]

7.1 ENVIRONMENTAL INFLUENCE ON PROPERTIES OF ENGINEERING CERAMICS

Ceramic materials are favorite materials for applications under severe conditions, such as to provide protection from an extensively corrosive environment, or provide strength at high temperatures, or provide resistance to rapid wear at high temperature, or provide electrical insulation in circuits. Hence the global spectrum of ceramics and their properties invites use of ceramic materials in extreme engineering environments. Certain important environments and properties are considered in this chapter, along with detailed explanations:

1. High-temperature strength (aerospace, liners, refractories, etc.)
2. Light weight (aerospace, automotive, material/fuel saving, etc.)
3. Wear resistance (aerospace, gears, articulating surfaces, etc.)
4. Corrosion resistance (biological, chemical, etc.)
5. High-temperature wear resistance (aerospace, turbine blades, motors, thermal barriers, etc.)
6. Electrical insulation (liners, heaters, dielectrics, etc.)
7. High-temperature oxidation resistance (aerospace, automotive, heaters, coatings, etc.)
8. High-temperature erosion resistance (rocket nozzle inserts, turbine blades, etc.)

The fundamental problem with ceramics resides in their inherently brittle nature. Ceramics often fracture like shattered glass upon an impact and do not show any plastic deformation. This leads to catastrophic failure without any prior warning. Hence, this limits the use of ceramic materials in engineering components that have to sustain impacts. Despite their high-temperature creep resistance being superior to that of metals, conditions in critical aerospace components such as jet engines limit the use of ceramics. Since efficiency improves with increasing operating temperature of the components, an insulating ceramic lining is necessary for improving efficiency of combustion. In addition, the presence of stress, notches, material impurities, a corrosive or oxidative environment, material processing, the existing microstructure, residual stresses, and so on can impart a complicated nature to the final mode of fracture or mode of defect accumulation.

Fracture of ceramics occurs due to crack initiation at defect sites such as pores, impurities, notches, and cracks; rapid loading extends the influence zone of these defects and causes failure of the component. An unstable crack initiates when the flaw size becomes greater than the critical size; under loading (cyclic or constant), this can cause fracture during service.

Certain high-temperature properties such as high refractoriness, chemical inertness, resistance to wear, resistance to high-temperature erosion, good oxidation resistance, low coefficient of thermal expansion, high or low thermal conductivity, and good creep properties are inherently required both for structural ceramics and for ultra-high-temperature ceramics engineered for re-entry space vehicles.[5] However, owing to the brittleness of ceramics, it becomes imperative to enhance their fracture toughness and flexural bend strength to move toward achieving their limitless applications.[5–11]

The processing, microstructure, and environment of an engineering component hold the key to deciding what damage will be incurred by a ceramic component during its service. Here, we consider only the effect of environment.

7.1.1 Oxidation Resistance

Oxide is generally the most stable phase of any element. Therefore members of the ceramic spectrum of borides, carbides, nitrides, silicides, and so on tend to get oxidized when exposed to an oxidizing atmosphere. Herein, oxide ceramics are the most stable. In general, ceramics are more stable than metals in terms of their oxidation. Oxide resistance is required when a surface is exposed to atmosphere at high temperatures. A few examples of oxidation-resistant materials include the following:

- Aluminum oxide
- Chromium oxide
- Zirconium oxide

7.1.2 Corrosion Resistance

In general, ceramics are known to possess good resistance to corrosion when exposed to chemicals. Household cutlery, pottery, century-old vases, and so on are made of ceramics. Corrosion resistance arises due to their chemical stability and the high covalent bonding typical of ceramics. Often HF, one of the strongest chemicals, is required to etch the microstructure of engineering ceramics.

7.1.3 Creep Resistance

Owing to their high melting points, ceramics often possess high creep resistance. Ceramics retain their strength up to >0.6 T_m, and their thermal insulation protects the core material and provides strength at high temperatures as well. Since the melting points (T_m) of ceramics approach close to 2300 K (compared with those of metals at >1300 K), increased strength can be expected for ceramics compared with metals. In addition, ceramics have higher elastic modulus compared with metals, and therefore remain stiff and introduce an extra dimension of strength retention even at higher temperatures (>1300°C).

7.1.4 Hard Bearing Surfaces

In locations where the material itself has to bear the wear on its surface with minimal lubrication to take away heat and reduce the friction of mating surfaces, the application of ceramic material wins such an engineering application. In addition, thermally sprayed ceramics are beneficial because of interlamellar porosity, which can hold lubricants. Examples include the following:

- Aluminum oxide
- Chromium oxide
- SiC
- WC/Co/Cr alloys

7.1.5 Thermal and Electrical Insulation

Usually oxide ceramics possess poor thermal and electrical conductivity, and their use as electrical insulation in high-voltage (~20 kV) transmission lines is not new.

Plasma-sprayed structure often induces porosity in the material, which adds to the thermally insulating nature of a coating. These materials are specifically advantageous in the thermal barrier coatings (TBCs) applied to turbine blades so that blades can take the load, and thermal insulation is provided by the TBC, so that blades can experience a lower temperature atmosphere and can retain strength at working temperature. Examples include the following:

- Yttria-stabilized zirconia (YSZ)
- Aluminum oxide

7.1.6 Abrasion-Resistant Ceramics

Ceramics are excellent candidates for abrasion resistance as well, mainly due to their high hardness. Once the hardness of a material is higher than its mating surface, it can render enhanced wear resistance. The following ceramics possess excellent wear resistance:

- Aluminum oxide, chromium oxide
- Chromium carbide, silicon carbide
- Titanium boride

7.1.7 Fretting Wear Resistance, Surface Fatigue, Impact Resistance

Coatings often require resistance to repeated sliding, rolling, impacting, or vibration in mechanically moving parts. Good fretting wear resistance requires a combination of high toughness and presence of compressive stresses, which restrict crack initiation, impede crack propagation, and tolerate cracking damage by blunting the crack tip. Usually a tough reinforcement (usually of metal) is required in ceramics to achieve high impact resistance as well, such as

- Cermet coatings WC/Co,
- CoCr/Ni-Cr, and
- Al_2O_3/Al.

7.1.8 Erosion and Cavitation Resistance

Erosion is highly complicated since the angle of impingement of the abrasive onto the ceramic surface is highly critical. For shallow contact angle attack (parallel to surface), higher hardness is preferred since the material loss is similar to that of abrasion. However, when the angle of attack is ~90° (perpendicular to surface), the impact transfer damages the material, and therefore the material's toughness is more important. Other than that, cavitation resistance and liquid impingement on a ceramic surface require good surface fatigue resistance since the explosion of bubbles leads to a cyclic state of compression and tension. Examples include the following:

- WC/Co
- Chromium carbide
- Aluminum oxide, chromium oxide

It is noted that mechanisms of material loss are defined at nanoscales and microscales, which later build up as macrodamage of the bulk component. Additionally, certain loss mechanisms prevalent at small length scales combine with contrasting mechanisms at large scales and result in overall damage to the bulk material. On one hand, small length scales involve interactions via single or a few asperity (or surface) contacts; on the other hand, macrodamage involves multiple asperity contacts or bulk interactions. Consequently, it becomes essential to bridge the gap between various length scales when estimating damage.

7.2 CLASSIFICATION AND ENGINEERING OF CERAMIC MATERIALS*

Structural ceramics can be broadly classified into non-oxide and oxide ceramics. Accordingly, the properties associated with the various classes of engineering ceramics also decide their performance in stringently constrained applications.

7.2.1 Non-Oxide Ceramics

Ceramics, in early years, involved techniques of adding various elements, compounds, and rare-earth oxides to achieve certain properties, such as a combination of hardness and toughness for applications in cutting tools, for example, AlN in SiC, Ni or Mo to TiB_2, and Co in WC. However, in later years, the use of pressing and sintering techniques (thermal, pressureless, liquid phase) and development of tough MAX-phase ceramics (see Chapter 12) to manipulate the microstructure, for example, in ZrB_2-$MoSi_2$, SiC, and SiAlON-SiC, have created a paradigm shift in realizing the ceramics for structural applications.[10,12–20]

Borides, carbides, nitrides, and silicides constitute the class of non-oxide ceramics. Ultra-high-temperature ceramics (UHTCs) such as HfC, TaC, ZrB_2, BN, HfN, TiN/TiB_2, and their composites have been used for rocket science and engineering applications that require structural integrity at temperatures in excess of 1800°C.[21] Toughening in the non-oxide ceramics can be achieved by grain refinement, addition of secondary reinforcements, high densification, and crack-healing agents (such as a glassy or amorphous phase, phase transformation, or presence of microporosity).[16,19] A summary of strengthening and toughening mechanisms in non-oxide structural ceramics is presented in Table 7.1. It can be observed that most of the research work and commercial development of non-oxide ceramics has been limited to SiC and Si_3N_4.[13,17,19,20]

*This section is mainly taken from the PhD thesis of Kantesh Balani (2007) Florida International University, Miami, FL.

Table 7.1. Fracture Toughening of Non-Oxide Structural Ceramics

Composition	Toughening mechanism	Fracture toughness	Other features/ comments	Reference
SiAlON–SiC	Liquid-phase sintering	~6.0 MPa $m^{1/2}$	Fine equiaxed grains and dense structure	13
SiC + AlN + rare earths	Smaller rare earth cations resulting in clean boundaries	~6.5 MPa $m^{1/2}$	Composition and microstructure of sintering additives affect toughness	17
WC–(0.5 wt%) Co–(0.25 wt%) VC	Abnormal grain growth reinforcement	7.34 MPa $m^{1/2}$	Toughening by crack-cutting elongated grains	18
Si_3N_4/SiC (5 wt% Y_2O_3 with up to 13 wt% SiC)	Inter- and intra-SiC nanoinclusions	5.8 MPa $m^{1/2}$ (at 8 wt% SiC)	Hardness increases with increase in SiC content	20
ZrB_2–$MoSi_2$	Uniform and fine microstructure	~2.6 MPa $m^{1/2}$	Flexural strength ~655 and 500 MPa at 1200 and 1500°C, respectively.	15
C/SiC	Presence of microcracks, carbon fiber pullout, homogeneity, and nondecomposition of matrix	–	Flexural strength ~124–287 MPa	22,23

Composites of Si_3N_4 with inter and intra SiC inclusions in wood-cutting ceramics provided enhancement of mechanical properties (fracture toughness increased from 5.1 to 5.8 MPa $m^{1/2}$ and microhardness increased from 16 to 19 GPa).[20] In addition, pressing and sintering techniques (thermal, pressureless liquid phase) used to manipulate microstructure in ZrB_2–$MoSi_2$, SiC, and SiAlON–SiC ceramics have proved to be significant in achieving toughened ceramic composites.[13,15,17] The role of utilizing abnormal grain growth in WC and the development of tough ternary-carbide-phase ceramics were the next big feats in the field of non-oxide ceramics.[10,12–20] C/SiC composites fabricated by chemical vapor deposition or infiltration and sol–gel techniques have emerged for applications such as aircraft brakes, re-entry shields, and rocket nozzles.[22,23]

Figure 7.2 demonstrates the change of fracture toughness during sintering with addition of rare-earth oxides to a SiC matrix.[17] Clean interphase boundaries without

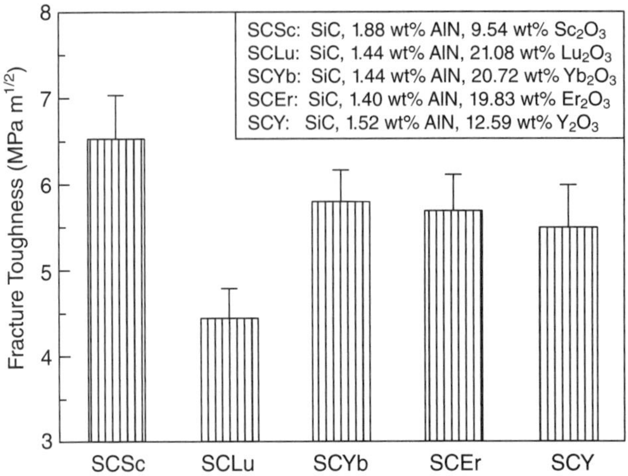

Figure 7.2 Fracture toughness of SiC ceramics sintered with Re_2O_3 and AlN.[17]

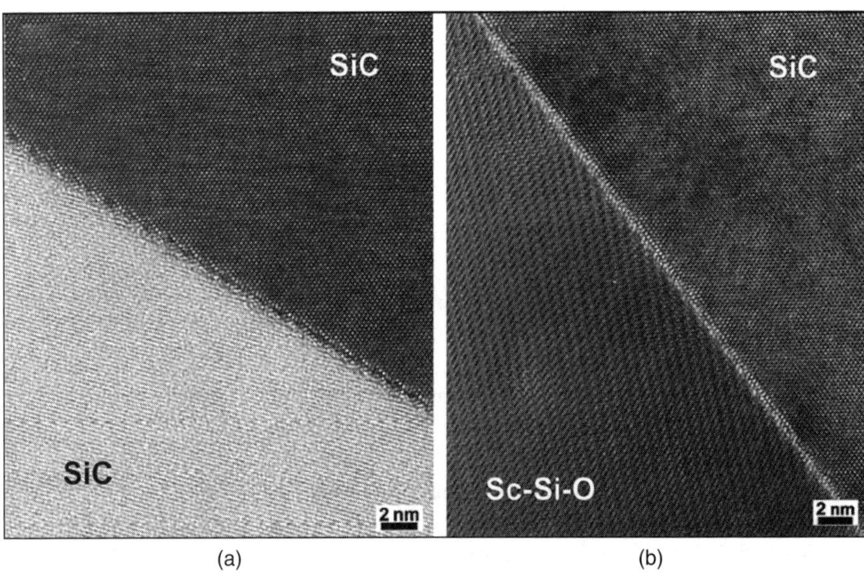

Figure 7.3 High resolution TEM micrographs revealing clean interfaces without the amorphous intergranular phase in SiC–AlN–Sc_2O_3: (a) SiC–SiC boundary and (b) SiC–junction phase boundary.[17]

an amorphous intergranular phase, as observed in Figure 7.3, result from the addition of smaller rare-earth cations.[17] Composition and microstructure of the sintering additives strongly aid the toughening and strengthening of SiC ceramics. Figure 7.4a features the increasing hardness of Si_3N_4/SiC nanocomposite with increasing SiC content.[20] Apart from increasing SiC content, microstructure refinement hinders dislocation movement within the Si_3N_4 matrix, enhancing hardness. Liquid-phase

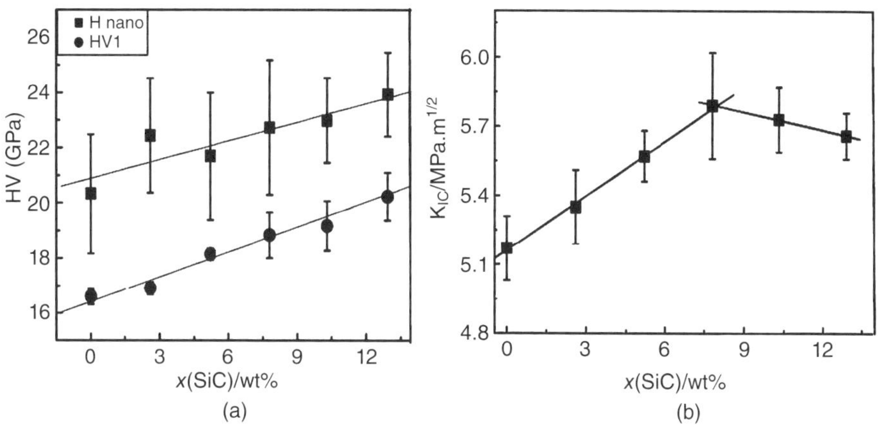

Figure 7.4 Comparison of (a) nanohardness and macrohardness and (b) fracture toughness of Si_3N_4/SiC nanocomposites.[20]

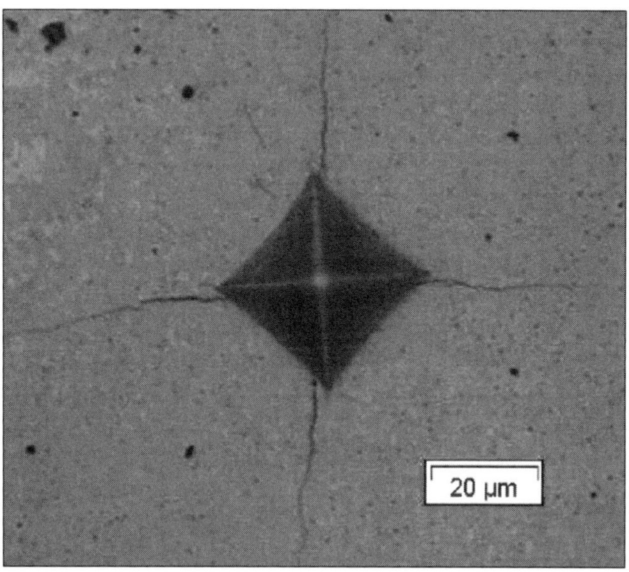

Figure 7.5 Vickers indentation on WC ceramic depicting origination of radial cracks.[18]

sintering resulted in percolation of SiC nanoinclusions in the intergranular regions, enhancing fracture toughness (Fig. 7.4b).

Figure 7.5 shows Vickers indentation generating radial cracks in cobalt-bonded WC. Vickers cracks generate due to residual stress relief upon unloading; the balancing relation of stress intensity ahead of the crack tip with crack termination can indirectly help evaluate the fracture toughness of the material. When Young's modulus for ceramics is known (taken as 390 GPa for bulk Al_2O_3), indentation fracture toughness (K) can be calculated by the semiempirical formula given by Anstis, Equation 7.1[24]:

$$K = \chi \left(\frac{E}{H_V} \right)^{1/2} \left(\frac{P}{c} \right)^{3/2}, \tag{7.1}$$

where $\chi = 0.016$ is a material-independent constant, E is Young's modulus, H_V is the Vickers hardness, P is the applied load, and c is the crack length. Oxide ceramics and their toughening mechanisms are reviewed in the next subsection.

7.2.2 Oxide Ceramics

Oxide ceramics are the most useful and common materials owing to ease of fabrication and their stability at high temperatures compared with those of nitrides, carbides, sulfides, and borides.[21,25] Oxide ceramics, such as Al_2O_3, ZrO_2, TiO_2, Cr_2O_3, SiO_2, and Y_2O_3, offer advanced technological applications owing to high hardness, resistance to corrosion, and high refractoriness apart from their high resistance to wear, fretting, cavitation, and erosion and their high dynamic modulus.[6,26–29] Though non-oxide ceramics display unique properties as potential candidates for extreme environments, they often require a protective oxide layer to create a diffusion barrier for oxidation protection.[21,30,31] Oxidation resistance becomes a natural requirement for the materials operating at elevated temperature in air. Aptly, oxide ceramics emerge as oxidation-resistant structural materials.[21,31,32] High-temperature applications introduce large volumetric phase transformations leading to structural instability. Also, the inherent brittleness of the ceramic oxides makes them more susceptible to thermal shock failures. Thereby, toughening of oxide ceramics becomes a prerequisite in their structural applications.

Conventionally, fracture toughening enhancement in oxides has been achieved through phase transformation or by introducing controlled or graded microstructure.[33–36] Incorporation of secondary phases to restrict crack propagation and nanocrystalline structure to enhance grain sliding have also been prominent in improving the fracture toughness of ceramic nanocomposites.[37–40] Table 7.2 reports various fracture toughening mechanisms in oxide ceramics.

Polished and thermally etched Y–ZrO_2–TiO_2 composite without TiO_2 addition (Fig. 7.6a) and with 10 vol% addition (Fig. 7.6b) at 1400°C for 4 hours increased the defects and pores in the ceramic composite.[33] Increased inhomogeneity and porosity resulted from the agglomeration tendency of TiO_2 particles. Though increased TiO_2 content resulted in deterioration of mechanical properties, an increase in fracture toughness and hardness was observed for higher sintering temperatures.[33]

Figure 7.7 shows hardness and crack propagation resistance of high-velocity oxy-fuel sprayed (HVOF) and (air plasma-sprayed (APS) TiO_2 coatings. Though hardness of all three processed coatings (viz. air plasma and HVOF sprayed TiO_2, and HVOF sprayed nano-TiO_2) is similar, nano-TiO_2 demonstrated extremely high fracture toughness compared with conventional processing (Fig. 7.7; also

Table 7.2. Fracture Toughness and Toughening Mechanisms for Oxide Ceramics

Oxide ceramics	Toughening mechanism	Fracture toughness	Other features/ comments	Reference
YSZ (yttria-stabilized tetragonal zirconia)	Transformation toughening (tetragonal to monoclinic ZrO_2)	4.5 MPa $m^{1/2}$	Additional toughening by $ZrTiO_4$ phase	33
Nano-TiO_2	Crack arrest by nanostructured zones	~27 MPa $m^{1/2}$	Isotropic crack propagation and distribution of agglomerated nanoparticles	37
SiO_2	Restricting cristobalite by AlN particles	Up to 2.96 MPa $m^{1/2}$ at 1400°C	Microcrack deflection and divergence	41
$MgAl_2O_4$	Restraining the grain growth by nano-Al_2O_3	7.79 MPa $m^{1/2}$	Low-temperature diffusional creep, thermally activated deformation	38
ZrO_2	Transformation toughening	10.1 MPa $m^{1/2}$	Addition of secondary toughening by WC, crack deflection	34

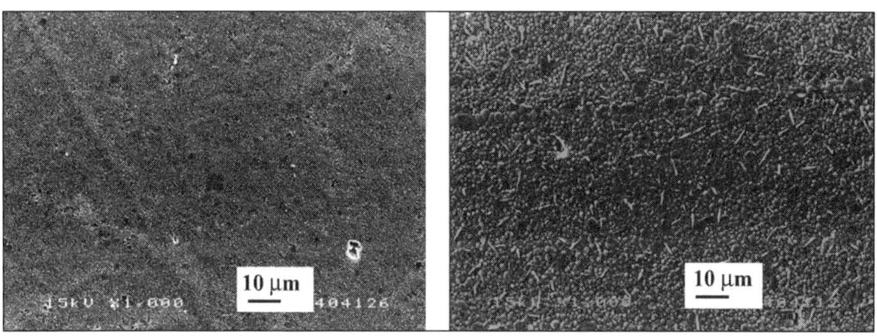

(a) (b)

Figure 7.6 SEM micrographs of the thermally etched surfaces of the Y–ZrO_2–$ZrTiO_4$ composites sintered at 1400°C with (a) 0 vol% and (b) 10 vol% TiO_2.[33]

see Table 7.2).[37] Enhanced fracture toughening was attributed to the arresting of cracks by nanostructured zones as observed in Figure 7.8.[37] Nanosized process zones impede the crack path and restrict the crack extension by absorbing the crack-propagation energy and arresting the crack, resulting in enhanced fracture toughness.

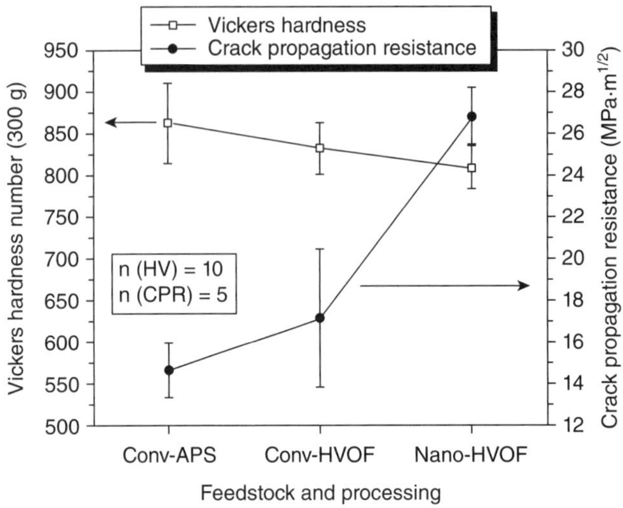

Figure 7.7 Vickers hardness and crack propagation resistance of coatings made from nanostructured and conventional TiO_2 feedstock sprayed via HVOF and APS. HV, Vickers hardness; CPR, crack propagation resistance.[37]

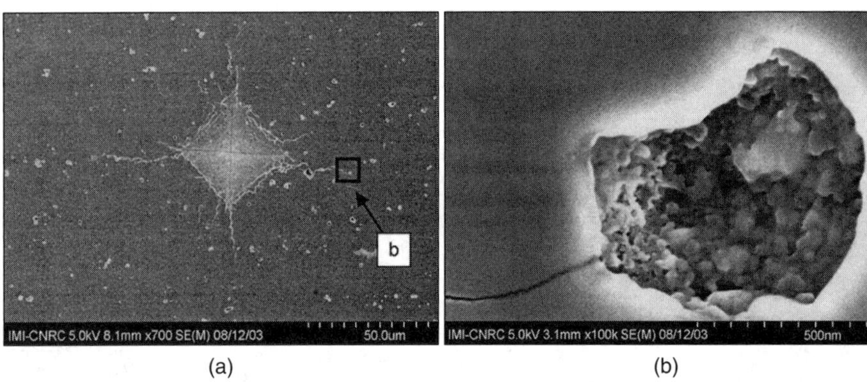

Figure 7.8 Vickers indentation impression (1 kgf) in the cross section of the (a) HVOF sprayed nanostructured TiO_2 coating and (b) the indentation crack tip being arrested by a zone of nanostructured particles.[37]

Engineering of ceramic materials requires a thorough knowledge of the environment in which the ceramic component will be used. Hence, it becomes essential to partake and understand the key strengths of ceramics, that is, low density compared with metals, high oxidation and corrosion resistance, high-temperature wear and erosion resistance, high fracture strength, and good thermal insulation properties. However, the major limitation of application of ceramics is their brittleness. Hence, understanding the various routes of energy dissipation during fracture becomes essential in order to make ceramics tougher!

REFERENCES

1 R. M. Jones. *Mechanics of Composite Materials*. Taylor & Francis Inc., Philadelphia, PA, 1999.

2 D. B. Miracle and S. L. Donaldson. *ASM Handbook: Composites*, Vol. 21. ASM International, Materials Park, OH, 2001, 3.

3 C. Suryanarayana and C. C. Koch. Nanostructured materials, in *Non-Equilibrium Processing of Materials*, C. Suryanarayana (Ed.). Pergamon, Oxford, UK, 1999, 313–372.

4 M. M. Schwartz. *Composite Materials: Processing, Fabrication and Application*. Prentice Hall PTR, Englewood Cliffs, NJ, 1997, 572.

5 K. Balani, A. Agarwal, and T. McKechnie. Near net shape fabrication via vacuum plasma spray forming. Trans. Indian Inst. Met. 59(2) (2006), 237–244.

6 G. R. Karagedov and N. Z. Lyakhov. Preparation and sintering of nanosized alpha-Al_2O_3 powder. Nanostruct. Mater. 11 (1999), 559–572.

7 Y. Katsuda, P. Gerstel, J. Narayanan, J. Bill, and F. Aldinger. Reinforcement of precursor-derived Si-C-N ceramics with carbon nanotubes. J. Eur. Ceram. Soc. 26 (2005), 3399–3405.

8 K. Balani, G. Gonzalez, A. Agarwal, R. Hickman, J. S. O'Dell, and S. Seal. Synthesis, microstructural characterization and mechanical property evaluation of vacuum plasma sprayed tantalum carbide. J. Am. Ceram. Soc. 89(4) (2006), 1419–1425.

9 K. Balani, R. Anderson, T. Laha, M. Andara, J. Tercero, E. Crumpler, and A. Agarwal. Plasma-sprayed carbon-nanotube reinforced hydroxyapatite coatings and their interaction with human osteoblasts in vitro. Biomaterials 28(4) (2007), 618–624.

10 M. Gu, C. Huang, B. Zou, and B. Liu. Effect of (Ni, Mo) and TiN on the microstructure and mechanical properties of TiB_2 ceramic tool materials. Mater. Sci. Eng. A 433 (2006), 39–44.

11 E. Turunen, T. Varis, T. E. Gustafsson, J. Keskinen, T. Fältc, and S.-P. Hannula. Parameter optimization of HVOF sprayed nanostructured alumina and alumina-nickel composite coatings. Surf. Coat. Technol. 200 (2006), 4987–4994.

12 C. H. Chen and H. Awaji. Temperature dependence of mechanical properties of aluminum titanate ceramics. J. Am. Ceram. Soc. 27 (2007), 13–18.

13 J. V. C. Souza, C. Santos, C. A. Kelly, and O. M. M. Silva. Development of α-SiAlON-SiC ceramic composites by liquid phase sintering. Int. J. Ref. Met. Hard Mater. 25 (2007), 77–81.

14 T. Zhang, Z. Zhang, J. Zhang, D. Jiang, and Q. Lin. Preparation of SiC ceramic by aqueous gel casting and pressureless sintering. Mater. Sci. Eng. A 443 (2007), 257–261.

15 D. Sciti, F. Monteverde, S. Guicciardi, G. Pezzotti, and A. Bellosi. Microstructure and mechanical properties of ZrB_2-$MoSi_2$ ceramic composite produced by different sintering techniques. Mater. Sci. Eng. A 434 (2006), 303–309.

16 F. Monteverde. Ultra-high temperature HfB_2-SiC ceramics consolidated by hot-pressing and spark plasma sintering. J. Alloys Comp. 428 (2007), 197–205.

17 Y. W. Kim, Y. S. Chun, T. Nishimura, M. Mitomo, and Y. H. Lee. High temperature strength of silicon carbide ceramics sintered with rare-earth oxide and aluminum nitride. Acta Mater. 55 (2007), 727–736.

18 T. Li, Q. Li, J. Y. H. Fuh, P. C. Yu, L. Lu, and C. C. Wu. Effects of AGG on fracture toughness of tungsten carbide. Mater. Sci. Eng. A 445–446 (2007), 587–592.

19 F. Eblagon, B. Ehrle, T. Graule, and J. Kuebler. Development of silicon nitride/silicon carbide composites for wood-cutting tools. J. Eur. Ceram. Soc. 27 (2007), 419–428.

20 M. Balog, J. Kečkéš, T. Schöberl, D. Galusek, F. Hofer, J. Křestan, Z. Lenčéš, J.-L. Huang, and P. Šajgalík. Nano/macro-hardness and fracture resistance of Si_3N_4/SiC composites with upto 13 wt.% of SiC nano-particles. J. Eur. Ceram. Soc. 27 (2007), 2145–2152.

21 K. Upadhya, J. M. Yang, and W. P. Hoffman. Materials for ultrahigh temperature structural applications. Am. Ceram. Soc. Bull. 76(12) (1997), 51–56.

22 K. Krnel, Z. Stadler, and T. Kosmac. Preparation and properties of C/C-SiC nano-composites. J. Eur. Ceram. Soc. 27 (2007), 1211–1216.

23 Y. Xu, Y. Zhang, L. Cheng, L. Zhang, J. Lou, and J. Zhang. Preparation and friction behavior of carbon fiber reinforced silicon carbide matrix composites. Ceram. Int. 33 (2007), 439–445.

24 G. R. Anstis, P. Chantikul, B. R. Lawn, and D. B. Marshall. A critical evaluation of indentation techniques for measuring fracture toughness: I, direct crack measurements. J. Am. Ceram. Soc. 64(9) (1981), 533–538.
25 M. Miyayama, K. Koumoto, and H. Yanagida. *Engineering Properties of Single Oxides. Engineered Materials Handbook®*, Vol. 4: Ceramics and Glasses. ASM International, Materials Park, OH, 1991, 748–757.
26 Y. Wang, S. Jiang, M. Wang, S. Wang, T. D. Xiao, and P. R. Strutt. Abrasive wear characteristics of plasma sprayed nanostructured alumina/titania coatings. Wear 237 (2000), 176–185.
27 H. Z. Wang, L. Gao, L. H. Gui, and J. K. Guo. Preparation and properties of intragranular Al_2O_3-SiC nanocomposites. Nanostruct. Mater. 10(6) (1998), 947–953.
28 W. Q. Li and L. Gao. Processing, microstructure and mechanical properties of 25%vol YAG-Al_2O_3 nanocomposites. Nanostruct. Mater. 11 (1999), 1073–1080.
29 Y. Yu, Y. Ma, C. Zhou, and H. Xu. Damping capacity and dynamic mechanical characteristics of the plasma-sprayed coatings. Mater. Sci. Eng. A A408 (2005), 42–46.
30 C. B. Bargeron, R. C. Benson, A. N. Jette, and T. E. Phillips. Oxidation of hafnium carbide in the temperature range 1400 to 2060 C. J. Am. Ceram. Soc. 76(4) (1993), 1040–1046.
31 J. B. B. Mattuck. High temperature oxidation. J. Electrochem. Soc. 114(10) (1967), 1030–1033.
32 M. W. Barsoum and T. El-Raghy. The MAX phases: Unique new carbide and nitride materials. Am. Sci. 89 (2001), 334–343.
33 X. Miao, D. Sun, P. W. Hoo, J. Liu, Y. Hu, and Y. Chen. Effect of titania addition on yttria-stabilised tetragonal zirconia ceramics sintered at high temperatures. Ceram. Int. 30 (2004), 1041–1047.
34 G. Anné, S. Put, K. Vanmeensel, D. Jiang, J. Vleugels, and O. Van der Biest. Hard, tough and strong ZrO_2-WC composites from nanosized powders. J. Eur. Ceram. Soc. 25(1) (2005), 55–63.
35 L. Xu, Z. Xie, L. Gao, X. Wang, F. Liana, T. Liu, and W. Li. Synthesis, evaluation and characterization of alumina ceramics with elongated grains. Ceram. Int. 31 (2005), 953–958.
36 K. A. Khor, Z. L. Dong, and Y. W. Gu. Plasma sprayed functionally graded thermal barrier coatings. Mater. Lett. 38 (1999), 437–444.
37 R. S. Lima and B. R. Marple. From APS to HVOF spraying of conventional and nanostructured titania feedstock powders: A study on the enhancement of the mechanical properties. Surf. Coat. Technol. 200 (2006), 3248–3437.
38 S. Bhaduri and S. B. Bhaduri. Microstructural and mechanical properties of nanocrystalline spinel and related composites. Ceram. Int. 28 (2002), 153–158.
39 H. V. Swygenhoven and J. R. Weertman. Deformation in nanocrystalline metals. Mater. Today 9 (2006), 24–31.
40 P. Bansal, N. P. Padture, and A. Vasiliev. Improved interfacial mechanical properties of Al_2O_3 13 wt%TiO_2 plasma-sprayed coatings derived from nanocrystalline powders. Acta Mater. 51 (2003), 2959–2970.
41 J. Wu, B. Li, and J. Guo. The influence of addition of AlN particles on mechanical properties of SiO_2 matrix composites doped with AlN particles. Mater. Lett. 41 (1999), 145–148.

Chapter 8

Thermal Spraying of Ceramics

Thermal spraying is defined as synthesis of a coating or freestanding structure via a high-velocity stream of molten or semimolten ceramic.[1,2] In thermal spraying, the primary source of energy during ceramic processing is "thermal energy," which causes material to change shape. Contrary to room-temperature plastic deformation processes, the material *flows* not because of applied mechanical stresses, but because of its reduced viscosity at high temperatures. Material is forced at high velocity toward a substrate at high temperatures, where its momentum at high temperature causes deformation of the ceramic to the desired shape. Often fine powders or wires are utilized as base material for processing; these break into fine droplets during spraying and are deposited as a coating. Later, the coating can be separated from the substrate and serve as a freestanding structure. There are several techniques that can be adopted for separation of substrate from the coating and these are discussed separately.

Thermal spraying has been widely used mainly for "reclamation" of worn surfaces, as thermal spraying can provide thick deposits in a very short time. Thicknesses on the order of a few millimeters can be easily repaired by thermal spraying. Hence thermal spraying of ceramics is highly utilized for wear protection, thermal insulation, corrosion and oxidation protection, surface repair or reclamation, electrical insulation, and more recently for biomedical coatings. A detailed chart of applications of thermal spray coatings is listed in Table 8.1.

8.1 MECHANISM OF THERMAL SPRAYING

Thermal spraying involves two parameters: the thermal energy of the plume, and the kinetic energy of the accelerated particles. The heat for softening the material is added to the momentum of material via forced carrier gas carrying the semimolten or molten particles. Successive impacts of semimolten or molten particles create layer-by-layer deposits that form the coating on the substrate. This phenomenon of deposition can be described in three stages:

Advanced Structural Ceramics, First Edition. Bikramjit Basu, Kantesh Balani.

Table 8.1. Thermal Spraying of Various Ceramics

Property	Examples	Details
Wear/abrasion resistance	Al_2O_3,WC–Co, Cr_2C_3–Ni–Cr, TiO_2, Al_2O_3–TiO_2,TiC, Mo, ZrO_2, SiC,	Resistance to abrasion, erosion, fretting, friction, sliding wear
Thermal barrier coatings	MCrAlY (M = Fe, Co, Ni), ZrO_2, Al_2O_3, YSZ	Oxidation, thermal insulation
Electrical insulation	Al_2O_3,Al_2O_3 + TiO_2	Insulating property
Electronic	TaN, In_2O_3, WN_2, TiN, WSi_2, YSZ, Co-fired ceramics (HTCC, LTCC)	Conductive ceramics
Biocompatibility	TiO_2, hydroxyapatite (HA) coatings, porcelain	Tissue growth propensity
Aesthetics	TiN, Al_2O_3	Jewelry, tiles, coatings, gemstones
Oxidation/corrosion resistance	Ti_3SiC_2, Al_2O_3, ZrO_2, SiO_2, SiO_2, MgO, Cr_2O_3	Resistance against oxidation/corrosion
Ultra-high-temperature resistance	W, W–Re, HfC, Al_2O_3, TaC	Resistance to ultra-high-temperature softening and wear

Stage 1: A coating gun heats up the source material (such as powder or wire) and makes the material plastic (molten or semimolten state). Now the material becomes easily plastically deformable.

Stage 2: Now these molten or semimolten droplets are accelerated to high velocities toward the substrate. This provides momentum to the stream of particles.

Stage 3: The molten or semimolten particles impact the substrate to form the coating. Deposition occurs due to successive impacts of particles. Deposition can be mechanically integrated or metallurgically bonded.

It has to be noted that the coating material should not decompose or sublime during the processing, and it should remain in a plastic condition. The kinetic impact (carrier gas velocity and particle mass) and the resulting viscosity (due to temperature and the material's inherent nature) and thermal characteristics (due to the particle's size and inherent nature and the substrate's thermal conductivity) decide the deposition of the coating and the evolution of microstructure. Consequent bonding between layers is often mechanical, but can be metallurgical if the material is superheated and/or if secondary postdeposition treatments are provided. Thermal spraying typically results in coatings greater than 50 μm thick.

Thermal spraying is a very complex process, with complexity arising from (i) the initial chemistry, size, and morphology of material, (ii) the thermal and kinetic profiles of particles in the thermal plume, and (iii) the interaction of molten/semimolten particles with the substrate. Additionally the selection of nozzle geometry, selection of gases, their purity, and the power rating also strongly affect the quality

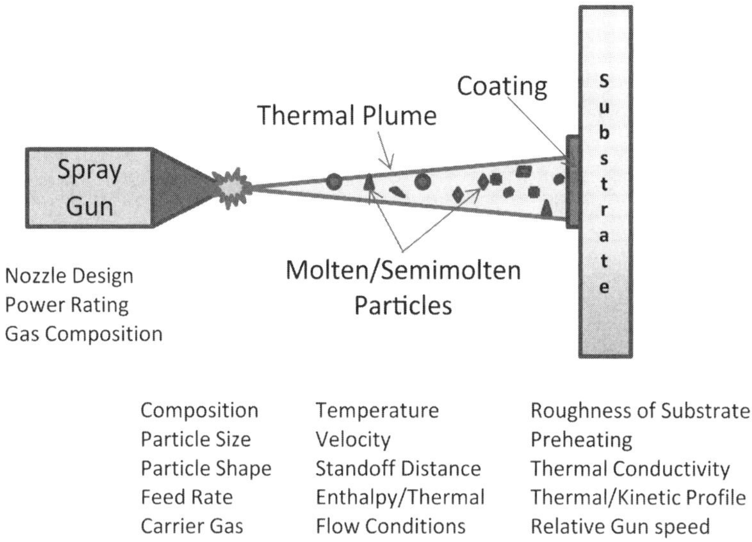

Figure 8.1 Complexity associated with thermal spraying of ceramics.

of ceramic deposition. A very high complexity is associated with thermal spraying (see Fig. 8.1), arising from the three stages of thermal spraying:

Complexity Associated with the First Stage. Certain parameters that affect the complexity in the process arise from the selection of material itself. The material composition, powder or wire size and shape, feed rate, density, and the carrier gas type and its flow, pressure, and velocity affect the manner in which the material is fed and the way it interacts with the thermal plume in the next stage.

Complexity Associated with the Second Stage. Most complexity-affecting parameters in this stage are related to the temperature and velocity of the stream of molten or semimolten particles. Additionally the spray distance decides the dwell time of particles in the plume. Enthalpy of the carrier gas decides the generated temperature, whereas thermal conductivity of particles decides the degree of their superheating. In addition, the flow turbulence of carrier gas dictates the heating of particles. The environment of the thermal stream restricts or assists oxidation of material at high temperatures.

Complexity Associated with the Third Stage. Here, the substrate on which deposition is to occur is vital. The surface preparation (in terms of roughness) and the purity, thermal conductivity, preheat temperature, and standoff distance of the surface decide the type and quality of coating that will be deposited. Additionally, in order to remove the substrate from the coating, several techniques need to be adopted (such as chemical leaching of the substrate, thermal-quench separation of the substrate, or simply machining off the substrate). Hence, it becomes highly critical to adopt the parameters that are required to achieve a perfect deposition.

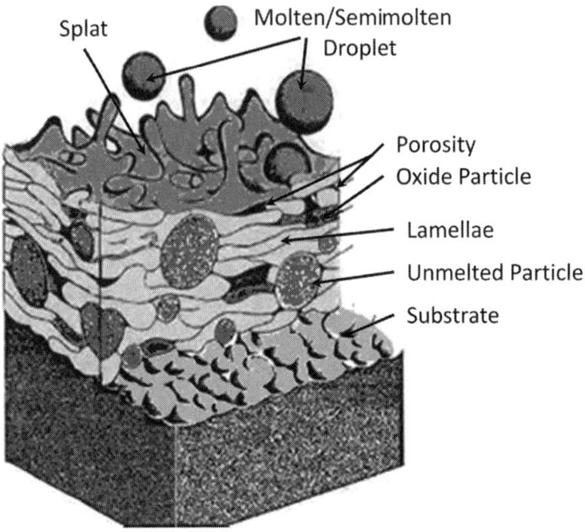

Figure 8.2 Typical microstructure of thermally sprayed coating (adapted from[1]).

Thermal spray coating is a layered deposition process, where splats build up by successive impact of solid particles or molten droplets as shown in Figure 8.2. The layered deposition can also entrap pores, oxides, and unmelted particles (Fig. 8.2), and overlapping splats lock onto one another to form a continuous coating layer. The microstructure of the resultant splats ultimately affects the properties of the coating. Furthermore, certain uncontrollable parameters, such as electrical fluctuations, wear and tear of the gun, choking of powders during feed, inhomogeneity in powder distribution, density of materials, and impurity of carrier gases, can also induce further complexity in controlling the final microstructure and properties of the deposited ceramics as shown earlier in Figure 8.1.

8.1.1 Advantages of Thermal Spraying

Thermal spraying is a highly versatile technique as it does not require a conventional container or crucible to "hold" the material. Hence it becomes easier to melt high-melting-point material in air and let it impact on the substrate to form a coating or freestanding structure. It encompasses various classes of deposition techniques, such as plasma spraying, combustion spraying, electric arc deposition, high-velocity oxy-fuel (HVOF) spraying, and cold spraying. Advantages include the following:

- A wide variety of materials can be deposited (metals, ceramics, composites, etc.).
- Higher rate of material deposition is achieved than with conventional coating processes.
- Minimal damage to substrate occurs; that is, coating can be done even on an apple with external cooling arrangement.

- Alloying of material is possible.
- Functionally gradient or porous materials can be produced.

Applications of thermal spraying are limited only by imagination.

8.1.2 Disadvantages of Thermal Spraying

Although selection of a process can be based on its economics and on certain properties of the particular processing technique, there are certain inherent disadvantages associated with the thermal spraying process, such as the following:

- It is a line-of-sight process.
- The process is very complex, and repeatability requires proper control.
- It cannot deposit thin coatings except for solution precursor thermal spraying.
- It produces a lamellar structure.
- Certain limitations regarding the oxidation or deformation might restrict use of particular materials under certain atmospheres.

8.2 CLASSIFICATION OF THERMAL SPRAYING

As shown in Figure 8.3, thermal spraying processes can be broadly classified into a few categories: (1) combustion, where the source of energy is combustion of gases; (2) electric arc, where energy is generated by creating an electric arc; (3) cold spray, where the material deformation is achieved by imparting high velocity via a gas; and (4) plasma, where the ionization recombination energy is utilized to create high temperature for material melting and is supplied with high-velocity carrier gas to impart kinetic energy for deposition.

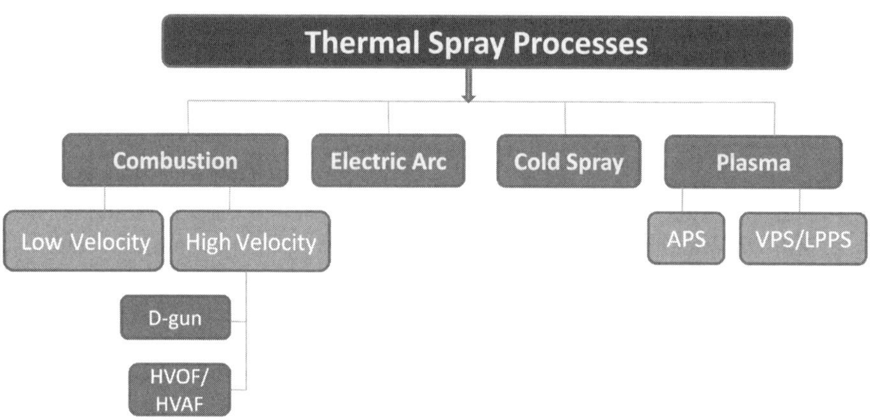

Figure 8.3 Classification of thermal spray processes. APS, air plasma spraying; VPS, vacuum plasma spraying; LPPS, low pressure plasma spraying; HVOF, high velocity oxy-fuel; HVAF, high velocity air-fuel.

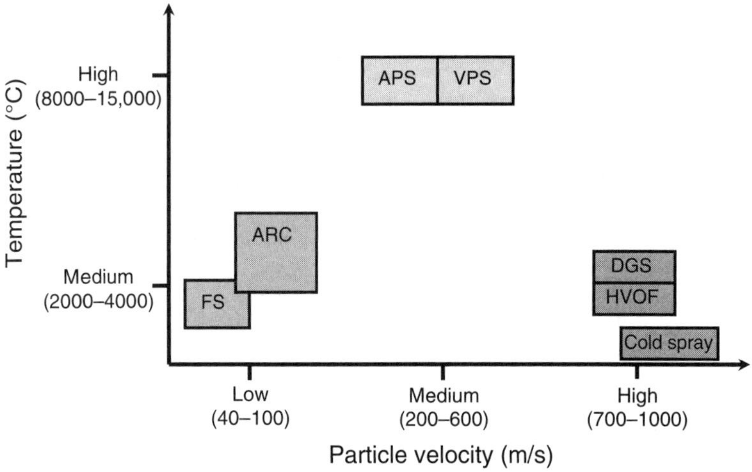

Figure 8.4 Schematic of temperatures and velocities typically attained in various thermal spray processes. ARC, arc spraying; FS, flame spraying; DGS, D-gun spraying.

The various temperatures and velocities attained in the thermal spraying processes are presented schematically in Figure 8.4. It must be noticed that selection of a thermal spraying process for a particular material broadly depends on the (1) inherent properties of the material, (2) performance aimed at, (3) availability of any alternative forming method, and (4) acceptable cost or economics.

The classification of thermal spray processes based on the different mechanisms of supplying heat is presented in Sections 8.2.1–8.2.4.

8.2.1 Combustion Thermal Spraying

8.2.1.1 Flame (Powder or Wire) Spraying

Flame spraying (powder or wire) or combustion spraying is one of the primitive coating techniques used on an industrial scale that is still widely used owing to its low cost and flexibility. Material to be deposited is in wire or powder form, and combustion of an oxy-fuel (OF) gas mixture provides the high temperature to melt it; the consequent molten or semimolten powder is carried by a compressed carrier gas to reach the final deposition surface. Flame temperatures are limited to 3000°C with in-flight velocities of 50 m/s.[1] This coating technique falls under the umbrella of low-velocity combustion thermal spraying.

In the case of limited batch work, and in cases where a wire cannot be formed from a material, a hopper is attached at the top of the torch for gravity feeding of powder (Fig. 8.5a). In some cases, fine powder can also be contained in plastic tubes and be fed as wire in a wire spray torch (Fig. 8.5b). In the wire spray torch, the wire is gradually fed into the flame of a specially designed torch—the wire tip continuously melts and a supply of compressed air blasts the molten tip to produce a fine molten metal spray that is subsequently deposited on the substrate (Fig. 8.5b).

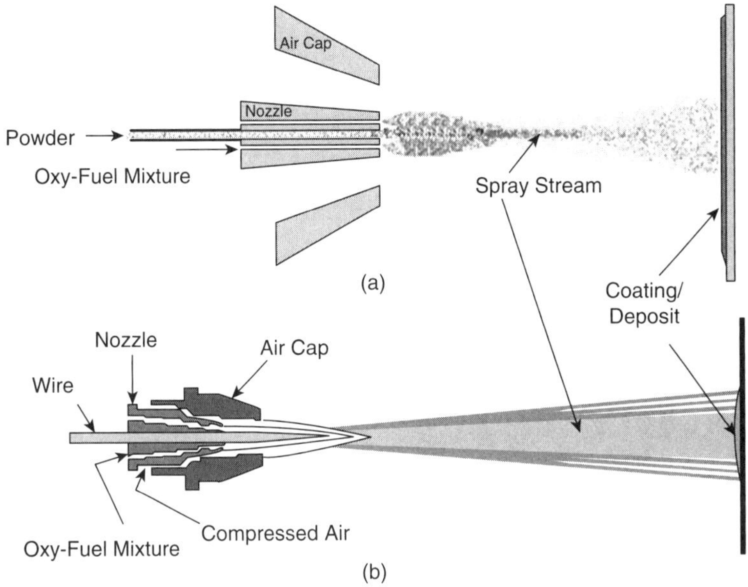

Figure 8.5 Schematic of combustion flame spraying: (a) powder and (b) wire (adapted from http://www.indiamart.com/plasmaapplication-processor-s/thermal-spraying.html).

Advantages:

- ***Simplicity and Economics.*** The combustion flame spray unit is exceedingly simple and this translates into a considerably cheaper equipment and lower capital investment.
- ***Limited Processing Parameters.*** The coating quality in combustion flame spraying depends mainly on the powder and powder injection variables and on the gun-to-job distance.
- ***Thermal and Kinetic Profiles.*** Furthermore, the velocity and temperature profiles in the combustion flame vary nowhere near as drastically with axial distance as they do in the case of a plasma flame; this makes the entire spraying process far easier to control with respect to *reproducibility* of coating quality.

Disadvantages:

- ***Limited Flame Temperature.*** A major limitation of the combustion flame technique is that the maximum available flame temperature is only around 3000°C, which is not sufficient to completely melt the coarse powders of many refractories and ceramics, but is enough to melt fine powders of metallic refractory metals.
- ***Workpiece Distortion.*** The long flame tends to heat the substrate—besides causing possible workpiece distortion, this precludes the use of low-melting-point substrates such as plastics.

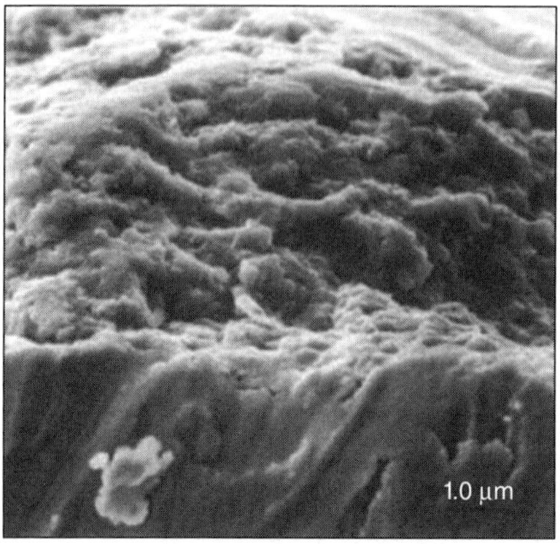

Figure 8.6 Flame sprayed hydroxyapatite on Ti–6Al–4V substrate showing a porous and layered hydroxyapatite coating.[3]

- ***Highly Porous with Oxidized Inclusions.*** The sprayed particles are accelerated to an impact velocity of only about 100 m/s; due to the oxidizing nature of the flame, this method can potentially yield coatings with a high oxide content and a high level of porosity, which may or may not be acceptable for the intended application of any given coating. However, use of compressed air as a propellant can substantially reduce the porosity. In some instances, using the torch to fuse the developed deposit can also increase the bond strength and coating density.

A typical high-velocity flame using propylene as fuel and oxygen as oxidizer was used to spray hydroxyapatite on a titanium-alloy (Ti–6Al–4V) substrate[3]; the flame temperature was $<3000°C$. The corresponding porous hydroxyapatite coating is shown in Figure 8.6.

8.2.1.2 High-Velocity Oxy-Fuel Spraying

High velocity oxy-fuel (HVOF) thermal spraying is regarded as the most significant development in the thermal spray industry since the plasma gun was first employed for coating applications. The growth of this technology since the introduction in 1982 of the first HVOF process, called JETKOTE (a registered trademark of Stoody Delero Stellite, Inc., San Diego, CA), has been phenomenal and numerous hypersonic combustion guns are now commercially available.

The HVOF processes rely on the continuous internal combustion of a fuel gas with oxygen to produce the high-temperature (~3000°C), high-velocity (~700–2000 m/s) exhaust gas stream essential for coating applications.[1] Propylene, propane, and hydrogen are the most commonly used fuel gases. While the flame temperature is close to 3000°C, the flame velocity is hypersonic (in the range 1500–2000 m/s), which causes severe plastic deformation (in metals) and refines grain size. It is reported that uniform heat input and acceleration is available over almost a 300-mm distance. The powder material to be used for coating is injected in a carrier gas (N_2,

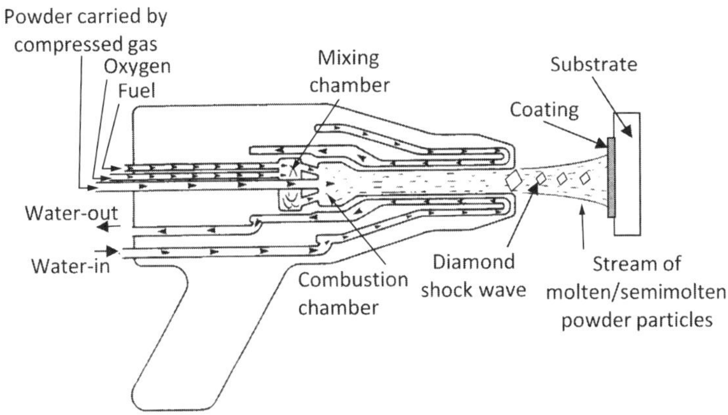

Figure 8.7 Schematic diagram of HVOF process.

He, or Ar) and, owing to the high back-pressures created by the combustion process, a pressurized powder feeder is required. Figure 8.7 shows the schematic diagrams of the gun chamber designs that have been commercially adopted.

Advantages:

- ***Dense Coatings.*** The powder is molten or semimolten and is propelled by the combustion gases to impact the substrate at very high speeds, providing well-adhered and dense coatings.
- ***Fine-Grained Microstructure.*** Owing to the high impact, the microstructure of the deposited coatings is often very fine-grained. This process is similar to mechanical working in terms of breaking the grain size.

Disadvantages:

- The relatively low gas temperatures available in the HVOF technique preclude the use of some refractory metals and many ceramics for spraying, which represents a significant shortcoming compared with the detonation spray systems.
- The severe impacts inherently dominant in the deposition process create residual stresses in the coating.
- High consumption of carrier gases is needed in achieving supersonic velocities.

8.2.1.3 Detonation Spray Technique

The detonation-gun (D-Gun) spray technique is known to produce coatings of superior quality for applications in various industries.[4,5] In particular, detonation spraying has been successfully utilized to obtain WC–12Co coatings, and in a recent study, an important process parameter, that is, OF ratio, was varied.[4] The extent of decarburization of WC has been found to depend strongly on the OF ratio, and the degree of decarburization varies from 4.4 to 45%, when the OF ratio is varied between 1.16

and 2.0, respectively. Higher hardness value was measured with the coatings deposited at an OF ratio of 1.50 (11.2 GPa) or 2.0 (11 GPa), compared with the coatings deposited at an OF ratio of 1.16 (9.2 GPa). Also, high fracture toughness (3–6 $MPa \cdot m^{1/2}$) was recorded with the coatings deposited at an OF ratio of 1.50. The abrasion properties of these coatings were also investigated.[5] The detonation spray coating equipment was originally developed and patented in the United States by Union Carbide Corporation in 1955 and independently in 1969 at the Institute of Materials Science (Kiev, Ukraine).[6] However, detonation spray equipment generally has been unavailable for widespread use in industry and research laboratories.

A typical detonation coating process consists of the following stages:

1. Flammable gas mixture (e.g., oxygen and acetylene) is fed into a tubular barrel closed at one end.
2. Simultaneously, a powder of the sprayed material is injected through a powder feeder before the explosion is triggered by a spark plug.
3. The detonation wave, formed instantaneously after ignition, obtains a velocity of around 2600–3500 m/s and a temperature of around 3500°C.[6,7]
4. This wave, followed by a rapid expansion of the reacted gas products, accelerates the powder particles so that they leave the open side of the barrel at a velocity of 700–1000 m/s and impinge on the surface being coated.[6,7]

Consequently, the coatings are dense and exhibit high bond strength. After each detonation, the barrel is purged with nitrogen gas and the process repeats several times per second. Purging with nitrogen avoids backfiring in the system as the detonation wave tends to propagate in all directions. Thus, the gas detonation process provides intermittent shock waves, which allows a lower deposition temperature at the substrate end and avoids undesirable thermal deformation of the substrate. Detonation gas spraying units can provide pulse rates of up to 10 cycles per second. The geometry of the detonation chamber must allow a steady detonation, which also depends on the mixture composition, temperature, and pressure. There is no upper limit for the barrel diameter; however, a critical minimum diameter is required for effective pressure coverage and particle-carrying capability. The most widely used design for gas-mixture explosion is a one-ended closed steel tube with approximately 25-mm internal diameter and 1350-mm length (Fig. 8.8).

The fundamental aspects of this technology are limited partly because of the phenomenon of gas detonation and nonequilibrium shock wave propagation on a millisecond scale. This complicates the study of the kinetics and thermodynamic interaction between the particle and detonation wave via both theory and experiments.

The key observations are summarized as follows:

- The velocity of the detonation wave varies between 1000 and 3000 m/s (~3–10 Mach), depending on the composition of the gas mixture.
- The velocity of the detonation is almost completely independent of the barrel material, the thickness of its walls, and its diameter, provided that it exceeds the critical diameter.

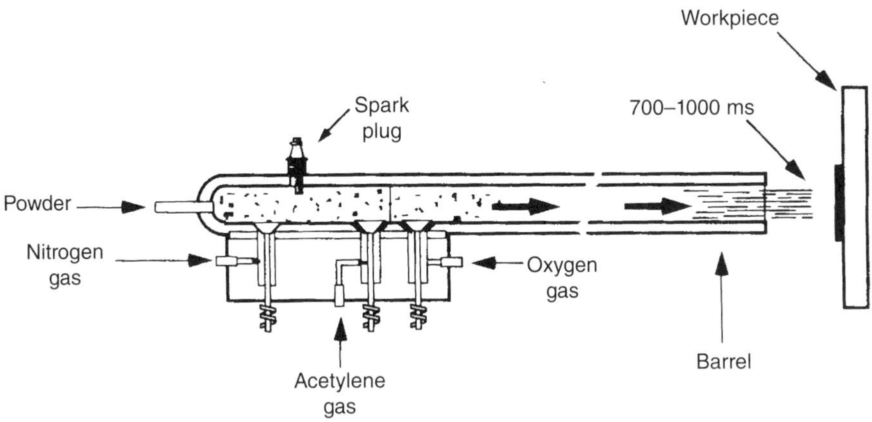

Figure 8.8 Schematic of detonation spray process.

- The velocity of the detonation does not depend on the conditions behind the detonation wave; that is, it matters not whether the ignition occurs at the open end of the barrel or at the closed end.
- The velocity of the detonation weakly depends on the temperature of the gas mixture.
- Increased pressure (density) of the gas mixture increases the velocity of the detonation wave; this increase is small at low pressures and becomes larger at higher pressures.
- Each gas mixture has an optimal composition that corresponds to the maximum speed of the detonation wave.
- The kinetics of particle acceleration by a detonation wave is distinctly different from particle acceleration in other methods of spraying, such as plasma and HVOF.[6,7]

For a pure oxygen/acetylene mixture, the velocity of detonation increases with an increased amount of acetylene; however, the presence of other gases (e.g., nitrogen) reduces the detonation wave velocity. During detonation spray coating, the full combustion reaction under stoichiometric conditions between acetylene and oxygen results in the formation of CO_2 and H_2O, which decide the resulting temperature and consequent velocity of the explosion. Water dissociation enhances, whereas CO formation reduces, the detonation velocity and temperature.

The advantages of the detonation process include the following:

1. Coatings are extremely dense and free from porosity.
2. Shock waves break the grains and microstructure becomes very fine.

The process, however, also suffers from some inherent disadvantages:

1. Deposited material should be able to take the impact without fracture. Hence brittle materials are often difficult to synthesize.

2. It requires controlled ignition for shock wave propagation, otherwise unsynchronized N_2 purging and ignition can lead to explosion.

8.2.2 Electric Arc Spraying

An electric arc is generated in a DC circuit by short-circuiting two electrodes; consequently arc-spraying is carried out in fashion similar to that of combustion spraying. The difference arises because the heat source is much more intense and temperatures attained in the electric arc are in excess of 5000°C. The material to be sprayed is fed as a rod or wire, which forms the consumable electrodes of the gun. As in the case of combustion flame wire spraying, the molten tips of the electrodes are atomized and blown against the substrate by a blast of compressed air. Figure 8.9 shows a representative arc spray torch. The temperature gradients in arc spraying are highly drooping as the molten or semimolten particles deflect from the axial trajectory. However, efficient heat transfer from the arc to melt the electrode allows deposition of marginally denser coatings compared with those of the combustion flame method. In addition, the higher temperature and power available for superheating the particles (>5000°C) enable the use of high spray rates and deposition of thick coatings.

Advantages:

1. High deposition rates up to 40 kg/h can be achieved.
2. Alloying can be done *in situ* as more than two electrodes can be utilized in depositing the coatings.
3. Extremely high temperatures can be attained, and high velocities allow dense coatings.
4. In addition, arc spraying does not heat the substrate too much.

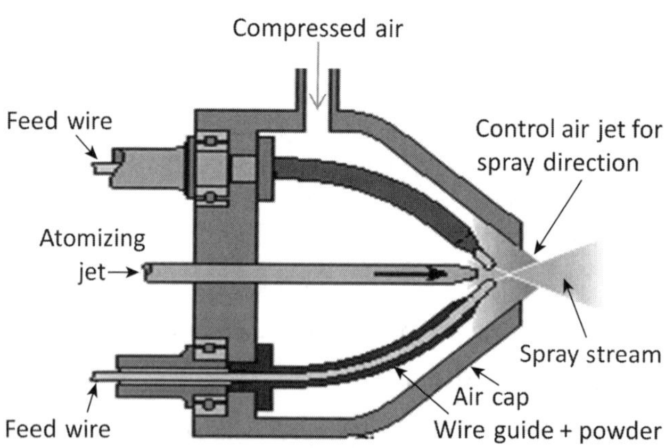

Figure 8.9 Schematic of electric arc spraying process (adapted from http://www.twi.co.uk/content/ksrdh002.html).

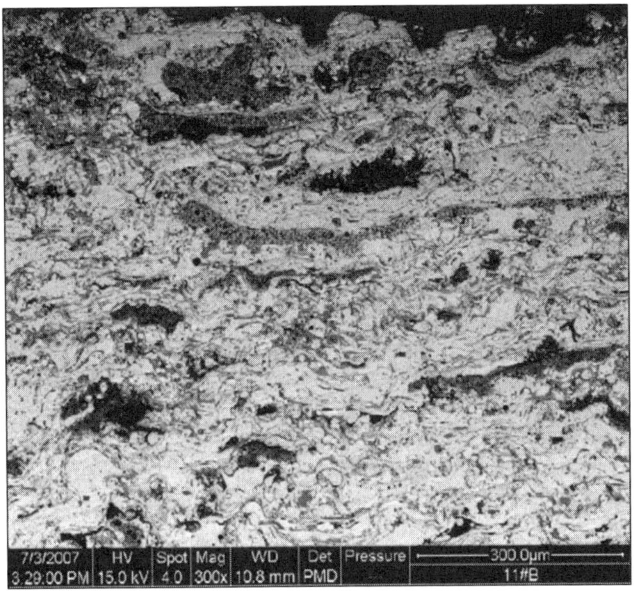

Figure 8.10 Flame sprayed coating of TiB_2 deposited using a Ni (Cr) cored wire.[8]

Disadvantages:

1. Deposition is limited to conductive metals since electrodes carry current and eventually get deposited as coatings.
2. Ceramics cannot be sprayed (as they are nonconductive).

Electric arc wire spraying utilizes sheathing ceramic particles (such as TiB_2 in this case) in a cored wire (such as Ni [Cr] and 304L steel).[8] Coating is dense and the dark spots seen in Figure 8.10 correspond to Al_2O_3.

8.2.3 Cold Spraying

Cold spraying invovles acceleration of powder particles to supersonic speeds at temperatures much below the melting point of the material (Fig. 8.11). A de-Laval-type (contracting–expanding) nozzle is utilized to accelerate powder particles to supersonic velocities. Hence, the deposition occurs via Helmholtz instability along the particle interface, followed by mere plastic deformation. Heating of the carrier gas is also often done (say up to 500 K), so the thermal energy component is not strong, and the supersonic velocities attained by easily deformable materials (such as Al, Cu) allow the particles to be deposited easily in their solid state. The lighter carrier gases (such as He) can render higher sonic velocities compared with heavier gases (such as N_2).

Advantages:

1. Solid-state deposition eliminates high-temperature oxidation, evaporation, melting, crystallization, gas release, and so on.

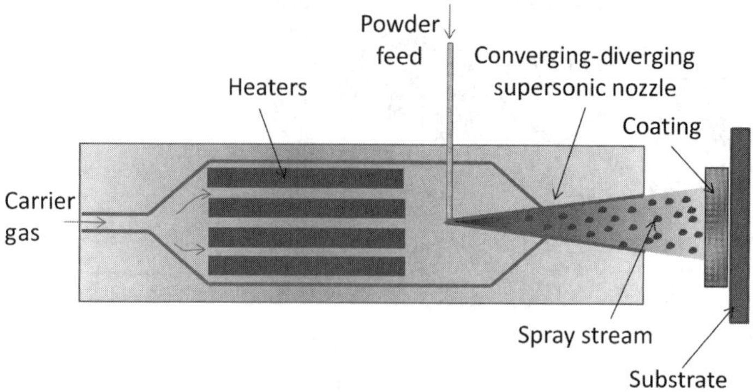

Figure 8.11 Schematic of cold-spraying process.

2. High density of coatings is usually achieved.
3. Fine, cold-worked microstructure is achieved, and it often posseses higher hardness.
4. Oxygen-sensitive materials can also be sprayed.
5. The process can work with dissimilar materials.
6. It can use fine particles (~5–10 μm) and deposit coatings less than 25 μm thick.

Disadvantages:

1. Brittle materials cannot be cold sprayed.
2. Residual stresses can be inherent as coatings are deposited primarily by plastic deformation. The strain hardening with enhaced yield strength often induces a higher degree of residual stresses.
3. Nozzle erosion is rapid due to the supersonic velocities inherent in cold spraying.
4. The process has a high consumption of carrier gases since they have to be thrown at high feed rates.

WC–Co coating deposited by cold spraying shows a highly dense coating in Figure 8.12a.[9] Helium was used as the carrier at a pressure of 1.7 MPa at a temperature of 823 K. Though cracking is observed due to residual stresses generated, nanocoating and utilization of pulsed gas cold dynamic spraying has been further shown to reduce the cracking effect in the deposited WC–Co coatings (Fig. 8.12b).[9]

8.2.4 Plasma Spraying

A plasma spraying gun typically consists of a cylindrical anode (usually made of copper) and a cone-shaped cathode (usually made of thoriated tungsten). A high-intensity arc struck between the two water-cooled electrodes ionizes the primary gas

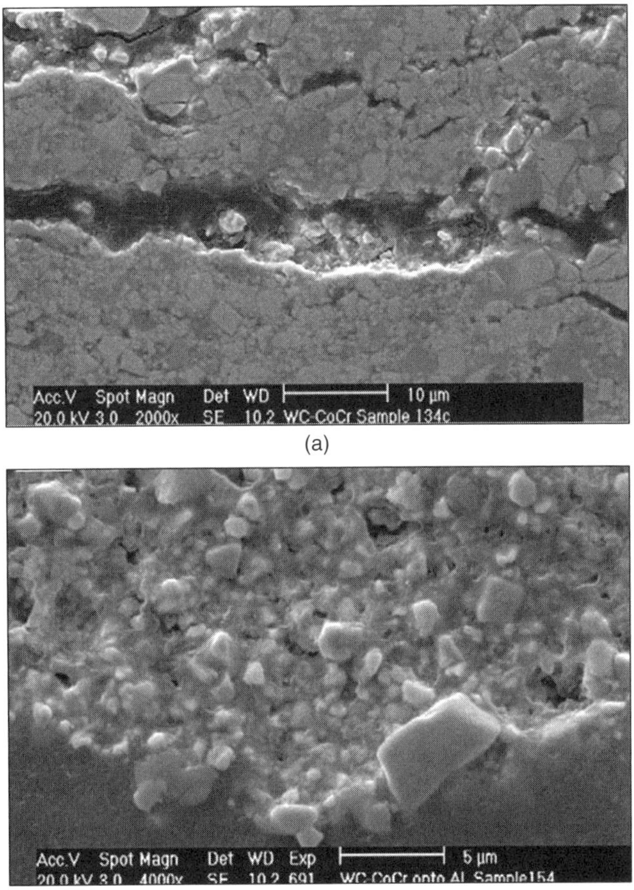

Figure 8.12 WC–Co coating deposited by (a) cold spraying and (b) pulsed gas dynamic spraying.[9]

(such as Ar) to form a plasma. Plasma is produced either by passing a plasma generating gas (such as Ar) through a high-intensity arc struck between two electrodes (arc plasma) or by high radio frequency excitation of plasma gas (RF plasma). Nontransferred DC arc plasma has also been conventionally used for coating and is most widely employed in the thermal spray industry. In a DC arc plasma torch, the arc is forced through a nozzle by its constriction thereby enhancing both the arc temperature and the plasma gas velocity. Powders are fed into the plasma stream via a carrier gas, which accelerates the particles to velocities of 100–500 m/s.[10,11] Successive impacts of molten or semimolten particles onto the substrate result in deposition of a thick (>25 µm) coating.

Nitrogen, hydrogen, argon, and helium are the commercially used plasma gases. In general, argon is the most suitable for routine applications, with hydrogen usually being added to enhance the enthalpy of the plasma gas and thereby improve the gas–particle heat transfer to aid particle melting. Since plasma can attain

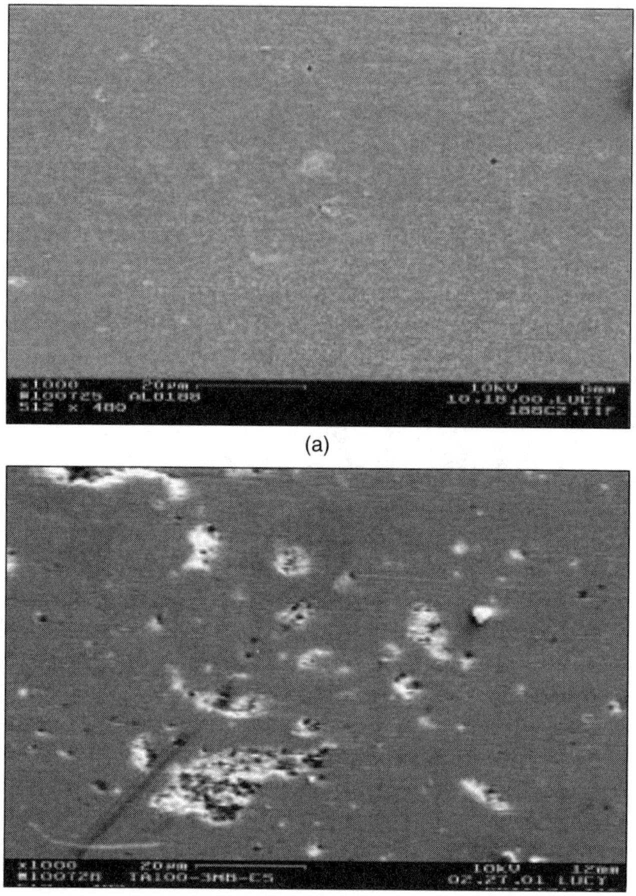

Figure 8.13 Cross-sectional images of (a) HVOF sprayed and (b) plasma-sprayed Al_2O_3–TiO_2 powders.[12]

temperatures in excess of 12,000°C, plasma spraying can melt any known high-melting-temperature material and deposit it as a coating.[10,11]

For the comparative densification of Al_2O_3–TiO_2 coatings, particle temperatures reach in excess of 2400 K at velocities of 800 m/s in HVOF spraying at 6–7 inches away from the nozzle, whereas temperatures in plasma spraying reach in excess of 10,000 K with much lower velocities. Therefore, coating is very dense in HVOF, shown in Figure 8.13a, compared with the plasma-sprayed coating seen in Figure 8.13b.[12]

8.2.4.1 Atmospheric Plasma Spraying

In atmospheric plasma spraying, the deposition environment is the atmosphere. Atmospheric plasma spraying deposits molten or semimolten particles using plasma;

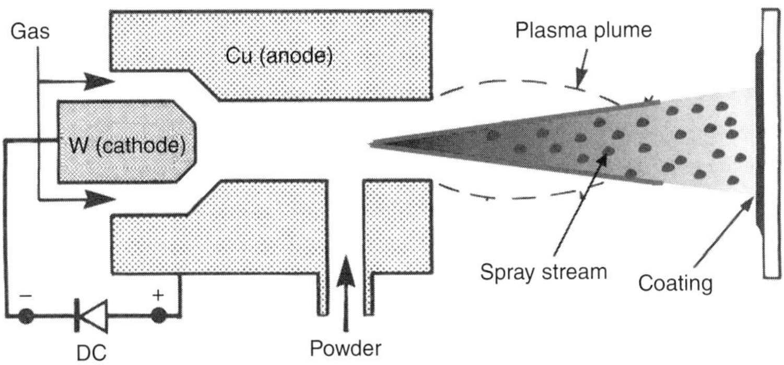

Figure 8.14 Schematic representation of atmospheric plasma spraying process.

the deposition occurs similar to that of other thermal spraying processes. There is no special environment for eliminating the negative effects of atmospheric gases. Primary and secondary gases are the sole shrouding gases for the deposition. A schematic diagram of atmospheric plasma spraying is shown in Figure 8.14. The plasma plume incurs steep temperature gradients: the temperature just outside the nozzle exit is higher than 15,000 K and drops off rapidly from the exit of the nozzle. Usually ceramic powders are introduced at this hottest part of the flame, whereas low-melting-temperature materials are introduced farther away. Powder feedstock has particles in the range of 10–40 μm in diameter to allow their melting and ease in carrying. Finer particles (<1–2 μm) are easy to melt, but they cannot overcome bow shock near the substrate to form a coating and are thrown off as overspray. On the other hand, bigger particles (>150 μm) do not melt and are reflected back without depositing on the substrate. Particle velocity is usually in the range of 100–300 m/s. Plasma spraying is a rapid solidification technique, where cooling rates are typically on the order of 10^6 K/s.

Advantages:

1. Microstructure is fine-grained, and equiaxed grains with columnar submicrostructure are observed.
2. Deposits are fairly uniform in composition along the thickness.
3. Functionally graded materials (with varying composition) can be deposited using the plasma spraying process.
4. High deposition rates of up to 4 kg/h can be achieved.
5. Plasma spray coatings are dense (>90% density), and clean.
6. The thermal spraying processes are highly versatile.

Disadvantages:

1. Complexity in processing (power fluctuation, electrode degradation, flowability of powders, etc.) makes it difficult to repeat the deposition even with the same parameters.
2. Equipment cost is high.

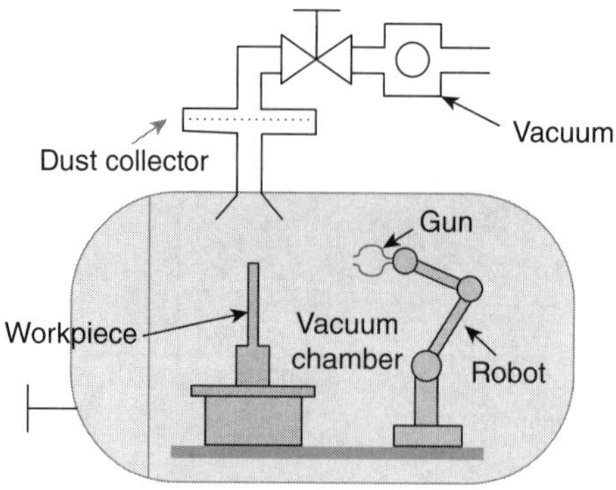

Figure 8.15 Schematic of vacuum plasma spraying system.

8.2.4.2 Vacuum Plasma Spraying

Vacuum plasma spraying (VPS) is the plasma processing technique (done in vacuum or low pressure) where metal or ceramic powders (10–50 μm in diameter) are fed into a plasma plume (~10,000 K), and the resulting molten or semimolten particles are accelerated to impact on a substrate to form a coating (Fig. 8.15). The spray chamber is filled with inert gas and maintained at low pressure (~100 mbar) during spraying, which allows processing of reactive materials without oxidation. In addition, 100% dense coatings are deposited owing to high impact velocities and longer plasma plume (owing to negligible air resistance). Since the deposition is done at low pressures (and is protected from atmosphere), oxidation of coatings is usually limited.[13]

Advantages:

1. The plume is longer and hotter, and hence highly dense coatings can be deposited. In addition, unmelted particles can be totally absent.
2. Low pressures avoid inclusion of oxides, and spraying of oxygen-sensitive materials can also be easily done.

Disadvantages:

1. The whole processing is done under vacuum, so cost automatically goes very high.
2. The part size can be limited owing to limited size of vacuum chamber.
3. The process is time consuming, as the chamber has to be evacuated before deposition can occur.

A typical TiC coating synthesized using VPS onto graphite substrate is presented in Figure 8.16.[14] The deposited coating is highly dense.

8.3 SPLAT FORMATION AND SPREAD

Splat inherently forms as the molten or semimolten droplet impacts on the substrate. The formation of the core and rim of the splat, shown in Figure 8.17, very much

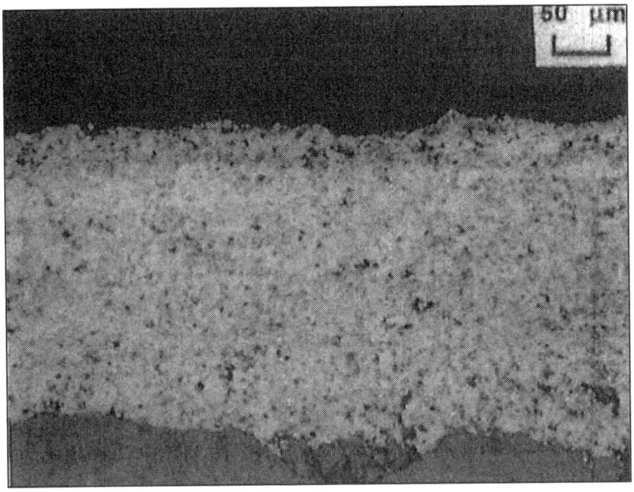

Figure 8.16 Vacuum plasma spray deposited TiC coating on graphite substrate.[14]

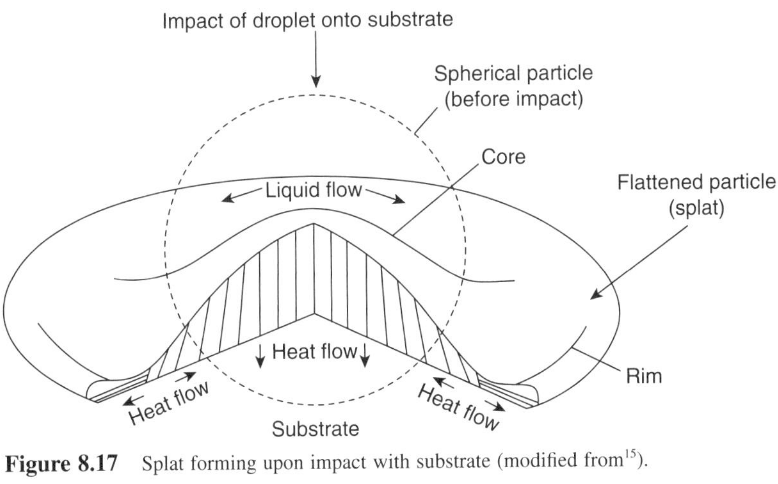

Figure 8.17 Splat forming upon impact with substrate (modified from[15]).

depends on the thermal history of the in-flight particle, the kinetics of thermal spraying, the thermal conductivity of the substrate, the particle's melting point, and its thermal or flow properties. Hence the following assist spreading of the splat:

1. ***Higher Plume Temperature.*** Higher temperature in the plume leads to superheating of the spray particle and decreases viscosity of the material.
2. ***Longer Dwell Time of Particles in Flight.*** More time in flight allows particles to reach higher temperature.

3. ***Absence of Air Resistance.*** Less resistance allows free flow of the material, and it avoids poor thermal transfer (which is otherwise present in the case of insulating air) between plume and material.
4. ***Low Thermal Conductivity of Substrate.*** Sudden cooling by the substrate freezes the droplet upon contact; hence poor thermal conductivity of the substrate will allow more spreading of the solidifying droplet.
5. ***Higher Thermal Conductivity of Sprayed Material.*** Once the droplet thins, it starts loosing heat dramatically as its surface area increases. If the core of material is able to keep supplying heat due to higher thermal conductivity of the material, more spreading will occur.
6. ***Successive Particle Impacts.*** If a successive particle impact occurs over a still-solidifying splat, further growth of the splat will be impeded and the splat will find it hard to spread further.

The "mushrooming effect" (thickening of the core without much spreading of the rim) also occurs when higher air resistance does not allow the molten droplet to spread, and the droplet freezes onto the substrate. The spreading of a splat is much greater in VPS compared with air plasma spraying (APS), because of its low air resistance and longer, hotter plasma plume. The spreading of a splat also depends on the cooling rate since this decides the time available for a splat to spread. In addition, cooling rates and solidification rates are highly critical in the evolution of microstructure. Highly refined grain sizes often occur in plasma-sprayed structures since the cooling rates are in excess of ~10^6 K/s. Cooling rate (Q) for Newtonian cooling (i.e., a thin layer of motionless fluid forms and solidifies) is given as

$$Q = h(T_t - T_s)/(\rho C_p s), \qquad (8.1)$$

where h is the heat transfer coefficient (W/m^2·K), T_s is the substrate temperature, ρ is the density of the splat, T_t is the melting point of the splat, C_p is the specific heat capacity of the splat (J/kg·K), and s is the splat thickness (in meters). Correspondingly, the solidification rate (R) is given as

$$R = h(T_t - T_s)/(\rho L_f), \qquad (8.2)$$

where L_f is the latent heat of fusion.

8.4 NEAR NET SHAPE FORMING

Near net shape (NNS) forming is the ability of a processing technique to synthesize ready-to-use engineering components. Thermal spraying can be tailored to produce an engineering component that can be put directly into service, with minimal or no postprocessing. Hence, NNS has gained popularity as the secondary processing operations can be avoided and the turnaround time and cost for making the component decrease. However, there are certain challenges associated with NNS:

1. ***Machining Negative Shapes.*** The first challenge is that a negative shape has to be created, onto which the coating has to be deposited to produce the

positive shape. Hence computer-aided machining is required to develop a negative mandrel onto which the thermal spraying can be done. Hence an additional step is introduced to achieve an NNS component.

2. ***Selection of Mandrel Material.*** The mandrel selection is highly critical as the mandrel has to be easily machinable, should be able to survive high spraying temperature, should not react chemically with the sprayed material, and finally, should be separable from the coating if required.
3. ***Thermal Stresses and Shrinkage.*** Thermal stresses have to be within limits since higher coating thickness might result in cracking of the coating. The aspect of thermal shrinkage has also to be designed for achieving the NNS. High-temperature spraying can produce larger inner diameter of bore and smaller surface grooves.
4. ***Mandrel Removal.*** If an NNS component is required, it might become necessary to remove the substrate or mandrel on which the coating was formed. Mandrel removal is one of the biggest challenges after thermal spraying has been done to achieve a desired coating. There are certain methods by which a mandrel can be removed: the mandrel can be machined off; it can be chemically etched; or it can be separated by deliberate induction of thermal stresses and the difference in coefficient of thermal expansion between the mandrel and coating, which upon careful control can result in mandrel removal.

8.5 OVERVIEW

Thermal spraying processes can have a variety of features in terms of (1) temperature and velocities attained, (2) deposition rates for productivity, (3) densification for a better quality, (4) adhesion to the substrate, and (5) relative cost of processing. A comparison of these parameters is provided in Table 8.2.

Table 8.2. Characteristics of Various Thermal Spraying Proceses

	Combustion spraying	HVOF	D-Gun	Arc spray	APS	VPS
Temperature (°C)	3000	3000	4000	5000	>12,000	>12,000
Particle Velocity (m/s)	40	800	800	100	200–400	300–600
Adhesion (MPa)	8	>70	>70	12	60–80	>70
Oxide content (%)	10–15	1–5	1–5	10–20	2–3	ppm
Porosity (%)	10–15	1–2	1–2	10	2–3	<0.5
Deposition rates (kg/h)	2–6	3–8	1	10–25	2–10	3–15
Relative costs[a]	Low	Moderate	High	Fair	High	Very high

[a]Cost: low < fair < moderate < high < very high.

Thus, it can be observed that a myriad of thermal spraying techniques are available to deposit ceramic coatings. In addition, NNS structures can also be engineered using thermal spraying methods. Thermal spraying deposition techniques eliminate the requirement of melting the ceramics in a container and thereby also reduce the cost of ceramic processing. Complexities are introduced in ceramic processing from initial powder feedstock through the powder purity, particle size, and distribution, followed by the spraying parameter, and the interaction of the molten or semimolten droplet with the substrate. Typically, lamellar structures are observed in thermally sprayed coatings, with the densification differing by the degree of heating (temperature), deformation (velocity), and the environment. Additionally, substrate properties, preheat or postheat conditions, deposition or standoff distance, flow rate of carrier gases, and so on also affect the finally developed microstructure.

REFERENCES

1 H. Herman and S. Sampath. Thermal spray coatings, in *Metallurgical and Ceramic Protective Coatings*, K. H. Stern (Ed.). Chapman & Hall, London, 1996, 261–289.

2 P. Fauchais, A. Vardelle, and B. Dussoubs. Quo vadis thermal spraying? J. Therm. Spray Tech. 10(1) (2001), 44–66.

3 H. Oguchi, K. Ishikawa, S. Ojima, Y. Hirayama, K. Seto, and G. Eguchi. Evaluation of high-velocity flame spraying technique for hydroxyapatite. Biomaterials 13(7) (1992), 471–477.

4 P. S. Babu, B. Basu, and G. Sundararajan. Processing-structure-property correlation and decarburization phenomenon in detonation sprayed WC-12Co coatings. Acta Mater. 56(18) (2008), 5012–5026.

5 P. S. Babu, B. Basu, and G. Sundararajan. Abrasive wear behavior of detonation sprayed WC–12Co coatings: Influence of decarburization and abrasive characteristics. Wear 268 (2010), 1387–1399.

6 E. Kadyrov and V. Kadyrov. Gas dynamical parameters of detonation powder spraying. J. Therm. Spray Tech. 4(3) (1995), 280–286.

7 E. Kadyrov. Gas-particle interaction in detonation spraying systems. J. Therm. Spray Tech. 5(2) (1996), 185–195.

8 J. J. Fang, Z. X. Li, and Y. W. Shi. Microstructure and properties of TiB_2-containing coatings prepared by arc spraying. Appl. Surf. Sci. 254 (2008), 3849–3858.

9 M. Yandouzi, E. Sansoucy, L. Ajdelsztajn, and B. Jodoin. WC-based cermet coatings produced by cold gas dynamic and pulsed gas dynamic spraying processes. Surf. Coat. Tech. 202 (2007), 382–390.

10 K. Balani, R. Anderson, T. Laha, M. Andara, J. Tercero, E. Crumpler, and A. Agarwal. Plasma-sprayed carbon nanotube reinforced hydroxyapatite coatings and their interaction with human osteoblasts in vitro. Biomaterials 28 (2007), 618–624.

11 K. Balani, S. R. Bakshi, D. Lahiri, and A. Agarwal. Grain growth behavior of aluminum oxide reinforced with carbon nanotube during plasma spraying and postspray consolidation. Int. J. Appl. Ceram. Tech. 7(6) (2010), 846–855.

12 Y. Liu, T. E. Fischer, and A. Dent. Comparison of HVOF and plasma-sprayed alumina/titania coatings—Microstructure, mechanical properties and abrasion behavior. Surf. Coat. Tech. 167 (2003), 68–76.

13 K. Balani, G. Gonzalez, and A. Agarwal. Synthesis, microstructural characterization, and mechanical property evaluation of vacuum plasma sprayed tantalum carbide. J. Am. Ceram. Soc. 89(4) (2006), 1419–1425.

14 D. J. Varacalle Jr., L. B. Lundberg, H. Herman, and G. Bancke. Titanium carbide coatings fabricated by the vacuum plasma spraying process. Surf. Coat. Tech. 86–87 (1996), 70–74.

15 S. Sampath and H. Herman. Rapid solidification and microstructure development during plasma spray deposition. J. Therm. Spray Technol. 5(4) (1996), 445–456.

Chapter 9

Coatings and Protection of Structural Ceramics

Coatings contribute as one of the important measures to provide surface resistance (to wear, corrosion, oxidation, erosion), insulation (thermal, electrical), cytocompatibility (bioactive, bioinert), aesthetics, failure (fracture, fatigue), and so on, as summarized in Figure 9.1. Hence to protect the base material from direct surface contact with the environment, the best alternative is to coat it with a superior material to face the harsh environment.

9.1 COATINGS

From the context of thermal spraying as the primary technique of commercial surface coatings, thermal spraying has been practiced since the early 1900s, when flame was utilized as the heat source for melting the material.[1] Since the development of the plasma spray torch by Thermal Dynamic Corp. in 1957, *plasma spraying* has established itself in depositing thick ceramic coatings (>50 μm). High-velocity oxy-fuel (HVOF) spraying, vacuum plasma spraying (VPS), detonation-gun (D-gun) spraying, and cold spraying are other processes in the family of thermal spraying. HVOF utilizes high-velocity carrier gas (3–5 Mach) and uses combustion as the source of thermal energy. Deposition of coatings in HVOF and D-gun spraying is achieved through plastic deformation with secondary assistance from thermal energy. Hence microstructure is usually observed to have minimal porosity (and high density) in HVOF and D-gun sprayed coatings. Plasma spraying utilizes thermal energy as the primary source of heat to melt the powders and deposit consolidated coatings. Thermal spraying has evolved as an effective processing tool to synthesize ceramic coatings with improved properties such as fracture toughness, indentation crack resistance, spallation resistance against a bend-and-cup test, adhesion strength, abrasive wear resistance, and sliding wear resistance.[2–11] Table 9.1 gives a list of ceramics produced using thermal spray processes. Conventional thermal spray processes such as wire arc spraying and flame oxy-fuel spraying are not considered

Advanced Structural Ceramics, First Edition. Bikramjit Basu, Kantesh Balani.

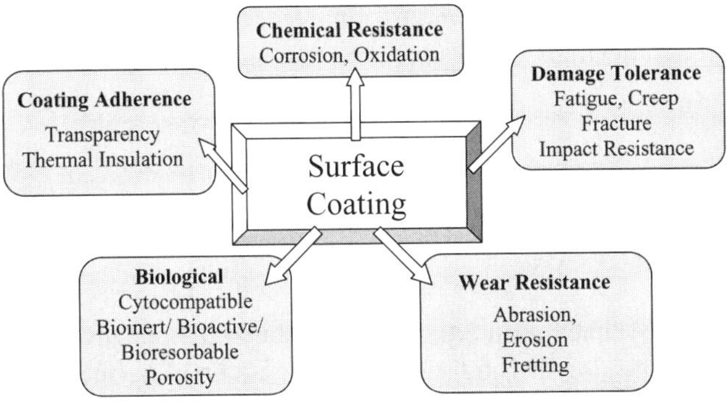

Figure 9.1 Role of surface coatings in rendering surface protection.

Table 9.1. Fabrication of Ceramics by Thermal Spraying

Ceramics	Thermal spraying processing	References
Non-oxides		
$MoSi_2$–Si_3N_4	VPS	12
TiC–Ni	HVOF	13
TaC	VPS	14
SiC/ZrB_2	Controlled APS	15
W	VPS	16
Oxides		
Al_2O_3	APS	17
Al_2O_3/SiC	HVOF, APS	18
Mo–MoO_2	APS	19
ZrO_2–Al_2O_3	APS	20
Al_2O_3–TiO_2	APS	21
TiO_2	HVOF, VPS	22–25
ZrO_2	APS	26
Al_2O_3–Ni	HVOF	27
HAp–ZrO_2	HVOF	28
HAp–CNT	APS	29
HAp	VPS, D-gun	30

VPS, vacuum plasma spraying; APS, air/atmospheric plasma spraying; HVOF, high-velocity oxy-fuel; D-gun, detonation-gun.

here, but these few specific examples provide the reader with an idea of the popularity of coating applications in academia, research, and industry.

Plasma spraying has been utilized by several researchers to synthesize nanocrystalline ceramic coatings.[2,31,32] With stringent material requirements, ceramic nanocomposites (such as WC–Co, Mo–Si–B, hydroxyapatite [HAp], FeAl, Y_2O_3–ZrO_2, ZrO_2–Al_2O_3, Al_2O_3–TiO_2) have been plasma sprayed to improve friction, wear, biocompatibility, oxidation resistance, and so on.[8,17,31,33–39] Nanostructured coatings also provide a solution to improving the fracture toughness of ceramics. Grain size refinement provides the required strength and improved toughness due to the Hall–Petch relationship.[40,41]

9.2 PROTECTIVE COATINGS

9.2.1 Biological Applications

Good biomaterials are required that can interface with the biological environment and assist the proper functioning of the body to provide a better lifestyle. These include stents for heart valves, ocular lenses, bone and dental implants, and supporting structural scaffolds, among others. Considering the example of bone implants, often a bone cement is applied over the metal implant to join it with surrounding bone *in vivo*, as shown in Figure 9.2. But this bone cement degrades with time and allows loosening of the body implant. Consequently, the whole implant loosens and requires additional surgery to fix this problem. In addition, the lifetime of implants is often ~10–15 years, so a young patient would require such surgeries twice or thrice during their lifetime. Hence the situation demands use of better materials and better coatings—ones that can have a lifetime of nearly 40 years.

The first strategy is that the implant should be able to take the loads and impacts usually encountered in daily life. Metals and alloys (e.g., Ti–6Al–4V) serve as the best candidates. But their poor compatibility with the body causes irritation; some-

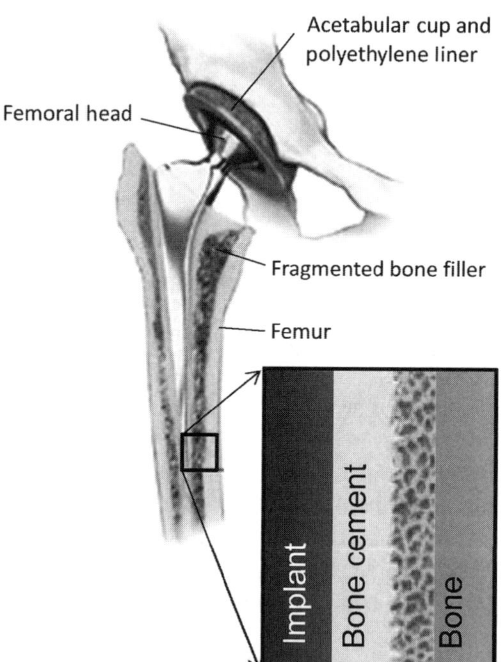

Figure 9.2 Incorporation of bone cement to join an implant with the natural bone. Adapted from http://www.zimmerindia.com/z/ctl/op/global/action/1/id/368/template/PC/prcat/P2/prod/y. See color insert.

times the metal ion release can be toxic as well. Hence the surface of the implant needs a coating. As we know that bone cement is not able to serve the purpose, a critical coating is required to bond the implant with the bone. So the idea is to create a biocompatible coating that can be deposited onto the implant material; the surface of the coating should have the following qualities:

1. ***Chemically Similar.*** So that the body does not differentiate the surface as foreign material and mount a hostile response. Difference in the chemical nature initiates a foreign-body response in the body.
2. ***Bioactive.*** To allow bone growth on its surface and act as surface on which bone cells can nucleate. This aspect further strengthens the integrity between the implant and surrounding tissues.
3. ***Porosity.*** So that bone cells can anchor themselves to the coating surface and form a strong bond. That will make the implant an inherent part of the newly formed bone, and will not loosen with time.
4. ***Permeability.*** To allow nutrients to reach the newly formed bone cells so that they survive and do not die out in the process of bone growth.
5. ***Chemically Inert.*** To withstand the corrosive environment of the body. Hence the material should not degrade uncontrollably under the *in vivo* environment.
6. ***Wear Resistant.*** To prolong the in-service life of the implant. In addition, the coating should not mechanically degrade and should not start releasing debris into the bloodstream.

With respect to above considerations, HAp has emerged as an ideal material owing to the similar of its chemical composition to that of bone and dentin mineral (Ca/P ratio of 1.67). The conventional method of depositing HAp on an implant surface has been plasma spraying since the process has high throughput and can provide the structural porosity needed for such an application.

9.3 ROCKET NOZZLE INSERTS

The stringent demands of aerospace applications require better and better materials to allow optimum fuel utilization, achieving maximum payload, and rendering full thrust to the aircraft. Hence the fuel must be burnt with maximum efficiency and at the highest rate. The propellant grain in the rocket motor provides such thrust via its oxidation (Fig. 9.3). However, the exit-gas temperatures exceed 2000°C with velocities reaching 2–3 Mach. This class of materials requiring temperature resistance in excess of 1800–2000°C is called ultra-high-temperature ceramics (UHTCs). These materials act as surface protectors for the underlying base material in terms of corrosion, oxidation, wear, erosion, high-temperature exposure, and so on. Therefore, exit nozzle inserts are required to protect the exit nozzle cone from excessive heat and high-temperature erosion. Exposure to high temperatures also leads to some severe problems:

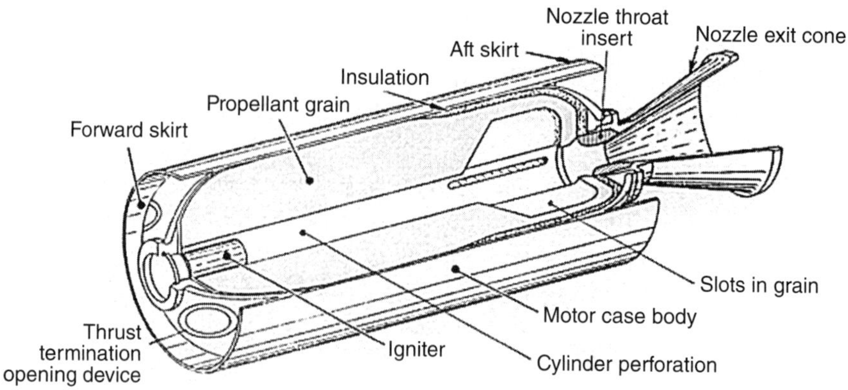

Figure 9.3 A sectional view of a solid-propellant rocket motor with conical exit nozzle.[42]

1. Easy oxidation of the material under open atmosphere
2. Material experiencing creep and losing its high-temperature strength
3. Material undergoing phase transformations and disruption of structural integrity
4. Thermal expansion and spalling off
5. Damage by high-velocity exit gases
6. Damaging the base material by not providing thermal insulation

Hence the material selection should arise from the material's possession of desired qualities: (1) highest melting point for resisting creep, (2) high-temperature strength for stability, (3) high-temperature hardness to withstand high-temperature wear, (4) freedom from phase transformation for structural integrity, (5) minimum mismatch of coefficient of thermal expansion with the base material for reducing interfacial stresses, (6) low thermal conductivity to provide insulation to the material underneath, (7) oxidation resistance, and (8) thermal shock resistance. An additional condition imposed on such materials is low density. Generally, high-melting-temperature materials posses very high densities (such as W, 3683 K, 19.25 g/cm^3, and Re, 3453 K, 21.04 g/cm^3). However, the advantages of the ceramic carbides are their comparatively lower density and even higher melting points than their compatriot base elements or oxides. A comparative schematic of the densities and melting points of various high-temperature ceramics is provided in Figure 9.4.

A few candidate materials are W, Re, SiC, ZrO_2, TaB_2, HfB_2, ThO_2, WC, C, ZrC, TiB_2, TiN, HfC, and TaC. The basic problem with the elements is their low resistance to oxidation. Hence their stability is a problem under oxidative conditions. Consequently, the oxides are highly brittle and have poor behavior under thermal shock. Borides face the problem of oxidizing to B_2O_3, which melts at ~500°C and evaporates at ~1500°C, hence resulting in rapid loss of strength and stability. Nitrides are also highly brittle. Therefore carbides emerge as potential materials for ultra-

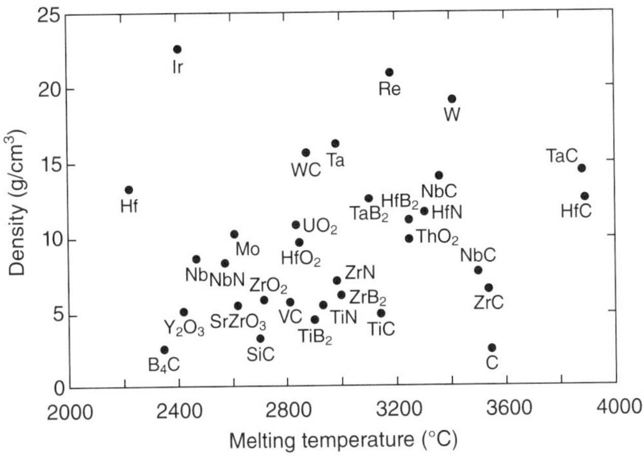

Figure 9.4 Melting points and density of various ultra-high-temperature materials.[43]

high-temperature applications of nozzle inserts owing to their high melting points and lower densities.

9.4 THERMAL BARRIER COATINGS

Thermal barrier coatings (TBCs), as the name suggests, provide protection to materials from high-temperature exposure. Since higher temperature operation increases the energy conversion efficiency of a system, engineering applications prefer conversion of energy under a highly efficient environment. However, as the operating temperature increases, material properties degrade and the material loses its strength and is no longer able to support high-efficiency conversion systems. Therefore, a TBC is given to the material (such as superalloy in a turbine blade), which restricts thermal exposure of the base material. Now the superalloy base does not see the high operating temperature and experiences much lower working temperatures (>200–300°C) than that of immediate environment, and thus retains its strength. In addition, the superalloy is protected from other aspects such as (1) oxidation from the environment, (2) corrosion from the flowing fluid and gases, and (3) wear from the high-velocity fluid containing dirt.

TBCs often utilize yttria-stabilized zirconia (YSZ) deposition on their surface due to the following qualities of YSZ materials:

1. High melting point
2. Lower density
3. Lower coefficient of thermal conductivity
4. Similar coefficient of thermal expansion
5. Good corrosion and oxidation resistance
6. Good wear resistance

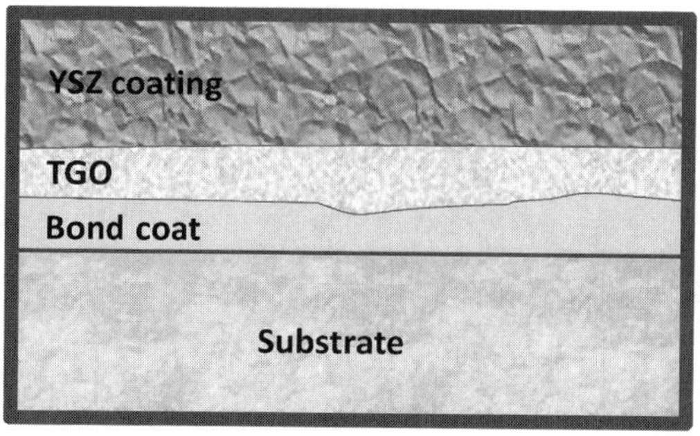

Figure 9.5 Schematic of thermal barrier coating.

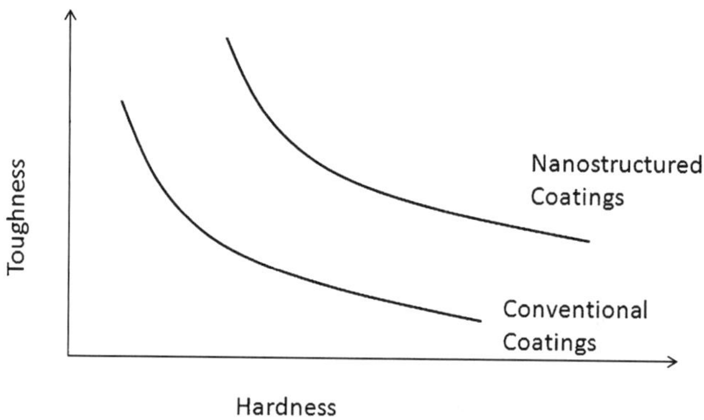

Figure 9.6 Schematic of toughening by nanocoatings compared with conventional coatings.

Typically the TBC consists of four layers: (1) the metal substrate, (2) a metallic bond coat, (3) a thermally grown oxide (TGO), and (4) a ceramic topcoat, usually of YSZ, as shown in Figure 9.5.

Incorporation of nano-YSZ particle sizes can have a valuable effect on fracture toughness (Fig. 9.6). Nanostructured YSZ allows termination of cracks since the crack-propagation energy can be released at the solid-state sintered nanoclustered regions.

9.5 WEAR RESISTANCE

Any mechanical structure in contact with a mating surface undergoes wear. Harder material bites into softer material and creates volume loss leading to instability of the structure. Ceramics often are very hard; therefore, preventing abrasion of ceramic surfaces is one of the prime facets in enhancing structural protection. High complexity of abrasion is associated with the following:

1. Interaction with mating surface
2. Surface roughness

3. Type and nature of loading
4. Available lubrication
5. Material properties and geometry of surfaces

It is to be noted that mechanisms of material loss are defined at nanoscales and microscales, which later build up as macrowear of the bulk component. Additionally, certain loss mechanisms prevalent at small length scales combine with contrasting mechanisms at large scales and result in overall damage to the bulk material. On one hand, small length scales involve interactions via single or a few asperity contacts; on the other hand, macrowear involves multiple asperity contacts. Consequently, it becomes essential to bridge the gap between various length scales when estimating wear.

A multiscale abridgment wear model has been developed by Balani et al., where bulk properties such as hardness (H) and fracture toughness (K) can be correlated with macrowear volume; consequently the critical pressure required to initiate a crack during wear can be estimated.[44] Macrowear volume (W_v) arising out of abrasion with applied pressure (p) and its corresponding hardness can be utilized to evaluate the wear constant (k) as[45]

$$k = \frac{W_v H}{p}. \tag{9.1}$$

The constant k can be calculated by fitting the bulk applied load impressed by an abrading pin to the resultant wear volume loss. The value of k represents the dependence of applied pressure in wear loss resisted by the inherent hardness of the material. Dependence of K and H at the macroscale with W_v can be further described by a modified equation, from which the fracture toughness exponent (a) can be calculated as[46]

$$a = \frac{-\ln(W_v + H^b)}{\ln K}, \tag{9.2}$$

where $b \sim 1.5$ for ceramics.[47] Consequently, the fracture toughness exponent a, determined from the wear volume loss can be parametrically fitted to match the aforementioned dependence in Equation 9.2. Later, fracture toughness (K) can be coupled with the friction (f) occurring at local-asperity to scale a critical contact pressure (P_{cr}), which initiates cracking in the coating during macrowear or nanowear. The contact pressure correlation to the previously calculated material properties (k and a) is then presented as[48]

$$P_{cr} = \frac{4.5}{1+10f} \frac{\left(k.p.W_V^{-\frac{b+1}{b}} \right)^{\frac{b}{a}}}{\sqrt{\pi a_0}}, \tag{9.3}$$

where a_0 was taken as wear debris size. Depending on the fracture toughness (K, macroproperty), hardness (H, macroproperty), and coefficient of friction, f

(nanoproperty), and inserting the calculated parameters (k and a) from Equations 9.1 and 9.2, the critical pressure (P_{cr}) that initiates cracking can be calculated. Consequently, a similar analogy can be drawn toward envisaging a map of strengthening mechanisms (in terms of E and H) at a multiscale level.

9.6 CORROSION PROTECTION BY CERAMICS

Corrosion is one of the most common ways by which metals and alloys degrade. But ceramic materials, being the stable form of elements (such as oxide, carbide, nitride, boride), form a stable low-energy phase. Ceramic tiles are often utilized as work tables to resist damage by chemical spills on a surface. They are resistant to high-temperature corrosion as well. Temperature plays a critical role in controlling reaction rates, and it is observed that each 10°C rise in temperature doubles the kinetics of a diffusion-controlled chemical reaction. Consequently, the presence of secondary parameters (such as humidity, saline environment, or galvanic coupling) can again alter the corrosion behavior of the base materials. However, resistance of ceramics arises from their electrically nonconducting nature, which does not allow formation of a complete galvanic cell.

The literature has plenty of data on the acidic resistance of ceramics, whereas data on resistance of ceramics in alkaline environments is limited. Weight loss of Al_2O_3 and SiC in various acidic and alkali media is provided in Table 9.2.

Often the measurement of weight loss by various researchers is based on the available chemicals (of different purity and concentration). Secondarily, different measurement methods and different environmental conditions add to the complexity of weight-loss measurement.[48] In addition, the impurities present in the base ceramic result in disparity in correlating the chemical resistance of various ceramics. Chemical resistance of various other ceramics is presented in Table 9.3.

Ideally, the higher melting point of ceramics renders a low ionic diffusion. The chemical composition of ceramics is highly stable, and corrosive conditions are well

Table 9.2. Weight Loss of Al_2O_3 and SiC Ceramic in Various Acid and Alkali Media[49]

Reactant	Mass content (%)	Treatment temperature (°C)	Weight loss (mg/(cm²·year))		
			Recrystallized SiC	Self-bound SiC (12 wt% Si)	Densely sintered Al_2O_3 (99 wt% Al_2O_3)
H_2SO_4	98	100	1.8	55.0	65.0
HNO_3	70	100	<0.2	0.5	7.0
H_3PO_4	85	100	<0.2	8.8	<1000
HCl	25	70	<0.2	0.9	72.0
HF	53	25	<0.2	7.9	20.0
HF + HNO_3	10 + 57	25	<0.2	<1000	16.0
NaOH	50	100	2.5	<1000	75.0
KOH	45	100	<0.2	<1000	6.0

Table 9.3. Chemical Resistance of Various Ceramics[48]

Material	Reference standard	Weight losses, % in determining	
		Acid resistance	Alkali resistance
Electromelted corundum	TU 3988-064-00224450–94	99.79	99.99
Electromelted periclase	TU 14-8-448–83	52.87	97.55
Electromelted mullite	TU 14-8-450–83	98.53	98.23
Electromelted zirconium dioxide	TU 48-4-489–87	83.54	99.99
Melted quartz (99.9% SiO_2)	TU 14-8-487–85	99.27	82.27
Silicon carbide	GOST 3647–80	99.99	98.62

withstood by ceramics. Chemical resistance of ceramics allows their use in chemical flooring, chemical cutting knives, body implants withstanding hostile environments (such as HAp, Al_2O_3, ZrO_2), and TBCs. This allows them to be in contact with the aggressive media and still survive the exposure for prolonged durations without damage.

9.7 OPTICALLY TRANSPARENT CERAMICS

Transparent ceramics come under the class of advanced applications for creating scratch-free panels and lenses for telescopes, armor and other impact-resistant materials, wear-resistant optical devices, windows with high-temperature thermal insulation for furnaces, and so on. Their high elastic modulus, strength, and high wear resistance can stand against high impact and damage. Their transmittance can be utilized in developing lasers, scintillators, peep-glasses, insulations, medical imaging, and so on. Light scattering by microstructural features (such as porosity and grain boundaries) limit the traverse of light through materials. Light is scattered when it encounters an obstacle whose size is similar to the wavelength of light. For visible light (wavelengths of 400–700 nm), submicron features arising at porosity sites and grain boundaries incoherently scatter the light either at the interfaces or at open porosity surfaces. Fine crystallites in a polycrystalline ceramic do not allow coherent scattering; thus light does not come out coherently and is absorbed via inelastic scattering losses. However, when these features are reduced to levels much below the size of 100 nm, scattering of light does not occur as light cannot encounter those features as obstacles anymore, resulting in a transparent ceramic.

9.8 CERAMIC POTTERY AND SCULPTURES

Ceramics have long been used in ancient history to make pottery. Their inert chemical nature and carefree maintenance have attracted people to archive them as vases, tiles, figures, sculptures, and so on, with their origin dating back to 30,000 B.C.

Chinese pottery has gained importance since A.D. 400, and beautiful ceramic sculptures are also observed in European modern history (since A.D. 1700). These include various colors obtained on the surface of ceramics by applying thin interference films, or different compositional layers, which provide different colors to the surface. Thereby, these surfaces can stay for thousands of years without any surface oxidation and retain their quality.

In summary ceramic coatings are widely utilized, mainly for resistance to high-temperature oxidation, corrosion, and wear and for damage tolerance. Further, ceramic coatings are also utilized for enhanced component life owing to their superior hardness and wear resistance compared with metals and polymers. Biological coatings have also come into widespread use in a variety of applications, and the advent of newer applications for optical materials, dielectrics, aesthetics, insulation, and so on has also enhanced the use of ceramic materials. The key to utilizing ceramic coatings is to expose the surface of a component to extreme conditions while the subsurface structure enjoys workable conditions. Thus, ceramic coatings enhance the value and lifetime of structural components while retaining the functionality of the entire system.

REFERENCES

1 P. Fauchais, A. Vardelle, and B. Dussoubs. Quo vadis thermal spraying? J. Therm. Spray Technol. 10(1) (2001), 44–66.

2 L. L. Shaw, D. Goberman, R. Ren, M. Gell, S. Jiang, Y. Wang, T. D. Xiao, and P. R. Strutt. The dependency of microstructure and properties of nanostructured coatings on plasma spray conditions. Surf. Coat. Technol. 130 (2000), 1–8.

3 E. H. Jordan, M. Gell, Y. H. Sohn, D. Goberman, L. Shaw, S. Jiang, M. Wang, T. D. Xiao, Y. Wang, and P. Strutt. Fabrication and evaluation of plasma sprayed nanostructured alumina-titania coatings with superior properties. Mater. Sci. Eng. A 301 (2001), 80–89.

4 L. T. Kabacoff. Nanoceramic coatings exhibit much higher toughness and wear resistance than conventional coatings. AMPTIAC Newsl. 6 (2002), 37–42.

5 H. Chen and C. X. Ding. Nanostructured zirconia coating prepared by atmospheric plasma spraying. Surf. Coat. Technol. 150 (2002), 31–36.

6 A. Kulkarni, Z. Wang, T. Nakamura, S. Sampath, A. Goland, H. Herman, J. ALlen, J. Ilavsky, G. Long, J. Frahm, and R. W. Steinbrech. Comprehensive microstructural characterization and predictive property modeling of plasma sprayed zirconia coatings. Acta Mater. 51 (2003), 2457–2475.

7 J. Gang, J. P. Morniroli, and T. Grosdidier. Nanostructures in thermal spray coatings. Scr. Mater. 48 (2003), 1599–1604.

8 Y. C. Zhu, K. Yukimura, C. X. Ding, and P. Y. Zhang. Tribological properties of nanostructured and conventional WC-Co coatings deposited by plasma spraying. Thin Solid Films 388 (2001), 277–282.

9 J. He, M. Lee, J. M. Schoenung, D. H. Shin, and E. J. Lavernia. Thermal stability of nanostructured Cr_3C_2-NiCr coatings. J. Therm. Spray Technol. 10(2) (2001), 293–300.

10 J. Karthikeyan, C. C. Berndt, J. Tikkanen, J. Y. Wang, A. H. King, and H. Herman. Nanomaterial powders and deposits prepared by flame spray processing of liquid precursors. Nanostruct. Mater. 8(1) (1997), 61–74.

11 A. Jadhav, N. P. Padture, F. Wu, E. H. Jordan, and M. Gell. Thick ceramic thermal barrier coatings with high durability deposited using solution-precursor plasma spray. Mater. Sci. Eng. A A 405 (2005), 313–320.

12 S. J. Hong, V. Viswanathan, K. Rea, S. Patil, S. Deshpande, P. Georgieva, T. McKechnie, and S. Seal. Plasma spray formed near-net-shape $MoSi_2$–Si_3N_4 bulk nanocomposites—structure property evaluation. Mater. Sci. Eng. A 404 (2005), 165–172.
13 N. Eigen, F. Gartner, T. Klassen, E. Aust, R. Bormann, and H. Kreye. Microstructures and properties of nanostructured thermal sprayed coatings using high-energy milled cermet powders. Surf. Coat. Technol. 195 (2005), 344–357.
14 V. Viswanathan, T. Laha, K. Balani, A. Agarwal, and S. Seal. Challenges and advances in nanocomposite processing techniques. Mat. Sci. Engg. R 54 (2006), 121–285.
15 M. Tului, G. Marino, and T. Valente. Plasma spray deposition of ultra high temperature ceramics. Compos. Sci. Technol. 201 (2006), 2103–2108.
16 H. Greuner, H. Bolt, B. Boswirth, S. Lindig, W. Kuhnlein, T. Huber, K. Sato, and S. Suzuki. Vacuum plasma-sprayed tungsten on EUROFER and 316L: Results of characterisation and thermal loading tests. Fus. Eng. Des. 75–79 (2005), 333–338.
17 A. Agarwal, T. McKechnie, and S. Seal. Net shape nanostructured aluminum oxide structures fabricated by plasma spray forming. J. Therm. Spray Technol. 12 (2003), 350–359.
18 U. Devi. Microstructure of Al_2O_3–SiC nanocomposite ceramic coatings prepared by high velocity oxy-fuel flame spray process. Scr. Mater. 50(7) (2004), 1073–1078.
19 R. Goswami, H. Herman, S. Sampath, X. Jiang, Y. Tian, and G. Halada. Plasma sprayed Mo–Mo oxide nanocomposites: Synthesis and characterization. Surf. Coat. Technol. 141 (2001), 220–226.
20 B. Liang, H. Liao, H. Ding, and C. Coddet. Nanostructured zirconia–30 vol.% alumina composite coatings deposited by atmospheric plasma spraying. Thin Solid Films 484 (2005), 225–231.
21 D. Goberman, Y. H. Sohn, L. Shaw, E. Jordan, and M. Gell. Microstructure development of Al_2O_3-13wt.%TiO_2 plasma sprayed coatings derived from nanocrystalline powders. Acta Mater. 50 (2002), 1141–1152.
22 R. S. Lima and B. R. Marple. High Weibull modulus HVOF titania coatings. J. Therm. Spray Technol. 12 (2002), 240–249.
23 R. S. Lima and B. R. Marple. Optimized HVOF titania coatings. J. Therm. Spray Technol. 12 (2003), 360–369.
24 R. S. Lima and B. R. Marple. From APS to HVOF spraying of conventional and nanostructured titania feedstock powders: A study on the enhancement of the mechanical properties. Surf. Coat. Technol. 200 (2006), 3248–3437.
25 Y. C. Zhu and C. X. Ding. Plasma spraying of porous nanostructured TiO_2 film. Nanostruct. Mater. 11(3) (1999), 319–323.
26 R. S. Lima, A. Kucuk, and C. C. Berndt. Integrity of nanostructured partially stabilized zirconia after plasma spray processing. Mater. Sci. Eng. A 313 (2001), 75–82.
27 E. Turunen, T. Varis, S.-P. Hannula, A. Vaidya, A. Kulkarni, J. Gutleber, S. Sampath, and H. Herman. Parameter optimization of HVOF sprayed nanostructured alumina and alumina-nickel composite coatings. Surf. Coat. Technol. 200 (2006), 4987–4994.
28 H. Li, K. A. Khor, R. Kumar, and P. Cheang. Characterization of hydroxyapatite/nano-zirconia composite coatings deposited by high velocity oxy-fuel (HVOF) spray process. Surf. Coat. Technol. 182 (2004), 227–236.
29 K. Balani, R. Anderson, T. Laha, M. Andara, J. Tercero, E. Crumpler, and A. Agarwal. Plasma-sprayed carbon nanotube reinforced hydroxyapatite coatings and their interaction with human osteoblasts in vitro. Biomaterials 28 (2007), 618–624.
30 H. C. Gledhill, I. G. Turner, and C. Doyle. In vitro fatigue behaviour of vacuum plasma and detonation gun sprayed hydroxyapatite coatings. Biomaterials 22(11) (2001), 1233–1240.
31 M. Gell, E. H. Ordan, Y. H. Sohn, D. Godberman, L. Shaw, and T. D. Xiao. Development and implementation of plasma sprayed nanostructured ceramic coatings. Surf. Coat. Technol. 146–147 (2001), 48–54.
32 Y. Wang, S. Jiang, M. Wang, S. Wang, T. D. Xiao, and P. R. Strutt. Abrasive wear characteristics of plasma sprayed nanostructured alumina/titania coatings. Wear 237 (2000), 176–185.
33 P. Bansal, N. P. Padture, and A. Vasiliev. Improved interfacial mechanical properties of Al_2O_3-13wt%TiO_2 plasma-sprayed coatings derived from nanocrystalline powders. Acta Mater. 51 (2003), 2959–2970.

34 K. A. Khor, Y. W. Gu, D. Pan, and P. Cheang. Microstructure and mechanical properties of plasma sprayed HA/YSZ/Ti-6Al-4V composite coating. Biomaterials 25 (2004), 4009–4017.
35 G. Skandan. Processing of nanostructured zirconia ceramics. Nanostruct. Mater. 5 (1995), 111–126.
36 Z. Mohannadi, A. A. Z. Moayyed, and A. S. M. Mesgar. Adhesive properties by indentation method of plasma-sprayed hydroxyapatite coatings. Appl. Surf. Sci. 253 (2007), 4960–4965.
37 N. Nomura, T. Suzuki, K. Yoshimi, and S. Hanada. Microstructure and oxidation resistance of a plasma sprayed Mo-Si-B multiphase alloy casting. Intermetallics 11 (2003), 735–742.
38 T. Grosdidier, G. Ji, and N. Bozzolo. Hardness, thermal stability and yttrium distribution in nanostructured deposits obtained by thermal spraying from milled-Y_2O_3 reinforced- or atomized FeAl powders. Intermetallics 14 (2006), 715–721.
39 A. L. Vasiliev, N. P. Padture, and X. Ma. Coatings of metastable ceramics deposited by solution precursor plasma spray: I. binary ZrO_2-Al_2O_3 system. Acta Mater. 54 (2006), 4913–4920.
40 K. J. Hemker. Understanding how nanocrystalline metals deform. Science 304 (2004), 221–222.
41 J. R. Vassen. Densification and grain growth of nano-phase ceramics. DKG 76(4) (1999), 19–22.
42 G. P. Sutton and O. Biblarz. *Rocket Propulsion Elements*, 7th ed. John Wiley & Sons, New York, 2001.
43 K. Upadhya, J.-M. Yang, and W. Hoffman. Materials for ultrahigh temperature structural applications. Am. Ceram. Soc. Bull. 76(12) (1997), 51–56.
44 K. Balani and A. Agarwal. Multiscale wear of plasma sprayed carbon nanotube reinforced aluminum oxide nanocomposite coating. Acta Mater. 56(20) (2008), 5984–5994.
45 S. D. Heintze, G. Zellweger, and G. Zappini. The relationship between physical parameters and wear of dental composites. Wear 263 (2007), 1138–1146.
46 G. Bolelli, V. Cannillo, L. Lusvarghi, and T. Manfredini. Wear behavior of thermally sprayed ceramic oxide coatings. Wear 261 (2006), 1298–1315.
47 K. H. Z. Gahr. *Microstructure and Wear of Materials*. Elsevier, Amsterdam, 1987.
48 H. Kong and M. F. Ashby. Wear mechanisms in brittle solids. Acta Metall. Mat. 40 (1992), 2907.
49 B. L. Krasnyi, V. P. Tarasovskii, E. V. Rakhmanova, and V. V. Bondar. Chemical resistance of ceramic materials in acids and alkalis. Sci. Ceram. Prod. 61(9–10) (2004), 337–339.

Section Four

Processing and Properties of Toughened Ceramics

Chapter 10

Toughness Optimization in Zirconia-Based Ceramics

This chapter discusses major toughening mechanisms in zirconia (ZrO_2) ceramics with particular attention to transformation toughening. Along with the physics and mechanics of transformation toughening, a comprehensive understanding of tetragonal ZrO_2 is provided in the context of stress-induced transformation toughening. Other toughening mechanisms, for example, microcracking and ferroelastic toughening, are also discussed. The coupling of various toughening mechanisms is also addressed.

10.1 INTRODUCTION

In the ceramics literature, toughening mechanisms[1] have been broadly classified into two categories: one involving a process zone around the crack tip, the other related to the bridging of crack faces by reinforcements (fibers, whiskers). The *process zone mechanism* is reported to enhance the material's intrinsic toughness; examples include *transformation toughening*[2] and microcracking. The second mechanism, that is, crack bridging, is useful in toughening ceramic matrix composites (CMCs). Among various engineering ceramics, ZrO_2 is known for an excellent combination of high strength (up to 1 GPa) and good fracture toughness (up to 10 MPa $m^{1/2}$), and such excellent properties are attributed to the transformation toughening mechanism. Hence, transformation toughening has attracted considerable attention during last few decades. The toughness obtainable with the transformation toughening mechanism is superior to that of the composites reinforced with particles (ZrO_2-toughened Al_2O_3 [ZTA]) or whiskers (e.g., Al_2O_3/SiC_W composites).[3] Various ZrO_2-based ceramic materials, including partially stabilized zirconia (PSZ), tetragonal zirconia polycrystals (TZPs), and fully stabilized zirconia (FSZ), as well as zirconia-toughened/dispersed ceramics (ZTC/ZDCs) are now considered as potential materials for structural, tribological, and biomedical applications.[4–7] From this perspective, progress is accomplished in terms of both understanding the physics of transformation toughening[3,4,8–14] and exploiting it to develop toughened materials.

Advanced Structural Ceramics, First Edition. Bikramjit Basu, Kantesh Balani.

The remainder of this chapter is organized into eight different sections. The characteristics of the tetragonal zirconia (t-ZrO_2)-to-monoclinic zirconia (m-ZrO_2) phase transformation are discussed in Section 10.2. This is followed by a short description on phase equilibria in the zirconia–yttria system in Section 10.3. Section 10.4 largely discusses the micromechanics of transformation toughening. Some theories explaining t-ZrO_2 stabilization are summarized in Section 10.5. Section 10.6 discusses the production and properties of Y-TZP ceramics. In Section 10.7, the role of various microstructural variables in transformation toughening is discussed. In Section 10.8, additional toughening mechanisms are mentioned. Finally, the coupling of different toughening mechanisms is addressed in Section 10.9.

10.2 TRANSFORMATION CHARACTERISTICS OF TETRAGONAL ZIRCONIA

The crystal structures of the three major zirconia phases have been shown earlier in Chapter 2 (Fig. 2.15) and the detailed crystallographic information on different ZrO_2 polymorphs can be found in Ref.[15] It can be reiterated here that the cubic zirconia has the ideal fluorite structure, whereas the other polymorphs (tetragonal and monoclinic) have a distorted fluorite structure.[16] It is known that undoped zirconia exhibits the following phase transitions:

$$monoclinic(m\text{-}ZrO_2) \underset{950°C}{\overset{1170°C}{\rightleftarrows}} tetragonal(t\text{-}ZrO_2) \xrightarrow{2370°C} cubic(c\text{-}ZrO_2) \xrightarrow{2680°C} liquid$$

The transformations among the three polymorphs need to be considered as far as the processing and mechanical properties (strength, toughness, etc.) of zirconia ceramics are concerned. It is recognized[3,4,6] that the t → m transformation in zirconia is a reversible thermal martensitic transformation, and such a transformation involves a large temperature hysteresis (~200°C), volume change (4–5%), and considerable shear strain (14–15%). Several dopants, for example, yttria and ceria, are added to stabilize the tetragonal and/or cubic phase at room temperature (RT).[4] Although c-ZrO_2, t-ZrO_2, and m-ZrO_2 are the commonly reported phases, other zirconia phases, for example, nontransformable tetragonal (t′-ZrO_2)[17] and rhombohedral (r-ZrO_2),[16,18] can be found under certain conditions. The extreme stability of t′-ZrO_2 phase is related to both higher dopant content and finer domain size (~0.1 μm). Importantly, t′-ZrO_2 is not reported to undergo phase transformation to m-ZrO_2; instead the toughening in the presence of t′-ZrO_2 is attributed to ferroelastic domain switching, as is discussed later.

It is known that the martensitic transformation of t-ZrO_2 to m-ZrO_2 is induced either during thermal cooling or by external stress application.[3,4] The stress-induced martensitic transformation of t-ZrO_2 enhances toughness, and this phenomenon is known as transformation toughening.

Mechanistically, the t→m ZrO_2 transformation typically takes place in two major stages.[16] In the first stage, the transition from tetragonal to monoclinic occurs by shear displacement of zirconium ions, and in the second stage the diffusional migration of oxygen ions to their respective sites in the monoclinic symmetry takes place. The displacement of the oxygen ions from the ideal fluorite positions has been experimentally confirmed by x-ray diffraction (XRD).[19] The reverse transition of the lattice structure from m symmetry to t symmetry and the migration of the Zr^{+4} and O^{-2} ions to their respective positions are controlled by the diffusional displacements of the respective ions.[20]

10.3 PHASE EQUILIBRIA AND MICROSTRUCTURE

An important feature of the ZrO_2–Y_2O_3 phase equilibria is the stabilization of the high-temperature cubic and tetragonal phases. Duwez et al. made an initial attempt to investigate the phase equilibria in the zirconia–yttria system.[21] A critical thermodynamic evaluation and a modified phase diagram were later proposed by Srivastava et al.[22] The widely accepted phase relationships were provided by Scott in 1975.[20] Using a large high-temperature XRD (HTXRD) experiment, it was confirmed that tetragonal phase could be retained on rapid cooling and in ZrO_2 ceramics with a grain size finer than a critical value (as is discussed later). The zirconia-rich part of the phase diagram is shown in Figure 10.1. The typical composition and the sintering temperature range for Y-TZP and Y-PSZ are also illustrated in Figure 10.1. The experimentally measured thermodynamic data as well as more analysis of phase stability in the ZrO_2–Y_2O_3 system is reported elsewhere.[23–25]

The microstructure and properties of TZP, PSZ, and ZTC can be found elsewhere.[26–29] TZP contains nearly 100% tetragonal ZrO_2 phase stabilized by yttria

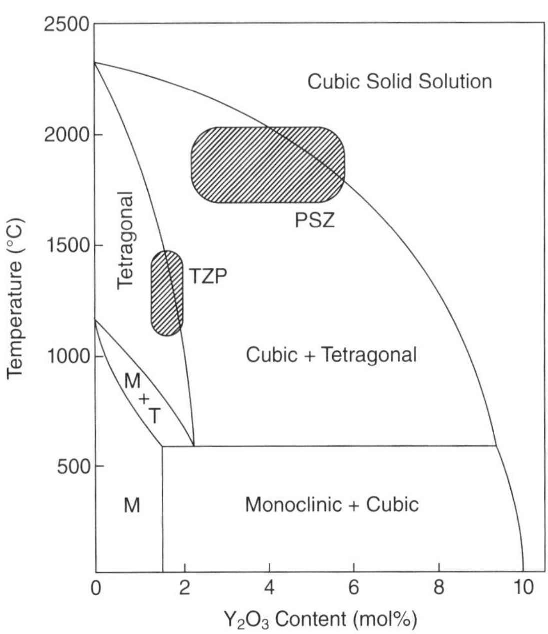

Figure 10.1 ZrO_2-rich part of the phase diagram in the ZrO_2–Y_2O_3 system (after Scott[20]). The shaded regions indicate the compositions and sintering temperatures commonly used for Y-TZP and Y-PSZ.

(a) (b)

Figure 10.2 Representative microstructures showing the distribution and morphology of tetragonal zirconia phase: (a) polygonal equiaxed tetragonal grains in a 3Y-TZP[102]; (b) lenticular-shaped tetragonal zirconia precipitates in a cubic matrix of a Mg-PSZ ceramic[5].

or ceria additions (Fig. 10.2a). Typical grain size of the TZPs is around 0.2–1 μm.[7] As far as nomenclature is concerned, TZP ceramics are often prefixed with "Ce-" or "CeO_2-" to denote ceria-stabilized ceramics, or with "Y-" or "Y_2O_3-" to denote yttria-stabilized ceramics. Also, a number in front of the acronym generally signifies the mole percent of dopant. For example, 3Y-TZP is the acronym for 3 mol% yttria-stabilized zirconia ceramic. The two most widely investigated TZP materials are Y-TZP and Ce-TZP, stabilized by yttria and ceria, respectively. Compared with TZP, PSZ is characterized by a coarse-grained microstructure with the t-ZrO_2 precipitates embedded in c-ZrO_2 matrix.[5] By annealing in the two-phase (c + t) field at high temperature (~1600°C), lens-shaped tetragonal phase is precipitated in a cubic zirconia matrix (see Fig. 10.2b).

10.4 TRANSFORMATION TOUGHENING

Transformation toughening refers to the phenomenon that t-ZrO_2 phase undergoes a phase transition to the stable monoclinic symmetry (Fig. 10.3a) in the tensile stress field around a propagating crack.[4] The concomitant volume expansion (4–5%) involved in such phase transition introduces a net compressive stress in the process zone around the crack tip.[30] This essentially reduces the local crack tip stress intensity factor as well as the driving force for crack propagation. The net result is an increase in the toughness (see Fig. 10.3b).

The mechanism of transformation toughening can be described by a change in the stress intensity factor,

$$\Delta K = K_{tip} - K_{\infty}. \tag{10.1}$$

Crack tip shielding occurs only when the crack tip stress intensity factor (K_{tip}) is less than that arising due to the applied stress (K_{∞}), that is, $\Delta K < 0$. The toughening is therefore expressed by the parameter ΔK_c, which is nothing but ($-\Delta K$).

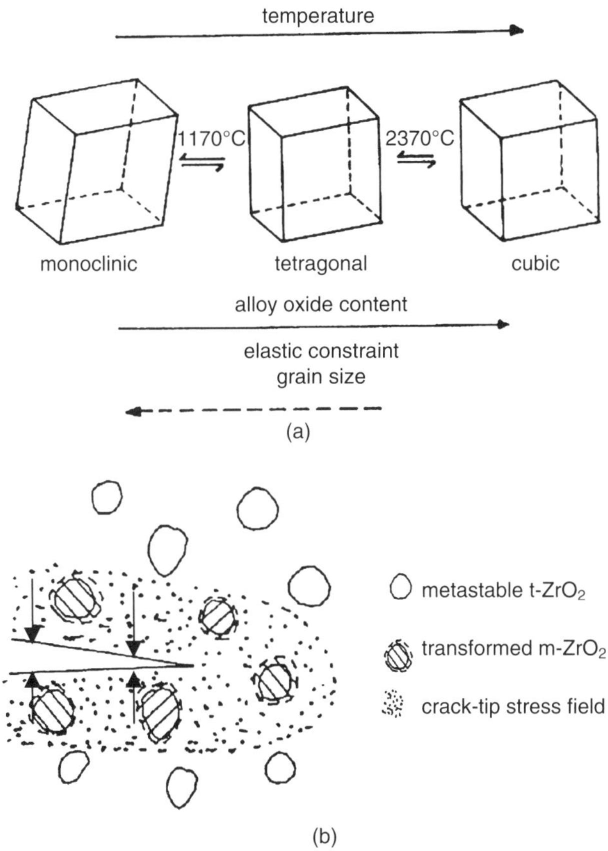

Figure 10.3 (a) The phase transformation in zirconia induced by thermal heating and cooling, addition of dopant oxides, or by elastic constraints and grain size. (b) Schematic showing the stress-induced phase transformation of metastable tetragonal zirconia particles in the crack tip stress field. The arrows in (b) indicate the generation of compressive residual stress due to the transformation-induced volume expansion and the microstructural constraint.[150]

For the effective contribution of transformation toughening, the retention of the maximum amount of t-ZrO_2 at RT or at the application temperature with optimum transformability is needed. In the transformation-toughening literature,[3,4,12,16,17] transformability is understood as the ease with which t-ZrO_2 transforms to m-ZrO_2 in the crack tip stress field.

10.4.1 Thermodynamics of Transformation

A brief thermodynamic analysis is presented next to describe the transformation toughening phenomenon. The thermodynamic approach was first proposed by Lange[31] and later modified by Becher.[32] The total free energy change per unit volume ($\Delta G_{t\rightarrow m}$) for constrained transformation can be described by the following expression:

$$\Delta G_{t\rightarrow m} = -\Delta F_{CH} + \Delta U_e + \Delta U_S - \Delta U_I, \tag{10.2}$$

where ΔF_{CH} is the chemical free energy change, ΔU_e is the strain free energy change associated with the transformation, ΔU_S is the change in surface free energy, and ΔU_I is the interaction energy density (related to the application of external stress).

From this, it is clear that total free energy change of transformation can be increased and the t-ZrO_2 can be retained in any or all of the following ways (see Fig. 10.3a):

- **(a)** A decrease in chemical free energy change by suitable dopants (yttria, ceria, calcia, magnesia, etc.)
- **(b)** An increase in strain free energy change by dispersing tetragonal phase in a constraining elastic matrix (such as alumina, cubic zirconia, etc.)
- **(c)** An increase in surface free energy change, for example, by reduction in tetragonal grain size

The energy balance approach essentially reveals that t → m transformation of constrained t-ZrO_2 particles is possible when the interaction energy density just balances the sum of the strain energy, surface energy, and the chemical free energy. Thus, one can arrive at the following definition of the critical transformation stress (σ_c):

$$\sigma_c = \Delta S \frac{(M_s - T)}{\varepsilon^t}. \tag{10.3}$$

where ΔS is the associated entropy change, M_S is martensitic transformation temperature, T is the temperature of interest, and ε^t is the transformation strain. Following the relationship in the preceding equation, the critical stress, at which stress-induced transformation initiates, is lowered as the M_s temperature approaches the temperature of interest.

10.4.2 Micromechanical Modeling

In last few decades, extensive effort has been put into developing a theoretical framework to predict the toughness enhancement due to stress-induced transformation.[4,17,33–37] There exist two mechanistic approaches: the *stress intensity approach* and the *conservation integral approach*.

A detailed mathematical treatment of these approaches is given elsewhere.[5,6,37] As mentioned earlier, enhanced transformation toughening is only realized during crack growth, that is, as the transformation zone develops around the crack tip (see Fig. 10.4)—a phenomenon demonstrated in the *resistance curve* (popularly known as the R curve).The evolution of an R curve can be described by three consecutive stages: the frontal zone, the partial zone, and the extended zone. If the long-range strain field is purely dilatational, no contribution from the frontal zone to the toughness increment could be realized ($\Delta K = 0$). In the partial zone, wherein only a finite

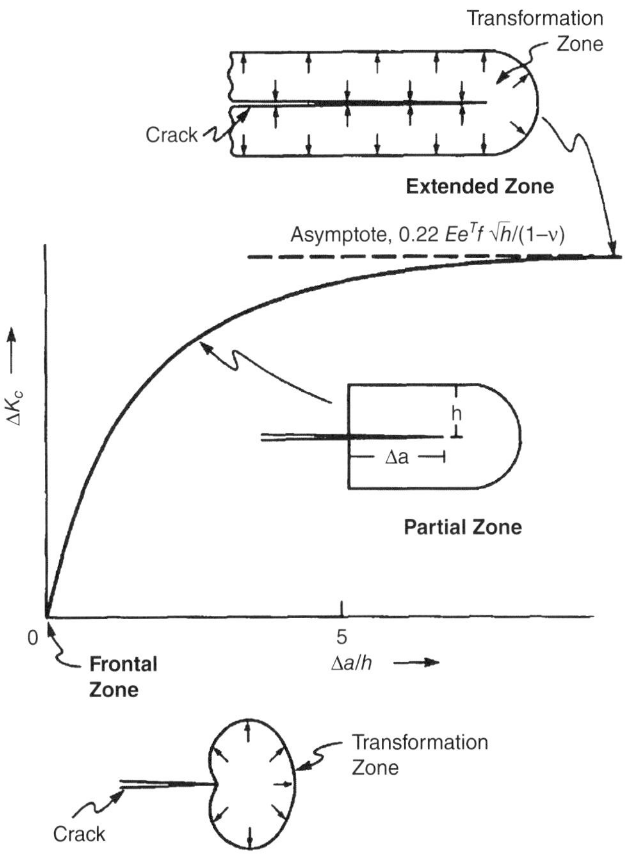

Figure 10.4 Schematic representation of the evolution of an R curve in transformation-toughened ceramics.[3]

fraction of the transformable tetragonal phase is transformed, the shielding of the crack tip takes place. Finally, crack tip shielding reaches its steady-state value in the fully developed extended zone, inside which the maximum amount of the tetragonal particles have already been transformed into monoclinic phase.

McMeeking and Evans[12] and Budiansky et al.[33] evaluated the transformation zone size with Mode I crack propagation, assuming that all the particles inside the process zone transform irreversibly. Budiansky et al.[33] came up with the following expression for the transformation zone size (h):

$$h = \frac{\sqrt{3}(1+\nu)^2}{12\pi}\left(\frac{K_\infty}{\sigma_c}\right)^2, \tag{10.4}$$

where ν is Poisson's ratio, K_∞ is the stress intensity factor due to the applied stress, and σ_c is the critical transformation stress. From this relationship, it is clear that a large transformation zone can develop when the critical transformation stress is low.

Both the analytical models formulated by McMeeking and Evans[12] and Budiansky et al.[33] can predict almost similar supercritical plane strain toughness increment (ΔK_c),

$$\Delta K_c = 0.22 f E \varepsilon^t \sqrt{h} /(1-\nu), \tag{10.5}$$

where f is the volume fraction of the tetragonal phase transformed within the transformation zone, E is the composite modulus, and ε^t is the dilatational strain involved in the transformation. Hence, the level of crack tip shielding due to stress-induced transformation can be linked to the size of the transformation zone and the volume fraction of the transformable t-ZrO_2 particles.

In majority of micromechanical models, it is assumed that stress-induced transformation is purely controlled by hydrostatic stresses. However, the transformation being martensitic in nature, shear stresses should play an important role when the transformation particularly involves deformation-induced twinning phenomena.[34–36] Evans and Cannon[2] considered the possible influence of shear strain and provided the following expression for toughness enhancement:

$$\Delta K_c = 0.38 f E \varepsilon \sqrt{h} /(1-\nu). \tag{10.6}$$

This expression indicates that a higher level of transformation toughening can be predicted when both shear and dilatational components of transformation strain are considered.

10.5 STABILIZATION OF TETRAGONAL ZIRCONIA

In the overall context of transformation toughening, the transformability of stabilized tetragonal phase plays an important role[12,16,17] and hence the transformability of t-ZrO_2 should be discussed in greater detail. Subbarao suggested that the size, charge, and amount of dopant cations influence the stabilization.[38] In experiments with various stabilizer additions, Kim observed[39] that the tetragonality, expressed by the "c/a" ratio, is considerably modified by the dopant addition. Morinaga et al.[40] postulated that the displacement of the oxygen anions from their ideal positions in the fluorite structure influences the stability.

Yoshimura studied the stability of the high-temperature zirconia phases from both thermodynamic and kinetic aspects.[41] From the thermodynamic aspect, the free energy of the t-ZrO_2 decreases with the increase in dopant content. Alternatively, the thermodynamically metastable phase can be retained at RT, if sufficient activation energy (ΔG^*) is not available to overcome the activation energy barrier of transformation.

Hillert[42] and others[43] reported that the oxygen vacancies also play an important role in the stabilization of the tetragonal phase. Adopting the Kröger–Vink notation,[1] the defect reaction for doping of zirconia doped with yttria can be written as

$$Y_2O_3 \xrightarrow{ZrO_2} 2Y'_{Zr} + 3O_o^x + V_o^{\cdot\cdot}, \tag{10.7}$$

where Y'_{Zr} indicates that a Y atom substitutes and occupies the Zr-lattice site, O_o^x represents an oxygen atom occupying a normal lattice site, and $V_o^{\cdot\cdot}$ expresses the vacancy formation in the oxygen lattice site. The increase in oxygen vacancies increases the disorder of the $ZrO_2–Y_2O_3$ system and enhances the stability of the tetragonal phase.[44]

It has been widely recognized that the t-ZrO_2 can only be stabilized at RT if a *critical grain size* is retained in the sintered microstructures. From the thermodynamic aspect,[45] one can finally arrive at the condition when t-ZrO_2 with a grain size $d < d_c$ will transform in the crack tip stress field:

$$-\Delta U_I = 6\Sigma\Delta S\left(\frac{1}{d}-\frac{1}{d_c}\right), \tag{10.8}$$

where $\Sigma\Delta S$ denotes the sum of all the interfacial energy terms, that is, $\Delta U_e + \Delta U_s$.

Garvie also proposed that there exists a critical size range (d_{cl}–d_{cu}), within which the retained t-ZrO_2 particles can transform in the applied stress field.[45] This reasoning clearly signifies that there exists a lower bound to this critical size (d_{cl}) below which a particle cannot transform even if the interaction energy density surpasses the dilatational strain energy density related to the transformation. The upper bound of the range (d_{cu}) essentially implies that when the particles are larger than d_{cu} they would spontaneously transform to m-ZrO_2 during the cooling from the sintering temperature. This would lead to the lowering of mechanical properties—in particular, toughness.

Finally, the critical factors, in the context of the stability of t-ZrO_2 embedded in a ceramic matrix (Mg-PSZ, ZTA, etc.), include matrix constraint, thermal residual stress, chemical composition, and the transformational nucleation barrier.[46,47] Garvie reported that the critical size range for Ca-PSZ and $Al_2O_3–ZrO_2$ composites is 62–95 nm and 0.38–0.45 μm, respectively.[45]

10.6 PRODUCTION AND PROPERTIES OF Y-TZP CERAMICS

It is well known that the strength and fracture toughness of structural ceramics is strongly dependent on the particle size, the chemistry of the starting powder, and the sintering parameters.[48] Since the mid-1990s, several new technologies have been developed for the production of submicron, ultrapure ZrO_2 powders with a narrow size distribution.[49] The different powder processing methods include hydroxide coprecipitation or alkoxide hydrolysis,[46] gel precipitation,[50] microemulsion techniques,[51] sol–gel synthesis,[50] hydrothermal synthesis,[52] and gas phase reaction–plasma coating,[53,54]. Only the first and the last methods are utilized in the large-scale production of zirconia powders for the fabrication of t-ZrO_2 ceramics. In the coprecipitation method,[55] the yttria-doped powders are obtained by ammonia leaching of $ZrOCl_2$ and $YCl_3/Y(NO_3)_3$ solution. In the plasma coating route, yttria is milled together with ZrO_2 base powders produced by plasma decomposition of $ZrCl_4$. More recently, Belgian researchers have developed a new route to synthesize Y_2O_3-doped powders using a "suspension drying" method.[55,56]

The state-of-the-art method of sintering and the new processing routes for the zirconia-based ceramics were reported in a 1996 paper.[53] Typically, the Y-TZP ceramics can attain full density by sintering in the temperature range 1400–1600°C for 1–2 hours in air. In this case, the sintering time and temperature need to be optimized to limit the grain growth in order to obtain a critical grain size ($d_{cl} < d < d_{cu}$). To attain the fully tetragonal microstructure in 3Y-TZP ceramics, the critical grain size must be below 0.8 μm.[31] Paek et al. studied the influence of sintering atmosphere (O_2, N_2) on the densification of 3Y-TZP.[57] Y-TZPs can be sintered by hot pressing (1400–1500°C for 1 hour in vacuum) or sinter–hot isostatic pressing (sinter-HIPing) to attain full density.[53] The use of advanced densification techniques, such as microwave sintering (MW)[58] and spark plasma sintering (SPS),[59–62] enables the densification of Y-TZPs with finer microstructure and improved properties. In the MW sintering route, the use of SiC rods (high dielectric constant) as susceptors in a MW cavity is adopted to facilitate the absorption of MW energy by ZrO_2 (low dielectric constant), which subsequently leads to densification typically at around 1400°C in 15–20 minutes. The SPS route is used widely to produce nanostructured ceramics.[63] Y-TZP ceramics with ~100-nm grain size have been reported to be sintered at 100–200°C lower temperature and in a shorter time (~5 minutes) than conventional sintering route.[63]

Concerning the mechanical properties, the hardness of Y-TZP lies in the range of 11–13 GPa. In an experimental study, Hannink measured a higher strength of 1300–1500 MPa for HIPed (1400°C, 150 MPa) Y-TZP samples.[64] High flexural strength (three-point bending) of around 1.7 GPa in 3Y-TZP was recorded after HIPing at 1500°C for 0.5 hour in 150 MPa.

The literature toughness data as measured by several researchers are summarized in Table 10.1. The reported toughness, varying between 2 and 20 MPa $m^{1/2}$, is dependent on the sintering parameters and the microstructural variables. The measured toughness values are also dependent on the measuring techniques, which can be broadly classified into (a) long crack and (b) short crack methods. The single edge notched beam (SENB) and single edge V-notched beam (SEVNB) techniques are classified in the first category, the details of which can be found in Ref.[63] In one small crack technique, the crack lengths (radial–median) around an indentation are measured to determine the toughness by adopting the established formulations proposed by Anstis et al.,[65] Kaliszewski et al.,[66] Niihara et al.,[67] or Shetty et al.[68] In another short crack method, the indentation strength in bending (ISB) method, an indented sample is fractured in bending and K_{Ic} is measured from the failure strength following the method proposed by Chantikul[69] or Cook and Lawn.[70] Assessing the data in Table 10.1, it can be noted that the indentation method has been used by many researchers in measuring toughness of transformation-toughened ceramics.

10.7 DIFFERENT FACTORS INFLUENCING TRANSFORMATION TOUGHENING

Various different parameters influencing the transformability of t-ZrO_2 are schematically presented in Figure 10.5. In the following subsections, the influence of some

Table 10.1. Summary of the Literature Results Reporting the Toughness of ZrO_2 Ceramics, Stabilized with Varying Yttria Content

Yttria (mol%)	Processing	Phases	Grain size (μm)	K_{Ic} (MPa $m^{1/2}$)	Toughness measurement	Reference
2.0	PS, 1500°C, 1 hour	Monoclinic 10%	0.51	5.0	ISB	80
	PS, 1500°C, 10 hours	Monoclinic 15%	0.89	6.8		
	PS ,1500°C, 80 hours	Monoclinic 20%	1.87	12.4		
2.5	PS, 1400°C, 30 minutes (unmilled powder)	78% monoclinic	0.29 ± 0.1	4.8 ± 0.36	Indentation	85
	PS, 1400°C, 30 minutes (milled, 1 hour)	60% monoclinic	0.35 ± 0.15	4.2 ± 0.32	Indentation	
1.8	HP	–	–	13.5 ± 0.5	Indentation	94
3.0	–	–	–	11.6	–	9
2.0	–	–	1.5	12	–	152
	–	–	0.5	7	–	
2.6	–	–	–	9.5	SENB	140
	–	–	–	5.6	SENB	
2.0	–	–	1.4	17.2	SENB (250-μm notch width)	141
	–	–	1.4	11.2	SENB (93-μm notch width)	
	–	–	1.3	12.3	SENB (130-μm notch width)	
3.0	PS, 1400°C, air, 15 minutes	–	–	6	Indentation	142
3.0	–	–	–	10.6	SENB	143
3.0	–	–	–	4.5 ($K_{plateau}$)	Indentation	144
4.0	–	–	–	3.1 ($K_{plateau}$)	Indentation	
2.0	–	–	–	4.4	SCF	145

(Continued)

Table 10.1. *(Continued)*

Yttria (mol%)	Processing	Phases	Grain size (μm)	K_{Ic} (MPa $m^{1/2}$)	Toughness measurement	Reference
2.0	Presintering at 1200°C, 12 hours	~35% tetragonal	–	5.0	Indentation	146
	Presintering at 1250°C,12 hours	~55% tetragonal	–	5.5	Indentation	
	1200°C, 12 hours and HP 1450°C, 2 hours	~99% tetragonal	–	11.0	Indentation	
	1250°C, 12 hour and HP 1450°C ,2 hours	~100% tetragonal	–	11.0	Indentation	
3.0	1450°C, 2 hours	25% c-ZrO_2, 75% t-ZrO_2	0.3 ± 0.1	5.0		79
3.0	1650°C, 3 hours, air, PS	Tetragonal	0.9	4.8 ± 0.2	Indentation	147
2.8	HP, 1400°C 40 minutes, Ar, 40 Mpa	–	–	4.6 ± 0.2	Indentation	148
2.0	HP, 1450°C, 1 hour, vac	–	0.4	5.9 ± 0.1	Indentation	59
3.0		–	0.3	2.5 ± 0.1		
3.0	PS, 1500°C, air	–	0.83	6.34	Indentation	149
				5.05	SEPB	
		–	0.62	5.8	Indentation	
		–	0.86	6.30	Indentation	
		–	1.14	8.2	Indentation	
2.0	PS, 1500°C, 2 hour, air	–	–	6.1 ± 0.2 ($K_{plateau}$)	–	153
1.5	1150°C, 2–5 hour, air	Fully tetragonal	–	11.9–13.9	Indentation	101
2.0			–	12.6–14.8		
3.0	HP, 1450°C, 1 hour, vac	Fully tetragonal	0.3	3.5 ± 0.1	Indentation	104,105
2.8			0.19	8.7 ± 0.3		
2.0			0.49	10.2 ± 0.5		
0	PS, 1100°C, 10 h	~92.2% monoclinic	0.133 ± 0.014	2.06 ± 0.04	SCF	151
0	PS, 1105°C, 5 h	~91.1% monoclinic	0.163 ± 0.022	2.00 ± 0.13		

Different processing routes: PS, pressureless sintering; HP, hot pressing. Various toughness measuring techniques: SENB, single edge notched beam; SEPB, single edge recracked beam; ISB, indentation strength in bending; SCF, surface crack in flexure.

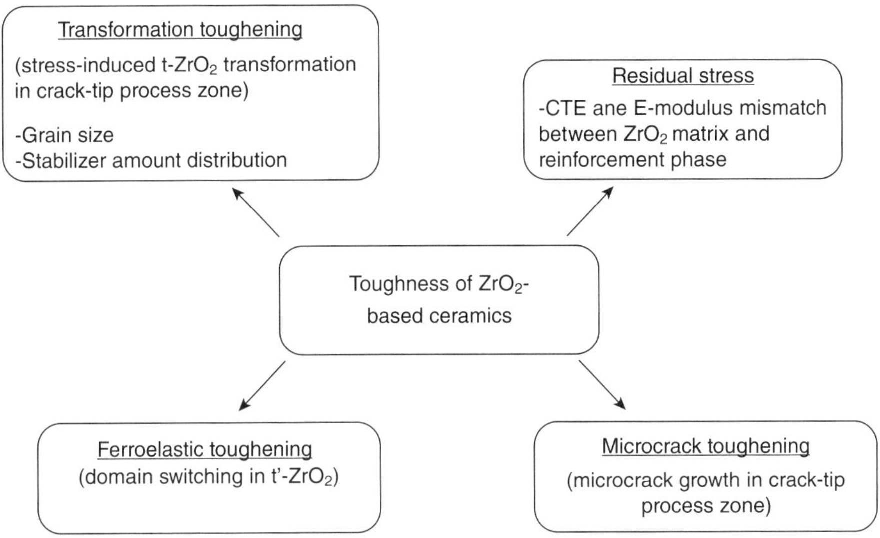

Figure 10.5 Summary of the different parameters influencing the transformation toughening and overall toughness of the monolithic zirconia ceramics.

important microstructural parameters on the properties of t-ZrO_2 and toughness is discussed.

10.7.1 Grain Size

As mentioned earlier, the grain size influences the transformability and toughness of tetragonal-zirconia ceramics.[71] The increase in the stability of t-ZrO_2 with decreasing grain size is due to the increased effectiveness of the grain boundary (GB) in constraining against the shape changes accompanying a martensitic transformation.[72] Another possible explanation is that the number of martensite units that can form decreases as the parent grain size decreases.[73] Thus, fewer accommodating variants can be allowed inside finer t-ZrO_2 grains. With lowering of the particle size, the strong shear component, in addition to the isotropic expansion, toward the overall transformation shape change dominates. The increased difficulty for the constraining matrix to accommodate both transformation strains explains the higher stability of smaller tetragonal grains.

Evans et al. reported that the twinned microstructure of transformed m-ZrO_2 was the origin of this size effect.[74] The shear strain involved in the martensitic transformations can be accommodated via one of the two lattice invariant deformation processes, twinning or slip. Clearly, the higher the twin density, the lower the strain energy change (ΔU_e)* or the larger the interfacial energy change (ΔU_s).[75] The

*The strain energy is given by $\Delta U_e = 0.4\mu V(\gamma_t)^2/\eta$, where μ is the shear modulus, γ_t is the strain energy, V is the particle volume, and η is the twin density, that is, the number of twins in the transformed particle (see Ref.[75]).

"size effect" could be further explained by considering the transformation thermodynamics (Eq. 10.2).

The competition between the two influencing factors (ΔU_s and ΔU_e) demands that there be an optimum size range in which t-ZrO_2 is retained. Within this optimum size range, a maximum volume fraction of tetragonal phase can be stabilized with a particular grain size. Additionally, all the energy terms in Equation 10.9 depend on the volume of the tetragonal particle, except the interfacial surface energy term (ΔU_s), which depends on the particle surface area. From simple energy considerations, one can therefore predict the size dependence of the tetragonal phase transformation.

Ruiz et al. reported the influence of heat treatment on the toughness of 3Y-TZP and explained the results in terms of the grain size.[76] By annealing a 2Y-TZP ceramic at 1500°C for varying timescale, Swain obtained a range of tetragonal grain sizes between 0.4 and 1.9 μm.[77] The transformation toughness linearly increased with grain size, with the highest toughness of around (12 MPa $m^{1/2}$) being measured at tetragonal grain size of 1.9 μm. While assessing the influence of yttria content on the critical grain size d_c, Lange observed that critical grain size d_c increased from 0.2 to 1 μm as the yttria content increased from 2 to 3 mol%, respectively.[78]

Lange also found the compositional dependence of the tetragonal grain size in Y-TZP materials with varying yttria content of 0.8–6.6 mol%.[78] In Table 10.1, literature data revealing the toughness–grain-size relationship are summarized. In general, the fracture toughness increases with grain size. One important observation is that high SENB toughness of 17 MPa $m^{1/2}$ can be obtained with 2Y-TZP grain size of 1.4 μm.[75] For sintered 3Y-TZP ceramics, the highest toughness (indentation method), 8.2 MPa $m^{1/2}$, was measured at a grain size of 1.14 μm.[79]

10.7.2 Grain Shape and Grain Boundary Phase

In addition to grain size, the grain morphology influences the transformability of t-ZrO_2. The grain morphology depends on the starting powder composition, type of stabilizer, sintering atmosphere, and so on.[80–85] For example, the grain shape changes in Ce-TZP depend on the sintering atmosphere.[80] Mecartney reported polygonal and faceted morphology of Y-TZP grains without an intentionally added second phase.[81] As is discussed later, the grain facets provide the heterogeneous nucleation sites for martensitic transformations. Thus, well-faceted grains can result in enhanced transformability and toughness. In contrast, the presence of an amorphous glassy phase results in rounded corners (see Fig. 10.6).

The formation of an aluminosilicate or a silica-rich glassy phase is favored in the presence of higher amounts of Al_2O_3[86] or SiO_2[86] in the starting powders, respectively. High resolution transmission electron microscopy–electron energy-loss spectrum (TEM-EELS) analysis confirmed the presence of a distinct amorphous phase in liquid-phase sintered Y-TZP ceramics.[83,86] Tsubakino et al. demonstrated that the use of 3 mol% Y-ZrO_2 powders with 5.12 wt% Al_2O_3 and 0.13 wt% SiO_2 leads to the formation of alumina–silicate phase with dissolved MgO at GB as well as at triple

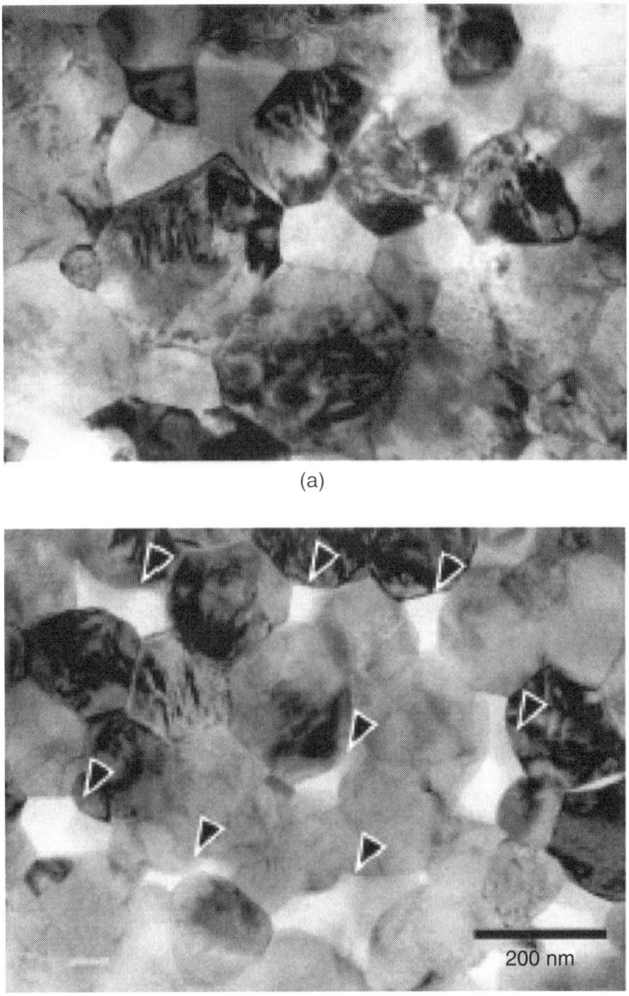

Figure 10.6 Bright field TEM images illustrating the change in grain shape of tetragonal grains in 3Y-TZP processed from starting powders containing varying amounts of SiO_2 addition: (a) sharply faceted (0 wt% SiO_2) to (b) grains with rounded edges (2.5 wt% SiO_2).[85] The presence of an intergranular glassy phase is indicated by arrows in (b).

junctions of grains.[86] The width of the GB phase is reported to increase from 2 nm to 50 nm with the increase in sintering temperature from 1550 to 1650°C. In another study, Morita and coworkers[85] obtained dense 2Y/3Y-TZP using Tosoh grade 3Y-ZrO_2 powder (Al_2O_3 < 50 ppm) with intentional SiO_2 addition (0–2.5 wt%) (Fig. 10.7). TEM images showing the existence of amorphous SiO_2-rich glassy phase at GB and triple pockets are illustrated in Figure 10.8. The change in grain shape from

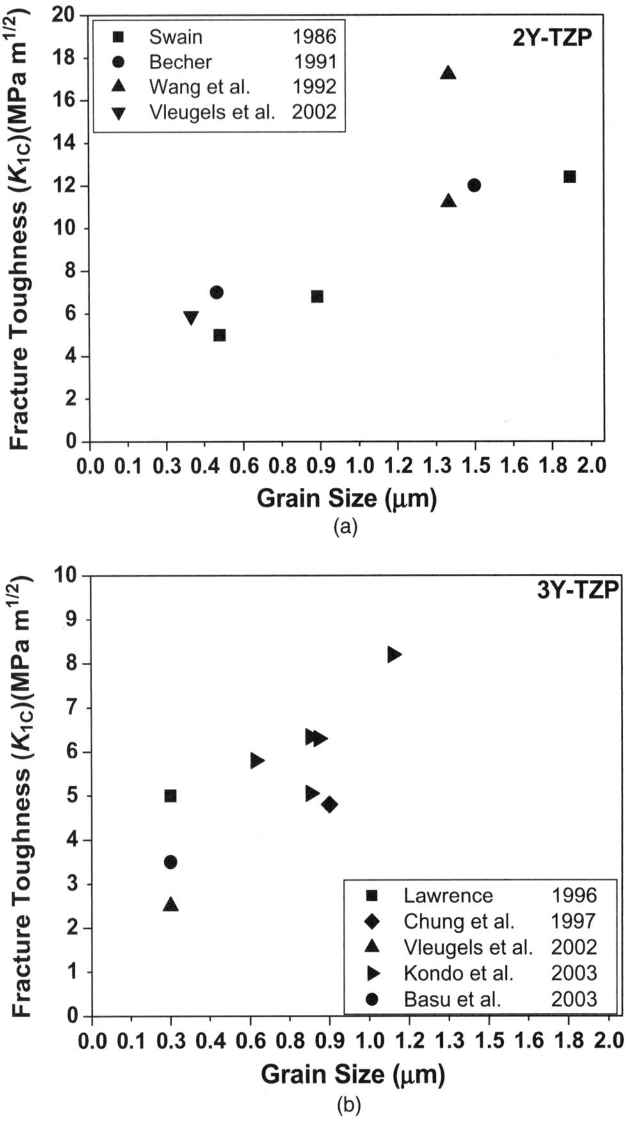

Figure 10.7 Literature data summary illustrating the grain size dependence of toughness for 2Y-TZP (a) and 3Y-TZP (b) ceramics.[1]

sharply faceted (0 wt% SiO_2) to rounded grain edges (2.5 wt%) can be observed. Energy-dispersive x-ray spectrometry (EDS) analysis further indicated the presence of SiO_2-rich glassy phase along with yttria segregation at triple pockets. However, the mechanical property (toughness) is not measured in any of the aforementioned studies. Stemmer et al. reconfirmed that the formation of amorphous GB phase is dependent on the impurity content of ZrO_2 starting powders.[83] Another observation

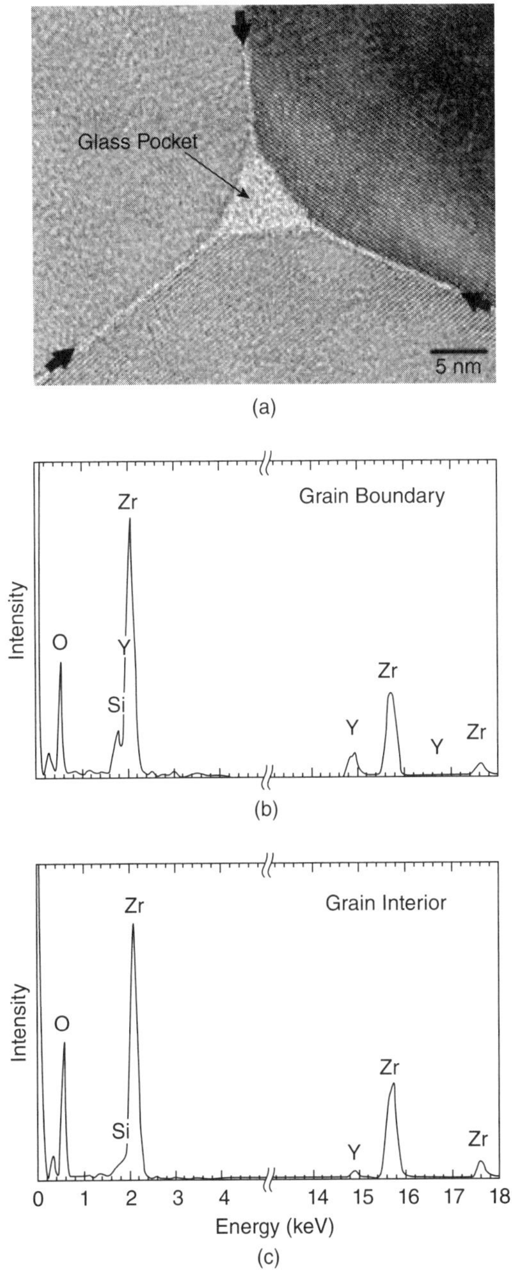

Figure 10.8 (a) HRTEM image of a grain boundary triple pocket revealing the formation of the glass phase in the 3YTZP doped with 0.3 wt% SiO2, sintered at 1673 K for 3 hours in air. The existence of the glassy phase along the grain boundary as a continuous phase is indicated by arrows. The obtained EDS spectra (b and c) confirm the formation of SiO_2-rich glassy phase and yttria segregation at the triple pockets.[85]

is that a strong yttria segregation at grain boundaries is recorded in the sintered 3Y-TZP, irrespective of grain sizes and impurity level.

The wetting of zirconia grains by a thin GB phase leads to the formation of an intergranular network. The presence of a continuous amorphous phase is reported to improve sinterability via liquid-phase sintering.[88] The repartition of the liquid phase among grain boundaries and multiple junction pockets has been related to the sintering temperature.[89] The experimental data to show the influence of GB glass phase on toughness of TZPs are, however, limited. In one study,[86] it was reported that 3Y-TZPs sintered using starting powders with higher impurity content (Daiichi grade, Al_2O_3, 0.72 wt%; and SiO_2, 0.08 wt%) exhibit higher indentation toughness (3.5 MPa $m^{1/2}$) compared with toughness (2.5 MPa $m^{1/2}$) measured with 3Y-TZPs processed from highly pure starting powders (Tosoh grade, Al_2O_3, <0.005 wt%; and SiO_2, 0.007 wt%). In future, a planned set of experiments need to be carried out to confirm the influence of varying the amount of alumina or silica addition on the toughness of Y-TZPs.

10.7.3 Yttria Content

The effectiveness of transformation toughening is determined by the nature and amount of the dopant cations.[87–94] Schubert and Petzow mentioned that the yttria content influences the chemical free energy term in two independent ways.[90] First, the undercooling required for stabilization decreases with increasing yttria content (660°C for 2Y-TZP and 360°C for 3Y-TZP). Second, coefficient of thermal expansion (CTE)-mismatch-induced residual stresses increase strongly with reduced yttria content.

In the pursuit of enhanced fracture toughness, Sakuma et al. measured a maximum toughness of 15 MPa $m^{1/2}$ in a sample with 1.8 mol% yttria.[93] Sakuma et al. also measured toughness as a function of the yttria content obtained in Y-TZP ceramics processed by conventional sintering and arc melting. The toughness peak lies at around 2 mol% yttria. Masaki et al. processed zirconia ceramics stabilized with yttria levels between 1.5 and 5.0 mol% using the hot isostatic pressing (HIP) route.[95] The fracture toughness of the obtained ceramics increased nonlinearly with decreasing yttria content from 2.5 to 2.0 mol%, with a maximum of 20 MPa $m^{1/2}$ for the 2Y-TZP. Importantly, the toughness values measured by Vickers indentation and ISB methods were comparable.

Matsui et al. reported that the difference in mechanical behavior of Y-TZP ceramics with equivalent average grain size can be attributed to the existence of critical yttria content (X_{cr}).[92] The X_{cr} for a TZP ceramic with 0.3-μm grain size lies between 4 and 5 wt% yttria content. For tetragonal ZrO_2 with yttria content lower than X_{cr}, a large driving force for the tetragonal-to-monoclinic transformation will be available during cooling from sintering. As a result, microcracking would result due to the thermally induced spontaneous tetragonal transformation. Gao et al. measured high strength of 1 GPa with a toughness of 14 MPa $m^{1/2}$ in a 2.1 mol% Y-TZP, with a grain size of 2 μm and 70% t-ZrO_2 retention.[96] Similarly, a critical grain size is required to retain 90% tetragonal phase.[94] For example, the critical grain size for 2Y-TZP is around 0.25 μm. This critical grain size also increases with the stabilizer content. The fracture toughness data, taken from the literature reports, are plotted against yttria content in Figure 10.9. In general, the trend is that toughness increases with decreases in yttria content. The maximum scatter in toughness is

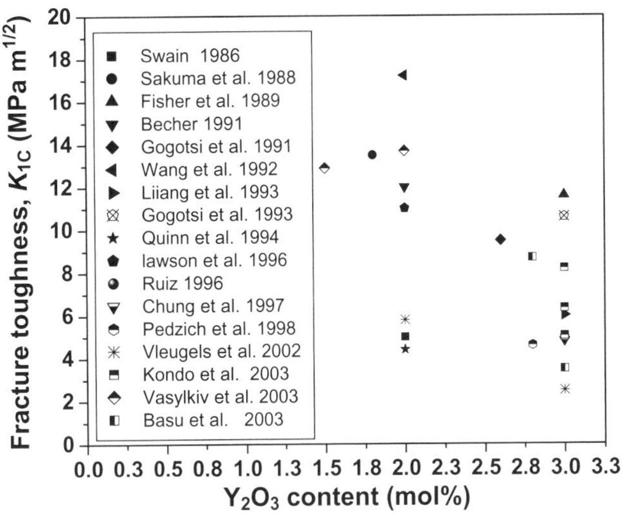

Figure 10.9 Influence of yttria content on the fracture toughness of Y-TZP ceramics.[1]

measured both at 3 mol% and 2 mol% yttria stabilization. The variation in toughness is largely related to the variation in grain size and the use of different toughness measurement techniques. High indentation toughness of larger than 10 MPa $m^{1/2}$ is achievable at lower yttria content of less than 2 mol%. An important observation has been that indentation toughness up to 14 MPa $m^{1/2}$ is obtainable with 1.5Y-TZP nanoceramics, sintered at 1150°C.[97]

10.7.4 Yttria Distribution

Apart from grain size and yttria content, the experimental results indicate that yttria distribution is another key factor influencing the t-ZrO_2 transformation[98] and fracture toughness of Y-TZP monoliths.[31,91,93,95,99–103] It is reported that Y-TZP materials, processed from yttria-coated ZrO_2 powders (titanium oxide powder, Tio3 grade) and Y-ZrO_2 coprecipitated powders display different fracture toughness trends.[31,57] The achievement of high toughness in coprecipitated ceramics is reported to be possible by growing the t-ZrO_2 grains by sintering at high temperature or by postsintering annealing for longer times at high temperature.[79,80]

Van Der Biest and coworkers reported an interesting approach to engineering the microstructure and tailoring the toughness of Y-TZP ceramics by means of mixing zirconia powders with varying yttria content (3 and 0 mol%).[99–103] By this route, it is possible that the toughness of TZP's can be considerably improved (up to 10 MPa $m^{1/2}$) compared with coprecipitated 3Y-TZP ceramics by tuning the starting powder composition. Detailed microstructural analysis suggested that the difference between the mean grain size of the coprecipitated 3Y-TZP (Tosoh grade, 0.3-μm grain size) and powder mixture based 2Y-TZP (hereafter referred to as TM2 grade ceramic, 0.5-μm grain size) ceramics is limited to less than 200 nm.[103] However, a significant difference in toughness between coprecipitated 3Y-TZP (2.5 MPa $m^{1/2}$) and the TM2 ceramic (10 MPa $m^{1/2}$) was measured. All these Y-TZP ceramics were hot pressed at 1450°C for 1 hour in vacuum.

Literature data[80] reveal that the fracture toughness for a coprecipitated powder based Y-TZP ceramic with an average grain size around 0.5 μm is about 6 MPa $m^{1/2}$. For coprecipitated Y-TZPs, the increased transformation toughening is also expressed in an increased M_s temperature as M_s temperatures of the coprecipitated 2Y-TZP and mixed grade 2Y-TZP (TM2) ceramics were measured to be at 390 and 312°C, respectively.[103] Based on the accepted theory, the M_s temperature for the 2Y-TZP ceramic (TM2) having overall yttria content of 2 mol% should have been higher because of its larger grain size. Another key observation is that the Tio3 ceramic exhibits excellent fracture toughness of 9 MPa $m^{1/2}$ at a mean t-ZrO_2 grain size of 0.19 μm (Fig. 10.10), a grain size at which coprecipitated t-ZrO_2 grains are hardly susceptible to transformation. This indicates that Y-TZPs based on mixing of 3Y-ZrO_2 powders with undoped ZrO_2 exhibit different toughening behavior.

The yttria distribution in the Tio3 ceramic (Fig. 10.10e) is broad and symmetric around 3 mol% yttria. However, a considerable number of areas analyzed with an electron probe microanalyzer (EPMA) were recorded with an yttria content below 3 mol%. These grains with low yttria content could be susceptible to transformation, since transformability increases with decreasing yttria content. Indeed, earlier experimental observations revealed a core–shell microstructure (Fig. 10.11) in the yttria-coated powder based Y-TZP[106] (Tio3 ceramic) and it was proposed that the enhanced transformability due to the characteristic core–shell microstructure is responsible for the high toughness of coated Y-TZP.[104] In 2000, Vleugels and coworkers reported high toughness in TZP ceramics processed from yttria-coated powders ("suspension drying" method).[58,59]

In contrast, TM2 ceramics display a coarser microstructure with mean grain sizes around 0.5 μm (Fig. 10.10b,d). Although the yttria distribution in the TM2 ceramic is characterized by a major frequency at 2 mol% yttria, a number of EPMA-analyzed areas having less than 2 mol% yttria were also recorded. The observation clearly indicates that the yttria is redistributed during sintering, resulting in a microstructure with heterogeneous yttria distribution. Furthermore, the high toughness of the TM2 ceramic can be ascribed to the enhanced transformability caused by the larger grains with lower yttria content (2 mol% or less). It can be noted here that the coprecipitated 3Y-TZP ceramic exhibits a narrow yttria distribution with highest frequency of yttria content around 3 mol%.[93,103,105] Therefore, the difference in yttria distribution can explain the observed toughness variation.

The exciting idea of mixing ZrO_2 powders with varying yttria content is further adapted in enhancing the toughness of 8Y-TZP ceramics.[105] Basu and coworkers measured high toughness up to 9–10 $MPa^{1/2}$ by mixing 8Y-ZrO_2 powders with a sufficient amount of undoped ZrO_2 to obtain overall yttria content of 2 mol%.[99–103] Additionally, EPMA measurements established the inhomogeneous yttria distribution in the high-toughness ceramics. From the preceding observations, an inhomogeneous yttria distribution, observed by using powders with widely different yttria

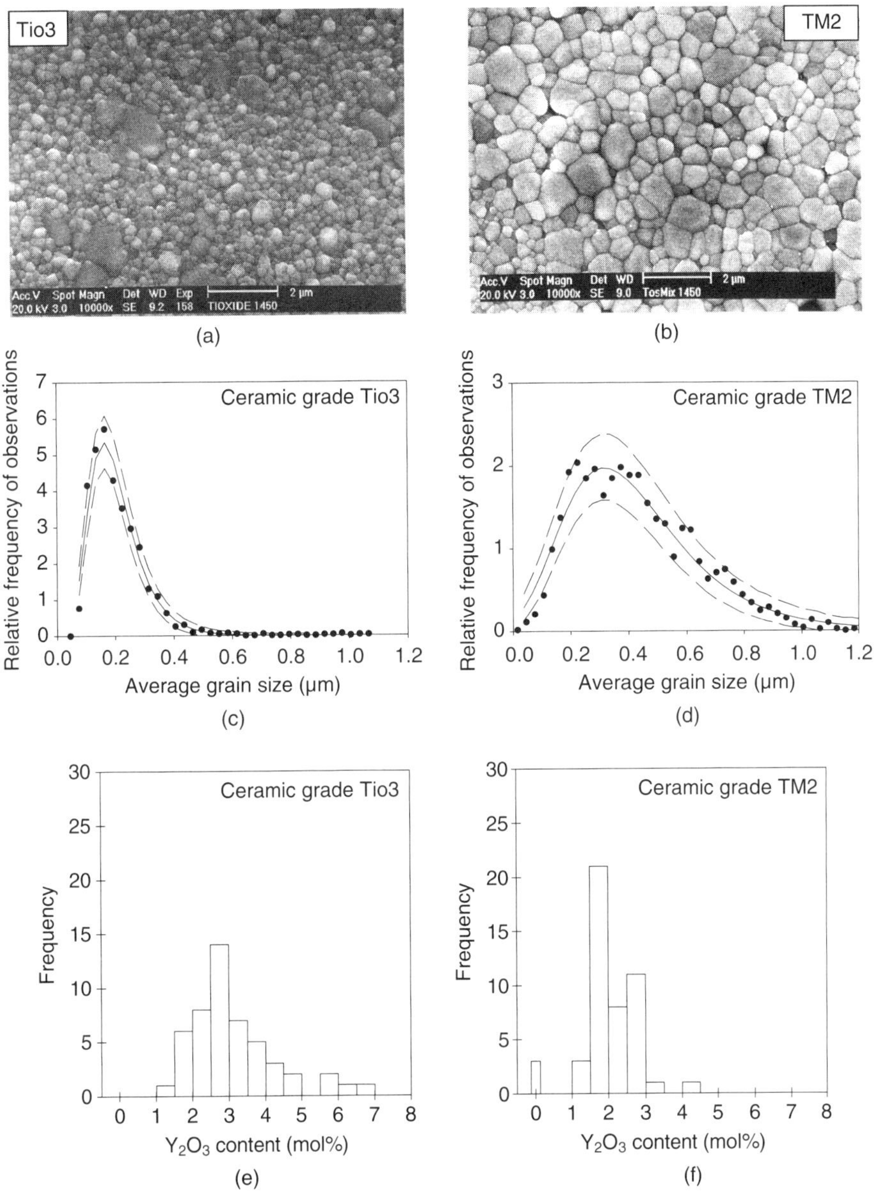

Figure 10.10 The sintered microstructures (a and b), grain size distributions (c and d), and EPMA yttria distribution (e and f) of two high-toughness Y-TZPs processed from 2.8 mol% Y-coated starting powder (Tio3mic) and mixture of 3 and 0 mol% Y-containing powders (TM2 ceramic).[103]

content, is identified as one of the key parameters in achieving high-toughness Y-TZPs. Based on these observations, the influence of grain size and yttria distribution on t-ZrO_2 transformability is explained (see Fig. 10.12). The presence of wider yttria and grain size distributions leads to high transformability and tougher Y-TZP ceramics. Lower-yttria-content grains with larger tetragonal grain sizes,

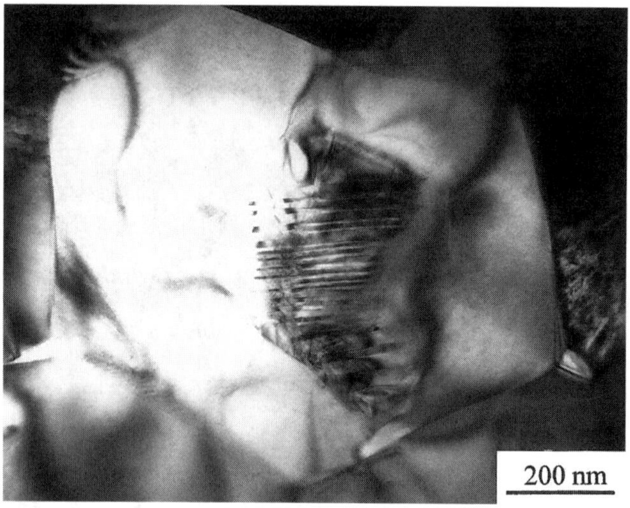

Figure 10.11 TEM image revealing the characteristic core–rim microstructure in a Y-TZP ceramic, processed from yttria-coated starting powders (after Bowen et al.[106]). The presence of yttria-poor core leads to high transformability and toughness.

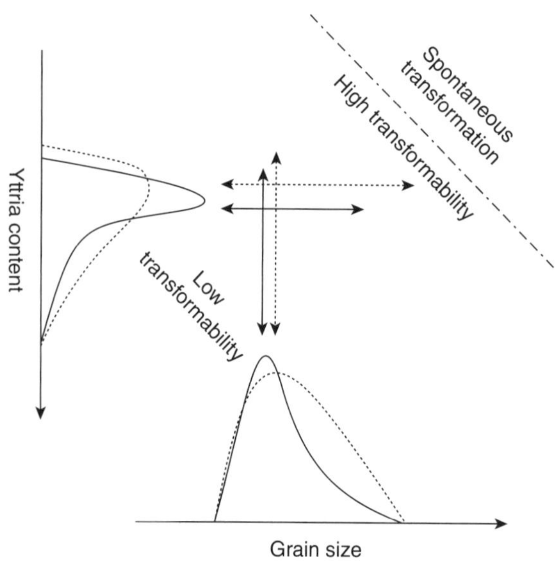

Figure 10.12 Schematic illustration showing the combined influence of grain size and yttria distribution on the t-ZrO_2 transformability in Y-TZP ceramics.

because of larger driving force, transform to monoclinic phase and subsequently the transformation in the higher-yttria-content grains occurs in an autocatalytic manner, a characteristic of martensitic transformation. More discussion of this plausible explanation on how the yttria and grain size distributions can contribute to enhanced toughness can be found in Refs.[59,93,103,107]

10.7.5 M_S Temperature

The onset temperature for the martensitic t–m transformation upon cooling is defined as the M_s temperature. Since the martensitic transformation in zirconia involves finite dilatational strain, researchers used dilatometry measurements to detect the M_s temperature.[108–110] Dilatometry experiments[105] can be performed to study the grain-size dependence of the M_s temperature in Ce-TZP and the kinetics of martensitic transformation in 2Y-TZP ceramics. Suresh et al. observed that the grain size has significant influence on the t–m transformation temperature.[107] M_s temperature data, summarized from literature, are plotted against grain size in Figure 10.13. The general trend is that M_s increases with grain size, independent of yttria content. Another important observation is that M_s increases with decrease in yttria content (Fig. 10.13). The following analytical expression can be achieved, illustrating the dependence of toughening on M_s temperature:

$$\Delta K_c = 0.08 fE\varepsilon^2(1+\nu)K_\infty / \Delta S(M_s - T)(1-\nu). \tag{10.9}$$

This expression implies that the higher the ($M_s - T$), the lower the transformation toughness. Clearly, it is desired that the M_s temperature is close to RT or the application temperature, so that the transformation-toughening contribution is maximized.

10.7.6 Transformation Zone Size and Shape

The size of the transformation zone is used as an input variable in analytical theories to model transformation-induced toughness. Rose considered both the

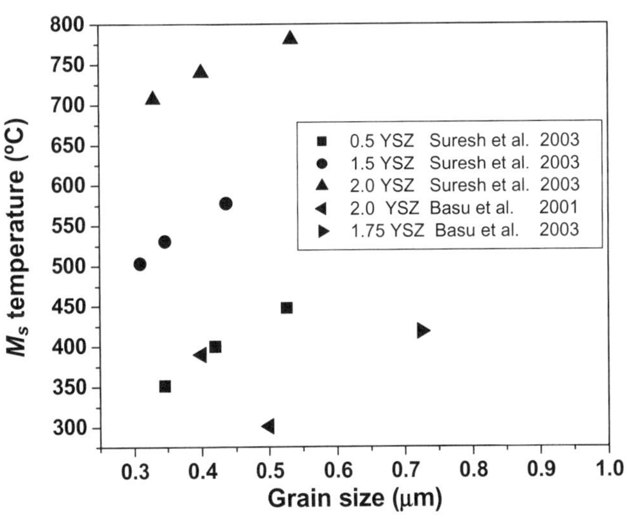

Figure 10.13 Grain size dependence of the martensitic transformation starting temperature for Y-TZP ceramics doped with varying yttria content.[1]

dilatational and shear strains while investigating the evolution of the shape and size of the transformation zone.[111] The assumption of plane stress conditions results in smaller zone width than does considering both the shear and dilatational components of the crack tip stress field. Based on the supercritical-transformation model, Stump and Budiansky proposed a refined model to account for crack growth.[13] The analytical calculations of their study revealed that the transformation strengthening, that is, the increase in strength due to phase transformation, occurs as a function of crack size during initial crack growth, with the increased transformation toughening only realized with the subsequent crack growth, that is, with long cracks (*R* curve).

Evans and Heuer[112] proposed an analytical relationship between the transformation zone size (h) and the critical transformation stress (σ_c):

$$h = \left[\frac{mK_0}{\sigma_c}\right]^2, \tag{10.10}$$

where m is an empirical constant and K_0 is the matrix toughness. Thus, increasing the transformation zone size (h) with corresponding decrease in critical transformation stress (σ_c) is the basis for achieving an enhanced contribution from transformation toughening. Different experimental techniques, for example, XRD,[113] transmission electron microscopy,[114] Raman spectroscopy,[115] acoustic emission,[116] atomic force microscopy,[117] and optical interference microscopy,[118] have been utilized to characterize the transformation zone. Clarke and Adar[115] reported a transformation zone size of 5 μm in a single-phase 3.5Y-TZP.

In Figure 10.14, literature data,[111] revealing the transformation toughness as a function of a composite microstructural parameter (product of volume fraction of transformable t-ZrO_2 [V_f] and square root of the transformation zone size h) are presented. All the investigated ceramics (Ce-TZP, Y-TZP, and Mg-PSZ) displayed a linear relationship. However, the slopes are different, revealing the difference in dopant-dependent linear relationship between K_{CT} and $V_f h^{1/2}$.

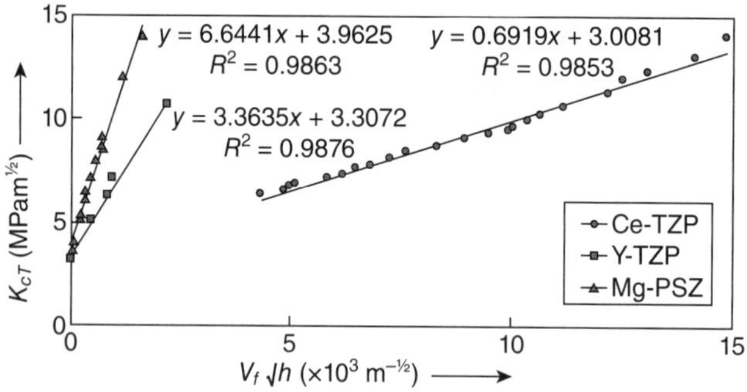

Figure 10.14 The interdependence of the transformation zone size, volume fraction of the transformable tetragonal phase, and the toughness of the stabilized zirconia ceramics.[1]

10.7.7 Residual Stress

The residual stress is reported to be an important parameter in optimizing the toughness of Y-TZP-based materials. Numerous investigations[92,118,119] have probed the influence of residual stresses on the transformability of the t-ZrO_2 phase and the concomitant toughness of TZP ceramics. In TZP monoliths, the thermal residual stresses (σ_r) arise due to two factors. The first one is due to the anisotropy in the CTE of the t-ZrO_2 phase ($\alpha_c = 11.4 \times 10^{-6}$°C^{-1} and $\alpha_a = 7.1 \times 10^{-6}$°C^{-1}).[120] Based on x ray lattice parameter measurements, Krell et al. reported a residual stress of 20–60 MPa in 3Y-TZP ceramics with grain sizes in the range of 0.5–1.0 μm.[121] Schubert determined the CTE of Y-TZP ceramics as a function of the yttria content.[119] The thermal expansion anisotropy of a 2Y-TZP ($\alpha_c/\alpha_a = 1.48$) is reported to be larger than that of a 3Y-TZP ($\alpha_c/\alpha_a = 1.18$). Finite element modeling (FEM) calculations showed that higher CTE anisotropy in 2Y-TZP results in larger residual stress as well as increased transformability and subsequently enhanced toughness.[122] Also, CTE mismatch (α_c/α_a) varies commensurately with the tetragonality (c/a) of t-ZrO_2 and a transition is reported to take place at about 4.5 mol% yttria with stabilization of the cubic structure.[120]

The second factor influencing the residual stress is the CTE mismatch between the GB phase (amorphous/crystalline) and the matrix. The intergranular phase (crystalline/amorphous) formed at the GB or the triple pocket typically has a lower CTE than the bulk zirconia. This results in tensile residual stress in the tetragonal grains. The formation of GB phase can be avoided with the use of high purity zirconia starting powders, in particular with lower silica and alumina content. Thus it is evident that only the first factor, that is, CTE-mismatch-(α_c/α_a)-induced residual stress will play an important role in TZP monoliths sintered from high purity starting powders.

The influence of the CTE-mismatch-induced residual stress on the tetragonal phase transformation in TZP with polygonal tetragonal grains is shown in Figure 10.15. The thermal expansion mismatch induced residual stress systematically decreases with increasing distance from the GB and typically is proportional to the "d/x" ratio, (d is the grain size and x is the distance from the GB). The larger strains at the grain facets or corners favor the heterogeneous nucleation of m-ZrO_2.

Concerning the limitations of transformation toughening, it can be commented that the effectiveness of transformation toughening strongly depends on the stability of the t-ZrO_2 phase. The experimental results reveal that the transformation-toughening contribution decreases linearly with increase in temperature.[4]

10.8 ADDITIONAL TOUGHENING MECHANISMS

The additional toughening mechanisms, for example, microcracking and ferroelastic toughening, are not temperature sensitive and offer useful toughening at elevated temperature, where transformation toughening is no longer effective. In the

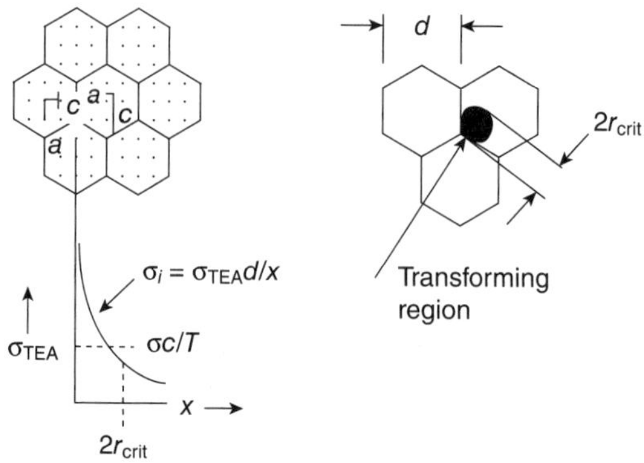

Figure 10.15 Schematic presentation illustrating the nucleation of the monoclinic zirconia phase induced by the local residual stress concentrations at the corners of polygonal tetragonal grains. The residual stress (σ_{TEA}) develops due to the anisotropy in thermal expansion coefficient (α_c/α_a) of the tetragonal zirconia phase. Note that the residual stress decreases with the distance (x) from the grain boundary in a tetragonal grain (d is grain size). The higher stress concentration at the grain facets triggers the tetragonal phase transformation, as illustrated by the development of a transformed zone (r_{crit}) near the grain boundary.[1]

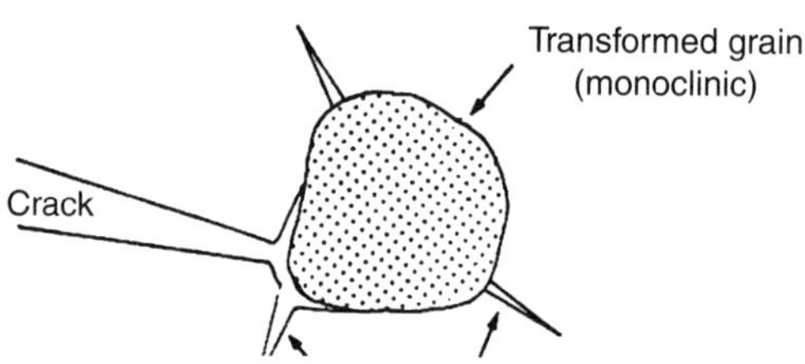

Figure 10.16 Microcrack toughening induced in the zirconia ceramics as a result of the formation and the subsequent growth of microcracks, as a consequence of stress-induced t-ZrO_2 phase transformation in the process zone.[1]

following subsections, these toughening mechanisms are presented with particular reference to their contribution to the toughness of zirconia monoliths.

10.8.1 Stress-Induced Microcracking

Conceptually, microcrack toughening involves the energy dissipation as microcracks nucleate and grow in size in the crack tip process zone.[120–138] Microcrack toughening is also realized in the presence of residual stresses, if the local tensile stress associated with the t–m zirconia transformation is sufficiently high. The shear strain involved in the martensitic t-ZrO_2 transformation is accommodated in the form of twins and/or microcracks, as shown in Figure 10.16. Ruhle et al. however argued that the radial microcracking may not accompany the stress-induced (t–m) ZrO_2 transformation.[47] They also commented that a given tetragonal grain either can transform or, if already transformed, can cause microcracking. In the presence of microcracks around a primary crack, toughness is enhanced by the growth of microcracks (dissipation of crack tip stress) and their interaction with the crack tip stress field. However, the presence of microcracks results in a reduction in stiffness (effective elastic modulus).

Following a continuum approach, Hutchinson investigated the effect of microcracking on the reduction in crack tip stress intensity, or shielding.[123] It was observed that the shielding contribution from microcracks is greater for steadily growing cracks than for stationary cracks. The strong resistance (R) curve behavior associated with microcracking arises mainly from the release of residual stress. Based on this theoretical work,[123] the half-width of the microcrack zone around a steadily growing crack is

$$h = 0.311\left(\frac{K}{\sigma_c}\right)^2 \tag{10.11}$$

and the toughness ratio is

$$\frac{K_{tip}}{K} = 1 - 1.42\varepsilon_s - 0.348E\theta\frac{\sqrt{h}}{K}, \tag{10.12}$$

where σ_c is the critical stress for the nucleation of the microcracks, θ is the dilatational strain involved with stress-induced microcracking, ε_s is the residual strain at the saturated level of microcracking, E is the composite elastic modulus, K_{tip} is the crack tip stress intensity factor, and K is the applied stress intensity factor. It was reported that, for $\varepsilon_s = 0.4$, a 50% reduction in K_{tip} could be achieved. It is evident that large residual strain is experimental in the presence of a larger microcrack density, which will evidently lead to a considerable decrease in elastic modulus. Hence, a larger contribution from microcrack toughening can be realized only at the expense of large modulus reduction.

Faber[138] reported a theoretical analysis to evaluate the microcrack toughening of zirconia-based ceramics. Faber formulated the toughness increment as a consequence of stress-induced zirconia phase transformation:

$$\Delta K^M = 0.25fE\theta\varepsilon_s\sqrt{h}, \tag{10.13}$$

where f is the microcrack density and explanations for the other terms could be found with Equation 10.13. Quantification of the toughness enhancement reveals that the toughness can only be increased as a consequence of stress-induced t-ZrO_2 transformation, and the magnitude of such an increase strongly depends on the relative increase in permanent strain (due to dilatational crack opening) compared with the decrease in modulus.

10.8.2 Ferroelastic Toughening

Ferroelastic domain switching is reported as an additional toughening mechanism in zirconia ceramics containing nontransformable t′-ZrO_2.[124–130] The tetragonal phase, formed as a transformation product of c-ZrO_2, is nontransformable and is known as t′-ZrO_2 (see also Section 10.2). Microstructural investigation revealed that

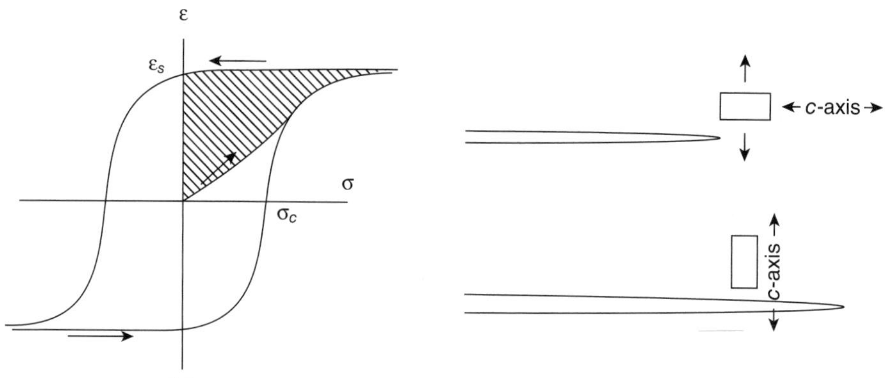

Figure 10.17 A schematic showing the stress–strain hysteresis loop (a) and the occurrence of the ferroelastic domain switching, as evidenced by reorientation of *c*-axis in tetragonal zirconia in the crack tip stress field (b), related to the ferroelastic toughening phenomenon (after Virkar and Matsumoto[127]). The shaded area in (a) is the mechanical energy absorbed due to domain switching of tetragonal zirconia phase. Two important parameters, the coercive stress (σ_c) and spontaneous strain (ε_s), are also indicated in (a).

t′-ZrO_2 has a polydomain structure, wherein each domain has a tetragonal symmetry with the *c*-axes along one of the three mutually orthogonal directions.

Ferroelastic toughening differs from stress-induced transformation in that the former does not involve any change in crystal structure, but a reorientation of the ferroelastic domains. Under stress, these domains are reoriented along *c*-axes to accommodate the strain. The ability to exhibit a permanent strain and its switchability result in a hysteresis loop between the strain (ε) and applied stress (σ), as illustrated in Figure 10.17.

Two important parameters relevant in the description of ferroelastic toughening mechanisms are coercive stress σ_c and spontaneous strain ε_s (Fig. 10.17). The area of the hysteresis loop is a measure of the mechanical energy dissipated in a single cycle. The dissipated energy eventually leads to toughening of the microstructure.

The application of a tensile stress on a t′-ZrO_2 grain in excess of σ_c along one of the *a*-axes will stretch it into *c*-axes and vice versa (see Fig. 10.17b). Similarly, the application of a compressive stress larger than σ_c along the *c*-axis will reorient it into an *a*-axis and one of the *a*-axes becomes a *c*-axis. Virkar and Matsumoto formulated the contribution of ferroelastic domain switching to the toughness,[131]

$$K_c = K_0\left[1+\frac{2hE}{(1-\nu^2)K_0^2}\oint\sigma_c d\varepsilon\right]^{0.5}, \tag{10.14}$$

where K_0 is the toughness in the absence of domain switching, E is Young's modulus, h is the height of the process zone around the crack where ferroelastic domain switching in tetragonal phase occurs, and ν is Poisson's ratio. The integral in the preceding equation is the shaded area shown in Figure 10.17a. Similar to the transformation-toughening mechanism, the process zone height h in this case is inversely proportional to σ_c^2 and directly proportional to K_c^2.

The first experimental evidence of ferroelastic toughening of t-ZrO_2 was reported by Virkar and Matsumoto.[127,131] Detailed XRD studies on TZP materials revealed that the surface grinding induced reversal of the peak intensities of (002), (200), and/or (113) and (131) is an evidence of the ferroelastic phenomenon.[129,131] Virkar and Matsumoto also fabricated a high-toughness ceria-doped zirconia ceramic (toughness 16 MPa $m^{1/2}$ as measured by the double cantilever beam technique).[131] *In situ* neutron diffraction on 12Ce-TZP materials confirmed the reversible nature of the ferroelastic toughening process.[130] Srinivasan and coworkers studied the ferroelastic phenomenon in 4 mol% yttria-doped zirconia single crystals both at RT and at high temperatures.[124] The experimental measurements at 1000°C revealed a toughness value as high as 8 MPa $m^{1/2}$ and a substantial domain switching behavior of the t-ZrO_2, as evidenced by the XRD study. Detailed microstructural observations of domain switching by Chan et al. indicated the domain reorientation around the fracture surface absorbs the energy from the loading system and thus enhances the fracture toughness of the zirconia materials.[129] High temperature *in situ* straining experiments on a high-voltage TEM provided evidences of crack tortuosity by the domain boundaries of the polydomain t′-ZrO_2 phase, which is a key factor resulting in ferroelastic toughening.[134–138]

10.9 COUPLED TOUGHENING RESPONSE

In the preceding sections, the microstructural factors in relation to the individual toughening mechanisms were discussed. Experimental observations and theoretical analysis confirmed that the combination of multiple toughening mechanisms results in tougher materials than those achieved by the sum of the individual processes.[132–138] Attempts have been made to explore concurrent transformation and microcrack toughening, which indicate that the coupling can indeed result in toughness augmentation.

10.10 TOUGHNESS OPTIMIZATION IN Y-TZP-BASED COMPOSITES

The toughness optimization of yttria-stabilized tetragonal zirconia (Y-TZP) by tailoring yttria distribution has been reported.[137] In this section, it is shown that a similar strategy can be used to develop high-toughness Y-TZP composites reinforced with hard TiB_2 particles. The experimental results confirmed that fully dense Y-TZP composites with 30 vol% TiB_2 can be obtained with a moderate hardness of 13 GPa, a high strength up to 1280 MPa, and an excellent indentation toughness up to 10 MPa $m^{1/2}$, when all were hot pressed at 1450°C for 1 hour in vacuum.[139] The toughness can be tailored between 4 and 10 MPa $m^{1/2}$ by carefully tuning the yttria stabilizer content of the ZrO_2 matrix between 3 and 2 mol%. As is mentioned later, the Y-TZP ceramic sintered from starting powers containing overall yttria content of 2.5 mol% will be referred as TM2.5 grade and the composite processed with TM2.5 matrix and TiB_2 (Starck grade E) will be designated as TM2.5E grade.

Representative bright field TEM micrographs revealing the morphology and size of the zirconia grains and TiB_2 phase in the TM2.5E composite are shown in Figure 10.18a. The ZrO_2 grains are retained with size of 0.3–0.4 μm. Occasionally,

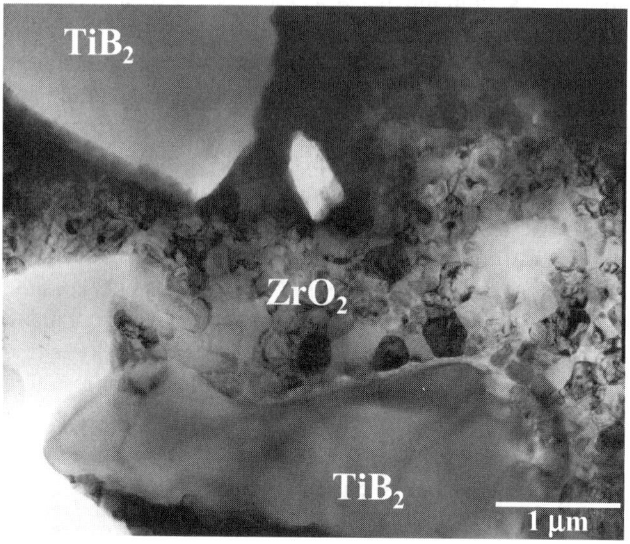

(a)

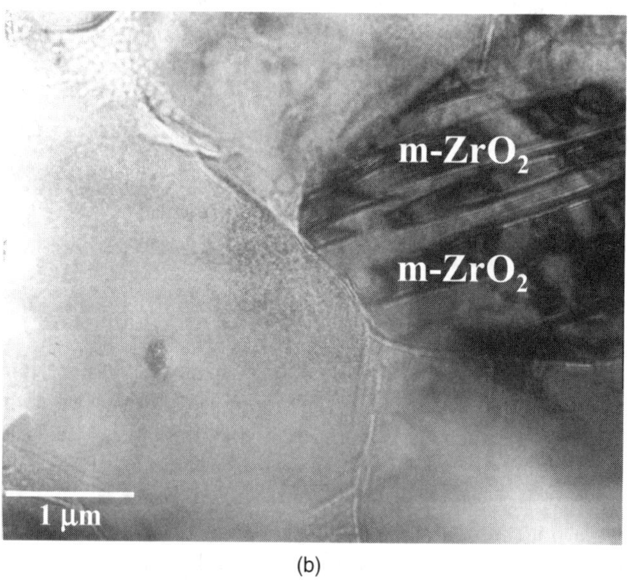

(b)

Figure 10.18 Bright field TEM images revealing the microstructure of 2.5Y-TZP–(30%)TiB_2 and coarser (2–3 μm) TiB_2 grains observed in a submicrometer-grain-sized ZrO_2 matrix (a). Typical martensite laths are observed in the transformed monoclinic ZrO_2 grains (b).[139]

some m-ZrO_2 grains display typical martensite laths, as shown in Figure 10.18b. The partial t-ZrO_2 transformation is caused either by the mechanical stress during the thin foil preparation or due to the thermal stress developed during the high-voltage electron beam irradiation.

In hot pressed ZrO_2–TiB_2, the boride particles contribute to the crack deflection toughening mechanism. The crack deflection model of Faber and Evans[138] predicts a toughness increase of about 15% for ceramic composites with 30 vol% of secondary phase, assuming an aspect ratio of 2. Looking at the relative toughness enhancement in the composite with respect to the matrix toughness, $\Delta K_{Ic}/K_m$, given in Figure 10.19,

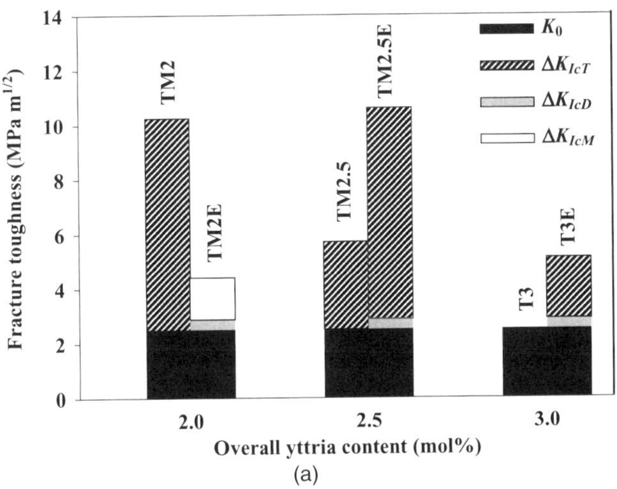

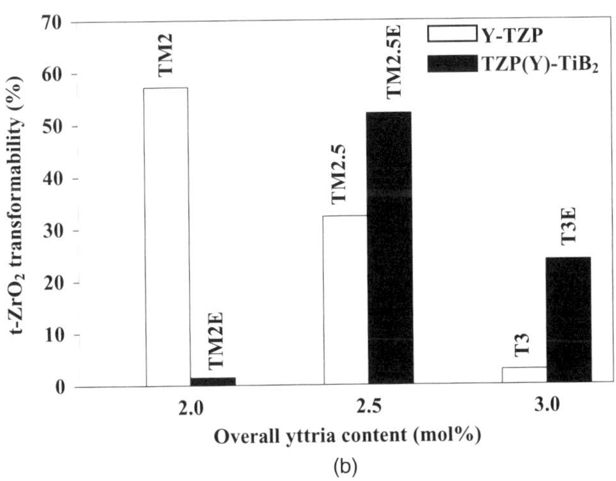

Figure 10.19 Tetragonal ZrO_2 phase transformability as determined by the difference in m-ZrO_2 content between fractured and polished surfaces (a) and the contribution of different toughening mechanisms to the overall toughness of the Y-TZP monoliths and ZrO_2–TiB_2 (70/30) composites (b). K_0 is the inherent matrix toughness, ΔK_{IcT} is the transformation toughening contribution, ΔK_{IcM} is the contribution from microcrack toughening, and ΔK_{IcD} is the contribution from crack deflection.[139]

and considering the previous estimate that crack deflection only accounts for a 15% increment in toughness of the composites, it is clear that enhanced transformation toughening contributes significantly to the overall toughness of the 3Y-TZP and 2.5Y-TZP based composites.

The overall toughness of a zirconia-toughened composite can be described by the following expression:

$$K_{Ic} = K_0 + \Delta K_{IcT} + \Delta K_{IcM} + \Delta K_{IcD}, \tag{10.15}$$

where K_0 is the inherent matrix toughness, that is, the zirconia matrix without or with negligible transformation toughening, ΔK_{IcT} is the transformation-toughening contribution, ΔK_{IcM} is the contribution from microcrack toughening, and ΔK_{IcD} is the toughening due to crack deflection by the TiB_2 phase. In the following analysis, the inherent matrix toughness, K_0 is taken as 2.5 MPa $m^{1/2}$.

In view of high transformability of the t-ZrO_2 phase in the 2Y-TZP material, the addition of TiB_2 facilitates spontaneous transformation of the t-ZrO_2 phase, resulting in microcracked 2Y-TZP–TiB_2 composites. This also indicates that microcrack toughening should be taken into account as an active toughening mechanism in these composites.

A comparison of Figure 10.19a,b reveals that the t-ZrO_2 transformability data correlate well with the transformation toughness in both the Y-TZP monoliths and composites. The higher the t-ZrO_2 transformability, the larger the transformation-toughening contribution. As is discussed in the following section, the transformation-toughening contribution toward the overall toughness of the composites is strongly dependent on the ZrO_2 matrix composition and the residual stress due to the presence of TiB_2.

10.10.1 Influence of Thermal Residual Stresses

The magnitude of the residual stress is determined by the mismatch in CTE and elastic modulus. If the residual stress is compressive, the critical stress necessary to initiate t-ZrO_2 phase transformation will increase. In contrast, the existence of tensile residual stresses will facilitate the t-ZrO_2 transformation at a much lower applied stress, since tensile stresses activate the t-ZrO_2 transformation in the crack tip stress field.[3]

Due to the lower CTE of TiB_2 ($\alpha_{300\text{–}800K} = 5 \times 10^{-6} K^{-1}$ and $\alpha_{950\text{–}2000K} = 9 \times 10^{-6} K^{-1}$)[139] compared with Y-TZP ($\alpha_{300\text{–}2000K} = 10 \times 10^{-6} K^{-1}$),[139] tensile residual stresses in the ZrO_2 matrix is produced during cooling from the hot pressing temperature. Several theoretical models to estimate residual stresses in multiphase ceramic composites have been reported in literature.[140,141] The estimated residual tensile stress in the zirconia matrix is 263 MPa. The toughness data in case of ZrO_2–TiB_2 composites reveal the influence of the tensile residual stresses in the ZrO_2 matrix on transformability. In fact, the transformation toughening of the ZrO_2 matrix is strong in the case of 3Y-TZP- and 2.5Y-TZP-based composites, contributing effectively to an increased transformation toughness, as shown in Figure 10.19.

In the presence of tensile residual stress (σ_r), the critical stress (σ_c) to initiate stress-induced transformation in the crack tip would be reduced (σ_c^m):

$$\sigma_c^m = \sigma_c - \sigma_r. \tag{10.16}$$

The reduction in critical stress results in a larger transformation zone (h) according to the formula proposed by Budiansky et al.:[134]

$$h = \frac{\sqrt{3}(1+\nu)^2}{12\pi}\left(\frac{K_\infty}{\sigma_c}\right)^2, \tag{10.17}$$

where ν is Poisson's ratio, K_∞ is the stress intensity factor due to the applied (external) stress, and σ_c is the critical transformation stress.

The increased transformation zone size can lead to enhanced transformation toughening contribution according to the established formula of Evans and Cannon[2]:

$$\Delta K_{IcT} = 0.38 fE\varepsilon^t \sqrt{h}/(1-\nu), \tag{10.18}$$

where f is the volume fraction of the t-ZrO_2 phase transformed within the transformation zone, ν is Poisson's ratio, E is the elastic modulus, and ε^t is the transformation strain.

The experimental data in Figure 10.19 reveal that the transformability of the ZrO_2 matrix should be affected in the presence of residual stress. The additional transformability due to the tensile residual stresses in a high-toughness 2Y-TZP matrix (highest tetragonal transformability among the monoliths) causes spontaneous transformation of the t-ZrO_2 matrix during postsintering cooling, resulting in the formation of microcracks and a degradation of the mechanical properties.

The influence of the residual stress is also reflected in a higher onset temperature (M_s) for the t-ZrO_2 to m-ZrO_2 transformation. The transformation data of all composites are reported in detail elsewhere.[142–153] The M_s temperature of the TM2.5E composite is measured at 330°C, whereas that of the TM2.5 monolith is below RT. The increased M_s temperature clearly reveals that the presence of TiB_2 reveals the t-ZrO_2 phase transformability in the investigated composites.

10.10.2 Influence of Zirconia Matrix Stabilization

The toughness increment ($\Delta K_{Ic}/K_m$) data presented in Figure 10.19 reveal that the TZP–TiB_2 (70/30) composites can be toughened by lowering the overall yttria content by tailoring the ZrO_2 matrix composition. In the 2–3 mol% yttria compositional window, the highest toughness is measured at an overall yttria content of 2.5 mol%, obtained by the addition of pure m-ZrO_2 to a 3 mol% Y_2O_3 coprecipitated ZrO_2 powder.

The experimental results are suggestive of the fact that overall yttria content and residual stress play a predominant role in optimizing the toughness of TZP–TiB_2 composites. It can be mentioned that yttria distribution can be tailored to develop ZrO_2–Al_2O_3, ZrO_2–ZrB_2, and ZrO_2–WC composites with high toughness up to 9 MPa m$^{1/2}$.[146–153]

10.11 OUTLOOK

The majority of the experimental work focused largely on the "grain size effect," and therefore this factor is critically considered in developing zirconia-based materials. The existing micromechanical models also considered the grain size effect. However, the yttria distribution is also focused as one of the crucial factors influencing the t-ZrO_2 transformability and toughness of the Y-TZP monoliths and composites. This observation strongly suggests that future micromechanical modeling work, in future, needs to consider this compositional aspect by establishing a mathematical framework in which transformability is linked to the distribution parameters of the yttria stabilizer and grain size.

Further work on the critical assessment of the stability of t-ZrO_2 particles is also suggested. Since the transformability of the t-ZrO_2 is governed by many parameters, a neural network algorithm can be developed to model the retained tetragonal stability. *Neural networks* are parameterized nonlinear models used for empirical regression, where conventional approaches are found to be weak. An implementation of the neural network analysis is therefore suggested to unravel the complex relationship between the various inputs (yttria content and distribution, grain size and distribution, residual stress, etc.) and outputs (t-ZrO_2 transformability, toughness). Such an analysis could be used to predict the transformation potential of t-ZrO_2 for different microstructures.

Since the yttria distribution is found to be an additional key factor controlling the toughness of Y-TZP monoliths, the distribution of other stabilizers, such as ceria, calcia, or magnesia, is also expected to play an equally significant role in optimizing the microstructure and toughness of Ce-TZP, Ca-PSZ, or Mg-PSZ, respectively. Careful experiments can be planned in future to examine this issue.

Despite the fact that considerable improvement in understanding transformation toughening has been accomplished since the 1980s, it remains an important aspect to effectively utilize transformation toughening in order to develop stronger and tougher ceramic components for structural applications. To this end, the coupling or interaction of transformation toughening with whisker and fiber reinforcement can be explored in technologically important ceramic systems and the findings critically analyzed to enhance our understanding of the interaction mechanism.

REFERENCES

1 B. Basu. Toughening of Y-stabilized tetragonal zirconia ceramics. Int. Mater. Rev. 50(4) (2005), 239–256.

2 A. G. Evans and R. M. Cannon. Toughening of brittle solids by martensitic transformations. Acta Mater. 34(5) (1986), 761–800.

3 M. Rühle and A. G. Evans. High toughness ceramic and ceramic composites. Prog. Mater. Sci. 33 (1989), 85.

4 R. C. Garvie, R. H. Hannink, and R. T. Pascoe. Ceramic steel? Nature 258 (1975), 703.

5 R. H. J. Hannink, P. M. Kelly, and B. C. Muddle. Transformation toughening in zirconia-containing ceramics. J. Am. Ceram. Soc. 83(3) (2000), 461–487.

6 (a) L. L. Hench. Bioceramics. J. Am. Ceram. Soc. 81 (1998), 1705–1728.(b) C. Piconi and G. Maccauro. Zirconia as a ceramic biomaterial. Biomaterials 20 (1999), 1-25.

7 T. E. Fischer, M. P. Anderson, and S. Jahanmir. Influence of fracture toughness on the wear reduction of yttria doped zirconium oxide. J. Am. Ceram. Soc. 72 (1989), 252–257.

8 R. H. J. Hannink and M. V. Swain. Progress in transformation toughening of ceramics. Annu. Rev. Mater. Sci. 24 (1994), 359–408.

9 I.-W. Chen. Model of transformation toughening in brittle materials. J. Am. Ceram. Soc. 74(10) (1991), 2564–2572.

10 I.-W. Chen. Model of transformation toughening in brittle materials. J. Am. Ceram. Soc. 74(10) (1991), 2564–2572.

11 J. B. Watchman. *Mechanical Properties of Ceramics: Chapter 11*. John Wiley & Sons, New York, 1996, 391–408.

12 R. M. Mcmeeking and A. G. Evans. Mechanics of transformation-toughening in brittle materials. J. Am. Ceram. Soc. 65(5) (1982), 242–246.

13 D. M. Stump and B. Budiansky. Crack-growth resistance in transformation toughened ceramics. Int. J. Solids Struct. 25 (1989), 635–646.

14 (a) P. M. Kelly and C. J. Wauchope. The tetragonal to monoclinic martensitic transformation in zirconia. Key Eng. Mater. 153–154 (1998), 97–124.(b) R. P. Ingel and D. Lewis III. Lattice parameters and density for Y_2O_3-stabilized ZrO_2. J. Am. Ceram. Soc. 69(4) (1986), 325–332. (c) I. R. Gibson and J. T. S. Irvine. Qualitative X-ray diffraction analysis of metastable tetragoal (t′) zirconia. J. Am. Ceram. Soc., 84(3) (2001), 615–18.

15 E. H. Kisi and C. J. Howard. Crystal structures of zirconia phases and their inter-relation. Key Eng. Mater. 153–154 (1998), 1–36.

16 D. P. Burke and W. M. Rainforth. Intermediate rhombohedral (r-ZrO_2) phase formation at the surface of sintered Y-TZP's. J. Mater. Sci. Lett. 16 (1997), 883–885.

17 A. H. Heuer, R. Chaim, and V. Lanteri. The displacive cubic → tetragonal transformation in ZrO_2 alloys. Acta Mater. 35(3) (1987), 661–666.

18 N. Mitra, K. Vijayan, B. N. Pramila Bai, and S. K. Biswas. Phase transformation induced by mechanical and chemical surface preparation of tetragonal zirconia polycrystals. J. Am. Ceram. Soc. 76 (1993), 533–535.

19 D. G. Lamas and N. E. W. D. Reca. Oxygen ion displacement in compositionally homogeneous, nanocrystalline ZrO_2–12 mol% Y_2O_3 powders. Mater. Lett. 41 (1999), 204–208.

20 H. G. Scott. Phase relationships in zirconia-yttria system. J. Mater. Sci. 10 (1975), 1527–1535.

21 P. Duwez, F. H. Brown Jun, and F. Odell. Zirconia-yttria system. J. Electrochem. Soc. 98 (1951), 356–362.

22 K. K. Srivastava, R. N. Patil, C. B. Choudhury, K. B. G. K. Gokhale, and E. C. Subbarao. Revised phase diagram of the system ZrO_2-$YO_{1.5}$. Trans. J. Br. Ceram. Soc. 73 (1974), 85–91.

23 T. Tojo, T. Atake, T. Mori, and H. Yamamura. Heat capacity and thermodynamic functions of zirconia and yttria-stabilised zirconia. J. Chem. Thermodyn. 31 (1999), 831–845.

24 J. Katamura, T. Seki, and T. Sakuma. The cubic-tetragonal phase equilibria in the ZrO_2-R_2O_3 (R=Y, Gd, Sm, Nd) systems. J. Phase Equilib. 16(4) (1995), 315–319.

25 (a) M. Hillert and T. Sakuma. Thermodynamic modeling of the c → t transformation in ZrO_2 alloys. Acta Mater. 39(6) (1991), 1111–1115. (b) H. M. Ondik and H. F. McMurdie (Eds.), *Phase Diagrams for Zirconia and Zirconia System.* American Ceramic Society, Westerville, OH, 1998, 35–66, 100–140, 249–277.

26 N. Claussen. *Science and Technology of Zirconia II. Advances in Ceramics*, Vol. 12, N. Claussen, M. Rühle, and A. H. Heuer (Eds.). American Ceramic Society, Columbus, OH, 1984, 325–351.

27 B. Basu, J. Vleugels, and O. Van Der Biest. Development of ZrO_2-ZrB_2 composites. J. Alloys Compd. 334(1–2) (2002), 200–204.

28 B. Basu, J. Vleugels, and O. Van Der Biest. Development of ZrO_2-TiB_2 composites: Role of residual stress and starting powders. J. Alloys Compd. 365(1–2) (2004), 266–270.

29 B. Basu, J. Vleugels, and O. Van Der Biest. ZrO_2-Al_2O_3 composites with tailored toughness. J. Alloys Compd. 372 (2004), 278–284.

30 D. J. Green, R. H. J. Hannink, and M. V. Swaini. *Transformation Toughening of Ceramics.* CRC Press, Boca Bacon, FL, 1989.

31 F. F. Lange. Transformation toughening: Part 1—Size effects associated with the thermodynamics of constrained transformations. J. Mater. Sci. 17 (1982), 225–234.

32 P. F. BECHER. Toughening behavior in ceramics associated with the transformation of tetragonal ZrO_2. Acta Metall. 34 (1986), 1885.

33 B. BUDIANSKY, J. W. HUTCHINSON, and J. C. LAMBROPOULOS. Continuum theory of dilatant transformation toughening in ceramics. Int. J. Solids Struct. 19 (1983), 337.

34 J. C. LAMBROPOULOS. Shear, shape, and orientation effect in transformation toughening. Int. J. Solids Struct. 22 (1986), 1083.

35 A. G. EVANS and A. H. HEUER. Review—Transformation toughening in ceramics: Martensitic transformations in crack-tip stress fields. J. Am. Ceram. Soc. 63 (1980), 241.

36 I. W. CHEN and P. E. ROYES MOREL. Implications of transformation plasticity in ZrO_2-containing ceramics. J. Am. Ceram. Soc. 69(3) (1986), 181–189.

37 P. E. ROYES MOREL and I. W. CHEN. Transformation plasticity of Ce_2O_3-stabilised tetragonal zirconia polycrystals: I, Stress assistance and autocatalysis. J. Am. Ceram. Soc. 71 (1988), 343.

38 (a) E. C. SUBBARAO. Zirconia—An overview, in *Advances in Ceramics 3, Science and Technology of Zirconia*, A. H. Heuer and L. W. Hobbs (Eds.). American Ceramic Society, Columbus, OH, 1981, 1–24. (b) E. C. SUBBARAO, H. S. MAITI and K. K. SRIVASTAVA. Martensitic transformation in zirconia. Phys. Status Solidi A 21(9) (1974), 9–40.

39 D.-J. KIM. Effect of Ta_2O_5, Nb_2O_5, and HfO_2 alloying on the transformability of Y_2O_3-stabilised tetragonal ZrO_2. J. Am. Ceram. Soc. 73(1) (1990), 115–120.

40 M. MORINAGA, H. ADACHI, and M. TSUKADA. Electronic structure and phase stability of ZrO_2. J. Phys. Chem. Solids 44(4) (1983), 301–306.

41 M. YOSHIMURA. Phase stability of zirconia. Am. Ceram. Soc. Bull. 67(12) (1988), 1950–1955.

42 M. HILLERT. Thermodynamic model of the cubic–tetragonal transition in nanostoichiometric zirconia. J. Am. Ceram. Soc. 74(8) (1991), 2005–2006.

43 S. P. S. BADWAL and N. NARDELLA. Formation of monoclinic zirconia at the anodic face of tetragonal zirconia polycrystalline solid electrolytes. Appl. Phys. A49 (1989), 13–24.

44 P. KOUNTOUROS and G. PETZOW. *Advances in Ceramics 3, Science and Technology of Zirconia*, A. H. Heuer and L. W. Hobbs (Eds.). American Ceramic Society, Columbus, OH, 1981, 202–216.

45 R. C. GARVIE. Thermodynamic analysis of the tetragonal to monoclinic transformation in a constrained zirconia monocrystal: Part 2: In the presence of an applied stress. J. Mater. Sci. 20 (1985), 3479–3486.

46 M. M. R. BOUTZ, R. J. M. OLDE SCHOLTENHUIS, A. J. A. WINNUBST, and A. J. BURGGRAAF. *British Ceramic Proceedings: Vol. 51: Nanoceramics*, R. Freer (Eds.). The Institute of Materials, London, 1993, 75–86.

47 A. H. HEUER, N. CLAUSSEN, W. M. KRIVEN, and M. RÜHLE. Stability of tetragonal ZrO_2 particles in ceramic matrices. J. Am. Ceram. Soc. 65(12) (1982), 642–650.

48 A. J. A. WINNUBST and M. M. R. BOUTZ. Sintering and densification: New techniques: Sinter forging. Key Eng. Mater. 153–154 (1998), 301.

49 M. J. MAYO. Processing of nanocrystalline ceramics from ultrafine powders. Int. Mater. Rev. 41(3) (1996), 85–115.

50 D. BURGARD, C. KROPF, R. NASS, and H. SCHMIDT. *MRS Symposium Proceedings*, Vol. 346, A. K. Cheetham, C. J. Brinker, M. L. Mecartney, and C. Sanchez (Eds.). The Materials Research Society, Pittsburg, CA, 1994, 101–107.

51 J. TARTAJ, J. F. FERNANDEZ, C. MOURE, and P. DURAN. Effects of seeding on the crystallisation kinetics of air-calcined yttria-doped hydrous zirconia. J. Eur. Ceram. Soc. 18 (1998), 229.

52 G. DELL'AGLI and G. MASCOLO. Hydrothermal synthesis of ZrO_2–Y_2O_3 solid solutions at low temperature. J. Eur. Ceram. Soc. 20 (2002), 139–145.

53 R. SINGH, C. GILL, S. LAWSON, and G. P. DRANSFIELD. Sintering, microstructure and mechanical properties of commercial Y-TZPs. J. Mater. Sci. 31 (1996), 6055.

54 W. BURGER, H. G. RICHTER, C. PICONI, R. VATTERONI, A. CITTADINI, and M. BOCCALARI. New Y-TZP powders for medical grade zirconia. J. Mater. Sci. Mater. Med. 8 (1997), 113–118.

55 J. VLEUGES, Z. X. YUAN, and O. VAN DER BIEST. Mechanical properties of Y_2O_3/Al_2O_3-coated Y-TZP ceramics. J. Eur. Ceram. Soc. 22(6) (2002), 873–881.

56 Z. X. YUAN, J. VLEUGES, and O. VAN DER BIEST. Preparation of Y_2O_3-coated ZrO_2 powder by suspension drying. J. Mater. Sci. Lett. 73(19) (2000), 359–361.

57 Y. K. PAEK, J. H. AHN, G. H. KIM, and S. J. L. KANG. Effect of nitrogen atmosphere on the densification of a 3-mol%-yttria-doped zirconia. J. Am. Ceram. Soc. 85(6) (2002), 1631–1633.
58 C. ZHAO, J. VLEUGELS, O. VAN DER BIEST, C. GROFFILS, and C. P. J. LUYPAERT. Hybrid sintering with a tubular susceptor in a cylindrical singular mode microwave furnace. Acta Mater. 48(14) (2000), 3795–3801.
59 W. LI and L. GAO. Rapid sintering of nanocrystalline ZrO_2(3Y) by spark plasma sintering. J. Eur. Ceram. Soc. 20 (2000), 2441.
60 B. BASU, J.-H. LEE, and D.-Y. KIM. Development of nanocrystalline wear resistant Y-TZP ceramics. J. Am. Ceram. Soc. 87(9) (2004), 1771–1774.
61 M. YOSHIMURA, T. OHJI, M. SANDO, and K. NIIHARA. Rapid rate sintering of nano-grained ZrO_2-based composites using pulse electric current sintering method. J. Mater. Sci. Lett. 17 (1998), 1389–1391.
62 M. YOSHIMURA, T. OHJI, M. SANDO, Y.-H. CHOA, T. SEKINO, and K. NIIHARA. Synthesis of nanograined ZrO_2-based composites by chemical processing and pulse electric current sintering. Mater. Lett. 38 (1999), 18–21.
63 B. LAWN. *Fracture of Brittle Solids*, 2nd ed. Cambridge Univ. Press, Cambridge, UK, 1992.
64 R. H. J. HANNINK. Significance of microstructure in transformation toughening zirconia ceramics. Mater. Forum 11 (1988), 43–60.
65 G. R. ANSTIS, P. CHANTIKUL, B. R. LAWN, and D. B. MARSHALL. A critical evaluation of indentation techniques for measuring fracture toughness: I, Direct crack measurements. J. Am. Ceram. Soc. 64 (1981), 533–538.
66 M. S. KALISZEWSKI, G. BEHRENS, A. H. HEUER, M. C. SHAW, D. B. MARSHALL, G. W. DRANSMANN, and R. W. STEINBRECH. Indentation dtudies on Y_2O_3-stabilized ZrO_2: I, Development of indentation-induced cracks. J. Am. Ceram. Soc. 77(5) (1994), 1185–1193.
67 K. NIIHARA, R. MORENA and D. P. H. HASSELMAN. Evaluation of K_{IC} of brittle solids by the indentation method with low crack-to-indent ratios. J. Mater. Sci. Lett. 1 (1982), 13–16.
68 D. K. SHETTY, I. G. WRIGHT, P. N. MINCER, and A. H. CLAUER. Indentation fracture of WC-Co cermets. J. Mater. Sci. 20 (1985), 1873–1882.
69 P. CHANTIKUL, G. R. ANSTIS, B. R. LAWN, and D. B. MARSHALL. A critical evaluation of indentation techniques for measuring fracture toughness: II, Strength method. J. Am. Ceram. Soc. 64(9) (1981), 539–543.
70 R. F. COOK and B. R. LAWN. A modified indentation toughness technique. J. Am. Ceram. Soc. 66(11) (1983), 200–220.
71 R. C. GARVIE and M. F. GOSS. Intrinsic size dependence of the phase transformation temperature in zirconia microcrystals. J. Mater. Sci. 21 (1986), 1253–1257.
72 W. C. JEONG, D. K. MATLOCK, and G. KRAUSS. Mater. Sci. Eng. A165 (1993), 1.
73 E. GIRAULT. Bainitic transformation in TRIP-assisted steels and its influence on mechanical properties. PhD thesis, KU Leuven, Belgium, March, 1999.
74 A. G. EVANS, N. BURLINGAME, M. DRORY, and W. M. KRIVEN. Martensitic transformations in Zirconia—Particle size effects and toughening. Acta Metall. 29 (1981), 447–456.
75 J. WANG, W. M. RAINFORTH, T. WADSWORTH, and R. STEVENS. The effects of notch width on the SENB toughness for oxide ceramics. J. Eur. Ceram. Soc. 10 (1992), 21–31.
76 L. RUIZ and M. J. READY. Effect of heat treatment on grain size, phase assemblage, and mechanical properties of 3 mol% Y-TZP. J. Am. Ceram. Soc. 79(9) (1996), 2331–2340.
77 M. V. SWAIN. Grain-size dependence of toughness and transformability of 2 mol% Y-TZP. J. Mater. Sci. Lett. 5 (1996), 1159–1162.
78 F. F. LANGE. Transformation-toughened ZrO_2: Correlations between grain size control and composition in the system ZrO_2-Y_2O_3. J. Am. Ceram. Soc. 69(3) (1986), 240–242.
79 H. KONDO, T. SEKINO, T. KUSUNOSE, T. NAKAYAMA, Y. YAMAMOTO, M. WADA, T. ADACHI, and K. NIIHARA. Solid-solution effects of a small amount of nickel oxide addition on phase stability and mechanical properties of yttria-stabilized tetragonal zirconia polycrystals J. Am. Ceram. Soc. 86(3) (2003), 523–525.
80 C. ZHAO. Functionally graded Ce-ZrO_2 based composites. PhD thesis, Katholieke Universiteit Leuven, Belgium, 1999.

81 M. L. MECARTNEY. Influence of an amorphous second phase on the properties of yttria-stabilised tetragonal zirconia polycrystals (Y-TZP). J. Am. Ceram. Soc. 70(1) (1987), 54–58.

82 T. MASAKI. Mechanical properties of Y-PSZ after aging at low temperature. Int. J. High Tech. Ceram. 2 (1986), 85–98.

83 S. STEMMER, J. VLEUGELS, and O. VAN DER BIEST. Grain boundary segregation in high-purity, yttria-stabilized tetragonal zirconia polycrystals (Y-TZP). J. Eur. Ceram. Soc. 18 (1998), 1565–1570.

84 H. TSUBAKINO, T. FUJIWARA, K. SATANI, and S. IOKU. Composition of grain boundary phase formed in zirconia-3 mol% yttria containing alumina J. Mater. Sci. Lett. 16 (1997), 1472–1475.

85 K. MORITA, K. HIRAGA, and B.-N. KIM. Effect of minor SiO_2 addition on the creep behavior of superplastic tetragonal ZrO_2 Acta Mater. 52 (2004), 3355–3364.

86 B. BASU. Zirconia-Titanium boride composites for tribological applications. PhD thesis, K. U. Leuven, Belgium, 2001.

87 S. RAMESH, C. GILL, and S. LAWSON. The effect of copper oxide on sintering, microstructure, mechancial properties and hydrothermal aging of coated 2.5Y-TZP ceramics. J. Mater. Sci. 34 (1999), 5457–5467.

88 R. M. GERMAN. *Liquid Phase Sintering*. Plenum Press, New York, 1985, 43.

89 S. LAWSON. The sintering, mechanical and aging properties of transition metal-oxide doped yttria-tetragonal zirconia polycrystals (Y-TZP). PhD thesis, University of Sunderland, UK, 1993.

90 H. SCHUBERT and G. PETZOW. Microstructural investigations on the stability of yttria-stabilised tetragonal zirconia; in Science and Technology of Zirconia III Edited by S. Somiya, N. Yamamoto and H. Yanagida. Adv. Ceram. 24A (1988), 21–28.

91 K. TSUKUMA, Y. KUBOTA, and T. TSUKIDATE. *Science and Technology of Zirconia II. Advances in Ceramics*, Vol. 12, N. Claussen, M. Rühle, and A. H. Heuer (Eds.). American Ceramic Society, Columbus, OH, 1984.

92 M. MATSUI, T. SOMA, and I. ODA. Stress-induced transformation and plastic deformation for Y_2O_3-containing zirconia polycrystals. J. Am. Ceram. Soc. 69(3) (1986), 198–203.

93 T. SAKUMA, H. EDA, and H. SATO. Science and technology of Zirconia III edited by S. Somiya, N. Yamamoto and H. Yanagida. Adv. Ceram. 24A (1988), 357–363.

94 F. F. LANGE and D. J. GREEN. *Science and Technology of Zirconia V*, S. P. S. Badwal, M. J. Bannister, and R. H. J. Hannink (Eds.), American Ceramic Society, Columbus, OH, 1981, 217–225.

95 T. MASAKI and K. SINJO. Mechanical properties of highly toughened ZrO_2-Y_2O_3. Ceram. Int. 13 (1987), 109–112.

96 L. GAO, T. S. YEN, and J. K. GUO. Science and technology of Zirconia III edited by S. Somiya, N. Yamamoto and H. Yanagida. Adv. Ceram. 24A (1988), 405–414.

97 O. VASYLKIV, Y. SAKKA, and V. V. SKOROKHOD. Low-temperature processing and mechanical properties of zirconia and zirconia–alumina nanoceramics. J. Am. Ceram. Soc. 86(2) (2003), 299–304.

98 T. C. LEI, G. Y. LIN, Q. L. GE, Y. ZHOU, and X. J. HE. Morphologies of monoclinic phase in ZrO_2 (2 mol% Y_2O_3) revealed by TEM in situ continuous observations. J. Mater. Sci. 32 (1997), 1105–1111.

99 B. BASU, J. VLEUGELS, and O. VAN DER BIEST. Toughness tailoring of yttria-doped zirconia ceramics. Mater. Sci. Eng. A 380 (2004), 215–221.

100 B. BASU, J. VLEUGELS, and O. VAN DER BIEST. Transformation behaviour of yttria stabilized tetragonal zirconia polycrystal-TiB_2 composites. J. Mater. Res. 16(7) (2001), 2158–2169.

101 B. BASU, J. VLEUGELS, and O. VAN DER BIEST. Y-TZP ceramics with tailored toughness. Key Eng. Mater. 206–213 (2002), 1185–1188.

102 B. BASU, J. VLEUGELS, and O. VAN DER BIEST. Microstructure-toughness-wear relationship of tetragonal zirconia ceramics. J. Eur. Ceram. Soc. 24(7) (2004), 2031–2040.

103 (a) B. BASU, J. VLEUGELS, and O. VAN DER BIEST. Transformation behaviour of tetragonal zirconia: Role of dopant content and distribution. Mater. Sci. Eng. A 366(2) (2004), 338–347. (b) B. BASU. Zirconia-titanium boride composites for tribological applications. Ph.D thesis, Kuleuven, Belgium, March 2001.

104 P. LIANG, H. L. LUKAS, S. G. FRIES, H. J. SEIFERT, and F. ALDINGER. Thermodynamic assessment of the Zr-O binary system. Proceedings of the 28th CALPHAD held at Grenoble, France, 2–7 May, 1999.

105 N. GUPTA, P. MALLIK, and B. BASU. Y-TZP ceramics with optimised toughness: New results. J. Alloys Compd. 379 (2004), 228–232.

106 C. R. BOWEN, A. W. TAVERNOR, J. LUO, and R. STEVENS. Microstructural design of sensor materials using the core-shell concept. J. Eur. Ceram. Soc. 19 (1999), 149–157.

107 A. SURESH, M. J. MAYO, W. D. PORTER, and C. J. RAWN. Crystallite and grain-size-dependent phase transformations in yttria-doped zirconia. J. Am. Ceram. Soc. 86(2) (2003), 360–362.

108 P. F. BECHER and M. V. SWAIN. Grain-size-dependent transformation behavior in polycrystalline tetragonal zirconia. J. Am. Ceram. Soc. 75(3) (1992), 493–502.

109 W. Z. ZHU and M. YAN. Effect of grain size on the kinetics of isothermal tetragonal to monoclinic transformation in ZrO_2 (2 mol%). J. Eur. Ceram. Soc. 17 (1997), 1729–1739.

110 P. F. BECHER, M. V. SWAIN, and M. K. FERBER. Relation of transformation temperature to the fracture toughness of transformation toughened ceramics. J. Mater. Sci. 22 (1987), 76–84.

111 L. R. F. ROSE. The mechanics of transformation toughening. Proc. R. Soc. Lond. A 412 (1987), 169–197.

112 A. G. EVANS and A. H. HEUER. Review—Transformation toughening in ceramics: Martensitic transformations in crack-tip stress fields. J. Am. Ceram. Soc. 63 (1980), 241–248.

113 J.-D. LIN and J.-G. DUH. The use of X-ray line profile analysis to investigate crystallite size and microstrain for zirconia powders. J. Mater. Sci. 32 (1997), 4901–4908.

114 D. L. PORTER and A. H. HEUER. Mechanisms of toughening partially stabilised zirconia (PSZ). J. Am. Ceram. Soc. 60(3–4) (1977), 183–184.

115 D. R. CLARKE and F. ADAR. Measurement of the crystallographically transformed zone produced by fracture in ceramics containing tetragonal zirconia. J. Am. Ceram. Soc. 65(6) (1982), 284–288.

116 A. TIEFENBACH, S. WAGNER, R. OBERACKER, and B. HOFFMANN. The use of impedence spectroscopy in damage detection in tetragonal zirconia polycrystals (TZP). Ceram. Int. 26 (2000), 745–751.

117 H. TSUBAKINO, Y. KURODA, and M. NIIBE. Surface relief associated with isothermal martensites in zirconia-3 mol% yttria ceramics observed by Atomic Force Microscopy. J. Am. Ceram. Soc. 82 (1999), 2921.

118 S. SCHMAUDER and H. SCHUBERT. Significance of internal stresses for the martensitic transformation in yttria-stabilised tetragonal zirconia polycrystals during degradation. J. Am. Ceram. Soc. 69(7) (1986), 534–540.

119 H. SCHUBERT. Aniostropic thermal expansion coefficients of Y_2O_3-stabilised tetragonal zirconia. J. Am. Ceram. Soc. 69(3) (1986), 270–271.

120 J. W. HUTCHINSON. Crack tip shielding by micro-cracking in brittle solids. Acta Metall. 35(7) (1987), 1605–1619.

121 A. KRELL, A. TERESIAK, and D. SCHLÄFER. Grain size dependent residual microstresses in submicron Al_2O_3 and ZrO_2. J. Eur. Ceram. Soc. 16 (1996), 803–811.

122 D. L. PORTER and A. H. HEUER. *Science and Technology of Zirconia II, Advances in Ceramics*, Vol. 12, N. Claussen, M. Rühle, and A. H. Heuer (Eds.). American Ceramic Society, Columbus, OH, 1984, 653–662.

123 J. W. HUTCHINSON. Crack tip shielding by micro-cracking in brittle solids. Acta Metall. 35(7) (1987), 1605–1619.

124 G. V. SRINIVASAN, J.-F. JUE, S.-Y. KUO, and A. V. VIRKAR. Ferroelastic domain switching in polydomain tetragonal zirconia polycrystals. J. Am. Ceram. Soc. 72(11) (1989), 2098–2103.

125 A. V. VIRKAR. Role of ferroelasticity in toughening of zirconia ceramics. Key Eng. Mater. 153–154 (1998), 183–210.

126 E. H. KISI, S. J. KENNEDY, and C. J. HOWARD. Neutron diffraction observations of ferroelastic domain switching and tetragonal-to-monoclinic transformation in Ce-TZP. J. Am. Ceram. Soc. 80(3) (1997), 621–628.

127 A. V. VIRKAR and R. L. K. MATSUMOTO. Ferroelastic domain switching as a toughening mechanism in tetragonal zirconia. J. Am. Ceram. Soc. 69(10) (1986), C-224–C-226.

128 G. A. GOGOTSI, V. P. ZAVADA, and M. V. SWAIN. Mechanical property characterisation of 9 mol% Ce-TZP ceramic material-II. Fracture toughness. J. Eur. Ceram. Soc. 16 (1996), 545–551.

129 C. J. CHAN, F. F. LANGE, M. RÜHLE, J.-F. JUE, and A. V. VIRKAR. Ferroelastic domain switching in tetragonal zirconia single crystals—Microstructural aspects. J. Am. Ceram. Soc. 74(4) (1991), 807–813.

130 L. S. PAN and S. HORIBE. Anelastic behavior of zirconia ceramics under monotonic and cyclic loadings. Acta Mater. 45(2) (1997), 463–469.

131 A. V. Virkar and R. L. K. Matsumoto. Toughening mechanism in tetragonal zirconia polycrystalline (TZP) ceramics, in *Advance in Ceramics, Vol. 24B, Science and Technology of Zirconia III*, S. Somiya, N. Yamamoto, and H. Yanagide (Eds.), American Ceramic Society, Westerville, OH, 1998, 653–662.

132 N. Claussen and G. Petzow. In *Tailoring of Multiphase and Composite Ceramics, Vol. 20, MRS*, R. E. Tressler, G. L. Messing, C. G. Pantano, and R. E. Newnham (Eds.). Plenum, New York, 1986.

133 Y. L. Cui. Interaction of fiber and transformation toughening. J. Mech. Phys. Solids 40 (1992), 1837–1850.

134 Y. L. Cui and B. Budiansky. Steady-state matrix cracking of ceramics reinforced by aligned fibers and transforming particles. J. Mech. Phys. Solids 41(4) (1993), 615–630.

135 P. F. Becher and T. N. Tiegs. Toughening behavior involving multiple mechanisms: Whisker reinforcement and zirconia toughening. J. Am. Ceram. Soc. 70(9) (1987), 651–654.

136 W.-H. Tuan and R.-Z. Chen. Interactions between toughening mechanisms: Transformation toughening versus plastic deformation. J. Mater. Res. 17(11) (2002), 2921–2928.

137 B. Basu, J. Vleugels, and O. Van der Biest. Toughness tailoring of yttria-doped zirconia ceramics. Mater. Sci. Eng. A380 (2004), 215–221.

138 K. T. Faber and A. G. Evans. Crack deflection processes-II. Experiment. Acta Metall. 31 (1983), 565–576.

139 B. Basu, J. Vleugels, and O. Van Der Biest. Processing and Mechanical properties of ZrO_2-TiB_2 composites. J. Eur. Ceram. Soc. 25 (2005), 3629–3637.

140 G. A. Gogotsi, A. V. Drozdov, V. P. Zavata, and M. V. Swain. Comparison of the mechanical behaviour of zirconia partially stabilized with yttria and magnesia. J. Aust. Ceram. Soc. 27 (1991), 37–49.

141 J. Wang, W. M. Rainforth, T. Wadsworth, and R. Stevens. The effects of notch width on the SENB toughness for oxide ceramics. J. Eur. Ceram. Soc. 10 (1992), 21–31.

142 M. Taya, S. Hayashi, A. S. Kobayashi, and H. S. Yoon. Toughening of a particulate-reinforced ceramic-matrix composite by thermal residual stress. J. Am. Ceram. Soc. 73(5) (1990), 1382–1391.

143 G. A. Gogotsi, E. E. Lomonva, and V. G. Pejchev. Strength and fracture toughness of zirconia crystals. J. Eur. Ceram. Soc. 11 (1993), 123–132.

144 G. W. Dransman, R. W. Steinbrech, A. Pajares, F. Guiberteau, A. D. Rodriguez, and A. H. Heuer. Strength and fracture toughness of zirconia crystals. J. Am. Ceram. Soc. 77(5) (1994), 1194–1201.

145 G. D. Quinn, R. J. Gettings, and J. J. Kubler. Indentation studies on Y_2O_3-stabilized ZrO_2: II, Toughness determination from stable growth of indentation-induced cracks. Ceram. Eng. Sci. Proc. 15 (1994), 252–257.

146 S. Lawson, K. Sing Tan, H. Gill, and P. Dransfield. *Sintering Technology*, R. M. German, G. L. Messing, and R. G. Cornwall (Eds.). Marcel and Dekker, New York, 1996, 473–480.

147 T.-J. Chung, H. Song, G.-H. Kim, and D.-Y. Kim. Microstructure and phase stability of yttria-doped tetragonal zirconia polycrystals heat treated in nitrogen atmosphere. J. Am. Ceram. Soc. 80(10) (1997), 2607–2612.

148 Z. Pedzich, K. Haberko, J. Piekarczyk, M. Farina, and L. Litynska. Zirconia matrix—tungsten carbide particulate composites manufactured by hot-pressing technique. Mater. Lett. 36 (1998), 70–75.

149 H. Kondo, T. Sekino, T. Kusunose, T. Nakayama, Y. Yamamoto, M. Wada, T. Adachi, and K. Niihara. Solid-solution effects of a small amount of nickel oxide addition on phase stability and mechanical properties of yttria-stabilized tetragonal zirconia polycrystals. J. Am. Ceram. Soc. 86(3) (2003), 523–525.

150 *Engineered Materials Handbook, Vol. 4, Ceramics and Glasses, ASM International*. The Materials Information Society, Materials Park, OH, 1991.

151 J. Eicher, U. Eisele, and J. Rodel. Mechanical properties of monoclinic zirconia. J. Am. Ceram. Soc. 87(7) (2004), 1401–1403.

152 P. F. Becher. Microstructural design of toughened ceramics. J. Am. Ceram. Soc. 74(2) (1991), 255–269.

153 A. H. De Aza, J. Chevalier, and G. Fantozzi. Slow-crack-growth behavior of zirconia-toughened alumina ceramics processed by different methods. J. Am. Ceram. Soc. 86(1) (2003), 115–120.

Chapter 11

S-Phase SiAlON Ceramics: Microstructure and Properties

In this chapter, the microstructure and properties of the newly developed nearly monophasic S-SiAlON ceramics, based on the composition of $Ba_2Si_{12-x}Al_xO_{2+x}N_{16-x}$ ($x = 2 \pm 0.2$) are discussed. It is shown that the sintering mechanism is based on liquid-phase sintering with formation of a Ba–Al silicate liquid (<5%) at intergranular pockets. The formation mechanism of the elongated-platelet morphology of S-phase SiAlON will also be discussed. Another focus of this chapter is to analyze and interpret the mechanical properties of hot-pressed Ba-doped S-SiAlON ceramics. Crack deflection by elongated S-phase grains in combination with crack bridging by β-Si_3N_4 needles has contributed to the observed high toughness.

11.1 INTRODUCTION

In view of moderately good toughness and high-temperature strength, silicon nitride (Si_3N_4)–based ceramics have been investigated as the potential candidate for various engineering applications, such as cutting tool inserts, valve seals, sealing rings, and cylinder liners, as well as for a variety of structural components in high-efficiency engines and other mechanical systems.[1] Among silicon nitride–based ceramics, SiAlON ceramics, in particular, have been developed for structural applications, because of their easier processability and high oxidation resistance.[1–8] In the ceramics literature, the solid solutions of α-Si_3N_4 and β-Si_3N_4 with Al and O are widely known as α-SiAlON and β-SiAlON, respectively. Of these two variants, α-SiAlON ceramics are characterized by a nearly single-phase microstructure with better thermal shock resistance, chemical stability, and high hardness over a large temperature range. This, coupled with lower density ($3.2\,gm/cm^3$), make these materials a potential candidate for bearing applications. However, the lower strength and fracture toughness remain bottlenecks for their successful application. Among many approaches attempted so far, "single-phase *in situ* toughened α-SiAlON" has been reported to exhibit an optimized combination of hardness and fracture toughness.[9–14]

Advanced Structural Ceramics, First Edition. Bikramjit Basu, Kantesh Balani.

Interestingly, a 2007 paper reported good cytocompatibility of rare-earth oxide-doped SiAlON ceramics, indicating potential biomedical application of SiAlON materials.[7] It needs to be mentioned here that an optimal combination of hardness, toughness, elastic modulus, and strength is desired for load-bearing implants.

Many of the engineering applications of β'/α' ceramics remain in the lower-temperature regime (generally less than 1000°C) and a greater market penetration is restricted largely by economic factors. From this perspective a continuing quest exists for novel SiAlON ceramics that could be sintered at lower temperatures and might be more tolerant of less-expensive starting powders. In M–Si–Al–O–N systems (where M is a cation in Group II of the periodic table), many phases have comparatively high oxygen-to-nitrogen ratios, with reduced covalency and intrinsic properties, and the phases with high nitrogen content are difficult to obtain with controlled phase content.[15,16] A relatively new SiAlON phase is the "S phase," having the composition $M_2Al_xSi_{12-x}N_{16-x}O_{2+x}$ with $x \approx 2$.[17,18] The Ba-containing S-phase ceramic is reported in this chapter with an aim to illustrate the main features of microstructural evolution, because it can be readily sintered to the nearly phase-pure state and it is less susceptible to evaporative loss of the stabilizing cation. A major focus is to discuss the relationship between microstructure and mechanical properties of S-SiAlON ceramics.

11.2 MATERIALS PROCESSING AND PROPERTY MEASUREMENTS

To sinter S-SiAlON ceramics, high purity (>99%) commercial powders were used: $BaCO_3$ (Aldrich), Si_3N_4 (Ube grade SNE10), AlN (Starck grade C), and Al_2O_3 (Alcoa). The precursor powders with a targeted S-phase composition of $BaAlSi_5O_2N_7$ were ball-milled in Si_3N_4 media for 24 hours and the dried powder mixture was hot pressed in BN-coated graphite dies in the temperature range 1600–1750°C for 2 hours in nitrogen environment.

The density, measured by the Archimedean method, was >97% of the theoretical density for all sintering temperatures; the microstructure was examined using x-ray diffraction (XRD) and a scanning electron microscope (SEM) and transmission electron microscope (TEM) equipped with energy-dispersive x-ray analysis spectrometers (SEM-EDS and TEM-EDS).

For all the hot-pressed ceramics the elastic modulus was measured using an ultrasonic pulse-echo technique. The hardness and indentation fracture toughness of S-SiAlON materials were determined with a Vickers hardness tester. It should be noted here that the use of microhardness measurements would cause a large variation in the observed hardness property of S-SiAlON materials (depending on the location of indents). In fact, Krell observed that the use of microloads needs to be avoided for determining hardness properties of various structural ceramics with complex microstructures.[19]

The indentation fracture toughness K_{IC} (MPa·m$^{1/2}$) was estimated using the formula proposed by Niihara et al.[20] for median cracks ($l/a \geq 1.5$):

$$K_{IC} = 0.0667(l+a)^{-3/2} \times P \times (E/H)^{2/5}. \quad (11.1a)$$

However, for the Palmqvist type of crack ($0.25 \leq l/a \leq 2.5$), a different formulation was used[21]:

$$K_{IC} = (0.0181/\sqrt{l})(P/a)(E/H)^{2/5}, \quad (11.1b)$$

where P is indentation load (N), $2a$ is the average indent diagonal length (μm), $2c$ is the crack length (from one crack tip to another), l is the difference of c and a (μm), E is the elastic modulus (GPa), and H is the hardness (GPa).

11.3 MICROSTRUCTURAL DEVELOPMENT

The properties of the hot-pressed S-SiAlON ceramics are presented in Table 11.1. Note that the highest sinter density (3.65 g/cm^3) was measured after hot pressing at 1750°C for 2 hours. While theoretical density (~3.6 g/cm^3) is achieved at relatively lower hot-pressing conditions (1600°C, 2 hours), the XRD traces (Fig. 11.1) reveal

Table 11.1. Summary of the Indentation Data—Crack Length (l), Indent Diagonal Length ($2a$)—for the Hot-Pressed S-SiAlON Ceramics[41]

Load	100 N			200 N			300 N		
Material	l (μm)	a (μm)	l/a	l (μm)	a (μm)	l/a	l (μm)	a (μm)	l/a
S1600	–	61.0	–	275.0	81.0	3.4	376.0	93.6	4.0
S1700	135.0	55.5	2.4	201.0	83.5	2.4	206.0	94.5	2.2
S1750	84.0	57.0	1.5	137.0	81.0	1.7	191.0	95.5	2.0

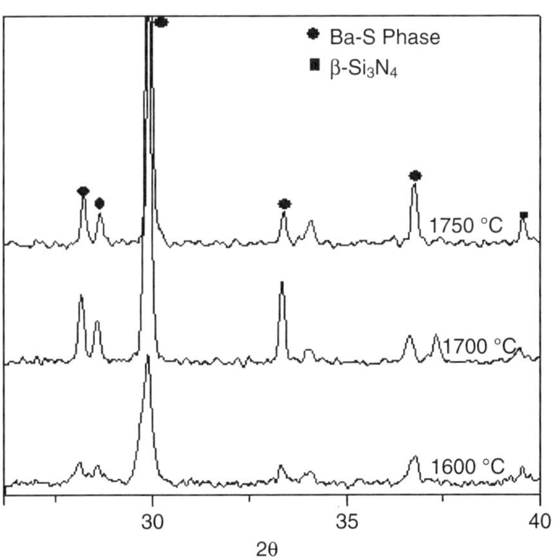

Figure 11.1 XRD spectra showing S-phase crystallization over a range of hot-pressing temperature. The sintering time at each temperature is 2 hours.[40]

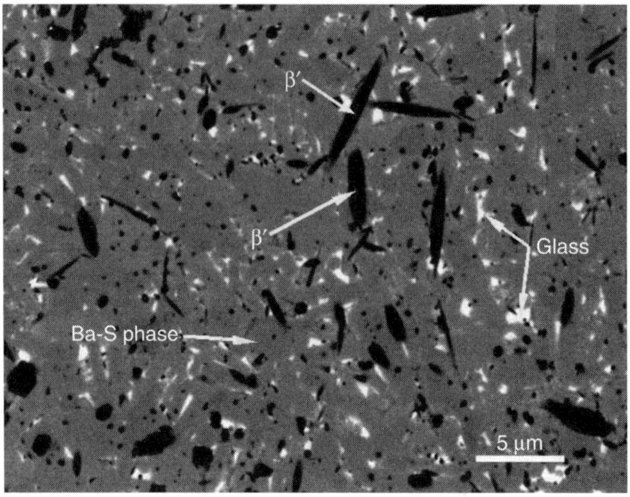

Figure 11.2 Backscattered electron (SEM) image showing minor phases (β′ and residual glass) in dark and light contrast relative to the major S-phase in an S-SiAlON ceramic, hot pressed at 1700°C for 2 hours.[40]

that the solution–reprecipitation reaction for S-phase recrystallization remains incomplete. XRD spectra also show a trace amount of β′-SiAlON and the partial crystallization of residual liquid to $BaAl_2SiO_6$. The acicular β′-phase and the residual silicate glass phase in an S-SiAlON ceramic (hot pressed at 1700°C for 2 hours) can be clearly distinguished in backscattered electron (BSE) SEM images (Fig. 11.2) in darker and lighter contrast, respectively. The S-phase has an intermediate Ba content, which results in the atomic number contrast. Careful phase analysis (semiquantitative) using SEM-EDS reveals that the hot-pressed microstructure contains (4–5%) residual glass and (6–8%) α/β-Si_3N_4, besides the characteristic S-phase.

To investigate the important features of the microstructural evolution, an S-SiAlON ceramic, hot pressed at 1700°C, was selected for detailed TEM analysis. As shown in Figures 11.3 and 11.4, bright field TEM images display a microstructure dominated by a contiguous array of elongated platelet crystals of S-phase with a minor intercrystalline liquid residue at triple-junction channels. The presence of fine β′ crystals within the S-phase are suggestive that they are formed early via sintering reaction and, subsequently, act as heterogeneous nucleants for the S-phase. The aspect ratios (ARs) of S-phase grains lie in the range 3.1–4.3, with an average value of 3.4. The width of S-phase grains (minor axis) varies in the range 0.6–1.4 μm, while the length (major axis) varies in the range 1–5 μm. The average width and AR of β-Si_3N_4 needles were estimated to be 0.57 μm and 7, respectively. In Figure 11.2, a typical acicular β-Si_3N_4 phase, with an AR of 7–8, is shown.

The analysis of the selected area diffraction patterns (SADPs) reveals that the S-phase crystals exhibit a preferred growth in the [001] orthorhombic axis with primary facets parallel to (010) and (100) planes (Fig. 11.3). TEM-EDS analysis

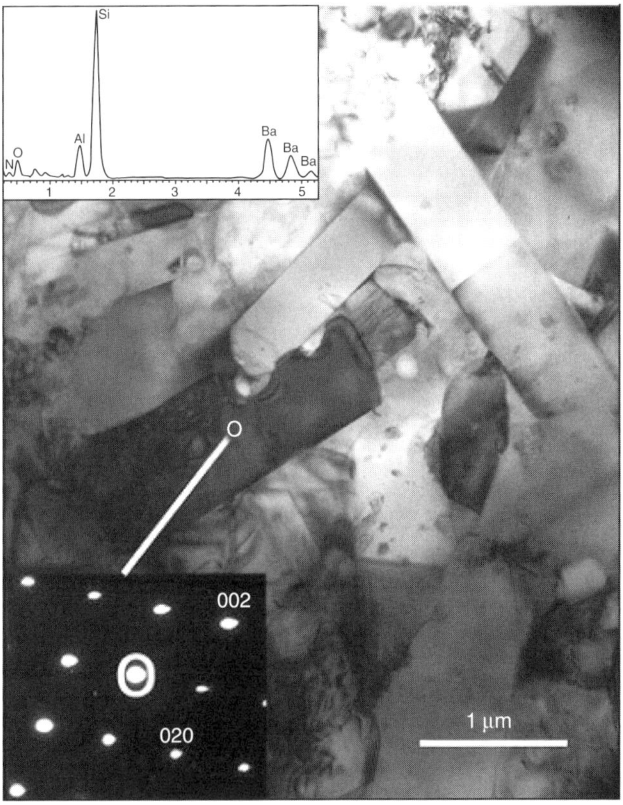

Figure 11.3 Bright field TEM image illustrating the elongated platelet morphology of S-phase in an S-SiAlON ceramic, hot pressed at 1700°C for 2 hours. A selected area diffraction pattern (SADP) and an energy dispersive X-ray (EDX) spectrum taken from the S-phase are also shown in insets.[40]

confirms an average composition for S-phase as $Ba_2Si_{12-x}Al_xO_{2+x}N_{16-x}$ ($x = 2.0 \pm 0.2$), but hot pressings with larger variations in x in the initial powder mixture are possible. The presence of the residual glass phase can be clearly observed in Figure 11.4a. TEM-EDS analysis of the glass, as shown in Figure 11.4b, reveals that the glass phase is Ba-rich aluminosilicate. Based on our analysis, the possible composition range of the triple-pocket glass is shown in Figure 11.5.

Many of the S-phase crystals are characterized by stacking-faults, imaged with characteristic fringe contrast, which have a weak crystallographic preference (Fig. 11.4). It is possible that the stacking-faults occur during the growth process and such defects terminate either on partial dislocations or on the crystal surface. Some representative bright field TEM images revealing the defect structure in the S-SiAlON ceramic are provided in Figures 11.6 and 11.7. The network of partial dislocations, appearing in fringes, is observed in Figure 11.6. In many of the investigated S-SiAlON grains, closely spaced antiphase boundaries (APBs) can be observed (Fig. 11.7).

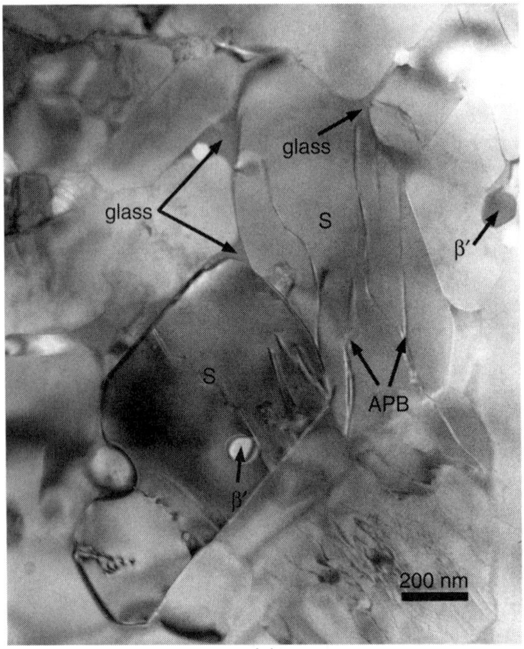

(a)

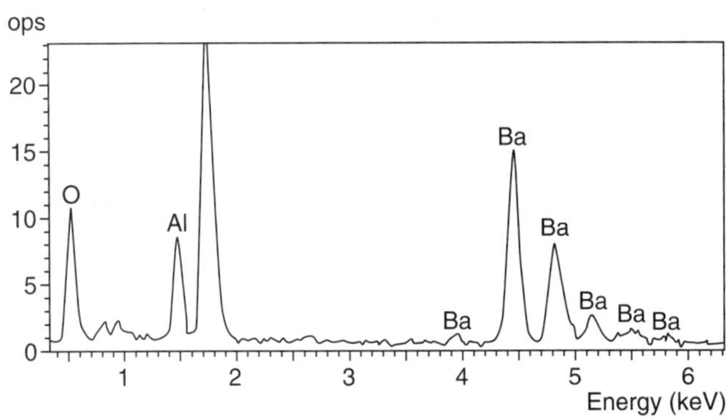

(b)

Figure 11.4 (a) Bright field TEM image of S-phase crystals with residual intercrystalline glass, included β′ and antiphase boundaries (APBs) in an S-SiAlON ceramic hot pressed at 1700°C for 2 hours. (b) TEM-EDS analysis of the glass phase.[40]

11.4 MECHANICAL PROPERTIES

The elastic modulus, as provided in Table 11.1, shows a value in the range 210–230 GPa, with the highest modulus of 228 GPa measured with the ceramic hot pressed at 1700°C. Some representative SEM images of the indented SiAlON surfaces are shown in Figure 11.8. A high-magnification SEM image (Fig. 11.8a) did

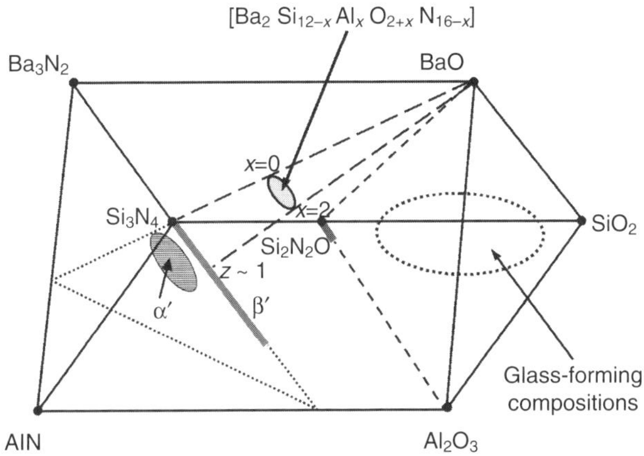

Figure 11.5 Phase equilibria of interest showing the possible composition range of the residual liquid phase, formed at the triple pocket during hot pressing of the investigated S-SiAlON ceramic.[40]

Figure 11.6 Bright field TEM image of S-phase revealing the presence of a network of partial dislocations in an S-SiAlON ceramic hot pressed at 1700°C for 2 hours.[40]

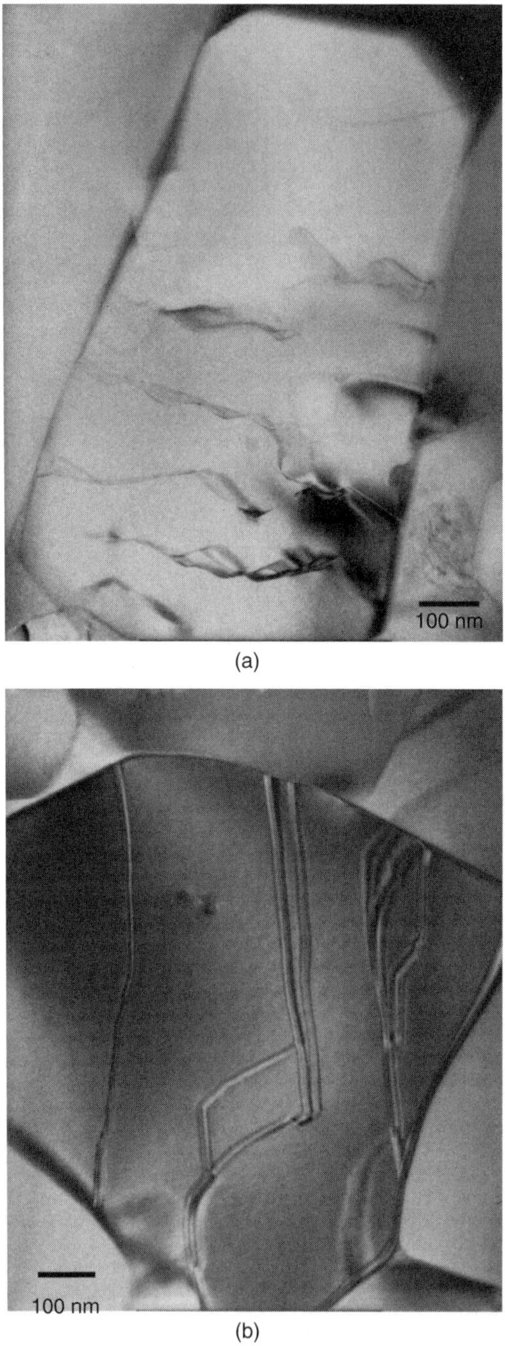

Figure 11.7 Bright field TEM images revealing the details of the defect structure within an S-phase grain in an S-SiAlON ceramic hot pressed at 1700°C for 2 hours. The defect structure is characterized by APBs (formed due to faults in the framework structure), infrequently observed to terminate at the partial dislocation (a) and also by the presence of APBs extending between the facets of an "S" crystal (b).[40]

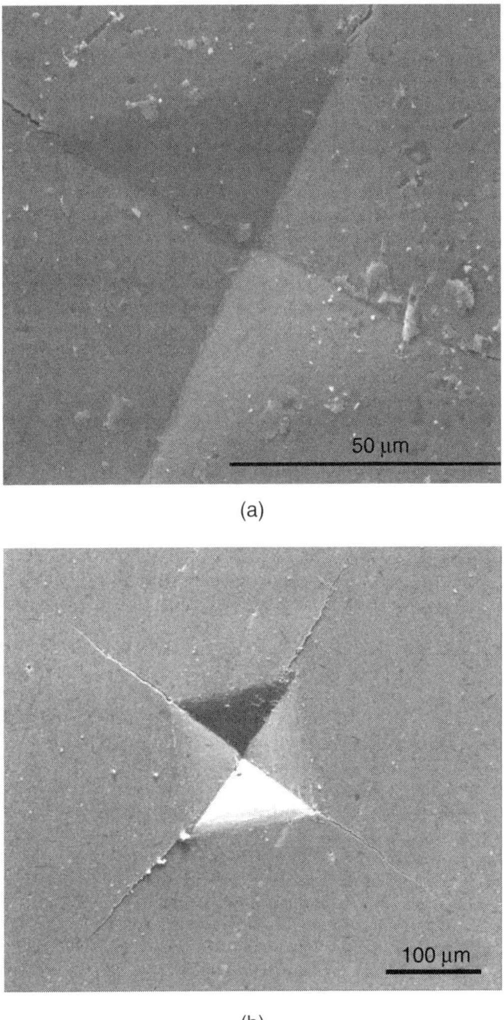

Figure 11.8 SEM topography images of the Vickers indents and indentation-induced radial crack pattern in the Ba-S-SiAlON ceramic hot pressed at 1750°C and indented at varying loads: 100 N (a) and 300 N (b).[41]

not reveal any significant cracking at the indent edges. Figure 11.8b also illustrates well-developed radial–median crack morphology, and multiple cracking from a single indent corner is generally not observed.

As seen in Table 11.2, the apparent hardness values lie in the range 8.6–16 GPa at varying loads of 50–300 N. A hardness-versus-load curve, plotted in Figure 11.9a, indicates that the apparent hardness modestly increases with indent load; such an observation can be attributed to behavior called the reverse indentation size effect (RISE).[18,22] In contrast to the normal indentation size effect (ISE), the indented material undergoes relaxation, which involves crack formation, dislocation activity, and/or elastic deformation of the tip of the indenter.[18,22] Therefore, it can be argued that,

Table 11.2. Summary of the Basic Mechanical Properties of S-SiAlON, Hot Pressed at 1600°C (S1600), 1700°C (S1700), and 1750°C (S1750), When Indented with Indent Load of 100, 200, and 300 N[41]

Load / Material	Density (gm/cm^3)	E (GPa)	Indentation at 50 N		Indentation at 100 N		Indentation at 200 N		Indentation at 300 N	
			H_v (GPa)	K_{IC} ($MPa{\cdot}m^{1/2}$)	H_v (GPa)	K_{IC} ($MPa{\cdot}m^{1/2}$)	H_v (GPa)	K_{IC} ($MPa{\cdot}m^{1/2}$)	H_v (GPa)	K_{IC} $MPa{\cdot}m^{1/2}$
S1600	3.62	215	8.6 ± 0.1	2.4 ± 0.2	12.6 ± 0.8	–	13.7 ± 0.9	5.9 ± 0.03	16.0 ± 2.0	5.5 ± 0.3
S1700	3.63	228	12.7 ± 0.4	3.9 ± 0.3	15.1 ± 1.3	7.4 ± 0.2	13.2 ± 0.8	8.5 ± 0.3	15.5 ± 0.2	11.1 ± 0.6
S1750	3.65	212	12.5 ± 0.4	3.7 ± 0.1	14 ± 0.7	11.6 ± 0.6	14.1 ± 1.0	12.0 ± 0.8	15.3 ± 1.8	11.6 ± 0.2

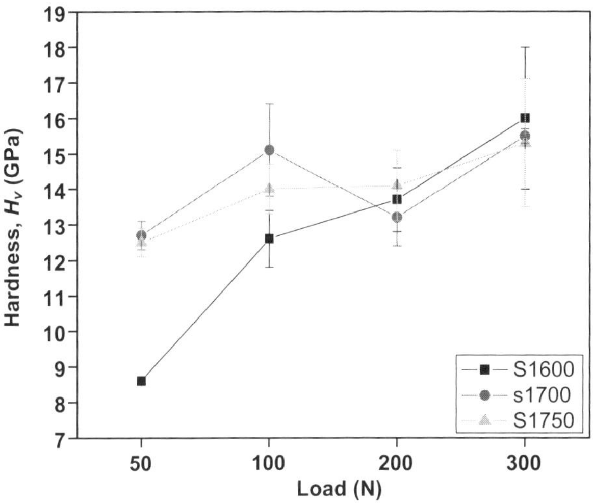

(a)

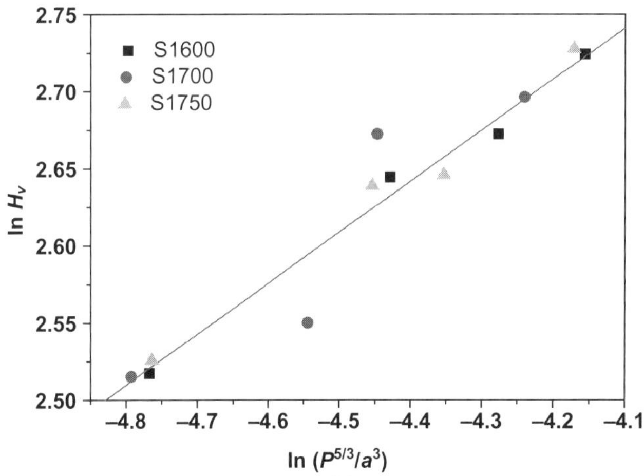

(b)

Figure 11.9 Plot of hardness versus load (a) and $\ln H_v$ against $\ln(P^{5/3}/a^3)$ (b), illustrating the reverse indentation size effect (RISE) exhibited by the S-SiAlON ceramics when indented with a varying load of 50–300 N.[41]

at low load, the dislocation plasticity can result in more deformation and therefore leads to reduced hardness in S-SiAlON materials. In contrast, the indentation response at higher load involved cracking.

It can be noted here that the RISE is observed at very low load (during micro-indentation and nanoindentation) for semiconductor materials, such as Si (111), GaAs (100), GaP (111), and InP (1000).[22–26] Also, the characteristic RISE phenomenon has also been reported for TiCN-based cermets with varying TiCN grain sizes at the low-load regime (1.47–40.67 N).[27]

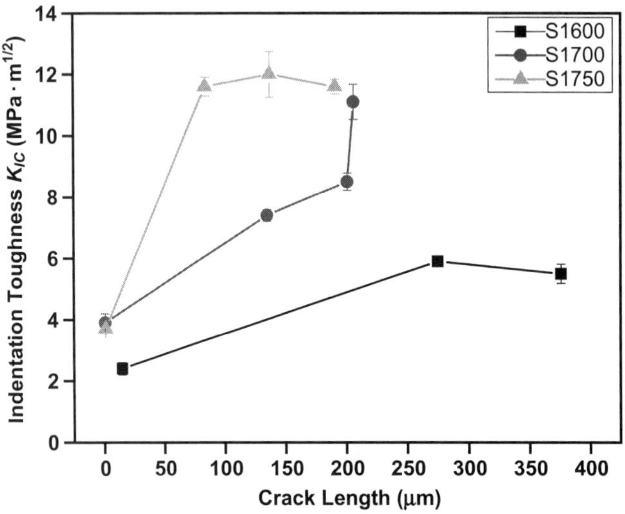

Figure 11.10 Plot of indentation toughness versus crack length for investigated S-SiAlON ceramics. A characteristic and a systematic increase in fracture toughness can be noted for all the investigated S-SiAlON ceramics.[41]

In Figure 11.10, a plot of fracture toughness against crack length reveals that, for S1600, toughness is low at small indent load and relatively higher toughness values are measured at 200 N or higher. It attains an almost steady-state toughness value of 5.5–5.9 MPa·m$^{1/2}$ (flat R curve). The relatively lower steady-state toughness of S1600 is attributed to the presence of unreacted silicon nitride phase, as noticed in Figure 11.1a. In contrast, a sharp increase in toughness from 8.5 to 11.1 MPa·m$^{1/2}$ is exhibited by S1700. More importantly, a systematic increase in K_c from 3.7 to 11.6 MPa·m$^{1/2}$ with increasing crack length (rising R-curve behavior) is measured in S1750. From Figure 11.10, it is evident that the pronounced "R-curve" behavior (systematic increase in toughness with crack extension) is exhibited by both S1700 and S1750 ceramics, over crack lengths up to 200 μm. In particular, a stable R-curve behavior and better toughness properties are achievable with S1750 ceramic.

11.4.1 Load-Dependent Hardness Properties

In the case of brittle ceramics, the indentation energy can be partitioned to the plastic deformation and fracturing of the indented material. At higher loads, the strain gradient plasticity and the elastic recovery effects can be neglected compared with plastic deformation and fracture. In the case of indentation cracking, the apparent hardness (H_{app}) can be analytically expressed as[25,26]

$$H_{app} = \lambda_1 k_1 (P/a^2) + k_2 (P^{5/3}/a^3), \tag{11.2}$$

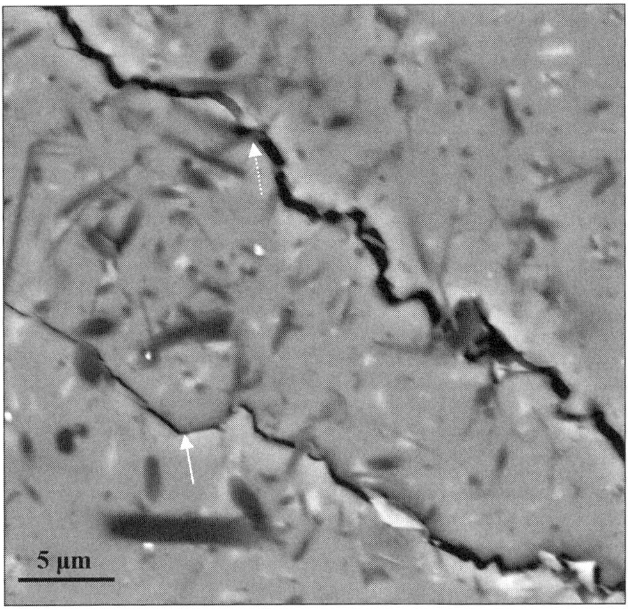

Figure 11.11 SEM image illustrating crack–microstructure interaction in S1750 ceramic. The single-pointed arrow indicates typical crack–S-phase interaction; the dotted arrow indicates the pullout of β-Si_3N_4 needles.[41]

where a is the indentation diameter, P is the indent load, and λ_1, k_1, and k_2 are constants.

For a linearly elastic brittle solid,

$$\lambda_1 = 0 \text{ and } H_{\text{app}} = K_2(P^{5/3}/a^3). \tag{11.3}$$

After extensive analysis of a large number of experimentally measured hardness data (0.4–22 GPa), Sangwal[23] proposed a modified relationship:

$$H_v = K(P^{5/3}/a^3)^m. \tag{11.4}$$

In Equation 11.4, both K and m are constants and the value of m is around 0.5–0.55.[24] Although deformation dominated the hardness response at lower load, the observation of indent-induced radial cracking can be made at intermediate and higher loads. In the case of S-SiAlON ceramics, the measured apparent hardness values are plotted against $P^{5/3}/a^3$ on a log-scale, in Figure 11.9b. It is clear from Figure 11.9b that the measured data can be fitted closely with an m value of 0.494, which is similar to an earlier report.[24] Here, it can be noted that earlier RISE effect was observed at extremely low loads (<1 N)[22–24] and in case of TiCN cermets at indent loads of less than 40 N.[27]

In the case of S- SiAlON ceramics, the plastic deformation is predominant at lower loads; as a result, much of the indentation energy is utilized in material flow and displacement, which lowers the apparent hardness. At higher load, cracking dominates, compared with plastic deformation; consequently, more energy is consumed for crack initiation and propagation. This is supported by the abrupt change in the *l/a* ratio from lower to higher load. The variation in *l/a* with load is suggestive of the dominance of cracking over deformation and is also an indirect measure of the partitioning of the relative energy for cracking and deformation. The fact that the characteristics of crack nucleation and propagation appear to be dependent on load essentially suggests that cracking and deformation at lower load are somewhat different from those at higher load. It also implies that the relative energy distribution for deformation and fracturing is dependent on load.

11.4.2 R-Curve Behavior

In analyzing the toughness data presented in Tables 11.1 and 11.2 as well as in Figure 11.10, some interesting observations emerge that imply a strong dependence of fracture toughness on microstructure. It can be recalled here that R curves are evaluated from Vickers indentation cracks for alumina,[28] Y-TZP,[29] Si_3N_4,[30] and SiC-based materials.[31] Although using the indentation technique to analyze the R-curve behavior of structural ceramics was criticized in a 2007 review article,[32] it has been reported that R curves determined from indentation cracks follow closely with those from the double cantilever beam (DCB) technique for silicon nitride ceramics.[33] Similar results have also been recorded for alumina ceramics.[34]

From Figure 11.10, three aspects R-curve behavior of S-SiAlON ceramics can be discussed: (1) initiation toughness (K_0), (2) steady-state toughness, and (3) characteristic bridging length (l_c), that is, the critical crack length needed to attain steady-state toughness. As far as the processing is concerned, S-SiAlON sintered at 1600°C has the lowest K_0, and the K_0 values of other S- SiAlON ceramics are comparable. Concerning the third aspect, the characteristic bridging length for S1600 is 97 μm, while that for S1700 and S1750 ceramics is around 200 μm.

To analyze the operating toughening mechanisms, the crack microstructure interaction is studied using SEM-BSE imaging mode. A typical, though not representative, case of multiple indentation cracking is shown in Figure 11.11 to provide evidence of important aspects of crack–microstructure interaction. Some important observations can be summarized as follows: (1) tortuous crack path, indicating better crack growth resistance; (2) evidence of grain bridging and pullout; and (3) frequent occurrence of debonding of elongated S-phase grains in the crack wake. Similar crack–microstructure interaction is commonly reported for other variants of SiAlON ceramics.[2,8] The increased resistance of S1700 and S1750 to the propagating crack, that is, rising R-curve behavior, is therefore attributed to the presence of elongated S-phase grains and whiskerlike β-Si_3N_4 grains, resulting in crack deflection and crack bridging, respectively. The large aspect ratio of interlocking S-phase grains (Fig. 11.2 and 11.3), in combination with the acicular β-Si_3N_4 needles, can potentially cause frictional and mechanical interlocking between the separated fracture surfaces by deflecting the crack. Therefore, more energy is required to separate the fractured surfaces, which in macroscopic terms results in increased toughness.

From the preceding discussion, it can be stated that crack bridging and grain pullout both contribute to the moderately high toughness of Ba-S-SiAlON ceramics. This observation can be analyzed in the light of a reported model that correlates fracture resistance with elongated ceramic grains.[35–37] Such a model can be used to predict frictional work, which is the major source of fracture energy. Also, the near tip elastic bridging is ignored. Another assumption is that the decohesion of a platelet-shaped grain, prior to pullout, preferentially occurs by intergranular fracture. On the basis of these assumptions, Becher and co-workers[35] proposed following expression:

$$K = C\sqrt{\sum_i V_i W_i (AR_i)^2}, \tag{11.5}$$

where V is the volume fraction of elongated grains, W is grain width, K is the fracture toughness, AR is the aspect ratio of elongated grains, and the suffix i denotes a group of ith-type elongated grains. Also, the value of C is taken to be the same for different elongated phases.

Since the Ba-doped S-SiAlON ceramics are characterized by two types of elongated phases: S-phase and β-Si_3N_4 needles; Equation 11.5 can be appropriately adapted as

$$K = C\sqrt{V_s W_s (AR_s)^2 + V_\beta W_\beta (AR_\beta)^2}, \tag{11.6}$$

where the suffixes s and b denote the S-phase and β-Si_3N_4, respectively. Following the original model,[35] the fitting parameter constant C in Equation 11.5 or Equation 11.6 can be found as

$$C = \sqrt{\frac{E\tau}{6(1-\nu^2)}}, \tag{11.7}$$

where E is elastic modulus, ν is Poisson's ratio, and τ is interfacial friction. It is also evident from Equations 11.5 and 11.7 that the AR of the platelet phase, compared with width or volume fraction, should have a dominant influence on toughness.

The set of Equations 11.5–11.7 was further used for S1700 ceramic, to illustrate the appropriateness of the toughening model. Average ARs of 3.44 and 7.0 and widths (W) of 1.14 and 0.57 μm for S-phase grains and β-Si_3N_4, respectively, were measured from SEM images. Considering the presence of 85% ($V_s = 0.85$) S-phase grains and 9% of β-Si_3N_4 needles ($V_\beta = 0.09$), the value of interfacial friction (τ) from Equations 11.6 and 11.7 has been computed to be 203.8 MPa. In our calculation, ν is taken as 0.3 and the toughness at 300 N load ($K = 11.1\ \text{MPa·m}^{1/2}$) was considered. Chen and co-workers estimated τ as 108 MPa in case of α-SiAlON materials (elongated grain morphology).[3] Therefore, the preceding calculation provides estimates of interfacial friction (τ) for S-SiAlON ceramics, which is in reasonable agreement with other SiAlON ceramics.

The following discussion assesses how the mechanical properties of S-SiAlON ceramics are comparable with the existing SiAlONs. Like other structural ceramics, the properties of various SiAlON ceramics are strongly dependent on microstructural characteristics, in particular grain shape and aspect ratio, intergranular phase, amount of seed crystals, ratio of α- to β-phases, and so on (e.g., see References 35–39). Table 11.3 summarizes the mechanical property data of some selected SiAlON ceramics and such data reveal a large variation in apparent hardness and toughness properties. Such a broad spectrum of properties can be uniquely described by two end members, that is, the equiaxed α-SiAlON material with high hardness (H_{v10} = 22 GPa) and low toughness (K_c = 3 MPa·$m^{1/2}$),[11] and β-SiAlON ceramic with high toughness (>6 MPa·$m^{3/2}$) and low hardness (~15 GPa). The advantages associated with α- and β-phases have been utilized in the development of "*in situ* self-reinforced multiphase α–β structure." Although, the toughness of S1750 (11.6 MPa·$m^{1/2}$) is much higher than that of several Si_3N_4-derived ceramics[6,8,12]; the apparent hardness of S-SiAlON (15.3 GPa) is much lower than that of many of the Si_3N_4-based ceramics.[5,8,13] Nevertheless, the hardness of Ba-doped S-SiAlON ceramics is comparable with some competing silicon nitrides.[35,37]

In summary, the S-SiAlON ceramics represent a new variant of SiAlON materials that can exhibit beneficial combinations of *E*-modulus (more than 200 GPa), apparent hardness (up to 16 GPa), and toughness (up to 12 MPa·$m^{1/2}$) properties. In addition, optimally processed S-SiAlON ceramics can exhibit R-curve behavior. It can be noted that R curves are exhibited by a wide range of SiAlON materials, for example, Y-α-SiAlON,[2] *in situ* toughened Y-α-SiAlON, and Ca-α-SiAlON.[39] Depending on the amount of seed crystals, a peak toughness of 10 MPa·$m^{1/2}$ was measured in these ceramics.

In view of the preceding observations, S-SiAlON ceramics appear to have potential for structural applications, provided the combination of hardness and toughness properties can be further enhanced by microstructural tailoring, for example, by incorporating Si_3N_4 seeds. An optimal balance of the amount of elongated S-phase grains and β-Si_3N_4 phase is desired. As an additional note, our 2007 tribological study on one of the investigated Ba-doped S-SiAlON ceramics (S1750) reveals friction and wear properties that are comparable with other competing SiAlON ceramics; more details can be found elsewhere.[36] In view of these observations, the newly developed Ba-doped S-SiAlON ceramic can be considered as a potential competitive SiAlON composition for various engineering applications. In future, gas pressure sintering with nitrogen gas or sinter-HIPing can be utilized to investigate the industrial-scale feasibility for the production of nitride ceramics, such as α-SiAlON ceramics. A detailed description of the microstructure and properties of S-SiAlON can be found elsewhere.[40,41]

11.5 CONCLUDING REMARKS

A typical hot-pressed microstructure of S-SiAlON ceramic is characterized by the anisotropic platelets of S-phase, residual glass (4–5%), and α/β-Si_3N_4 (6–8%). TEM-EDS analysis suggests the S-phase composition to be slightly Al-rich $Ba_2Si_{12-x}Al_xO_{2+x}N_{16-x}$ ($x = 2 \pm 0.2$). Detailed TEM analysis suggested "S" is the preferred anisotropic growth of phase direction along the *c*-axis of the orthorhombic unit cell and the facets of the platelets have a crystallographic orientation along (100) and/or (010) primary crystal planes. Careful analysis of the indentation data reveals

Table 11.3. A Comparative Study of the S-SiAlON Ceramic, Hot Pressed at 1750°C, with Previously Investigated Competing Si_3N_4-Based Materials, in Terms of Hardness and Indentation Toughness Properties

Material under study	Starting material and/or method of fabrication	Hardness H_{v10} (GPa)	Fracture toughness K_c (MPa·m$^{1/2}$)
Y-α-SiAlON ($Y_{0.5}Si_{9.3}Al_{9.3}O_{1.2}N_{14.8}$)	α-Si_3N_4, AlN, Al_2O_3, Y_2O_3: HP at 1900°C for 1 hour, in N_2	21.0	12.0
Nb-Stabilized α-SiAlON	β-Si_3N_4, AlN, Al_2O_3, Si_3N_4: HP at 1950°C and heat treated for 12 hours at 1650°C	21.7	6.3
Lu + 5% Lu_2SiO_5	α-Si_3N_4, Al_2O_3, AlN, SiO_2, Lu_2O_3: HP at 1950°C for 2 hours in N_2	18.9	4.4
Dy-α-SiAlON	α-Si_3N_4, AlN, Al_2O_3, Dy_2O_3: PLS at 1800°C and then GPS at 1900°C for 1 hour	18.8	6.3
Y-α-SiAlON	By post-sintering the nitrided compact at 1900°C for 3 hours	18.5	5.1
Si_3N_4 + 20 vol% (9:1 AlN/Y_2O_3)	α-Si_3N_4, with 20% AlN, Al_2O_3, AlN:Y_2O_3 in 9:1 molar ratio: GPS at 1900°C, 2 hours in N_2	18.5	5.1
Y-Yb-α-SiAlON	α-Si_3N_4, AlN, (Y, Yb)O_3: HP at 1600–1750°C in vacuum	18.9	4.6
Si_3N_4	HIPed Si_3N_4 + MgO	17.0	6.0
S-SiAlON	α-Si_3N_4, AlN, Al_2O_3, $BaCO_3$: HP at 1750°C, 2 hours in N_2 atmosphere	14.0	11.6
β-SiAlON	$Si_{6-z}Al_zO_zN_{8-z}$, $z = 0.3$	15.5	4.7
(Ce-Y) α-SiAlON	α-Si_3N_4, AlN, Al_2O_3 with CeO_2 (SHS-ed) CeO_2:Y_2O_3 in 1:0:0.375 molar ratio: HPat 1750°C, for 1 hour in N_2	15.2	4.9
α-SiAlON–(10%)β-SiAlON	α-SiAlON 76.92 wt% Si_3N_4, 13.46 wt% AlN, 5.77 wt% Y_2O_3, 3.85 Al_2O_3, reinforced with 10% β-SiAlON fiber	14.8	5.9
α-SiAlON (equiaxed)	Starting powders α-Si_3N_4, Al_2O_3, AlN, and Y_2O_3	22.0	3.0
Y_2O_3–Si_3N_4	Hot pressed (1650°C, 1 hour)	15.2	7.1–8.1
Yb_2O_3–Si_3N_4	Hot pressed (1650°C, 1 hour)	13.9	7.4–8.5

Different processing routes mentioned: HP, hot pressed; PLS, pressureless sintering; HIP, hot isostatic pressing; GPS, gas pressure sintering; SHS, self-propagating high-temperature synthesis.[41]

the RISE in S-SiAlON ceramics. The RISE effect has been discussed in terms of the predominance of cracking over deformation at higher load, resulting in an increase in apparent hardness compared with that at lower load. Another important result has been that the hot-pressed S-SiAlON ceramics display pronounced R-curve behavior—a systematic increase in toughness with increase in crack length. A maximum toughness of 11–12 MPa·m$^{1/2}$ was recorded for S-SiAlON, hot pressed at 1750°C. The increased toughness arises from noticeable contribution from crack deflection, crack bridging, and crack branching, imparted by elongated S-phase grains as well as β-Si_3N_4 needles. The use of a theoretical toughening model based on predominant toughening by platelet-shaped S-phase grains and β-Si_3N_4 needles provided an estimate of the interfacial friction of around 200 MPa, which is in reasonable agreement with other competing SiAlON ceramics.

REFERENCES

1 S. Kurama, M. Herrmann, and H. Mandal. The effect of processing conditions, amount of additive and composition on the microstructures and mechanical properties of α-SiAlON ceramics. J. Eur. Ceram. Soc. 22 (2003), 109–119.

2 M. Zenotchkine, R. Shuba, J.-S. Kim, and I.-W. Chen. Effect of seeding on the microstructure and mechanical properties of alpha-SiAlON: I, Y-SiAlON. J. Am. Ceram. Soc. 85(5) (2002), 1254–1259.

3 M. Zenotchkine, R. Shuba, and I.-W. Chen. Effect of seeding on the microstructure and mechanical properties of alpha-SiAlON: III comparison of modifying cations. J. Am. Ceram. Soc. 86 (2003), 1168–1175.

4 Y. Kaga, M. I. Jones, K. Hirao, and S. Kanzaki. Fabrication of elongated α-SiAlON via a reaction-bonding process. J. Am. Ceram. Soc. 87(5) (2004), 956–959.

5 C. Santos, K. Strecker, S. Rebeiro, J. V. C. de Souza, O. M. M. Silva, and C. R. M. da Silva. α-SiAlON ceramics with elongated grain morphology using an alternative sintering additive. Mater. Lett. 58 (2004), 1792–1796.

6 J. Jiang, P. Wang, W. He, W. Chen, H. Zhuang, Y. Cheng, and D. Yan. Study on the stability of Ce α-Sialon derived from SHS-ed powder. J. Eur. Ceram. Soc. 24 (2004), 2853–2860.

7 C. Santos, S. Ribeiro, J. K. M. F. Daguano, S. O. Rogero, K. Strecker, and C. R. M. Silva. Development and cytotoxicity evaluation of SiAlON ceramics. Mater. Sci. Eng. C 27 (2007), 148–153.

8 H. Mandal. New developments in α-Sialon ceramics. J. Eur. Ceram. Soc. 19 (1999), 2349–2350.

9 S. Turan, F. Kara, and H. Mandal. Transmission electron microscopy of SrO containing Multi-Cation doped α-SiAlON ceramics. Mater. Sci. Forum 383 (2002), 37–42.

10 N. Acikbas Calis, E. Suvaci, and H. Mandal. Fabrication of functionally graded SiAlON ceramics by tape casting. J. Am. Ceram. Soc. 89 (2006), (10)3255–3257.

11 I. Wei-Chen and A. Rosenflanz. A tough SiAlON ceramic based on α-Si_3N_4 with a whisker like microstructure. Nature 389 (1997), 701.

12 F. Ye, M. J. Hoffmann, S. Holzer, Y. Zhou, and M. Iwasa. Effect of amount of additives and post heat treatment on the microstructure and mechanical properties of Yttrium α-SiAlON ceramics. J. Am. Ceram. Soc. 86 (2003), (12)2136–2142.

13 C. Zhang, K. Komeya, J. Tatami, and T. Meguro. Inhomogeneous grain growth and elongation of Dy-α-SiAlON ceramics at temperatures above 1800°C. J. Eur. Ceram. Soc. 20 (2000), 939–944.

14 C. J. Hwang, D. W. Susintzky, and D. R. Beaman. Preparation of multication α-SiAlON containing strontium. J. Am. Ceram. Soc. 78(3) (1995), 588–592.

15 J. Grins, Z. Shen, M. Nygren, and T. Ekstrom. Preparation and crystal structure of $LaAl(Si_{6-z}Al_z)N_{10-z}O_z$. J. Mater. Chem. 5 (1995), 2001–2005.

16 M. H. Lewis, C. J. Reed, and N. D. Butler. Pressureless-sintered ceramics based on the compound Si_2N_2O. Mater. Sci. Eng. 71 (1985), 87–94.

17 Z. Shen, J. Grins, S. Esmaeilzadeh, and H. Ehrenberg. Preparation and crystal structure of a new Sr Containing SiAlON phase $Sr_2Al_xSi_{12-x}N_{16-x}O_{2+x}$ (x~2). J. Mater. Chem. 9 (1999), 1019–1022.

18 C. J. Hwang, D. W. Susnitzky, and D. R. Beaman. Preparation of multication {alpha}-SiAlON containing strontium. J. Am. Ceram. Soc. 78 (1995), 588–592.

19 A. Krell. A new look at the influences of load, grain size, and grain boundaries on the room temperature hardness of ceramics. Int. J. Refractory Met. Hard Mater. 16 (1998), 331–335.

20 K. Niihara, R. Morena, and P. H. Hasselman. Evaluation of K_{Ic} of Brittle solids by the indentation methods with low crack to indent ratios. J. Mater. Sci. Lett. 1 (1982), 13–16.

21 S. Palmqvist. Occurrence of crack formation during Vickers indentation as a measure of the toughness of the hard materials. Arch. Eisenhuttenwesen 33 (1962), 629–633.

22 N. K. Mukhopadhyay and P. Paufler. Micro- and Nano-indentation techniques for Mechanical characterization of materials. Int. Mater. Rev. 51(4) (2006), 1–37.

23 K. Sangwal. On the reverse indentation size effect and microhardness measurement of solids. Mater. Chem. Phys. 63 (2000), 145–152.

24 K. Sangwal and B. Surowska. Study of indentation size effect and microhardness of $SrLaAlO_4$ and $SrLaGaO_4$ single crystals. Mater. Res. Innov. 7 (2003), 91–104.

25 H. Li and R. C. Bradt. The effect of indentation induced cracking on apparent microhardness. J. Mater. Sci. 31 (1996), 1065–1070.

26 C. Hays and E. G. Kendall. An analysis of knoop microhardness. Metallography 6 (1973), 275–282.

27 J. Gong, H. Miao, Z. Zhao, and Z. Guan. Load-dependence of the measured hardness of Ti(C, N)- based cermets. Mater. Sci. Eng. A303 (2001), 179–186.

28 A. M. Thompson, H. M. Chan, M. P. Harmer, R. E. Cook. Crack Healing and Stress Relaxation in Al_2O_3-SiC "Nanocomposites." J. Am. Ceram. Soc. 78 (1995), 567–571.

29 R. M. Anderson and L. M. Braun. Technique for the R-curve Determination of Y-TZP using indentation technique. J. Am. Ceram. Soc. 73 (1990), 3059–3062.

30 C. W. Li, D. J. Lee, and S. C. Lui. R-curve and strength for in-situ reinforced silicon nitrides with different microstructures. J. Am. Ceram. Soc. 75 (1992), 1777–1785.

31 R. Choi and J. A. Salem. Strength, toughness and R-Curve behavior of SiC Whisker-reinforced composite Si_3N_4 with reference to monolithic Si_3N_4. J. Mater. Sci. 27 (1992), 1491–1498.

32 D. Munz. What can we learn from R-Curve measurements? J. Am. Ceram. Soc. 90(1) (2007), 1–15.

33 C. W. Li and J. Yamanis. Super tough silicon nitride with R curves behavior. Ceram. Eng. Sci. Proc. 10 (1989), 632–645.

34 R. F. Cook, E. R. Liniger, R. W. Steinbrech, and F. Deuerler. Sigmoidal indentation strength characteristics of polycrystalline alumina. J. Am. Ceram. Soc. 77 (1994), 303–314.

35 P. F. Becher, H. T. Lin, S. L. Hwang, M. J. Hoffmann, and I.-W. Chen. The influence of microstructure on the mechanical behavior of silicon nitride ceramics. Mater. Res. Soc. Symp. Proc. 287 (1993), 147–158.

36 Manisha and B. Basu. Tribological properties of a hot pressed Ba-doped S-Phase SiAlON ceramic. J. Am. Ceram. Soc. 90(6) (2007), 1858–1865.

37 Y. S. Zheng, K. M. Knowles, J. M. Vieira, A. B. Lopes, and F. J. Oliveira. Microstructure, toughness and flexural strength of self-reinforced silicon nitride ceramics doped with yttrium oxide and ytterbium oxide. J. Microsc. 201 (2001), 238–249.

38 Z. Y. Deng, Y. Inagaki, J. She, Y. Tanaka, Y. F. Liu, M. Sakamoto, and T. Ohji. Long crack R-curve of aligned porous silicon nitride. J. Am. Ceram. Soc. 88(2) (2005), 462–465.

39 M. Zenotchkine, R. Shuba, J. S. Kim, and I. W. Chen. R-curve behavior of in situ toughened α-SiAlON ceramics. J. Am. Ceram. Soc. 84(4) (2001), 884–886.

40 B. Basu, M. Lewis, M. E. Smith, M. Bunyard, and T. Kemp. Microstructure development and properties of novel Ba-doped S-phase sialon ceramics. J. Eur. Ceram. Soc. 26 (2006), 3919–3924.

41 B. Basu, Manisha, and N. K. Mukhopadhyay. Understanding the mechanical properties of hot pressed Ba-doped S-phase sialon ceramics. J. Eur. Ceram. Soc. 29 (2009), 801–811.

Chapter 12

Toughness and Tribological Properties of MAX Phases

Ceramics form the top of the chart when it comes to low-weight high-temperature applications. They are highly robust due to excellent resistance to oxidation and corrosion, retention of high-temperature strength, and resistance to wear and erosion even at high temperatures while providing the benefit of their low density. But, owing to the brittleness of ceramics, it has become a challenge in ceramic engineering to enhance a material's toughness by a few times. In addition, the poor thermal shock resistance and toughness of ceramics usually limit their use in common engineering applications.

12.1 EMERGENCE OF MAX PHASES

An engineering material showing superior high-temperature wear and corrosion resistance similar to that of a ceramic while displaying toughness similar to that of metals is often sought for enhanced performance during service. The so-called MAX phases show nanolaminate structure during deformation and possess better wear resistance and toughness compared with those of conventional ceramics. Their typical stoichiometry is of the form $M_{n+1}AX_n$ phases (termed MAX for short), where M=metals (early transition metals), A=Group A elements, and X=carbon and/or nitrogen.[1,2] A typical combination of MAX-phase materials is presented in Figure 12.1. These materials will allow development of new technologies involving high-efficiency engines and motors, durability in extreme conditions of thermal conductivity, damage-tolerant thermal protection systems, enhanced fatigue resistance, and retention of rigidity at high temperatures. Recently, this class of nanolaminate structural ceramics has been engineered by researchers.

A mere 1° increase in the working temperature of jet engines could save \$1 billion in a year, and a meager 3 mi/gal (~1.3 km/L) increase in fuel efficiency of automobiles could save a million barrels of oil per day. But, the problem lies in the fact that no material exists to survive the heat of high-temperature operation while

Advanced Structural Ceramics, First Edition. Bikramjit Basu, Kantesh Balani.

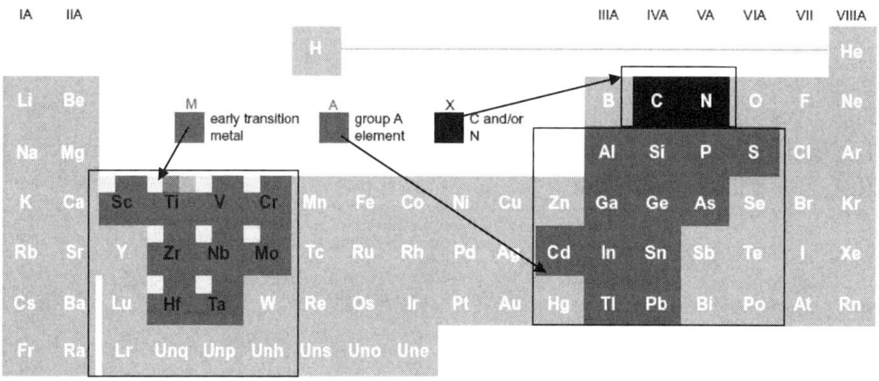

Figure 12.1 The MAX phase depicted by early transition metals (M), group A elements (A), and C and N (X) in the periodic table (adapted from Reference[3]).

spinning at thousands of revolutions per minute (rpm). Similarly, if ceramic chambers are able to withstand heat, the radiative cooling can be discarded (along with fans, cooling water, etc.) to achieve a lighter structure that would be worth more miles in every gallon of fuel.

High machinability of MAX phases along with their high stiffness and electrical- and thermal-conductance makes them a highly attractive class of ceramics. Some of the MAX phases are damage tolerant and can withstand high thermal shock while maintaining high resistance to oxidation, corrosion, and creep.[1,2] A schematic comparing MAX phases with other engineering alloys is presented in Figure 12.2.[3]

12.2 CLASSIFICATION OF MAX PHASES

MAX phases are classified based on the stoichiometric ratio of each M, A, X phase constituting the tough ceramic, namely, 211, 312, and 413. This class of materials naturally forms three groups (based on the stoichiometry of MAX elements): 211, 312, and 413 materials (Figure 12.3). The MAX carbides often possess hexagonal crystal structure (space group $P6_3/mmc$) with two formula units per unit cell showing near close-packed layers of M, with interlayers of A, and C atoms filling the octahedral sites within the M layers.

Every category contains a set of compounds, the classification of which is made on the basis of the ratio of each element in the compound. To date, more than 50 MAX phases are known. A list of a few MAX phases based on the Group A elements are presented in Table 12.1. Additionally, the mechanical properties of a few MAX phases are listed in Table 12.2. Difference in the bulk modulus arises because of the difference of atomic density in the laminate structure and different lattice parameters (along the *a* and *c* directions) of the MAX phase (Table 12.3).

A combination of C and N can alter the lattice parameter of the MAX phase depending on their concentration as follows[4]:

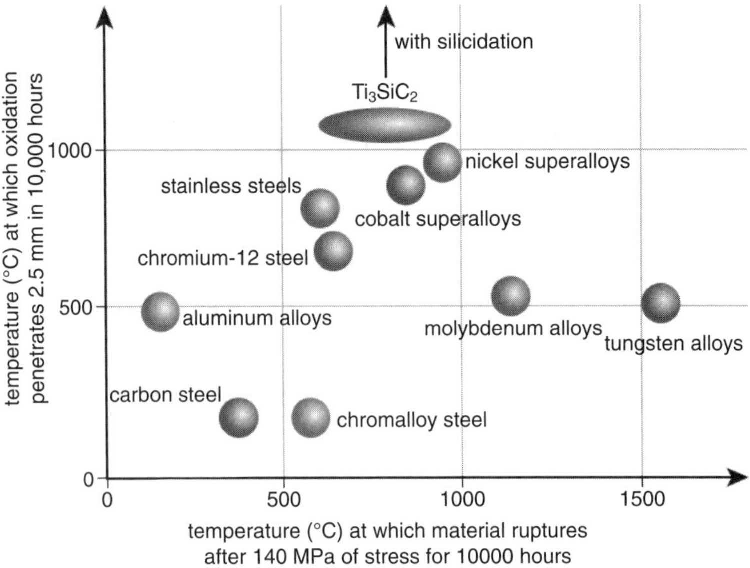

Figure 12.2 Position of Ti_3SiC_2 showing high-temperature oxidation and strength retention. Materials on the top can sustain a higher oxidation environment before they are ruptured at high temperatures. Titanium silicon carbide appears to be superior, compared with nickel superalloys, in oxidation resistance and behaves similarly in high-temperature strength requirements.[3]

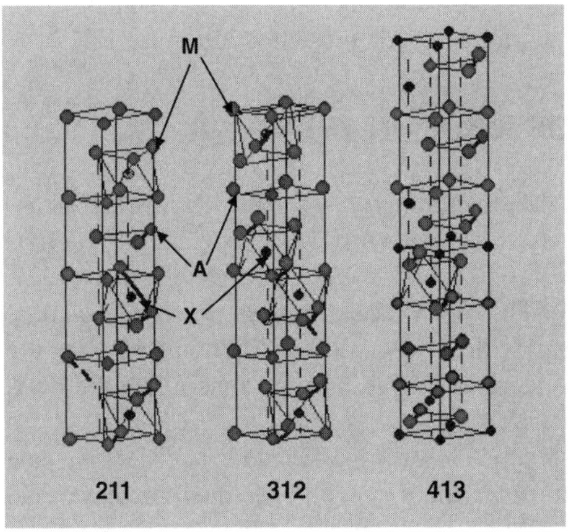

Figure 12.3 Classification of the MAX phases (adapted from Reference 3).

Table 12.1. Various MAX Phases Based on the Group A Elements[a]

Al	Si	P	S
Ti_2AlC, Ti_2AlN, $Ti_2Al(C_{0.5}, N_{0.5})$, Nb_2AlC, $(Nb,Ti)AlC$, Ta_2AlC, V_2AlC, Cr_2AlC, Ti_4AlN_3, Ti_3AlC_2	Ti_3SiC_2	V_2PC, Nb_2PC	Ti_2SC, Zr_2SC, Nb_2SC, Hf_2SC
Ga	**Ge**	**As**	**Se**
Ti_2GaC, V_2GaC, Cr_2GaC, Nb_2GaC, Mo_2GaC, Ta_2GaC, Ti_2GaN, Cr_2GaN, Cr_2GaN	Ti_2GeC, V_2GeC, Cr_2GeC, Ti_3GeC_2	V_2AsC, Nb_2AsC	
In	**Sn**	**Sb**	**Te**
Sc_2InC, Ti_2InC, Zr_2InC, Nb_2InC, Hf_2InC, Ti_2InN, Zr_2InN	Ti_2SnC, Zr_2SnC, Zr_2SnC, Hf_2SnC, Hf_2SnN		
Tl	**Pb**	**Bi**	**Po**
Ti_2TlC, Zr_2TlC, Hf_2TlC, Zr_2TlN	Ti_2PbC, Zr_2PbC, Hf_2PbC		

[a]http://www.materials.drexel.edu/max/MAX%20properties/properties.html.

Table 12.2. Mechanical Properties of a Few MAX Phases[1,2,8,23–25]

Compound	Elastic modulus (GPa)	Shear modulus (GPa)	Poisson ratio	Bulk modulus *M* (GPa)	Bulk modulus *C* (GPa)
Ti_2AlC	277	118	0.19	144	186
V_2AlC	235	116	0.20	152	201
Cr_2AlC	245	102	0.20	138	166
Nb_2AlC	286	117	0.21	165	208
Ti_3AlC_2	297	124	0.20	165	226
Ti_4AlN_3	310	127	0.22	185	216
Ti_3SiC_2	339	139	0.20	190	206
Ti_3GeC_2	340	142	0.19	169	179

$$a(\text{Å}) = 3.074 - 0.07\frac{[N]}{[N]+[C]}, \qquad (12.1)$$

with correlation coefficient of 0.98. Similarly,

$$c(\text{Å}) = 18.537 - 0.245\frac{[N]}{[N]+[C]}, \qquad (12.2)$$

with correlation coefficient of 0.79.

Replacing C with N in the MAX phases (Ti_2AlC and Ti_3AlC_2) results in a decrease of the bulk modulus (from 185 GPa in TiAlC to a predicted ~175 GPa in $Ti_3Al(CN)$, and from 230 GPa in Ti_3AlC_4 to 220 GPa in $Ti_3AlC_2N_2$). This decrease

Table 12.3. Lattice Parameters and Density of Various MAX Phases[1–3,8]

Compound	Lattice parameter (a)	Lattice parameter (c)	Density (g/cm^3)
Ti_2AlC	3.056	13.623	4.11
V_2AlC	2.916	13.13	4.07
Cr_2AlC	2.860	12.82	5.24
Hf_2SnC	3.320	14.388	11.80
Nb_2AlC	3.103	13.83	6.50
Nb_2GaC	3.13	13.56	7.73
Nb2InC	3.17	14.37	8.30
Ta_2AlC	3.07	13.80	11.82
Ti_2GaC	3.07	13.52	5.53
Ti_3AlC_2	3.08	18.58	4.50
Ti_4AlN_3	2.99	23.37	4.76
Ti_2GeC_2	3.07	12.93	5.68
Ti_3SiC_2	3.07	17.67	4.52
Ti_3GeC_2	3.09	17.76	5.55
V_2PC	3.08	10.91	5.38
Zr_2InC	3.34	14.91	7.10

is contrary to the fact that unit cell volumes decreased with addition of N in MAX phases.[4]

A simple transition from Ti_3SiC_2 to TiC can be done by replacing Si by C atoms, as shown in Figure 12.4. However, replacing Si by C results in high twinning of the regions (Fig. 12.4a), and the inserted C-layer becomes a mirror plane of the twinned region (Fig. 12.4b).[5] Detwinning requires rotation of the C-layer to produce the (110) plane of TiC (Fig. 12.4c). Since the transformation of Ti_3SiC_2 occurs with 15% volume shrinkage, this process becomes highly exciting to engineers working toward utilizing shrinkage with microcracks and strains in the matrix in these engineered MAX phase ceramics.[5]

12.3 DAMAGE TOLERANCE OF MAX PHASES

Low hardness ($H \sim 4$ GPa), high elastic modulus ($E \sim 320$ GPa), and easy machinability of Ti_3SiC_2 make it a very attractive ceramic. The basal slip assisted with grain buckling and sliding is rendered in Ti_3SiC_2 because of its low H/E ratio (~0.013), more representative of soft metals. Hardness is strongly dependent on the indentation load (as the contact dimension increases) because of resistance from grains. As the contact dimension increases to more than the grain size, basal slip planes align and produce enhanced deformation and reduced hardness values (Fig. 12.5).[6]

Hertzian indentation produces a distinct surface depression, corroborating the higher degree of deformation being accommodated in the MAX-phase ceramic. Subsurface deformation extends well below the contact area, and the damaged zone

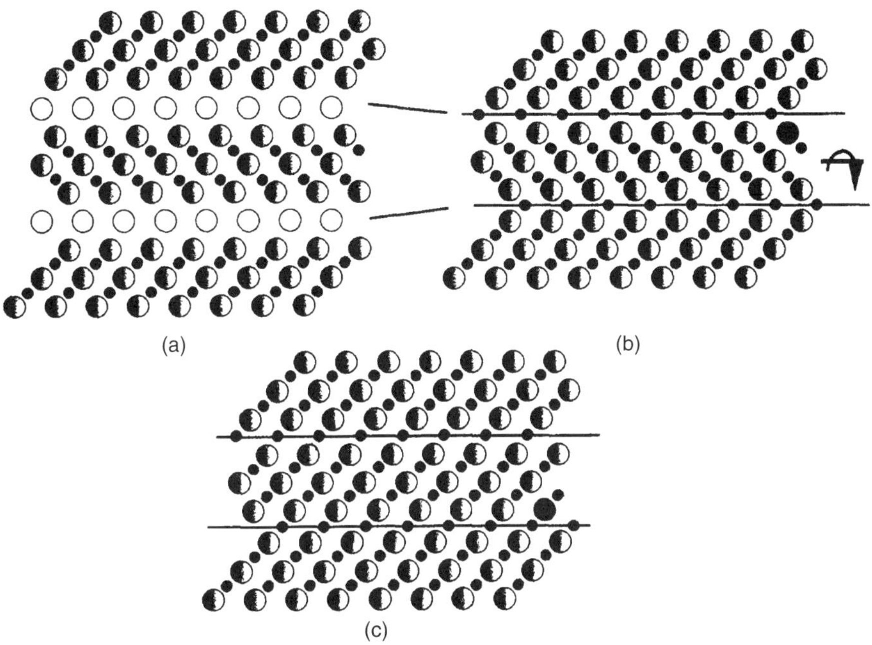

Figure12.4 **(a)** Schematic of (110) plan in Ti_3SiC_2. (b) Twinning in TiC structure by replacement of Si by C atoms. (c) Relaxation in TiC by rotation around axis (b) (reprinted with permission from Reference[5]).

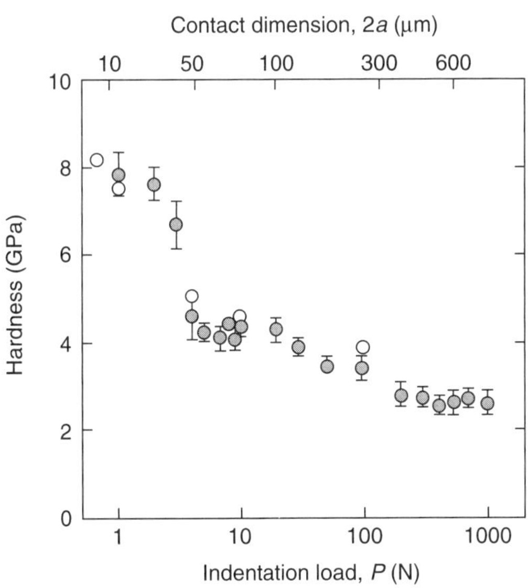

Figure 12.5 Hardness of Ti_3SiC_2 as a function of load.[6]

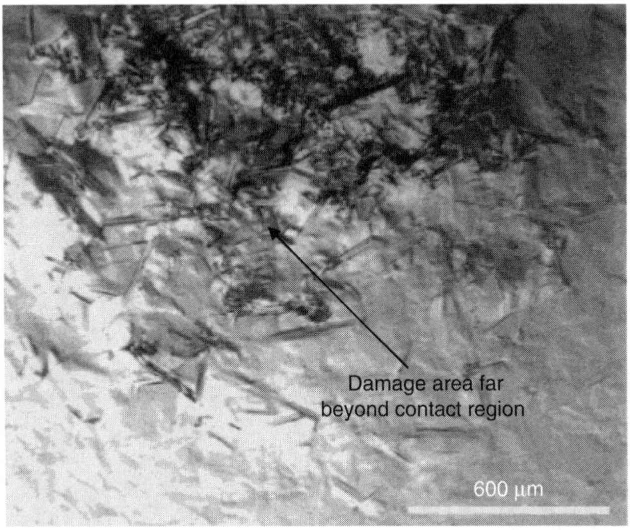

(a)

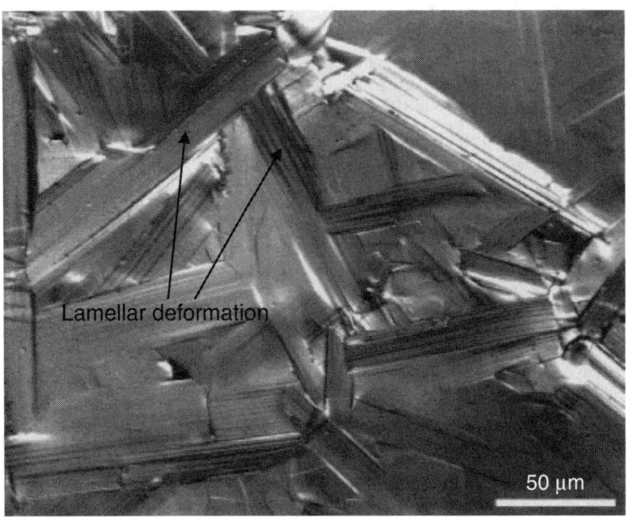

(b)

Figure 12.6 Micromechanical damage showing (a) damage beyond contact region and (b) lamellar deformation along grain boundaries in the subsurface region.[6]

is presented in Figure 12.6a.[6] The presence of lamellar structure showing slip along basal planes and deformation along grain boundaries is confirmed via scanning electron microscopy (SEM) images (Fig. 12.6b).[6] The bulk deformation of the MAX phases represents very much that of a kink band (Fig. 12.7a)[3] that generates in a pack of cards under extreme pressures (Fig. 12.7b).[3]

Once the tough class of nanolaminated ceramics is achieved, the characteristic toughness of these must be evaluated via resistance curves (R curves). A typical R

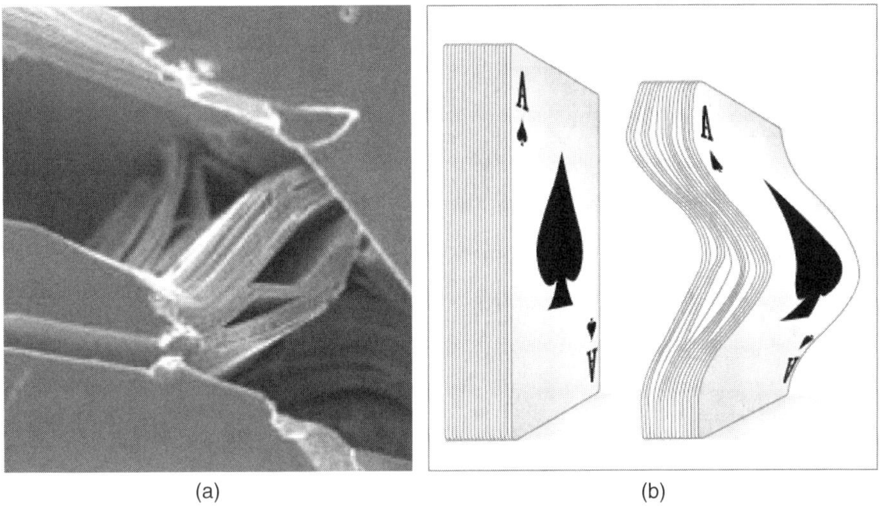

Figure 12.7 (a) Deformation of MAX phase. (b) Representative deformation of MAX phase shown as similar to that of a deck of cards.[3]

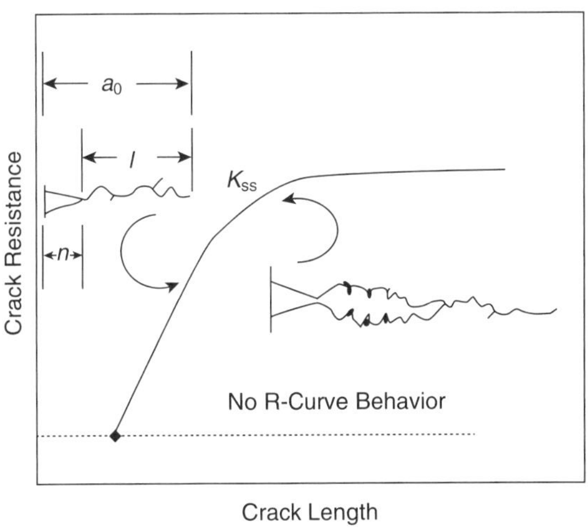

Figure 12.8 Functional dependence of crack resistance with the crack front length. The R curve shows more damage tolerance than the materials not showing R-curve behavior.[7]

curve of toughened ceramic is presented in Figure 12.8, showing a process zone developing around the crack front.[7]

Activation of basal slip planes in the MAX phases results in damage tolerance as observed in the Vickers indentation showing absence of cracking even at 50-N load (Fig. 12.9a).[8] However, as the grain size is decreased (2–4 μm) compared with that of 10–20 μm in coarse-grained MAX phase, Vickers indentation cracks have appeared. Crack emanation implies that energy dissipation occurs mainly by kinking of laminates, followed by buckling in coarse-grained samples, whereas this kinking gets limited in fine-grained MAX phase and leads to cracking (Fig. 12.9b).[8]

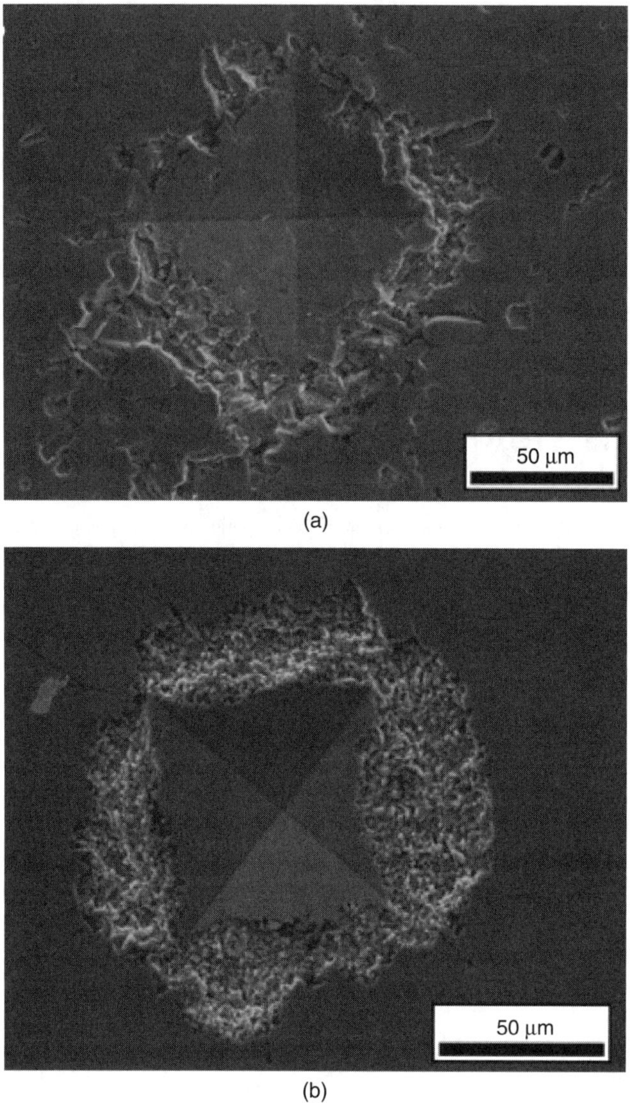

Figure 12.9 Scanning electron microscope (SEM) micrographs of Vickers indentations of (a) coarse-grained and (b) fine-grained Ti_3SiC_2, Vickers indented at a load of 50 N.[8]

An important aspect of mechanical behavior of Ti_3SiC_2 ceramics is that Ti_3SiC_2 exhibits R-curve behavior. Precracking for the four-point bend test can also have strong implications for the resulting fracture toughness of the Ti_3SiC_2 MAX phase. Consequently, the effect of precracking shows a variation in the fracture toughness, indicating that Ti_3SiC_2 follows the characteristic R-curve behavior (Fig. 12.10).[7]

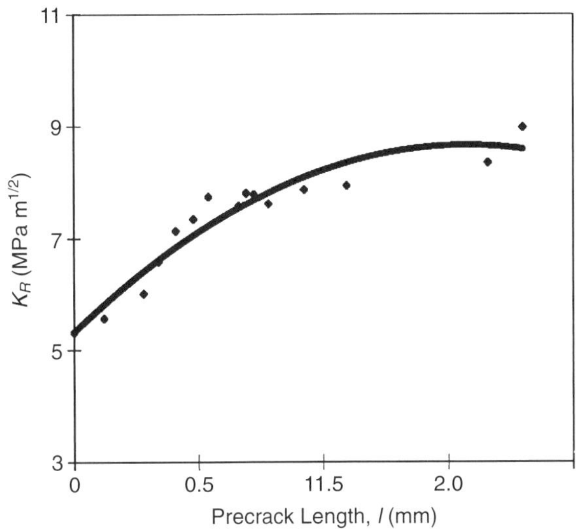

Figure 12.10 R-curve behavior of Ti_3SiC_2 showing change in fracture toughness with change in precrack length.[7]

The experimentally measured values of the mode I critical stress intensity factor (K_{IC}), a measure of fracture toughness, summarized in Figure 12.10 were obtained using the following equations[9,10]:

$$K_{IC} = g \cdot \frac{P_{max} S}{BW^{3/2}} \frac{3(a/w)^{1/2}}{2(1-a/w)^{3/2}} \tag{12.3}$$

where

$$g = \frac{1.99 - (a/W)(1-a/W)[2.15 - 3.93(a/W) + 2.7(a/W)^2]}{1 + 2(a/W)}, \tag{12.4}$$

where a = average precrack length, W = thickness of the beam (nominally 4 mm), g is a geometric factor based on the average crack length and specimen thickness, P_{max} = maximum load required for fracture, $S = (S_0 - S_i)$, with outer span length S_0 = 30 mm and inner span length S_i = 10 mm, B = width of the beam (nominally 3 mm). It should be noted that Equations 12.3 and 12.4 are valid for (S/W) ratio of 4–5 and (a/W) ratio varying between 0.35 and 0.70.

Following a theoretical analysis,[11] R-curve behavior can be described by the following functional dependence of toughness on crack size:

$$K_R(c) = K_{IO} + A\left[1 - \exp\left\{\frac{-(c - c_0)}{B}\right\}\right], \tag{12.5}$$

where $K_R(c)$ is crack growth resistance with a particular precrack length (MPa m$^{1/2}$), K_{IO} is fracture toughness without a precrack (MPa m$^{1/2}$), A and B are constants for a particular system, with A having units of MPa m$^{1/2}$ and B in millimeters, $c_0 = 0$ (initial flaw size, only V-notch, without any precrack length, i.e., n), and c is precrack length (l), in millimeters.

In the case of hot-pressed Ti_3SiC_2, the R-curve behavior can be described as

$$K_R(c) = 5.32 + 3.186\left[1 - \exp\left\{\frac{-(c - c_0)}{0.56}\right\}\right]. \tag{12.6}$$

A significant bridging effect is expected in the coarse-grained Ti_3SiC_2 structure, which leads to higher plateau toughness.[7] The fracture toughness of Ti_3SiC_2, measured by compact tension (CT) method was reported to be 9.5 MPa m$^{1/2}$ and 16 MPa m$^{1/2}$ for fine- (3–10 μm) and coarse- (50–200 μm) grained, Ti_3SiC_2, respectively.[12]

R-curve behavior results from the unique interaction between a growing crack and microstructural characteristics. Ti_3SiC_2 is characterized by relatively weak bonding between the silicon layer and the TiC octahedra along the basal plane. Usually, the dislocations of Ti_3SiC_2 are reported to be mobile and multiply at room temperature.[1] The dislocation movement is restricted on either the basal plane or kink boundaries. In general, the deformation behavior Ti_3SiC_2 is unusual for carbides and is caused primarily by the layered structure and the metallic nature of the bonding.[13] In addition to regular slip, the mechanisms for room-temperature plastic deformation in Ti_3SiC_2 involve the readjustment of local stress fields from kink band (boundaries) formation, buckling, and delamination of individual grains.[13] Specifically, delamination along the weaker basal planes facilitates the formation of microlaminae contained within a single grain; consequently, the deformation and distortion of such laminae contribute to toughening.

The R-curve characteristic cracking-behavior essentially implies that a crack path is highly tortuous in nature and allows crack deflection along the weak interface (Fig. 12.11a).[7] The fluctuating crack tip needs to overcome the path transition barrier and follow an extended path, resulting in high absorption of fracture energy. In addition, the buckling of the basal planes allows the breakage of a few tile-like features to block the crack completely, as in Figure 12.11b,[7] and generate an entirely new crack from that region.[7] Thereby, the fracture toughness of Ti_3SiC_2 is enhanced because of bifurcation of the crack path as well.

In addition, fracture energy is also absorbed due to crack branching and crack delamination at the weak interface. This relaxes the stress at the crack tip and improves the fracture resistance as well as damage resistance. Such bridging processes provide evidence for both elastic-ligament bridging and frictional pullout, similar to bridging processes observed in Al_2O_3/SiC_w composites.[11] In addition, crack propagation causes more transgranular and/or translamellar cracking. In particular, this severely diminishes the propensity for grain bridging in the crack wake in the fine-grained (5–20 μm) microstructure. This enhanced crack growth resistance can be attributed to the metallic nature of the bonding and the absence of strong in-plane Si–Si bonds in case of Ti_3SiC_2.[1] This bonding characteristic is highly unusual in ceramic systems and accounts for high steady-state fracture toughness by promoting crack bridging.

12.4 WEAR OF Ti_3SiC_2 MAX PHASE

The tribological properties of Ti_3SiC_2 was evaluated to a limited extent. Myhra et al.,[14] using lateral force microscopy, recorded extremely low kinetic friction

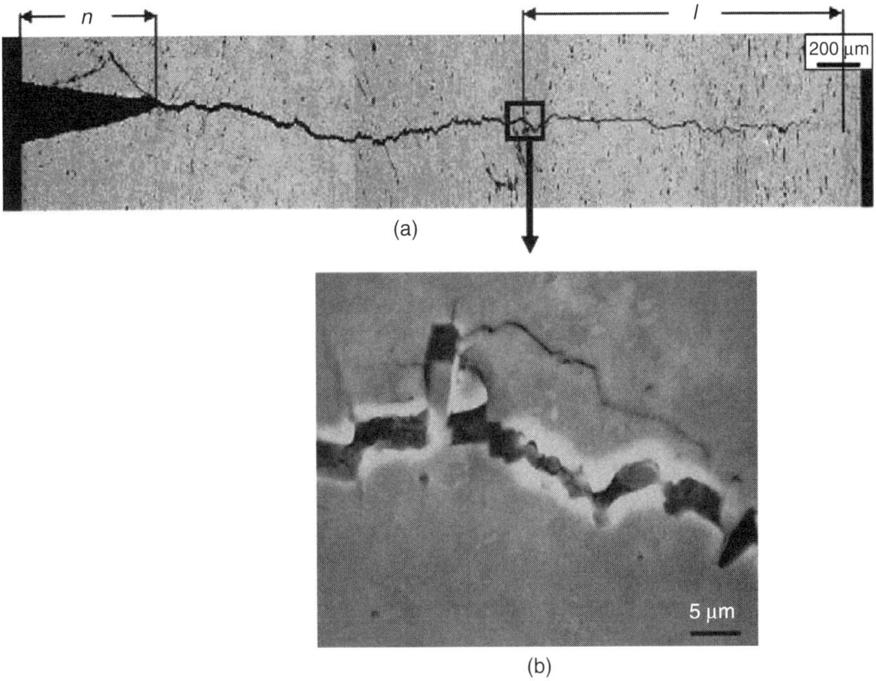

Figure 12.11 Fracture surface of Ti_3SiC_2 after four-point bending, showing (a) tortuous crack path and (b) crack bifurcation and impediment.[7]

coefficient (μ) of 0.002–0.005 at 1000–20,000 nN normal force for the basal planes of Ti_3SiC_2 single crystal. Barsoum and coworkers measured a rather high coefficient of friction (COF) of 0.8 for a Ti_3SiC_2/steel tribocouple at 5-N load.[15] The frictional characteristics, as measured using a pin-on-disk tribometer, were found to be independent of Ti_3SiC_2 grain size (5–100 μm). Zhang et al.[16] investigated the friction and wear behavior of a self-mated Ti_3SiC_2 tribocouple and Ti_3SiC_2/diamond pair using a pin-on-flat tribometer. The COF of the former is 1.16–1.43, but that of the latter is below 0.1 for varying loads of 0.49–9.8 N. The low COF of Ti_3SiC_2 against diamond was attributed to the formation of a lubricating film. Chen et al.[17] reported that a ternary metal silicide, $Cr_{13}Ni_5Si_2$, alloy could exhibit excellent wear resistance compared with hardened steel and 1.0%C–1.5%Cr–containing tool steel at loads of 98–196 N, measured using a ball-on-wheel tribometer. The tribological study of MAX phases in contact with Ni-alloy reported COFs of ~0.5–0.6.[18] High wear rate was found to be due to the release of massive wear particles during the abrasion.[18] Apparently, there is not much difference in the COF reported for various MAX phases (see Table 12.4 and Fig. 12.12).[18]

During wear at high temperatures (550°C), tribofilms form on both of the contact surfaces. It is reported that this tribofilm is amorphous or nanocrystalline composed of partially oxidized superalloy, which smears over the MAX surfaces, making it wear resistant compared with phenomena occurring at room temperature.[19]

Table 12.4. Wear Rate (WR), and Coefficient of Friction (μ) of Various MAX Phases with Respect to Ni-Superalloy at a Load of 3 N at Ambient Environment[18]

Specimen	Dynamic partner	WR of MAX ($mm^3/N{\cdot}m$)	μ mean
Ti_2AlC	Inc718	$\approx 5.5 \times 10^{-4}$	0.5 ± 0.1
Ti_2AlN	Inc718	$\approx 3 \times 10^{-2}$	0.8 ± 0.15
Ti_4AlN_3	Inc718	$\approx 3 \times 10^{-2}$	0.8 ± 0.15
Ti_3SiC_2	Inc718	$\approx 2.5 \times 10^{-2}$	0.6 ± 0.15
Cr_2AlC	Inc718	$\approx 1.2 \times 10^{-3}$	0.6 ± 0.1
Ta_2AlC	Inc718	$\approx 1.5 \times 10^{-2}$	0.5 ± 0.1
Cr_2GeC	Inc600	$\approx 1.2 \times 10^{-3}$	0.5 ± 0.1
Cr_2GaC	Inc600	$\approx 4 \times 10^{-3}$	0.4 ± 0.1
Ti_2SnC	Inc600	$\approx 8 \times 10^{-3}$	0.63 ± 0.1
Nb_2SnC	Inc600	$\approx 1.5 \times 10^{-2}$	0.63 ± 0.1

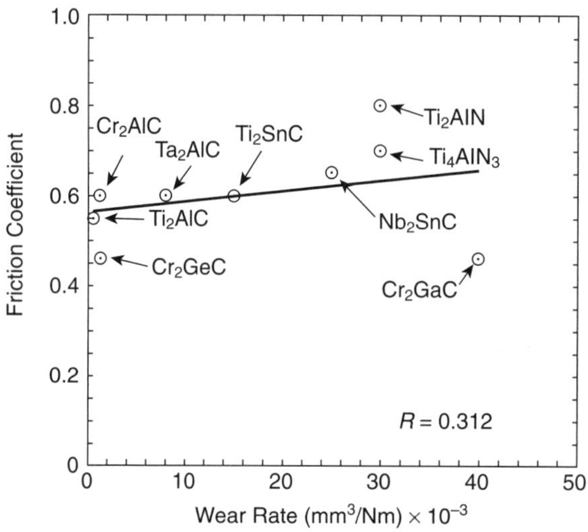

Figure 12.12 Wear rate of MAX phases with a fitted linear regression line ($R = 0.312$).[18]

This phenomenon reduces friction and wear rate, as shown in Table 12.5.[18] The smoother surface of the postwear surface (tested at 550°C) is shown in Figure 12.13.[18] The postsliding root mean square (RMS) roughness of the Ta_2AlC was 350 nm (Fig. 12.13).[18] Similarly the postwear roughness of Cr_2AlC, Ti_3SiC_2, and Ti_2AlC (not shown) surfaces were 325, 540, and 272 nm, respectively. From the preceding results, it appears that Ti_3SiC_2 can be used at moderately high temperature.

All the aforementioned results were obtained using a pin-on-disk wear tester. Since the wear mechanisms and tribological data typically depend on the contact

Table 12.5. Wear Rate (WR), and Coefficient of Friction (μ) of Various MAX Phases with Respect to Ni-Superalloy of 550°C with 3-N Load

Static partner	Dynamic partner	WR_s (mm^3/N·m)	WR_d (mm^3/N·m)	μ
Ta_2AlC	Inc718	$\geq 1 \times 10^{-6}$	$\sim 10^{-5}$	≈0.4
Ti_3SiC		$\geq 1 \times 10^{-6}$		≈0.4
Cr_2AlC		$\geq 1 \times 10^{-6}$	$\sim 10^{-5}$	≈0.3
Ti_2AlC		$\geq 1 \times 10^{-6}$		≈0.4
Cr_2GeC	Inc600	$\approx 5 \times 10^{-4}$		≈0.5
Cr_2GaC	Inc600	$\approx 6 \times 10^{-6}$		≈0.35
Ti_2AlN	Inc600	$\approx 3 \times 10^{-5}$		≈0.4
Ti_4AlN_3	Inc600	$\approx 1 \times 10^{-3}$		≈0.6

WR_s, specific wear rate of the static tribopartner; WR_d, specific wear rate of the dynamic tribopartner.[18]

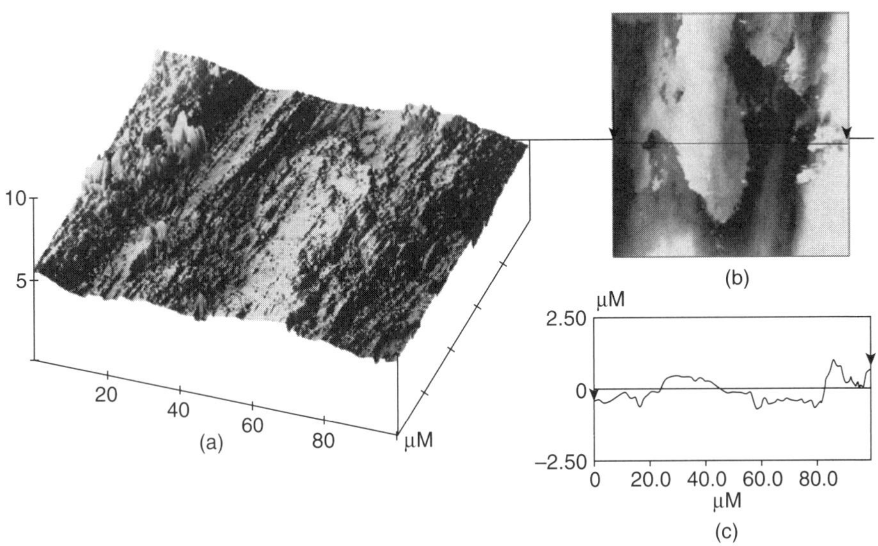

Figure 12.13 Postwear atomic force microscopy (AFM) analysis of 100 × 100 μm Ta_2AlC surface against Inconel 718 for 2-km sliding at 550°C: (a) 3D isometric view, (b) top view, and (c) side view profile of the region marked by arrows in top view.[18]

configuration, some studies were also conducted using a fretting wear tester, which can provide the opportunity to study the friction and wear properties under small-amplitude oscillatory reciprocatory motion. In the following, the fretting wear results[20] obtained with hot pressed dense Ti_3SiC_2 (~99% density) are summarized. In fretting experiments, the stroke length, oscillation frequency, and total number of fretting cycles were fixed at 100 μm, 8 Hz, and 100,000, respectively (Fig. 12.14).[20] Low loads (~1–8 N) were used to evaluate the resulting COF and damage of the MAX phase. The wear rate was observed to increase with load from 1 to 2 N

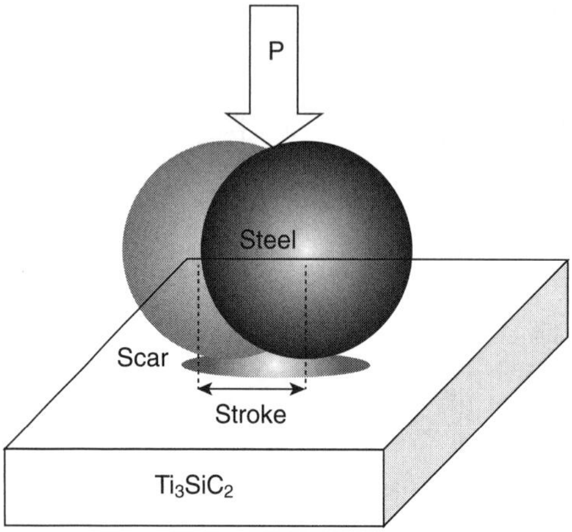

Figure 12.14 Schematic of ball-on-flat fretting configuration.[20]

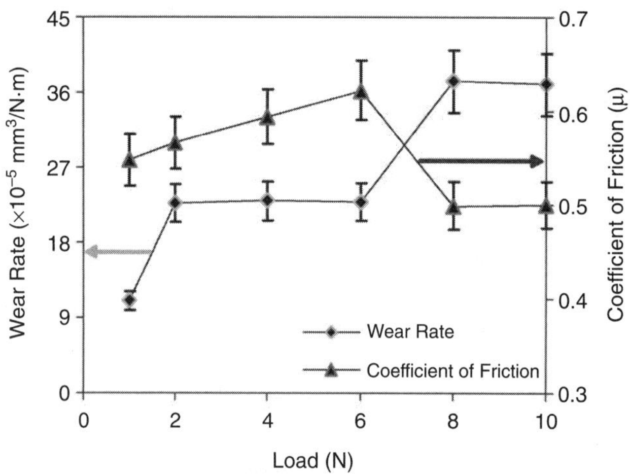

Figure 12.15 Wear rate and coefficient of friction for Ti_3SiC_2 fretted against steel.[20]

(~10 × 10^{-5} mm^3/N·m), whereas it appeared constant for the load range 2–6 N (Fig. 12.15).[20] However, a transition was observed in the wear rate (from 25 × 10^{-5} mm^3/N·m to ~37 × 10^{-5} mm^3/N·m) when load was increased from 6 to ~8 N.[20] A typical illustration of the increasing damage with increasing load is depicted in Figure 12.16.[20] At 10-N load, the transverse wear scar diameter is maximum and around 500–510 μm.

To see the topographical features of the worn surfaces after fretting in the low-load regime, some representative wear scar images are presented in Figure12.17.[20] The deeper abrasive scratches of groove width around 2–3 μm are observed even at the lowest load (1 N). The presence of a transfer layer and wear debris particles (brighter contrast) are found to adhere on the abrasive scratches (Fig. 12.17a). The formation of a tribochemical layer is significant at 4-N load, as shown in Figure 12.17b. However, the tribochemical layer is nonprotective and spalls off due to

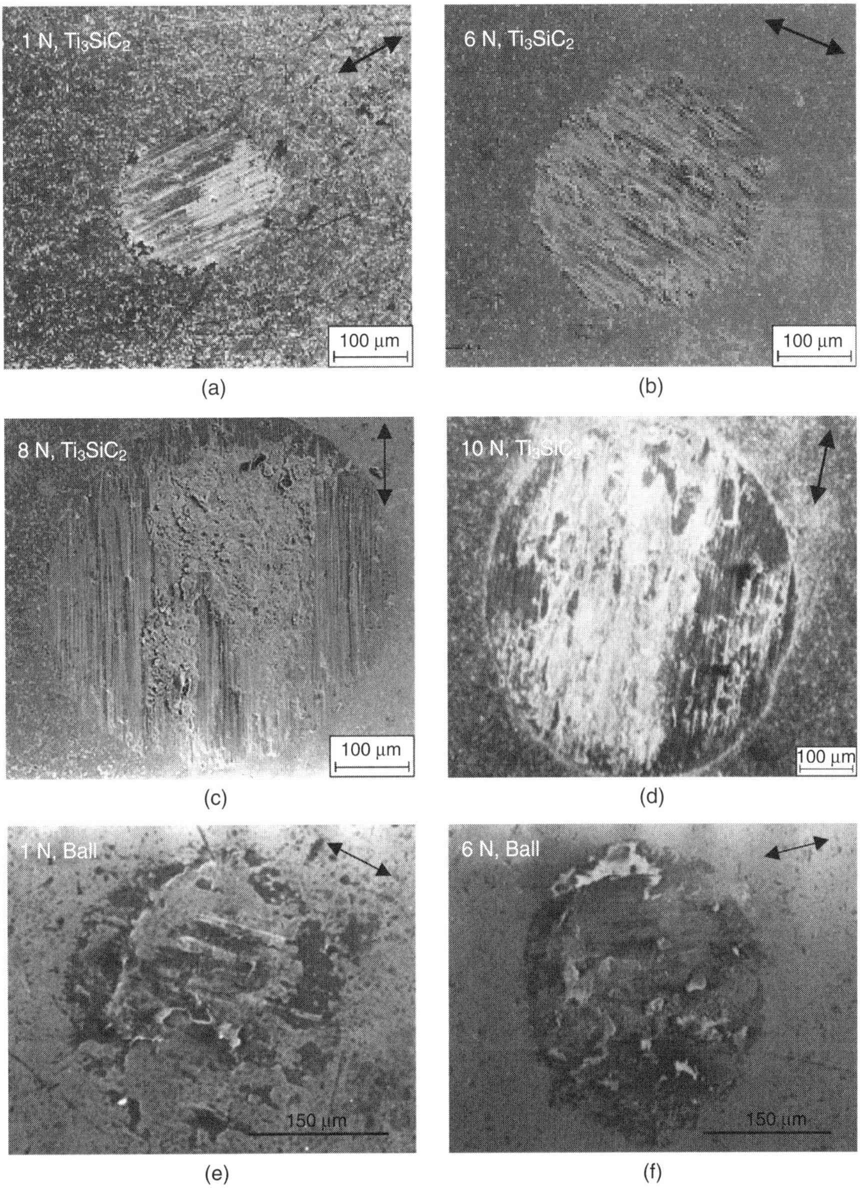

Figure 12.16 Postfretting wear topography of Ti3SiC2 flat (a–d) as well as steel counterbody (e and f), for 100,000 cycles at varying loads. The double-pointed arrow indicates the sliding direction (adapted from Reference[20]).

propagation of cracks, primarily in the direction perpendicular to the fretting motion. Also, at 6-N load, a plastically deformed layer contributes to the material removal (Fig. 12.17c).

A clear change in worn surface topography is observed at higher loads of 8 and 10 N (see Fig. 12.18[20]). The evidence of extensive plastic deformation is clear at 8-N

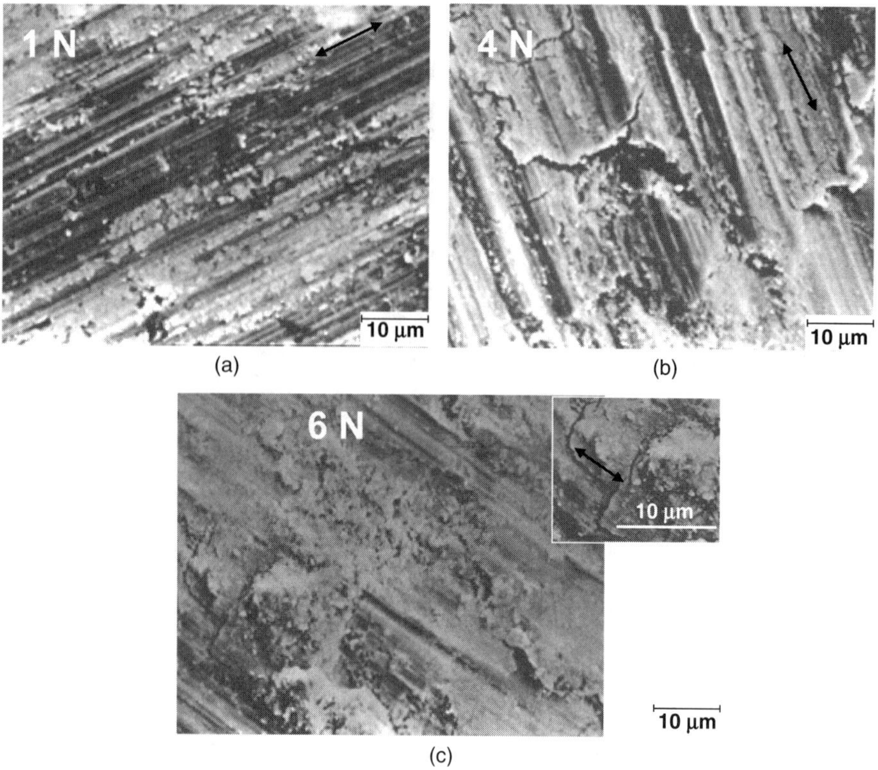

Figure 12.17 SEM images showing deeper abrasive scratches of groove width around 2–3 μm (a), spalling of nonprotective tribochemical layer due to propagation of cracks (b and c) on Ti_3SiC_2 worn surface after fretting against bearing steel for 100,000 cycles under varying loads, as indicated on the individual micrographs.[20]

load (Fig. 12.18a) and the topographical features are more like that of worn surface, commonly observed with metallic materials. The fracture of elongated Ti_3SiC_2 grains on the fretted surface is also observed and the cracks are found to propagate along the edges of elongated grains. At the highest load, 10 N, the contribution of plastic deformation is significant and the presence of cracks is also observed (Fig. 12.18b).

To characterize the tribochemical layer, Raman spectroscopy analysis of worn Ti_3SiC_2 is shown in Figure 12.19.[20] A comparison with literature reports[19,21] reveals the formation of Fe_2O_3, TiO_2, and SiO_2. It is therefore evident that the steel ball is severely oxidized and the transfer of oxidized metallic debris takes place during fretting. Also, it is believed that residual TiC is oxidized to TiO_2 during fretting and Ti_3SiC_2 oxidizes to form TiO_2 and SiO_2. Therefore, the tribochemical oxidation reaction can be expressed as follows:

$$2Fe + (3/2)O_2 = Fe_2O_3; \qquad (12.7)$$

$$TiC + 2O_2 = TiO_2 + CO_2\ (g); \qquad (12.8)$$

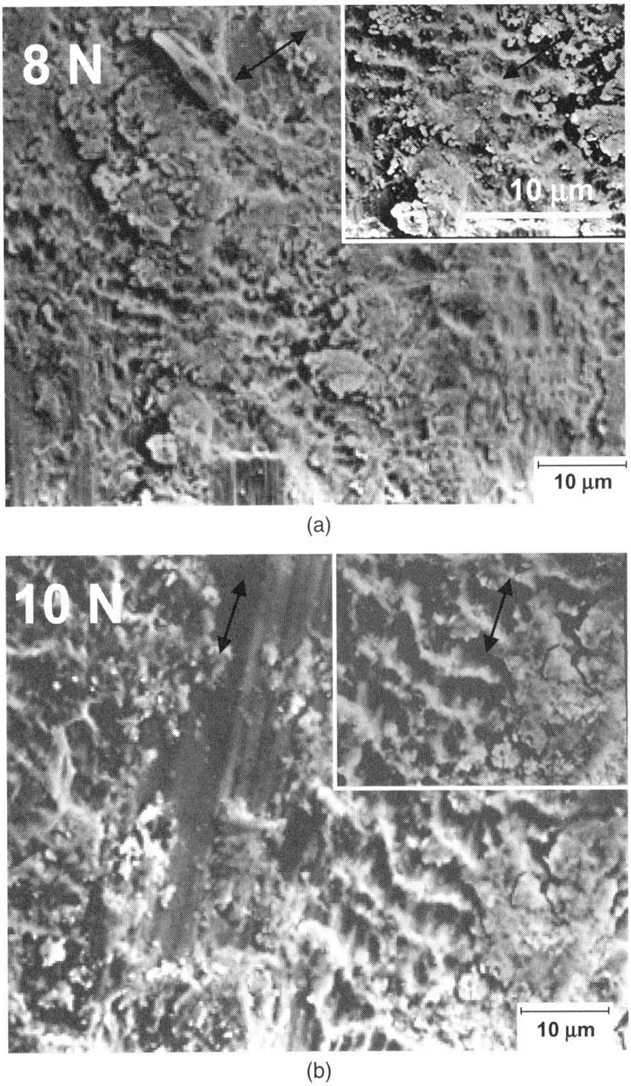

Figure 12.18 SEM images showing extensive plastic deformation (a and b) on the fretted surface of Ti_3SiC_2 after fretting against steel for 100,000 cycles at high loads (8 and 10 N). The details of the deformed tribolayer are also shown in the inset of (a) and (b). The double-pointed arrow indicates the sliding direction.[20]

$$Ti_3SiC_2 + 6O_2 = 3TiO_2 + SiO_2 + 2CO_2\ (g). \quad (12.9)$$

The literature report[22] confirms that Reactions 12.7–12.9 initiate at 900°C and the parabolic oxidation behavior in air between 900 and 1400°C facilitates the formation of distinct rutile and silica layers. The activation energy of such reactions is 370 ± 20 kJ/mol. From the preceding discussion, the major mechanisms contributing

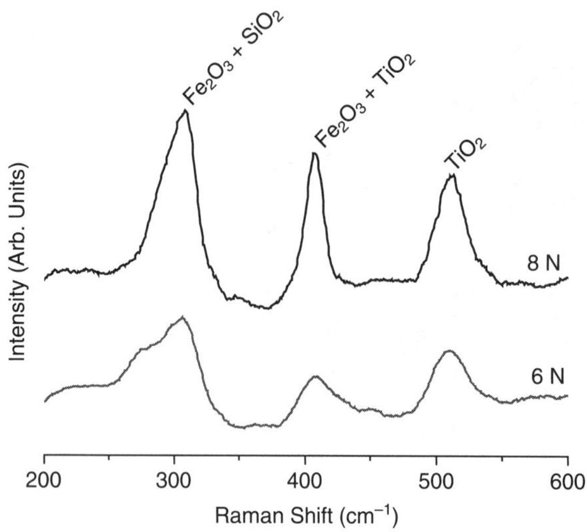

Figure 12.19 Raman spectra obtained from the fretted surface on Ti_3SiC_2, worn against steel for 100,000 cycles for different loads, as indicated against the individual spectra.[20]

to the process of friction and wear of Ti_3SiC_2 can be summarized as (1) abrasion, (2) tribochemical layer formation, and (3) plastic deformation.

Before discussing the mechanisms of material removal from tribosurfaces, it is imperative to distinguish between abrasion and tribochemical wear. When the surface asperities (irregular protuberances of the surface profile) of the harder mating solid slide on a softer surface, the deformation- or plowing-induced damage of the softer material is known as abrasion. In the case of contact between two mating solids, it is known as two-body abrasion. Alternatively, if the harder wear debris particles abrade the softer of the mating solids, then it is known as three-body abrasion. In contrast, tribochemical wear is described as a material removal process due to the chemical reactions either between a sliding solid and the surrounding environment or between two moving solids, resulting in the formation of a distinct tribolayer. The chemical reactions are triggered at tribological contacts due to frictional energy. Friction and wear with tribochemical layer formation largely depends on the stability or properties of the layer itself.

As reported,[11] the steady-state COF of a Ti_3SiC_2/steel couple increased from 0.55 to 0.6 as load was increased from 1 N to 6 N. The increase in COF is presumably due to more severe abrasion with increased load, as revealed in Figure 12.17. However, a decrease in COF from 0.62 to 0.5 was recorded with the increased load from 6 to 8 N and COF remains constant (~0.5) at 10-N load (Fig. 12.15). This decrease in COF can be explained as follows. At the higher load (>6 N), the formation of tribochemical reaction products and wear debris takes place to a larger extent. These debris (third body) particles, while being entrapped between Ti_3SiC_2 and steel (first two bodies), tend to roll during sliding motion, thereby decreasing the friction. Hence, a transition from two-body to three-body abrasion takes place at higher load (>6 N), as is also observed in Figure 12.15.

Another interesting observation is that Ti_3SiC_2, despite having a characteristic chainlike unit cell structure, experiences higher COF (0.5–0.6), compared with other ceramics with layered structures. Typically, solid lubricants such as graphite and $MoSi_2$ are strongly anisotropic in their mechanical properties, with less resistance to shear deformation in the basal planes than in other directions. Displacement of the layers occurs by easy slippage, leading to low COF under ambient humidity (~40–50% relative humidity [RH]) and temperature (23–25°C). For example, graphite and $MoSi_2$ experience low COF of ~0.2 at room temperature, and the COF increases to 0.8 between 400 and 600°C.[26] It is also reported that h-BN has similarly low COF (0.2), which is maintained even at higher temperature (850°C). The fact that Ti_3SiC_2 has higher COF indicates that a similar lubrication mechanism does not operate. This is essentially because of the inherent bond structure, as shown in Figure 12.3. It has been reported[27] that the interatomic bond length in Ti(1)–Si is around 2.69 Å, which is lower than the interplanar van der Waals bond length in graphite (3.40 Å). Because of the smaller bond length, the bond strength is expected to be higher in Ti_3SiC_2 than in other lamellar solids (MoS_2, graphite), and this causes the difficulty of slippage in the Ti–C–Ti–C–Ti–Si network. This also contributes to a reasonably higher COF (0.5–0.6) of Ti_3SiC_2. It can be mentioned here that mica, having a characteristic lamellar structure, does not deform easily under shear force and Ti_3SiC_2 also exhibits higher COF than other solid lubricants (WS_2, graphite, MoS_2, and polytetrafluoroethylene [PTFE]).[28]

As far as the characteristics of Ti_3SiC_2 wear behavior are concerned, severe abrasion on Ti_3SiC_2, even at the lowest load (1 N), is primarily due to the difference in hardness between mating counterfaces. At intermediate loads (4 and 6 N), cracking of the tribolayer is significant and the nonprotective nature of the tribolayer increases wear of Ti_3SiC_2. A change in wear mechanism is recorded at high loads (8–10 N). It should be noted here that, although the tribochemical wear remains an active wear mechanism at load > 6 N, plastic deformation appears to be significant. At 8-N load, severe plastic deformation contributes to increase in wear rate. The deformation of Ti_3SiC_2, as explained in the existing literature,[2,4,8,12] can be used to elucidate the observed deformation-induced wear. In Ti_3SiC_2, two adjacent covalent bond chains of Ti–C–Ti–C–Ti–Si form a chain couple with length equal to the cell dimension in the c-direction (see Figs. 12.3 and 12.4). The chains are bonded together with strong metallic Ti layers, which were found to be inhomogeneous in the free charge density distributions along the c-axis. Therefore, deformation of Ti_3SiC_2 at the worn surfaces can also be expected from the metallic-like nature of its bonds. Also, the interaction between Ti planes is mediated by hexagonal layers of C and Si atoms. Previous investigations reported that the Ti(1,2)–C interaction has a stronger covalent p–d nature compared with the Ti(1)–Si interaction. However, a rigid interaction exists between Si–Si atoms inside Si monolayers.[8,12] Additionally, the polar character of the directional bonding reveals the presence of ionic bonding in Ti–C and Ti–Si interactions. This anisotropy of metallic–covalent–ionic bonding was thought to be responsible for the Ti_3SiC_2 plasticity. It is worth noting here that, although Young's modulus (E) of the investigated polycrystalline Ti_3SiC_2 is quite high (316 GPa), the ratio of modulus to hardness (~63) lies in the regime of ductile materials.[12] According to Barsoum and co-workers, plastic deformation of Ti_3SiC_2 was attributed to delamination and kink band formation at temperatures above 1200°C and at room temperature if grains are oriented. Also, the plastic deformation occurring only at higher load (8 N) indicates that the plasticity of the chainlike structure under fretting motion requires a critical contact pressure. The calculated Hertzian contact pressure at 8-N load is around 800 MPa. Also, a considerable fraction of frictional energy is dissipated as heat energy, which is partitioned at the fretting contact between two mating solids. This evidently increases the contact temperature. Therefore, the combined effect of high contact pressure (at load ≥8 N) and high temperature results in the observed plasticity on the worn surfaces of Ti_3SiC_2.

12.5 CONCLUDING REMARKS

Summarizing the discussion of this chapter, it is clear that ternary carbides, such as Ti_3SiC_2, are possibly among the few monolithic ceramics (such as SiAlONs) that can exhibit a strong crack growth resistance. It has been reasoned that the characteristic layered structure of Ti_3SiC_2 is the origin of the rising R-curve behavior. This was further confirmed by the crack tip shielding by grain bridging in the crack wake.

Interestingly, Ti_3SiC_2 exhibits interesting tribological behavior at varying loads under fretting contacts. A transition in friction and fretting wear rate is critically observed in the low-load region (<10 N). Such a transition is a result of the interplay among abrasion (two-body vs. three-body), tribochemical reactions, and deformation wear.

Special introduction of evaluating toughness and tribological properties of MAX phases is essential as a separate section because their failure is more in the laminated form (like the compression of a deck of cards along its longitudinal direction), providing them some degree of toughness. This tiling nature of MAX phases separates them from other ceramics since they retain their high wear and abrasion resistance while demonstrating higher toughness. In addition, the machining of MAX phases was shown to be quite easy, as reported by the Barsoum group.[1–3] Hence, the shaping of MAX-phase ceramics also does not constrain the selection of ceramics for certain applications.

REFERENCES

1 M. W. Barsoum, D. Brodkin, and T. El-Raghy. Layered machinable ceramics for high temperature applications. Scrip. Met. Mater. 36 (1997), 535–541.

2 M. W. Barsoum and T. El-Raghy. Synthesis and characterization of a remarkable ceramic: Ti_3SiC_2. J. Am. Ceram. Soc. 79 (1996), 1953–1956.

3 M. W. Barsoum and T. E. Raghy. The MAX phases: Unique new carbide and nitride materials. Am. Sci. 89 (2001), 334–343.

4 B. Manoun, S. K. Saxena, G. Hug, A. Ganguly, E. N. Hoffman, and M. W. Barsoum. Synthesis and compressibility of $Ti_3(Al,Sn_{0.2})C_2$ and $Ti_3Al(C_{0.5},N_{0.5})_2$. J. Appl. Phys. 101 (2007), 113523.

5 M. W. Barsoum. The $M_{N+1}AX_N$ phases: A new class of solids: Thermodynamically stable nanolaminates. Prog. Solid State Chem. 28 (2000), 201–281.

6 I. M. Low, S. K. Lee, B. R. Lawn, and M. W. Barsoum. Contact damage accumulation in Ti_3SiC_2. J. Am. Ceram. Soc. 81(1) (1998), 225–228.

7 D. Sarkar, B. Basu, M. C. Chu, and S. J. Cho. R-curve behavior of Ti_3SiC_2. Ceram. Int. 33 (2007), 789–793.

8 S. Amini, M. W. Barsoum, and T. El-Raghyy. Synthesis and mechanical properties of fully dense Ti_2SC. J. Am. Ceram. Soc. 90(12) (2007), 3953–3958.

9 H. Tada. *The Stress Analysis of Cracks Handbook*, 2nd ed. Paris Productions Inc, St Louis, MO, 1985.

10 Standard test methods for the determination of fracture toughness of advanced ceramics at ambient temperatures. ASTM Provisional Standard Designation No. PS070-97. American Society for Testing and Materials, West Conshohocken, PA.

11 B. R. Lawn, *Fracture of Brittle Solids*, 2nd. Edition. Cambridge University Press, Cambridge, UK, 1998.

12 D. Chen, K. Shirato, M. W. Barsoum, T. El-Raghy, and R. O. Ritchie. Cyclic fatigue-crack growth and fracture properties in Ti_3SiC_2 ceramics at elevated temperatures. J. Am. Ceram. Soc. 84(12) (2001), 2914–2920.

13 M. W. Barsoum, L. Farber, T. El-Raghy, and I. Levin. Dislocations, kink bands and room temperature plasticity of Ti_3SiC_2. Metall. Mater. Trans. A 30 (1999), 1727–1738.

14 S. Myhra, J. W. B. Summers, and E. H. Kisi. Ti_3SiC_2—A layered ceramic exhibiting ultra-low friction. Mater. Lett. 39 (1999), 6–11.

15 T. E. Raghy, P. Blau, and M. W. Barsoum. Effect of grain size on friction and wear behavior of Ti_3SiC_2. Wear 238 (2000), 125–130.

16 Y. Zhang, G. P. Ding, Y. C. Zhou, and B. C. Cai. Ti_3SiC_2—A self-lubricating ceramic. Mater. Lett. 55 (2002), 285–289.

17 M. Chen, K. Kato, and K. Adachi. Friction and wear of selfmated SiC and Si_3N_4 sliding in water. Wear 250 (2001), 246–255.

18 S. Gupta, D. Filimonov, V. Zaitsev, T. Palanisamy, and M. W. Barsoum. Ambient and 550°C tribological behavior of select MAX phases against Ni-based superalloys. Wear 264 (2008), 270–278.

19 U. Serincan, G. Kartopu, A. Guennes, T. G. Finstad, R. Turan, Y. Ekinci, and S. C. Bayliss. Characterization of Ge nanocrystals embedded in SiO_2 by Raman spectroscopy. Semicond. Sci. Technol. 19 (2004), 247–251.

20 D. Sarkar, B. V. M. Kumar, and B. Basu. Understanding the fretting wear of Ti_3SiC_2. J. Eur. Ceram. Soc. 26 (2006), 2441–2452.

21 S. C. Tjong. Electron microscope and Raman characterization of the surface oxides formed on the Fe–Cr alloys at 400–850°C. Mater. Character. 26 (1991), 29.

22 M. W. Barsoum, T. E. Raghy, and L. Ogbuji. Oxidation of Ti_3SiC_2 in Air. J. Electrochem. Soc. 144 (1997), 2508–2516.

23 T. E. Raghy, A. Zavaliangos, M. W. Barsoum, and S. R. Kalidindi. Damage mechanisms around hardness indentation in Ti_3SiC_2. J. Am. Ceram. Soc. 80 (1997), 513–516.

24 J. Etzkorn, M. Ade, D. Kotzott, M. Kleczek, and H. Hillebrecht. Ti_2GaC, Ti_4GaC_3 and Cr_2GaC—Synthesis, crystal growth and structure analysis of Ga-containing MAX-phases Mn+1GaCn with M ¼ Ti, Cr and n ¼ 1, 3. J. Solid State Chem. 182 (2009), 995–1002.

25 A. Onodera, H. Hirano, T. Yuasa, N. F. Gao, and Y. Miyamoto. Static compression of Ti_3SiC_2 to 61 GPa. Appl. Phys. Lett. 74 (1999), 3782–3784.

26 R. Deacon. Lubrication by lamellar solids. Proc. Roy. Soc. 243A (1957), 464–482.

27 N. I. Medvedeva, D. L. Novikov, A. L. Ivanovsky, M. V. Kuznetsov, and A. J. Freeman, Electronic properties of Ti_3SiC_2-based solid solutions. Phys. Rev. B, 58 (1998), 16042–16050.

28 B. Bhushan, *Principles and Applications of Tribology*, John Wiley & Sons 1999, 413.

Section Five

High-Temperature Ceramics

Chapter 13

Overview: High-Temperature Ceramics

The last few decades have witnessed the development of various boride-based materials and such wider efforts are particularly due to their combination of properties, which include high hardness, elastic modulus, abrasion resistance, and superior thermal and electrical conductivity. The targeted applications include high-temperature structural materials, cutting tools, armor material, electrode materials in metal smelting, and wear parts. In this overview chapter, a review of the current state of knowledge in the development of bulk boride-based materials is presented, with particular emphasis on consolidation–microstructure–property relationships.

13.1 INTRODUCTION

The boride-based structural ceramics, because of their refractoriness and high-temperature strength, are well suited for applications at high temperature.[1] Among the structural ceramics, titanium diboride (TiB_2) is considered as the base material for a range of high-technology applications.[2,3] TiB_2 is a refractory material with a combination of attractive properties, including exceptional hardness (≈25–35 GPa at room temperature, more than three times harder than fully hardened structural steel), which is retained up to high temperature. It has a high melting point (>3000°C), good creep resistance, good thermal conductivity (~65 W/m·K), high electrical conductivity, and considerable chemical stability. TiB_2 also has properties similar to TiC, an important base material for cermets and many of its properties, namely, hardness, thermal conductivity, electrical conductivity, and oxidation resistance, are better than those of TiC.[4–7] The unique combination of properties enables TiB_2 to be a candidate material for heavy-duty wear applications, particularly at elevated temperatures.

With respect to chemical stability, relevant in high-temperature machining applications, TiB_2 is more stable in contact with pure iron than are WC and Si_3N_4 and, therefore, TiB_2-based materials can be preferred over WC-based materials for high-temperature applications. The good electrical conductivity of TiB_2 (electrical

Advanced Structural Ceramics, First Edition. Bikramjit Basu, Kantesh Balani.

resistivity ≈13 × 10^{-8} Ω·m) makes it an excellent candidate for special electrical applications, for example, cathodes used in aluminum electrosmelting or vaporizing elements for vacuum metal deposition installations.[8] The relatively low fracture toughness of monolithic TiB_2 (≈5 MPa·$m^{1/2}$) is the bottleneck for engineering applications.[9] This has been a major driver for considerable research efforts to improve both toughness and sinterability.[10,11]

One of the avenues to improved sinterability and toughness of boride and carbide materials has led to the development of cermets,* wherein a metallic binder is used to obtain dense bulk materials.[12–14] For example, the TiB_2-based cermets typically contain a metallic binder phase (Co, Ni). Cemented borides are developed in the TiB_2–Fe system.[15] These materials could be a novel lower density (having potentially higher hardness) substitute for the WC/Co system. In a different approach, the addition of TiB_2 to an Al_2O_3 or Si_3N_4 matrix considerably increases its hardness, strength, fracture toughness, and electrical conductivity.[16,17] Such composites have found applications as wear parts, cutting tools, and heat exchangers. Additionally, these electroconductive toughened ceramics can be suitably shaped by electrodischarge machining (EDM) to obtain complex-shaped components, greatly increasing the number of industrial applications of these ceramic materials.

13.2 PHASE DIAGRAM AND CRYSTAL STRUCTURE

Titanium diboride is a stable intermetallic compound in the Ti–B system (see Fig. 13.1). The Ti–B binary system is characterized by three intermetallic phases: orthorhombic TiB; orthorhombic Ti_3B_4; and hexagonal TiB_2, which has a stoichiometric composition range of 28.5–30.0 wt% boron. Due to its stability and high melting point, TiB_2 is a candidate material for high-temperature structural applications. In a TiB_2 lattice, Ti atoms form a hexagonal close-packed (HCP) structure in TiB_2. The hexagonal unit cell of single-crystal TiB_2, having space group P6/mmm ($a = b = 3.029$ Å, $c = 3.229$ Å; $\alpha = \beta = 90°$, $\gamma = 120°$) is illustrated in Figure 13.2a. The boron (B) atoms are located interstitially between the A-layers, leading to a strong network structure (see Fig. 13.2b). The high hardness and elastic modulus of TiB_2 as well as its chemical resistance are primarily due to its inherent crystal structure and atomic bonding.

Table 13.1 summarizes the important physical and mechanical properties of various engineering ceramics with those of TiB_2. From Table 13.1, TiB_2 is superior to other advanced structural ceramics, such as Al_2O_3, B_4C, and SiC, in terms of its better hardness and toughness and its electrical and thermal conductivity. In particular, in regards to hardness, TiB_2 (35 GPa) is better than TiC (32 GPa), WC (24 GPa), and Si_3N_4 (25 GPa), except B_4C (47 GPa). Since densification of TiB_2 remains a major challenge, the sintering of monolithic borides will be discussed in the following section.

*Theoretically, all ceramic materials with a metal binder are classified as cermets, including the cemented carbides. However, the cutting tool industry considers only the TiC-, Ti(C,N)-, and TiB_2-based materials to be cermets, while the WC-based materials are classified as cemented carbides.

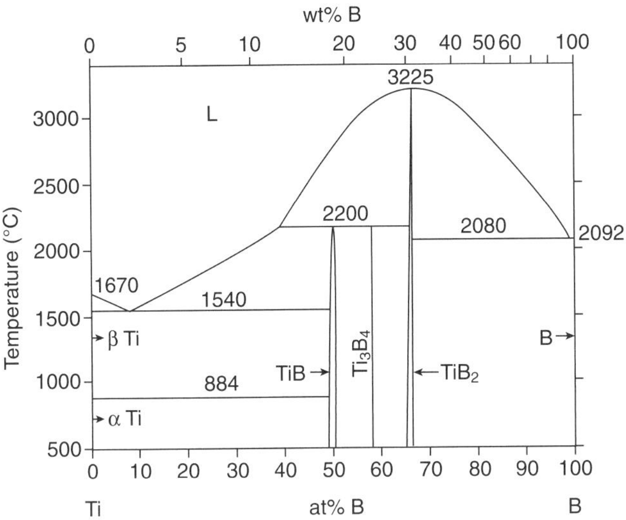

Figure 13.1 Ti–B binary equilibrium phase diagram, indicating the possibility of formation of three intermetallic compounds, that is, TiB, Ti_3B_4, and TiB_2. Note that TiB_2 has the highest melting point (3225°C) and little off-stoichiometric compositional variation.[1]

13.3 PROCESSING, MICROSTRUCTURE, AND PROPERTIES OF BULK TiB_2

13.3.1 Preparation of TiB_2 Powder

Titanium diboride powder can be synthesized using a variety of high-temperature methods, such as the direct reactions of titanium or its oxides or hydrides, with elemental boron over 1000°C; carbothermal reduction of titanium oxide and boron oxide; or hydrogen reduction of boron halides in the presence of the metal or its halides. Among various synthesis routes, electrochemical synthesis and solid-state reactions[18]are widely used to obtain finer titanium diboride in large quantity. An example of solid-state reaction is the borothermic reduction:

$$2TiO_2 + B_4C + 3C \rightarrow 2TiB_2 + 4CO. \quad (13.1)$$

Typical oxygen and carbon content of optimally synthesized TiB_2 is around 0.5 and 0.6 wt%, respectively. Synthesized TiB_2 powders are characterized by finer sizes, with D_{50} around 1.1 μm (see Fig. 13.3). It is important to mention that it is possible to produce large quantities (kilogram scale) of phase-pure TiB_2 powders on both laboratory scale and commercial scale using the borothermic reduction process. Nevertheless, the aforementioned synthesis routes cannot produce nanosized powders. Bates and co-workers[19] produced nanocrystalline TiB_2 of sizes 5–100 nm using a solution-phase reaction of $NaBH_4$ and $TiCl_4$, followed by annealing the obtained amorphous precursor at 900–1100°C. Axelbaum et al. reported a gas-phase combustion process that directly yielded nonagglomerated, low-oxygen-content

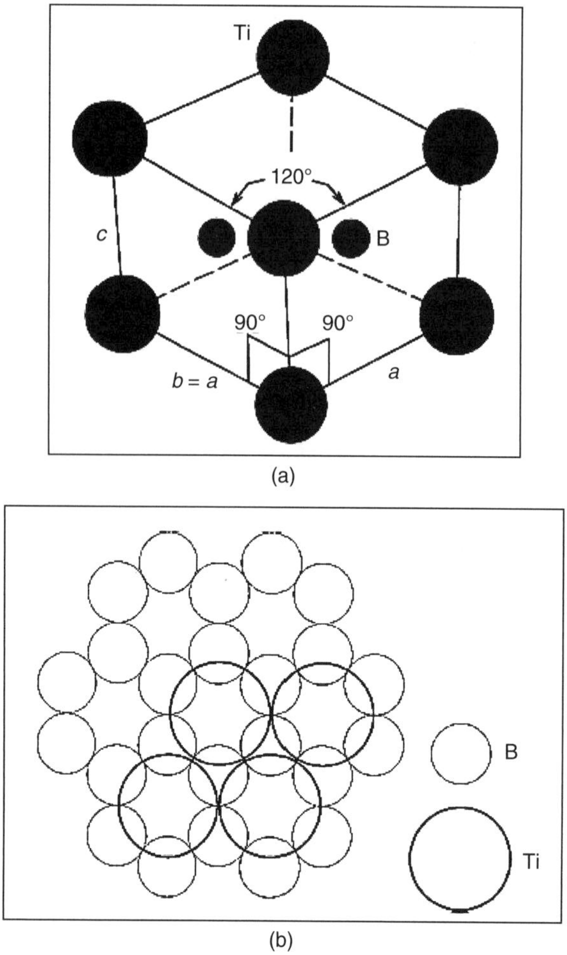

Figure 13.2 (a) The hexagonal unit cell of single-crystal TiB_2, $a = b \neq c$ ($a = b = 3.029$ Å, $c = 3.229$Å), $\alpha = \beta = 90°$, $\gamma = 120°$, 1 formula unit per cell, with Ti at (0,0,0) and B at (1/3,2/3,1/2) and (2/3,1/3,1/2). (b) Illustration of the hexagonal net of boron atoms; the Ti atoms are situated half-*c*-axis above and below the boron network. The *c*-axis is perpendicular to the paper.[1]

TiB_2 nanoparticles by the reaction of sodium vapor with $TiCl_4$ and BCl_3.[20] However, the as-synthesized powders were reported to be contaminated with metallic titanium and titanium oxide.

Another synthesis approach to obtain submicrometer-sized TiB_2 powder is mechanical alloying of the mixture of elemental Ti and B powders.[21] No amorphization occurs during the process, because of the negative heat of formation of TiB_2. It has been reported that the size of the transition metal and the heat of formation of borides influenced the mechanical alloying time, while producing finer-sized TiB_2. Ultrafine TiB_2 powder could also be produced using a self-propagating high-temperature synthesis (SHS) process, involving addition of varying amounts of NaCl.[22] TiB_2 powders with the finest particle size (26 nm) were obtained in the case of 20 wt% NaCl addition. The ignition temperature for the stoichiometric mixture of TiO_2, H_3BO_3, and Mg was as low as 685°C.

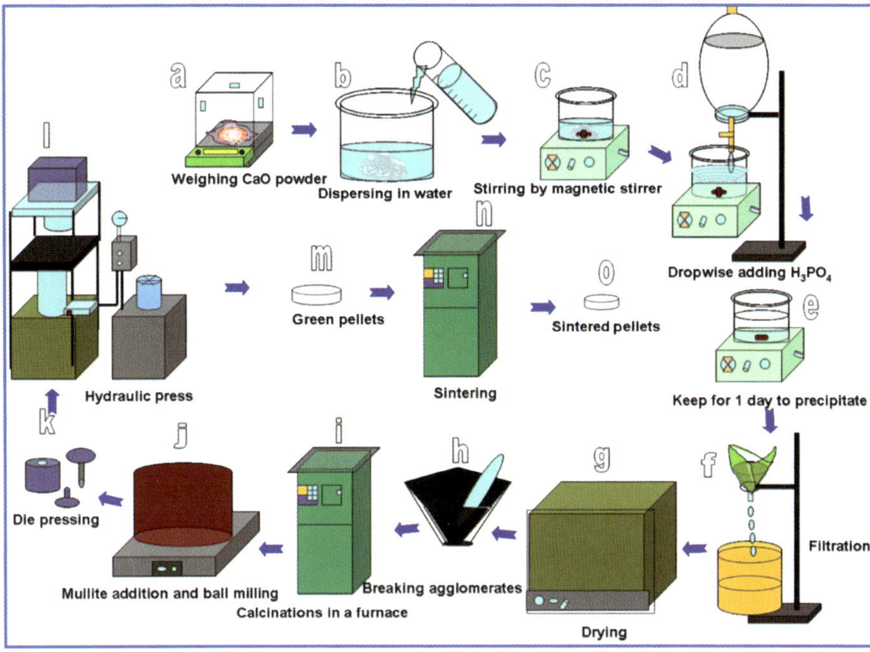

Figure 4.2 (a) Illustration of the synthesis of HAp powders and subsequent cold pressing to obtain green pellets.

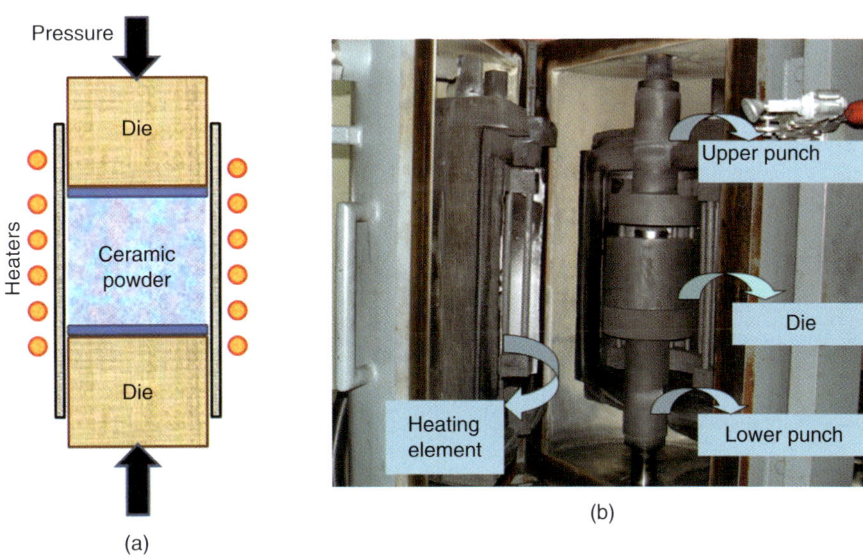

Figure 6.2 (a) Schematic and (b) industrial hot-pressing setup.

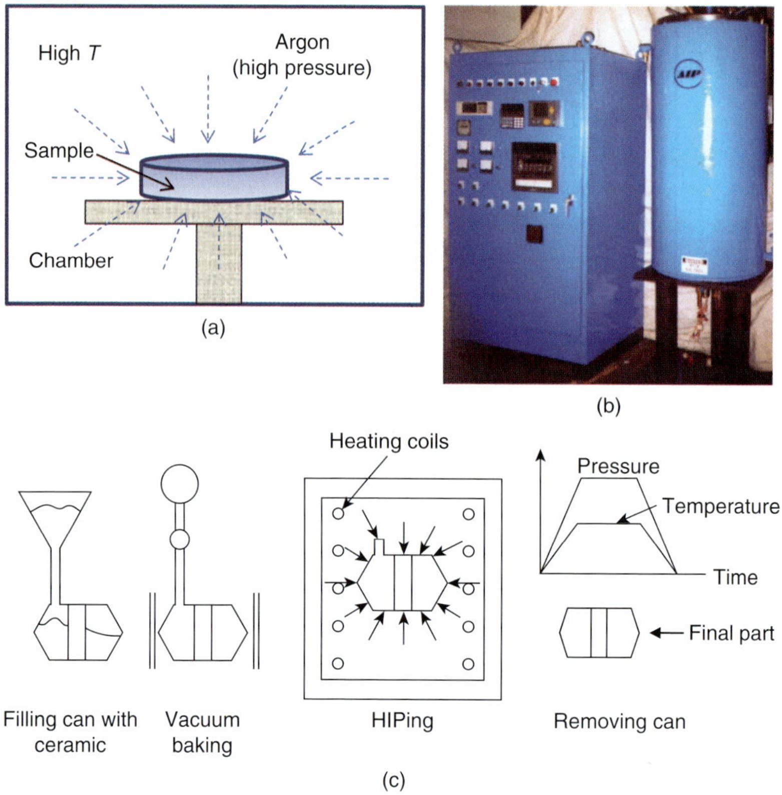

Figure 6.7 Schematic showing (a) the principle involved, (b) a commercial unit for hot isostatic pressing (HIPing), and (c) the sequence of HIPing.

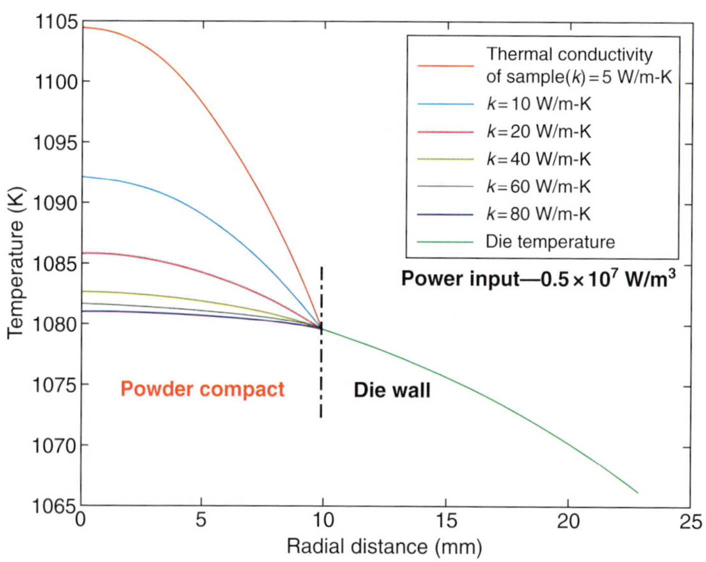

Figure 6.15 Effect of thermal conductivity in generating temperature gradient from surface to core of the ceramic powders inside a graphite die.[14]

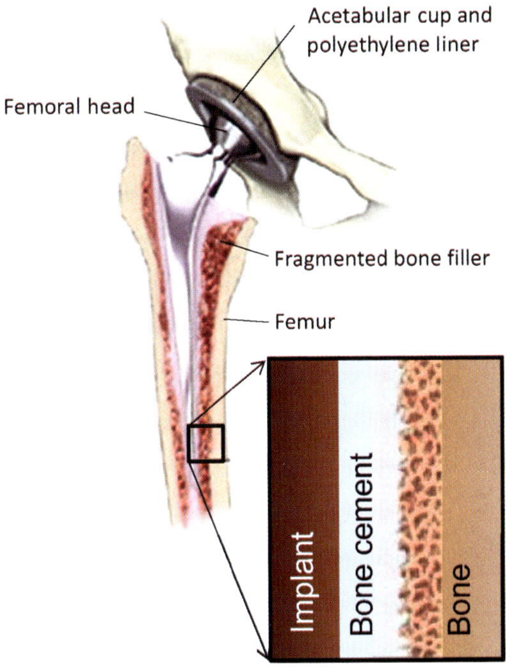

Figure 9.2 Incorporation of bone cement to join an implant with the natural bone. Adapted from http://www.zimmerindia.com/z/ctl/op/global/action/1/id/368/template/PC/prcat/P2/prod/y.

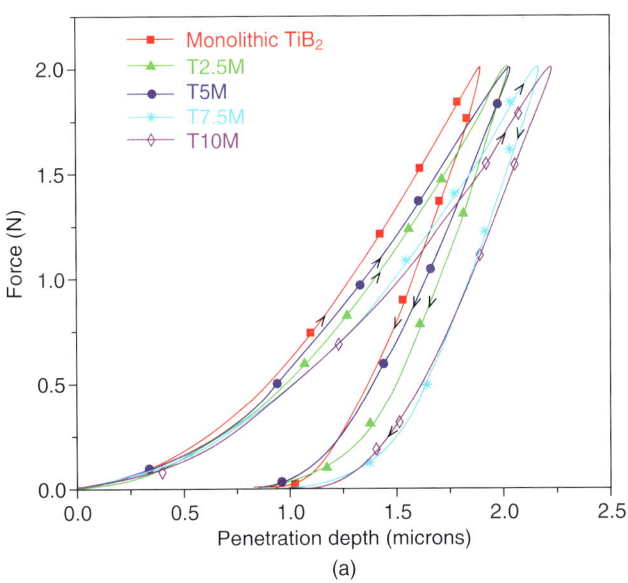

Figure 14.4 (a) Plots of load versus penetration depth (*P–d*), recorded for monolithic TiB_2 (HP, 1800°C) and TiB_2-based composites (HP, 1700°C) during instrumented Vickers indentation at a peak load of 2 N.[29]

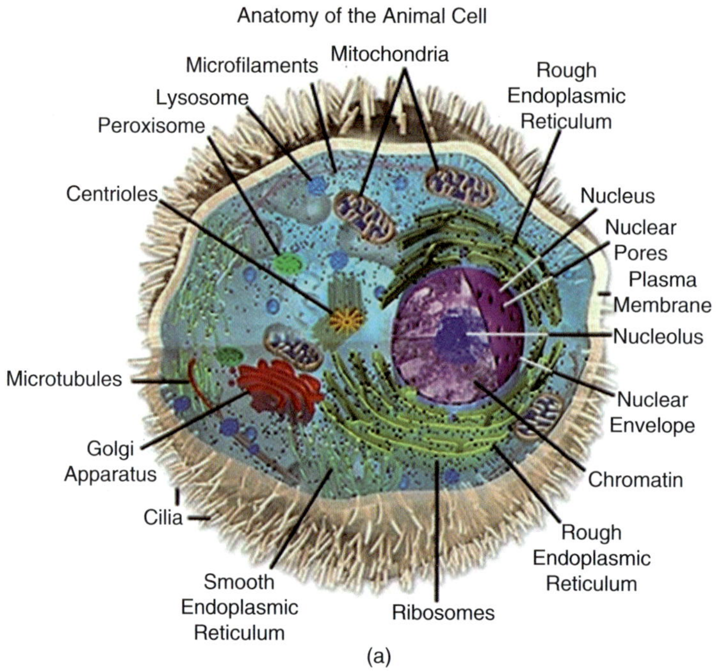

Figure 19.4 (a) Schematic illustration showing the anatomy of a eukaryotic animal cell.

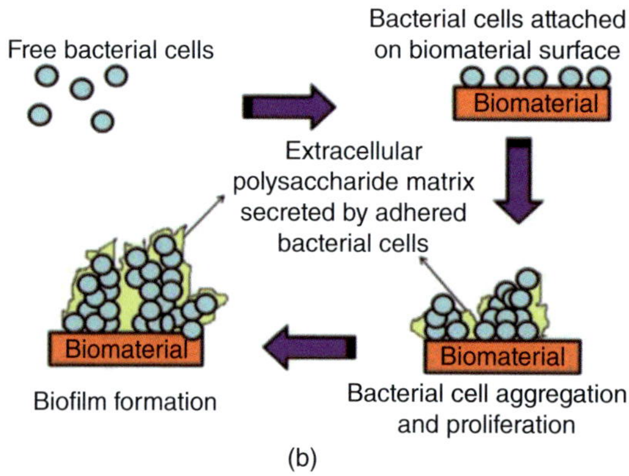

Figure 19.7 (b) Schematic representation of biofilm formation on material substrate.

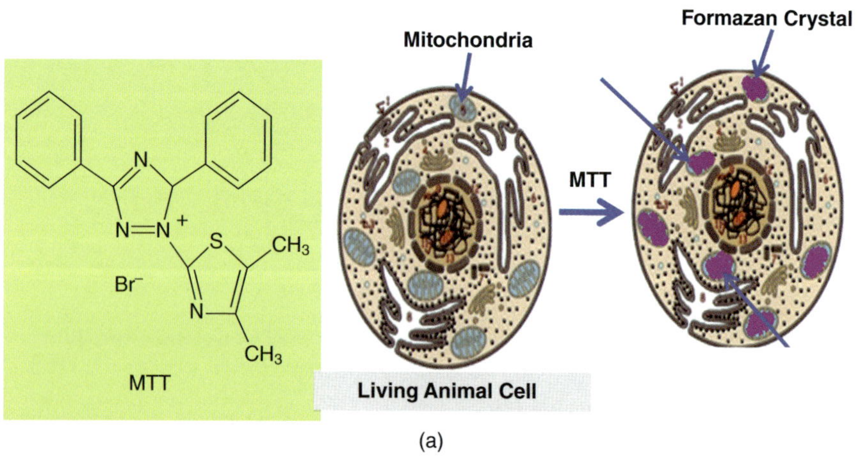

Figure 19.9 (a) Chemical structure of MTT (left, taken from http://en.wikipedia.org/wiki) as well as fundamental principle involved in MTT assay to investigate cell viability).

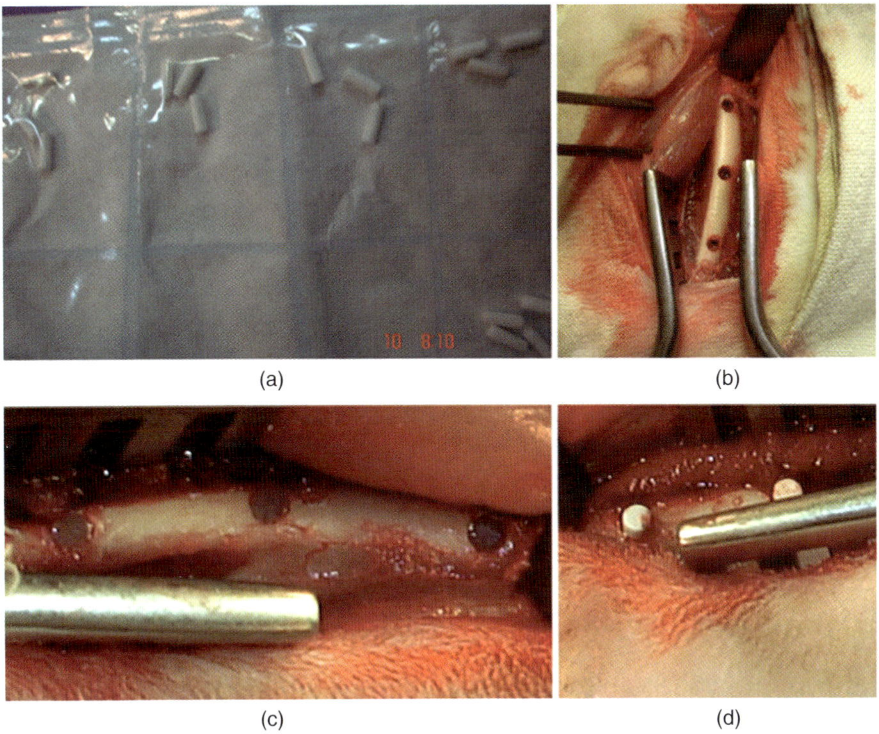

Figure 19.11 See text for full caption.

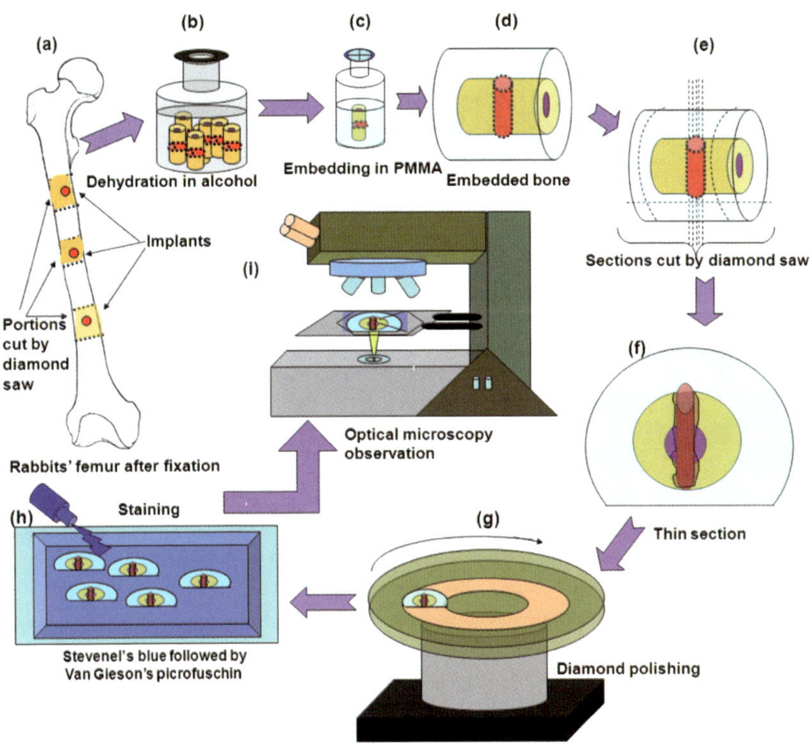

Figure 19.12 See text for full caption.

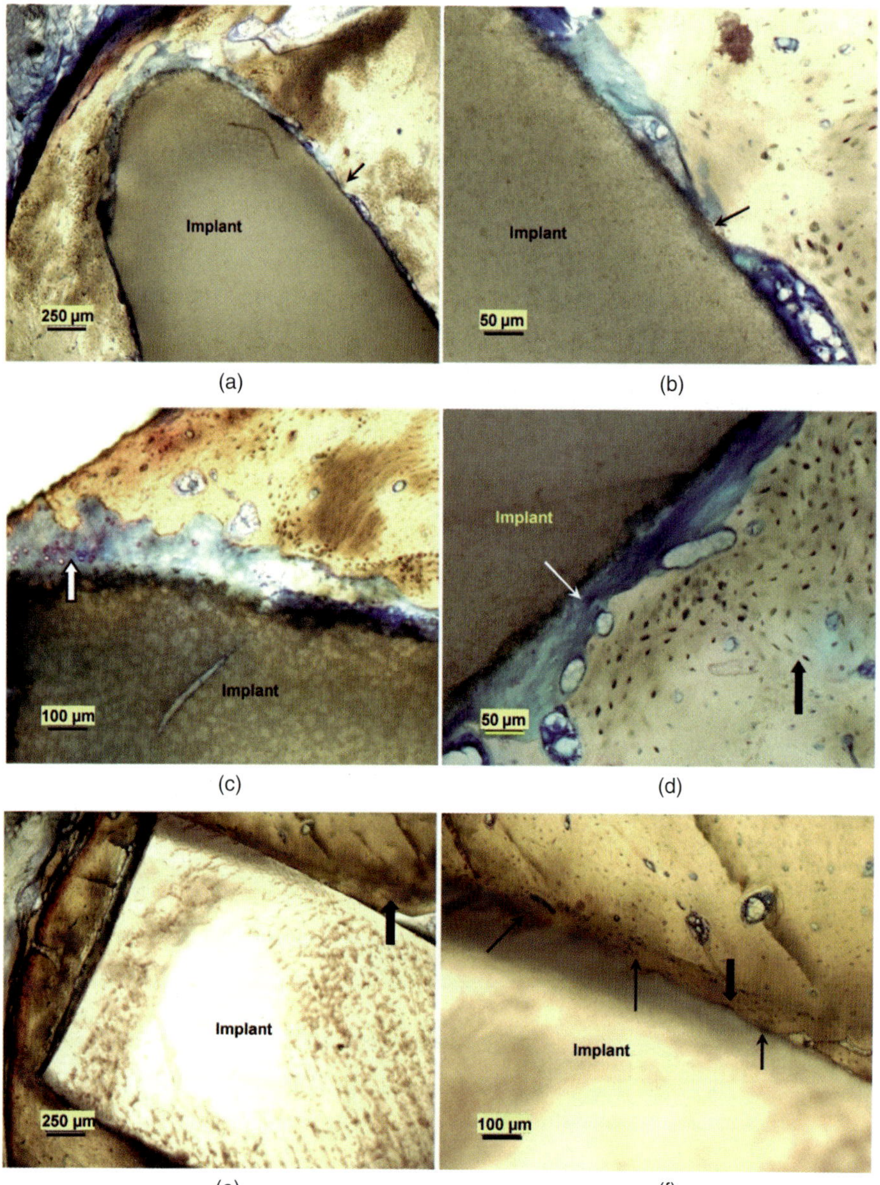

Figure 20.6 Photomicrographs of host bone–implant interface (a–d) BCP–(20 wt%)mullite and (e and f) UHMWPE after 12 weeks postimplantation: (a) bone formation around implant with focal direct contact (arrow) with implant; (b) higher magnification of focal bone–implant contact; (c) focal chondrogenesis (arrow) at interface; (d) osteocytes (white arrow) at implant–host new bone (black arrow) interface; (e) bone formation (arrow) around implant; (f) woven bone (thick arrow) at implant interface (thin arrows).[78]

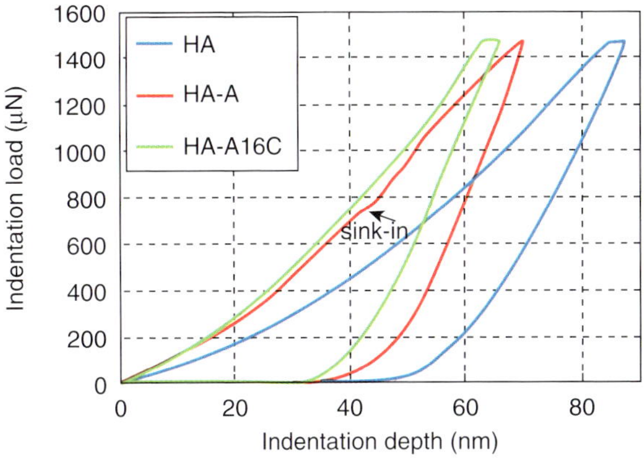

Figure 21.9 Load–depth curve of monolithic HAp (HA), HAp–Al_2O_3 (HA–A), and HAp–Al_2O_3–CNT (HA–A16C) coatings. Reprinted with permission from Reference 25.

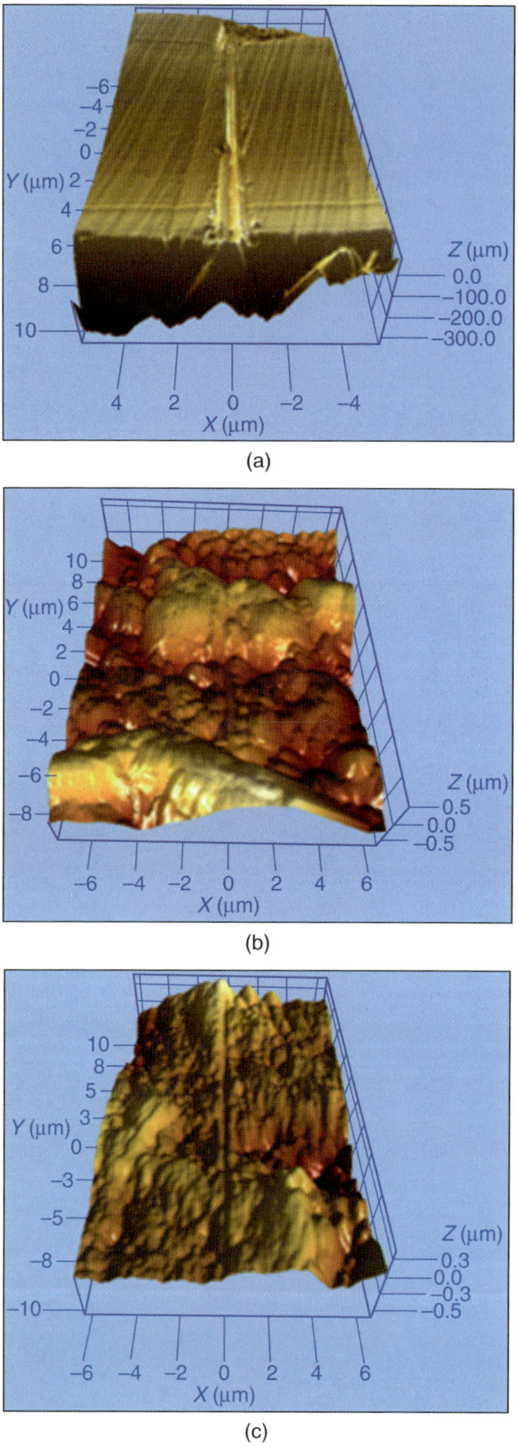

Figure 21.10 Nanoscratched surface topography of plasma-sprayed (a) monolithic HAp (HA), (b) HAp–Al_2O_3 (HA–A), and (c) HAp–Al_2O_3–CNT (HA–A16C) coatings. Reprinted with permission from Reference 25.

Table 13.1. Summary of Important Physical and Mechanical Properties of TiB_2 and the Other Important Ceramics

Property	TiB_2	ZrB_2	B_4C	SiC	WC	TiC	Si_3N_4	Al_2O_3	$MoSi_2$
Crystal structure	hex	hex	rhom	hex	hex	fcc	hex	hex	tet
Melting point (°C)	3225	3000	2450	2200	2600	3067	1900	2043	2050
Density (g/cm^3)	4.52	6.1	2.52	3.2	15.7	4.93	3.44	3.99	6.3
Linear thermal expansion, α (10^{-6}/K)	α_a- 6.6 α_c- 8.6	6.83	4.5	5.68	5.2–7.3	7.42	2.4	8.0	8.4
Thermal conductivity (W/m·K)	60–120	23.03	27.63	15–155	29–121	17–32	20–24	30.1	50–221
Electrical resistivity (10^{-6} Ω·cm)	10–30	9.2	10^6	>10^5	17	52	10^{18}	10^{20}	21
Fracture toughness, K_{IC} (MPa·m$^{1/2}$)	5–7	–	3–3.5	2.5–6	–	4	4–6	2.5–4	2–2.5
Elastic modulus (GPa)	560	350	450	480	720	400	210	400	384
Hardness (GPa)	25–35	22–26	37–47	20–35	20–24	24–32	14–25	18–21	13
Three-point flexural strength (MPa)	700–1000	–	300	300–800	480–830	240–270	1000–1200	323	–
Enthalpy (kJ/m)	324.1	326.6	72	71.6	35.2	183.8	750.5	1580.1	108.9
Oxidation resistance (°C)	1100	1100	1100	1400	800	1200	1200	>1700	1700

fcc, face-centered cubic; hex, hexagonal; rhom, rhombohedral or trigonal; tet, tetragonal (taken from Reference 1).

Figure 13.3 (a) TEM image and (b) selected area diffraction pattern (SADP) of the nanocrystalline TiB_2 powders, synthesized by reaction between $TiCl_4$ and $NaBH_4$ for 12 hours at 600°C. (c) Scanning electron microscopy (SEM) image of the micron-sized TiB_2 powders, produced by borothermic reaction among TiO_2, B_4C, and C.[1]

In an effort to synthesize nanocrystalline titanium diboride, Gu and co-workers[23] used the solvothermal reaction of metallic sodium with amorphous boron powder and $TiCl_4$ at 400°C. Such a synthesis route can be described by the following reaction:

$$TiCl_4 + 2B + 4Na \xrightarrow{\text{Benzene at } 400^\circ\text{C}} TiB_2 + 4NaCl. \quad (13.2)$$

The benzene is used as reaction medium to control the rate of reaction and particle size. Chen et al.[24] also prepared nanocrystalline TiB_2 by the reaction of $TiCl_4$ with

$NaBH_4$ at 500–700°C for 12 hours in an autoclave. As shown in Figure 13.3, the nanocrystalline TiB_2 had a size range of 10–20 nm.

From the preceding discussion, it should be evident that a few laboratory-scale synthesis routes are available to synthesize micron- or submicron-sized TiB_2 powders. From the sintering–consolidation point of view, it is necessary to obtain finer TiB_2 powders with a narrow size distribution as well as with limited agglomeration. From classical Herring's scaling law, it can be theoretically predicted that a decrease in particle size by an order of magnitude can reduce sintering time by three to four orders of magnitude. The presence of agglomerates causes a decrease in sinterability along with microcracks and macrocracks. Besides the size of TiB_2 starting powders, their purity, in terms of oxygen content, largely controls sinterability. For example, TiB_2 powders containing ≥1 wt% oxygen can be sintered up to 90% of theoretical density (ρ_{th}), even at higher sintering temperature.[25]

13.3.2 Densification and Microstructure of Binderless TiB_2

It can be reiterated here that the densification of phase-pure ceramics of transition metal borides, such as TiB_2, is inherently difficult because of three characteristics of these compounds: high melting point, low self-diffusion coefficient, and the comparatively high vapor pressure of the constituents. To obtain optimum densification, sintering temperatures exceeding 70% of the melting temperature (in degrees Kelvin) are applied during consolidation. This implies that TiB_2 ($T_m \sim 3250°C$) requires sintering temperatures of 1800–2300°C, such that appreciable grain boundary and volume diffusion-induced material transport can enable attainment of more than 95% theoretical density. However, borides undergo abnormal grain growth at high temperatures.[2] Even at relatively lower temperatures, evaporation–condensation induces growth of faceted crystals. The occurrence of microcracking at the grain boundaries is favored with the increase in grain size. Thus, it is extremely difficult to achieve uncracked dense TiB_2 by pressureless sintering, as no shape accommodation occurs in the absence of external pressure and the large pores tend to coarsen during high-temperature sintering.[2]

It was mentioned earlier that oxygen in TiB_2 starting powders is an important factor as far as their sinterability is concerned. A thin oxygen-rich layer (mainly TiO_2 and B_2O_3) has been reported to be present on the surface of TiB_2 powder, irrespective of the powder synthesis route. Baik and Becher[25] investigated the adverse effect of oxygen contamination, which gets introduced during synthesis and/or subsequent processing, on the densification of TiB_2. To achieve higher density while inhibiting abnormal grain growth, it has been suggested that the total oxygen content of the powder must be limited to less than 0.5 wt% or strong reducing additives need to be used to remove TiO_x below 1600°C.

Table 13.2 summarizes the data on sinter density and material properties, obtained with monolithic TiB_2 (without any sinter-additive). A variation in sintering condition can evidently produce varying material properties. With the exception of B_4C, other ceramics have lower hardness than TiB_2. Also, the fracture toughness of

Table 13.2. Summary of Properties of TiB_2-Based Ceramics, Sintered without Any Sinter-Additive under Various Processing Conditions

Material composition (in wt%)	Processing conditions	Sintered density (% ρ_{th})	Microstructural phases	Hardness (GPa)	Fracture toughness, K_{IC} ($MPa \cdot m^{1/2}$)	Flexural strength MPa
TiB_2 – 0%	PS at 2150°C, Ar	93.1	TiB_2	. . .	5.4	. . .
sinter-additive	HIP at 1500°C, 196 MPa, 2 hours, Ar	97.6	Equiaxed grains of 1.8 µm	. . .	3.7	650 (3-P)
	HIP at 1600°C, 200 MPa, 2 hours, Ar	. . .	Grain growth: TiB_2 2 µm → 4–6 µm	26	3.5	450(3-P)
	HP at 1600°C, 1 hour, Ar	91.0	Equiaxed TiB_2 grains of 4.8 µm	. . .	4.1	305 (3-P)
	HP at 1700°C, 1 hour, Ar	95.0	Equiaxed TiB_2 grains of 5.0 µm	. . .	4.3	498(3-P)
	HP at 1800°C, 0.5 hour, Ar	96.0	Equiaxed TiB_2 grains of 5.3 µm	. . .	5.3	545 (3-P)
	HP at 1800°C, 1 hour, Ar	97.0	Equiaxed TiB_2 grains of 8.1 µm	. . .	5.7	558 (3-P)
	HP at 1800°C, 1.5 hours, Ar	. . .	TiB_2 grains of 10.5 µm	. . .	6.3	538 (3-P)
	HP at 1800°C, 2 hours, Ar	. . .	TiB_2 grains of 12.7 µm	. . .	6.8	475 (3-P)
TiB_2 – 0%	HP at 1900°C, 1 hour, Ar	. . .	TiB_2 grains of 12.2 µm	. . .	6.2	521 (3-P)
sinter-additive	HP at 2000°C, 1 hour	. . .	. . .	26.7DPH	4.8	. . .
	HPS at 1977°C, 5 minutes, 3 GPa	94.6	Equiaxed TiB_2 grains of 1.4 µm	19.2	. . .	. . .
	HPS at 2227°C, 5 minutes, 3 GPa	97.1	. . .	21.7	2.8	. . .
	HPS at 2477°C, 5 minutes, 3 GPa	98.0	. . .	24.5	3.2	. . .
Ti + B (1:2)	HPCS at 1977°C	97.8	. . .	23.6	3.5	. . .
Ti + B (1:2)	HPCS at 2227°C	98.0	. . .	24	3.8	. . .

Ti + B (1:2)	HPCS at 2477°C	98.0	. . .	23.9	3.6	. . .
Ti + B (1:2)	HPCS at 1977°C, 5 minutes	98.7	. . .	24.2	3.9	. . .
Ti + B (1:2)	HPCS at 2227°C, 5 minutes	98.9	. . .	24.5	3.8	. . .
Ti + B (1:2)	HPCS at 2477°C, 5 minutes	99.2	. . .	24.6	4.5	. . .
Ti + B + C	HPCS at 1977°C	98.5	. . .	22.7	4.3	. . .
Al_2O_3	HP at 1550°C, 45 minutes	100	2 μm	19.4	4.6	420 (4-P)
B_4C	HP at 2150°C, 65 minutes	95.0	6–10 μm	29	2.5	220 (3-P)
β-SiC	HP at 1950°C, 2 hours	–	0.8–3 μm	–	2.5	394 (4-P)
Si_3N_4	HIP at 1650°C, 2 hours, 18 MPa, Ar	100	–	15.5	4.6	–
WC	PPC at 1600°C, 4 minutes, 35 MPa	–	5.5 μm	19.2	8.2	–
ZrB_2	PS at 2150°C, 540 minutes	98	9.1 μm	14.0	–	444(4-P)
ZrB_2	HP at 1900°C, 45 minutes	98	4.0 μm	23.0	–	565(4-P)

PS, pressureless sintering; HP, hot pressing; HIP, hot isostatic pressing; HPS, high-pressure sintering; HPCS, high-pressure self-combustion synthesis; PPC, plasma pressure compaction; 3-P, three-point bending; DPH, Vickers hardness in GPa at 1-kg load (taken from Reference 1).

TiB_2 is comparable with or better than other high-temperature ceramics, with the exception of WC. From Table 13.2, it is clear that, while hot pressing at or above 1800°C can produce more than 95% theoretical density, hot isostatic pressing (HIPing) at 1500–1600°C enables the attainment of similar density with good mechanical properties (hardness ~26 GPa and three-point flexural strength ~450 MPa). Additionally, high-pressure sintering (HPS) at temperature greater than 1900°C and at a pressure of 3 GPa has demonstrated its capability to produce 95% or higher density TiB_2 with good hardness (see Table 13.2). As mentioned in Table 13.2, the elemental powder mixture of Ti and B in 1:2 weight ratio can be used to obtain 95% or higher theoretical density via a high-pressure self-combustion synthesis (HPCS) route with carbon addition.

Figure 13.4 illustrates the influence of sintering temperature on the relative density and the grain size of TiB_2. With an increase in sintering temperature, the density increases rapidly to theoretical density at 2173 K. At high sintering temperature, TiB_2 grains grow rapidly. An illustration of increase in TiB_2 grain size with the increase in hot-pressing time is presented in Figure 13.5. The experimental data in Figure 13.5 indicate that lower sintering temperature in combination with shorter sintering time can be used to limit the grain growth of TiB_2. Plastic flow and diffusion creep are the main sintering mechanisms during the initial and final stages of sintering, respectively. The hot-pressing parameters need to be tailored to prevent abnormal grain growth.[26]

In an important paper, Baumgartner and Steiger[27] reported almost full densification (99% ρ_{th}) of titanium diboride powder by pressureless sintering. The high purity micron-sized titanium diboride powder was synthesized in the presence of excess hydrogen in arc plasma heating, according to the following reaction:

$$TiCl_4\ (g) + 2BCl_3\ (g) + 5H_2\ (g) \rightarrow TiB_2\ (s) + 10HCl\ (g). \qquad (13.3)$$

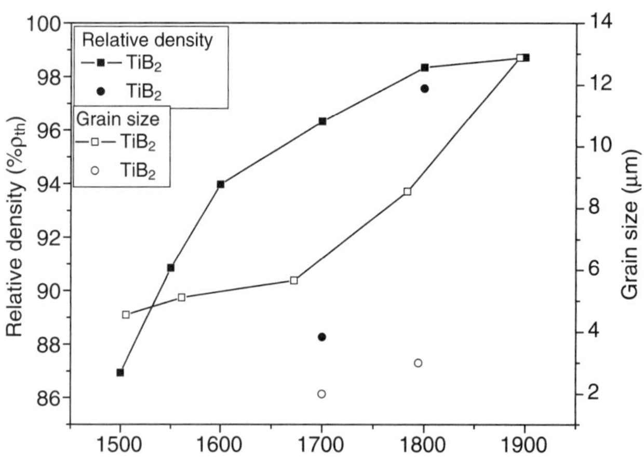

Figure 13.4 Plot showing the influence of sintering temperature on relative density and average grain size of monolithic TiB_2, sintered without any sinter-additive.[1]

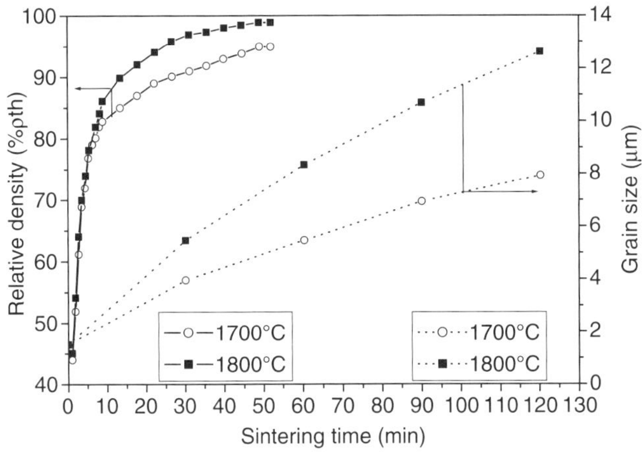

Figure 13.5 Densification and grain growth kinetics of monolithic TiB_2 (without any sintering aid) as function of sintering time at 1700 and 1800°C. It may be noted here that grain size of >15 µm can potentially promote the occurrence of microcracking during postfabrication cooling.[1]

It can be mentioned here that the difficulties in achieving fully dense titanium diboride by the pressureless sintering of carbothermic powder are attributed to rapid grain growth (activation energy ~1.02 MJ/mol, i.e., 244 kcal/mol) prior to complete densification.

13.4 USE OF METALLIC SINTER-ADDITIVES ON DENSIFICATION AND PROPERTIES

In this section, the influence of metallic binders on the densification and properties of TiB_2 is discussed.[28–30] Table 13.3 summarizes available literature reports, including some of the important cermets, which reveal the influence of metallic binders on the microstructure and mechanical properties of TiB_2-based materials. It is evident from Table 13.3 that high-density TiB_2 (>99% ρ_{th}) can be sintered with the use of smaller amounts (1–2 wt% or even lesser amounts) of metallic binders, such as Fe, Cr, and Ni. Also, very high hardness (23–31 GPa) can be achieved in combination with moderate toughness (4–6 MPa·m$^{1/2}$). Additionally it is to be noted that a modest four-point flexural strength value (>500 MPa) has been measured for dense TiB_2, sintered using metallic binder; compared with cermets, the hardness of TiB_2 (sintered with metallic additive) is better than 18 GPa, whereas the hardness of ZrB_2-, TiC-, and WC-based materials is modest (14 GPa or lower). Also, a TiB_2-based cermet has high toughness of 9.2 MPa·m$^{1/2}$, which is comparable with TiC-based cermets.

Earlier sintering experiments, using metallic additives such as nickel, iron, cobalt, stainless steel, and manganese, indicated that 99% of theoretical density can be achieved by liquid-phase sintering (LPS). More than 99% theoretical density was obtained in TiB_2–Ni by a hot-pressing route (1425°C). It was proposed that the transition metals (Ni, Co, Cr) react with TiB_2, forming various metal borides with a

Table 13.3. Overall Summary Illustrating the Influence of Metallic Additives on Microstructure and Mechanical Properties

Material composition (in wt%)	Processing conditions	Sintered density (% ρ_{th})	Microstructure	Vickers hardness, H_v (GPa)	Indentation toughness ($MPa \cdot m^{1/2}$)	Flexural strength (MPa)
TiB_2–(0.014%)Ni	HP at 2250°C, 30 Mpa, 20 minutes	97.9	Equiaxed TiB_2 Grains of 1.5-μm traces of Ni_4B_3, Ni_3B	23.3	5.8	716(4-P)
TiB_2–(0.7%)Ni	HP at 1550°C, 1 hour, Vacuum	97.9	..	23.3	5.1	716(4-P)
TiB_2–(1.4%)Ni	HP at 1425°C	>99	. . .	. . .	6.4	670(4-P)
TiB_2–(7.9%)Ni	HP at 1425°C	>99	. . .	. . .	4	420(4-P)
TiB_2–(0.017%)Fe	HP at 1700°C, 1 hour, Ar	99	Equiaxed TiB_2 grains of 6.4 μm	. . .	6.6	520(3-P)
TiB_2–(0.5%)Fe–(0.5%)Cr	PS at 1800°C, 2 hours, Ar	97.6	. . .	27.0	6.2	506(4-P)
TiB_2–(0.5%)Fe–(0.5%)Cr	PS at 1900°C, 2 hours, Ar	98.6	..	31.3	5.9	262(4-P)
TiB_2–Ti	HP at 1900°C, 2 hours, Ar	99.6	Needle-like fiber of TiB_2	19.0	4.5	360(3-P)
TiB_2–(14.4%)Fe–(6.1%) Ni-(8%)$TiAl_3$	Sinter-HIP at 1500°C, Ar	–	TiB_2 grains of 5.0 μm	17.8	9.2	1019(4-P)
ZrB_2–(4.0%)Ni	HP at 1850°C, 30 minutes	98.0	–	14.4	2.8	371(4-P)
WC–(6.5%)Co	PS at 1550°C, 4 hours	–	2.5 μm	15.8	10	–
TiC–(10%)Mo_2C–(25%)Ni	PS at 1550°C, 4 hours	–	4.2 μm	15.1	6.1	–
TiC–(10%)TiN–(10%) Mo_2C–(25%)Ni	PS at 1550°C, 4 hours	–	2.4 μm	12.7	10.0	–

PS, pressureless sintering; HP, hot pressing; 4-P, four-point bending; 3-P, three-point bending (taken from Reference 1).

low melting point (approximately 900–1100°C) and these borides also exhibit good wetting behavior. In the case of Ni-bonded TiB_2, a ternary τ-phase $Ni_{21}Ti_2B_6$ forms by the dissolution of TiB_2. Typical metal contents used to obtain optimum LPS of TiB_2 are 5–25 wt% (2–12 at%) of either Ni or Co. In using metallic additives, the sintering temperatures have been decreased from 2100 to 1400°C.

In LPSed TiB_2, the boride particles form a rigid skeleton of faceted crystals, as a result of reaction with sinter-aids, such as iron, nickel, or cobalt.[17] Depending upon the wetting behavior, round pores accumulate at particle–matrix interfaces or close to triple junctions. Moreover, the evaporation of Fe-, Co-, or Ni-borides results in entrapped gas pores. Hence, hot pressing is necessary for a homogeneous distribution of the liquid phase, particle rearrangement, and complete removal of the residual porosity. In contrast to hardmetals based on Co, the matrix phase is brittle; for example, the K_{IC} of Ni_3B is 1.4–1.9 MPa·m$^{1/2}$,[2] and hence the toughness of borides, sintered using metallic additives, remains inferior.

Pressureless sintering experiments with nickel, nickel boride, and iron additives were also attempted. Relatively small additions (1–5 wt%) of nickel, nickel boride (NiB), and iron facilitated LPS of titanium diboride (TiB_2). High density (>94% of theoretical density) was recorded at temperatures greater than or equal to 1500°C. Exaggerated grain growth was observed in TiB_2 sintered with binders such as Ni, NiB, and Fe during sintering at 1700°C, and this was closely related to the oxygen content of the samples and to sintering temperature. An example of LPSed microstructure (TiB_2–(1.5 wt%)Ni) is presented in Figure 13.6a. Although the TiB_2 grains are around 4–5 μm or coarser, most of the triple pockets appear to contain sintering liquid residue.

In a different approach, simultaneous addition of 0.5 wt% Cr and Fe could enhance the densification of TiB_2 up to 98.8%.[29] Typical microstructure of TiB_2–(0.5%)Cr–(0.5%)Fe shows equiaxed grain morphology with TiB_2 sizes between 2 and 10 μm (see Fig. 13.6b). In Figure 13.6c, a high-resolution transmission electron microscopy (TEM) image reveals the existence of (Fe,Cr)-rich sintering liquid. In a case of B_4C addition along with 0.5 wt% Fe, abnormal grain growth was restricted remarkably and an increase in sintered density was recorded up to 95%.[30] Microstructural observations revealed the existence of Fe-rich phase at the triple junction and at grain boundaries. From the perspective of high-temperature applications, nonmetallic additives are preferred for improving sinterability of TiB_2 without promoting grain growth.

13.5 INFLUENCE OF NONMETALLIC ADDITIVES ON DENSIFICATION AND PROPERTIES

Various nonmetallic additives, such as aluminum nitride (AlN), SiC, Si_3N_4, CrB_2, B_4C, and TaC, were used as sinter-aid to attain densification of TiB_2 with good mechanical properties.[31–49] Table 13.4 presents a summary of the literature results to show the influence of minor amounts of nonmetallic additives on densification and mechanical properties. A critical comparison of Tables 13.3 and 13.4 indicates that

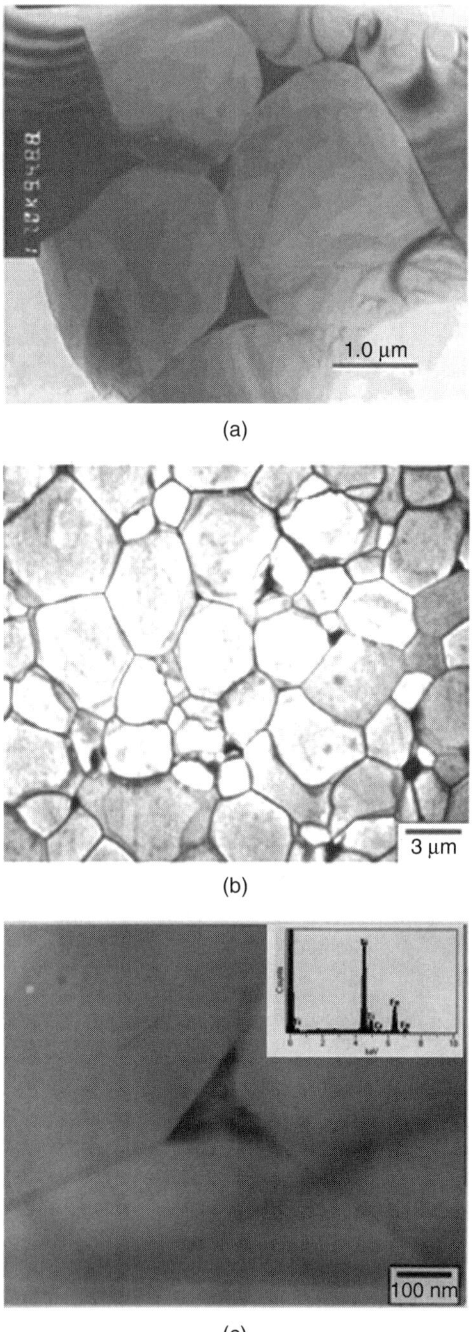

Figure 13.6 (a) Bright field (BF) TEM image showing the wetting behavior of the liquid phase in TiB_2–(1.5 wt%)Ni, pressureless sintered at 1700°C for 1 hour in argon atmosphere; (b) SEM image of TiB_2–(0.5 wt%)Fe–(0.5 wt%)Cr, pressureless sintered at 1800°C for 2 hours; and (c) BF TEM image along with energy-dispersive x-ray spectrometry (EDS) analysis revealing the presence of Fe, Cr in sintering liquid residue at the triple pocket in a sintered TiB_2–(0.5%) Fe–(0.5%)Cr.[1]

Table 13.4. Summary of Research Results Illustrating the Effect of Different Nonmetallic Sinter-Additives on Microstructure and Mechanical Properties of TiB_2

Material composition (in wt%)	Processing conditions	Sintered density (% ρ_{th})	Microstructural phases	Hardness (GPa)	Fracture toughness, K_{IC} ($MPa{\cdot}m^{1/2}$)	Flexural strength (MPa)
TiB_2–(0%)AlN	HP at 1800°C, 1 hour, 30 MPa, Ar	89.0	TiB_2 only	12.5	4.5	360 (4-P)
TiB_2–(2.5%)AlN	HP at 1800°C, 1 hour, 30 MPa, Ar	94.0	TiB_2, BN, TiN, Al_2O_3	16.1	5.0	500 (4-P)
TiB_2–(5.0%)AlN	HP at 1800°C, 1 hour, 30 MPa, Ar	98.0	TiB_2, BN, TiN, Al_2O_3, unreacted AlN	22.0	6.8	650 (4-P)
TiB_2–(10.0%)AlN	HP at 1800°C, 1 hour, 30 MPa, Ar	88.5	TiB_2, BN, TiN, Al_2O_3, unreacted AlN	14.0	5.2	500 (4-P)
TiB_2–(20.0%)AlN	HP at 1800°C, 1 hour, 30 MPa, Ar	87.5	TiB_2, BN, TiN, Al_2O_3, unreacted AlN	12.0	4.6	400 (4-P)
TiB_2–(0%)SiC	PS at 1700°C, Vacuum + HIP at 1600°C, 200 MPa of Ar	62.0	TiB_2 only	. . .	3.3	450 (3-P)
TiB_2–(2.5%)SiC	PS at 1700°C, Vacuum + HIP at 1600°C, 200 MPa of Ar	99.0	Amorphous SiO_2, liquid phase along grain boundaries	. . .	4.3	660 (3-P)
TiB_2–(5.0%)SiC	PS at 1700°C, Vacuum + HIP at 1600°C, 200 MPa of Ar	93.0	TiB_2, SiC, and TiC	. . .	4.9	850 (3-P)
TiB_2–(0%)TaC	HP at 2000°C, 30 minutes	90.0	. . .	96.0 (RN 15)	. . .	. . .
TiB_2–(1.0%)TaC	HP at 2000°C, 60 minutes	98.0	(Ti, Ta)B_2 and (Ta,Ti)(C,N) solid solution	97.8 (RN 15)	. . .	. . .

(*Continued*)

Table 13.4. *(Continued)*

Material composition (in wt%)	Processing conditions	Sintered density (% ρ_{th})	Microstructural phases	Hardness (GPa)	Fracture toughness, K_{IC} (MPa·m$^{1/2}$)	Flexural strength (MPa)
TiB_2–(5.0%)TaC	HP at 2000°C, 60 minutes	99.4	(Ti, Ta)B_2 and (Ta,Ti)(C,N) solid solution	98.2 (RN 15)	. . .	. . .
TiB_2–(0%)Si_3N_4	HP at 1800°C, 1 hour, Ar	90.0	TiB_2 grains (3–7 μm)	23.0	5.8	380 (4-P)
TiB_2–(2.5%)Si_3N_4	HP at 1800°C, 1 hour, Ar	99.0	TiB_2 TiN, BN	27.0	5.1	810 (4-P)
TiB_2–(5.0%)Si_3N_4	HP at 1800°C, 1 hour, Ar	97.5	TiB_2 TiN, BN	21.0	4.8	510 (4-P)
TiB_2–(5.0%)Si_3N_4	HP at 2000°C, 30 minutes	86.0	. . .	87.0 (RN 15)		
TiB_2–(10.0%)Si_3N_4	HP at 1800°C, 1 hour, Ar	96.0	TiB_2 TiN, BN	20.0	4.4	400 (4-P)
TiB_2–(0%) sinter-additive	HP at 2000°C, 30 minutes	90.0	. . .	96.0 (RN 15)	. . .	. . .
TiB_2–(5.0%)TaN	HP at 2000°C, 30 minutes	96.8	. . .	98.2 (RN 15)	. . .	. . .
TiB_2–(5.0%)TiC	HP at 2000°C, 30 minutes	96.5	. . .	98.0 (RN 15)	. . .	. . .
TiB_2–(5.0%)WC	HP at 2000°C, 30 minutes	95.9	. . .	94.3 (RN 15)	. . .	. . .
TiB_2–(10.0) B_4C– (0.5)Fe	PS, 2000°C, 30 minutes	95.0	. . .	5.5	. . .	. . .
TiB_2	HP, 1850°C, 1 hour	96.8	. . .	24.4	5.4	. . .
TiB_2–(10 vol%) B_4C–(0.7 vol%)Ni	HP, 1600°C, 15 minutes	97.9	. . .	19.3	6.2	. . .
TiB_2–(20 vol%) B_4C–(1 vol%)Ni	HP, 1600°C, 20 minutes	97.9	. . .	24.7	6.4	. . .

TiB_2–(10 vol%) B_4C–(10 vol%)2Y–ZrO_2–(0.7 vol%) Ni	HP, 1650°C, 60 minutes	95.6	. . .	15.9	5.7	. . .
TiB_2–(10 vol%) B_4C–(10 vol%)2Y–ZrO_2–(1.2 vol%) Ni	HP, 1750°C, 20 minutes	97.0	. . .	15.0	6.8	. . .
TiB_2–(5.0%)TiN	HP at 2000°C, 30 minutes	95.0	. . .	98.1 (RN 15)	. . .	. . .
TiB_2–(5.0%)ZrN	HP at 2000°C, 30 minutes	83.9	. . .	92.8 (RN 15)	. . .	. . .
TiB_2–(5.0%)ZrB_2	HP at 2000°C, 60 minutes	79.6	. . .	88.4 (RN 15)	. . .	. . .
TiB_2–(5%)ZrO_2 (2Y)–(5%)SiC	HIP at 1600°C	98.0	SiC acts as a grain growth inhibitor	23.1 GPa (H_v)	5.2	1280 (3-P)
TiB_2–(3%)CrB_2	MW, 1900°C, 30 minutes, Ar	95.0	. . .	28.9 (DPH)	6.2	. . .
TiB_2–(3%)CrB_2	MW, 2100°C, 30 minutes, Ar	98.0	. . .	27.0 (DPH)	6.1	. . .

HP, hot pressing; HIP, hot isostatic pressing; HPCS, high-pressure self-combustion synthesis; HPS, high-pressure sintering; MW, microwave sintering; 4-P, four-point bending; 3-P, three-point bending; H_V, Vickers hardness; RN, Rockwell superficial (N scale); Ar, argon atmosphere (taken from Reference 1).

a relatively larger amount of nonmetallic additive, more than 5–10 wt%, is typically added to densify TiB_2, whereas a much smaller amount of metallic additive, even less than 2 wt%, is used to obtain dense TiB_2. Such comparison also reveals that, similar to the use of metallic binder, a combination of high hardness (~20–27 GPa) and moderate toughness (~4–7 $MPa{\cdot}m^{1/2}$) is achievable with the use of a large variety of nonmetallic sinter-additive (added in an appropriate amount).

It can be noticed in Figure 13.7a that monolithic bulk TiB_2, without any sinter-additive, is characterized by coarse grains of around 10 μm. The pores are reduced in size with the addition of 2.5 wt% Si_3N_4 (Fig. 13.7b). Almost fully dense microstructure with finer TiB_2 grain sizes of 5 μm or less has been reported when AlN was used as sinter-additive, with an optimal amount of 5 wt% (Fig. 13.7c). Similarly, Torizuka et al.[33,34] reported the formation of grain boundary liquid phase (amorphous SiO_2), when SiC was used as an additive. According to Murata et al.,[37] TaC and TaN can be quite effective for densification of TiB_2. In their work, (Ti, Ta)B_2 and (Ta, Ti)(C,N) solid solutions were identified in sintered microstructures after hot pressing at 2000°C, when TaC was added to TiB_2.

The influence of sinter-additive content on the relative density, grain size, and mechanical properties (hardness, fracture toughness, and flexural strength) is demonstrated in Figures 13.8–13.10. In the following, important experimental observations are summarized.

The presence of different reaction products, such as TiN and BN, has been reported for TiB_2–Si_3N_4 composite system.[38] The relative density of pure TiB_2 was only 90% ρ_{th} with average grain size of ~7 μm, as shown in Figure 13.8a. The sintered density remarkably increased to >99% ρ_{th}, and fine microstructure (~3 μm) was recorded, when 2.5 wt% of Si_3N_4 was added. The hardness of TiB_2 can be increased with the addition of 2.5 wt% of Si_3N_4, as shown in Figure 13.9a. In contrast to the strength or hardness, the fracture toughness decreased steadily with an increase in the amount of Si_3N_4 (Fig. 13.9b). In particular, crack deflection along grain boundaries was reported to contribute to the high fracture toughness.

In another work, Park et al.[38] investigated the effect of hot-pressing temperature (1500–1800°C) in the case of TiB_2–(2.5 wt%)Si_3N_4 composition. Unlike the density, the average grain size increased steadily with the sintering temperature. In densifying TiB_2, the removal of the surface oxide layer is necessary and the formation of a liquid phase during sintering is critically important. The hardness and fracture toughness of the specimens, shown in Figure 13.10b, illustrate that TiB_2 with a good combination of mechanical properties can be densified at temperatures as low as 1600°C with the addition of a small amount of Si_3N_4 sinter-additive.

AlN,[32] another nonmetallic sinter-additive, when added in a small amount (≤5 wt%), eliminated the rutile phase (TiO_2), present on the TiB_2 powders, by the following reaction with AlN to form TiN and Al_2O_3:

$$3TiO_2\ (s) + 4AlN\ (s) \rightarrow 3\ TiN\ (s) + 2Al_2O_3\ (s) + (1/2)N_2\ (g), \tag{13.4}$$

$$TiB_2\ (s) + (3/2)N_2\ (g) \rightarrow TiN\ (s) + 2BN\ (s). \tag{13.5}$$

The elimination of TiO_2 enhanced the sinterability and, consequently, the mechanical properties of TiB_2. The evidence of BN formation at grain triple pockets is provided

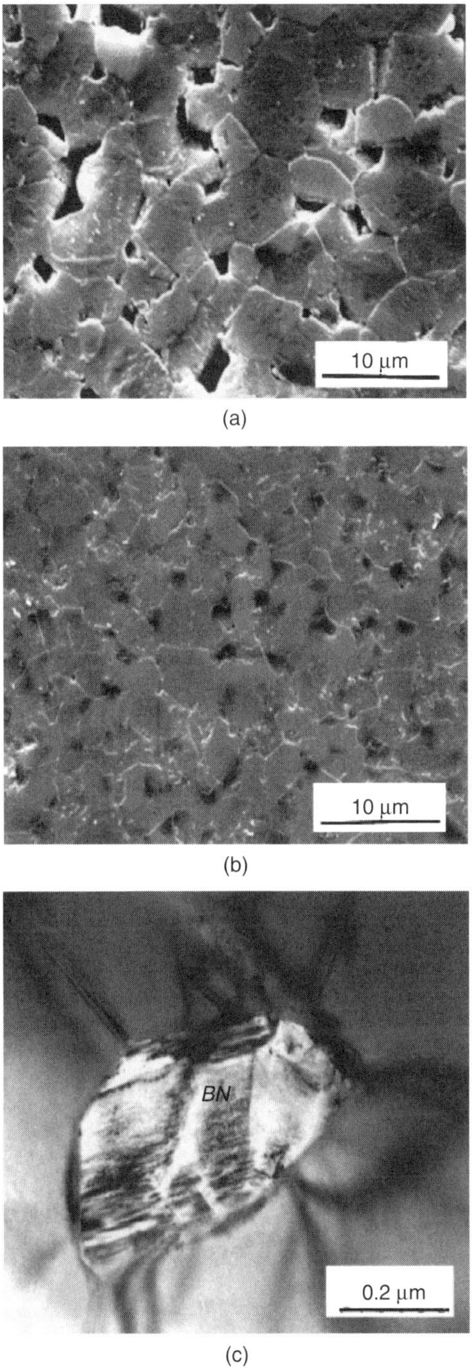

Figure 13.7 SEM images of TiB_2 specimens, hot pressed at 1800°C for 1 hour containing (a) 0 wt% sinter-additive, (b) 2.5 wt% Si_3N_4, and (c) BF TEM image revealing the presence of BN phase, formed due to sintering reaction, when 5 wt% AlN is added as sinter-additive to densify TiB_2.[1]

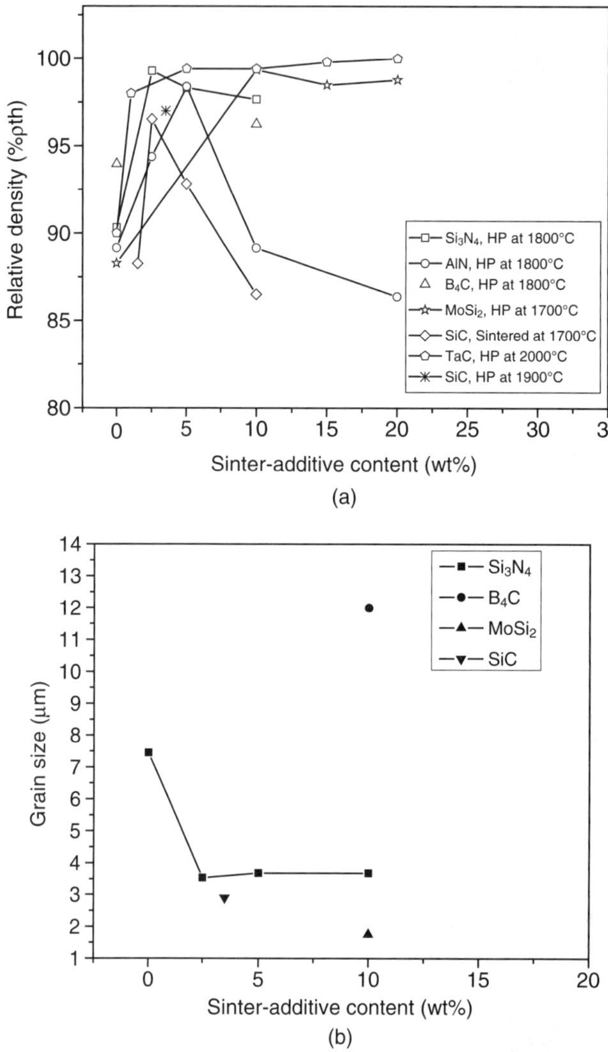

Figure 13.8 Plot revealing the influence of various sinter-additives, in varying amounts, on (a) sintered density and (b) grain size of hot-pressed TiB_2 ceramics.[1]

in Figure 13.7c. However, when a large amount of AlN was added (≥10 wt%), the sinterability and the mechanical properties degraded, apparently due to remaining unreacted AlN.

The effect of SiC and ZrO_2 on mechanical properties of titanium nitride, titanium carbonitride, and titanium diboride was reported by Torizuka et al.[33] Although TiB_2 and TiB_2–(20%)ZrO_2 lacked sinterability, the addition of SiC was effective in improving the sintered density. For example, the density of TiB_2–(19.5%)

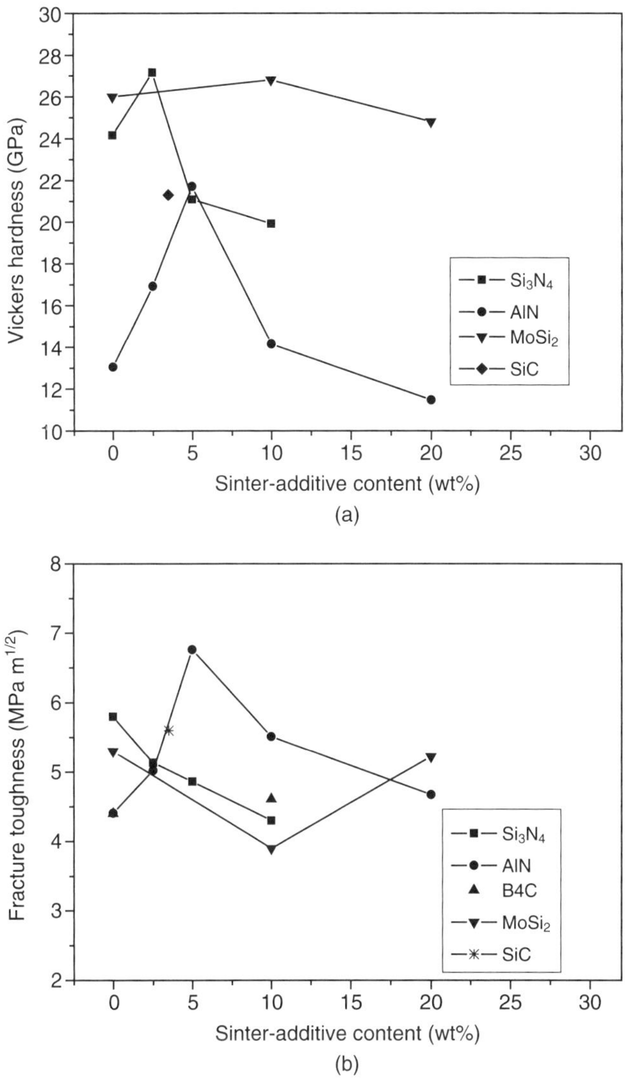

Figure 13.9 Variation of room-temperature mechanical properties of hot-pressed TiB_2 with sinter-additives: (a) Vickers hardness; (b) fracture toughness.[1]

ZrO_2–(2.5%)SiC was 97% ρ_{th}. It was reported that TiO_2 reacts with SiC and formed TiC and SiO_2, according to the following reaction:

$$TiO_2 + SiC \rightarrow TiC + SiO_2. \quad (13.6)$$

In the case of TiB_2–(2.5 wt%)SiC sintered compacts, about 3.5 vol% of SiO_2 was reported to form as a result of Reaction 13.6. The existence of the amorphous SiO_2 phase in the microstructure is suggestive that the densification is enhanced by LPS.[44]

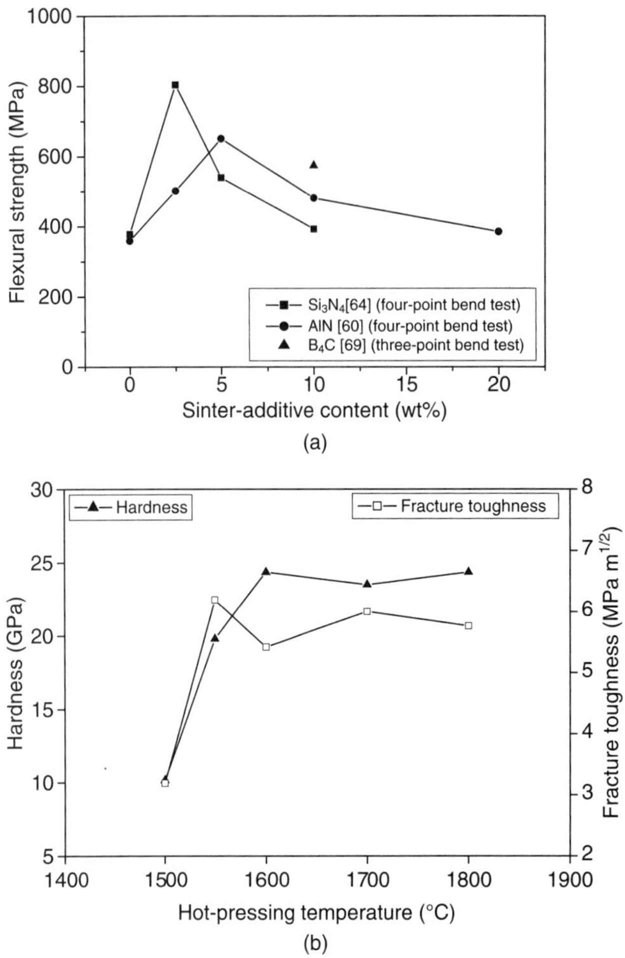

Figure 13.10 (a) Room-temperature flexural strength of hot-pressed TiB_2 ceramics containing various amounts of sintering additives and (b) Vickers hardness and fracture toughness—both measured at room temperature—of TiB_2–(2.5)Si_3N_4 (wt%) ceramics, hot pressed at different temperatures.[1]

Besides hot pressing and pressureless sintering, limited efforts have been invested in using advanced sintering techniques, such as microwave sintering.[41] Using a 2.45-GHz, 6-kW microwave furnace adapted for inert gas sintering, titanium diboride (TiB_2) was rapidly sintered to >90% theoretical density at 1900–2100°C with soaking time of 30 minutes or less. A comparison with conventional sintering revealed that microwave sintering of TiB_2–(3 wt%)CrB_2 took place at 200°C lower temperature and yielded material with significantly improved hardness, grain size, and fracture toughness.

13.6 IMPORTANT APPLICATIONS OF BULK TiB_2-BASED MATERIALS

The property requirements for armor applications include low density, superior hardness, and high compressive strength, which enable "defeat of the projectiles."[50] Other desirable characteristics for armor materials comprise the combination of high compressive yield strength or hardness, high tensile spall strength, high fracture toughness, and high Poisson's ratio. The spall strength of an armor material can be defined in different ways. The stress or strain at which a further increase in stress or strain causes the material to deform inelastically is the Hugoniot elastic limit (HEL), a measure of spall strength. Alternatively, the spall strength can be described as the stress at which a material loses its cohesiveness when placed under shock-induced tension.

Since 1980, the ballistic performance of various ceramics (AlN, Al_2O_3, B_4C, SiC, TiB_2, WC, and ZrO_2) has been studied.[51–54] As far as armor applications are concerned, TiB_2 displays favorable properties, such as high impact velocity for dwell–penetration transition and deformation-induced hardening. Dandekar et al.[51] reported the strength properties of TiB_2 under plane shock wave loading in terms of its spall threshold and the shear stress under application of a shock compressive stress of 60GPa. The HEL value for TiB_2 is 13–17GPa. The spall strength of TiB_2 decreases with increasing impact stress and becomes negligible at the HEL. More recently, two-phase ceramics of titanium diboride and alumina have been developed.[52] For armor applications, these materials exhibited a wide range of fracture toughness values, higher than monolithic TiB_2/Al_2O_3.

As far as other applications are concerned, TiB_2 is an attractive material for the aluminum industry, because of its easier wettability and low solubility in molten aluminum as well as good electrical conductivity.[55–57] Watson and Toguri[58] reported the wettability of pure TiB_2 and TiB_2/carbon composites by aluminum in cryolite melts. The wettability of the composite material increased with TiB_2 content. The hot-pressed TiB_2 was reported to be completely wet by aluminum with contact angle of zero in a cryolite melt at 980°C. Fang and Knodler[59]fabricated porous TiB_2 electrodes for the alkali metal thermoelectric converter (AMTEC). The electrical performance of these new electrodes was superior to that of other electrodes, such as TiN or Mo.

Other important applications of TiB_2 have also been reported, for example, as electrical contact barrier for Si in the semiconductor industry.[60] However, boron diffusion from TiB_2 into underlying silicon was observed above 1000°C, restricting its wider application.

13.7 CONCLUDING REMARKS

As a concluding note, a summary of various issues and aspects concerning the development of ultra-high-temperature ceramics is provided in Figure 13.11. While the microstructural stability is a major issue with respect to ensuring a good

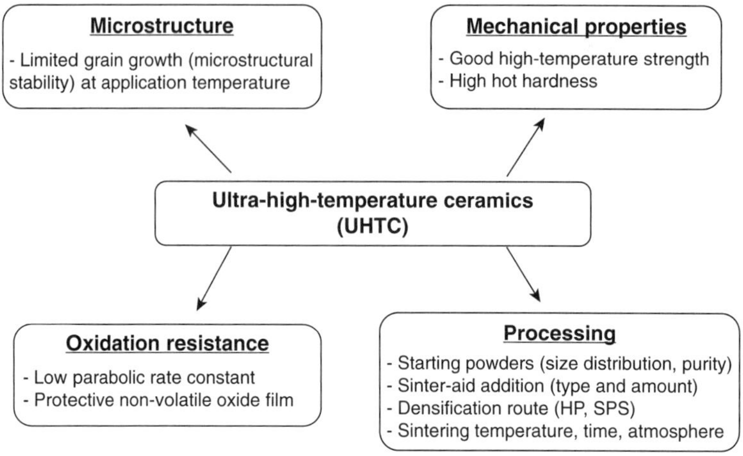

Figure 13.11 Various aspects of processing and properties of ultra-high-temperature ceramics.

combination of high-temperature strength and hardness properties, the oxidation resistance controls the performance at high temperature. The type and amount of sinter-aid in combination with tailoring of the sintering parameters need to be considered while developing boride-based high-temperature ceramics.

It should be evident from the discussion in this chapter that extensive research activity has been invested in the use of several sinter-additives and optimizing sintering conditions in order to improve sinterability and mechanical properties of bulk TiB_2. Densification of TiB_2, more than 99% theoretical density (ρ_{th}), was achieved by LPS in the presence of metallic sinter-aid. Several ceramic sinter-aids, such as AlN, SiC, Si_3N_4, CrB_2, B_4C, and TaC, were also used to obtain dense TiB_2 with better mechanical properties.

In regards to the fabrication routes, it has been noted that hot pressing and pressureless sintering are employed to densify borides. Future work can be pursued to use advanced processing techniques such as spark plasma sintering (SPS) in order to develop TiB_2-based materials with refined microstructure and improved properties. It would be interesting to assess the influence of electric field assisted sintering on grain growth of conductive ceramics, such as TiB_2.

Considering oxidation at elevated temperature, the oxidation resistance of TiB_2 could be improved by coating a protective amorphous SiO_2 layer on the surface of borides.[61] Since TiB_2 materials can experience thermal cycling during high-temperature applications, future studies should focus on oxidation resistance behavior in terms of the stability and protectiveness of oxide scale during cyclic oxidation behavior. The study of aqueous corrosion behavior of TiB_2 also requires attention. Although ceramics are generally known to be corrosion resistant, the presence of metallic binder can degrade the corrosion resistance properties of TiB_2.

To produce near-net-shape materials, EDM can be used to produce complex and intricate shapes of TiB_2. Although bulk TiB_2 with little sinter-additive has the potential to be used as an electrode tool in EDM, further EDM study needs to be performed to evaluate the material removal rate (MRR), percentage of tool wear rate (TWR), and surface roughness (R_a) of EDMed tools. It should be emphasized here that the results of such a study would widen the scope of application of a better EDMable TiB_2 in different sizes and shapes.

REFERENCES

1 B. Basu, G. B. Raju, and A. K. Suri. Processing and properties of TiB_2-based materials: A review. Int. Mater. Rev. 51(6) (2006), 352–374.
2 R. Telle. Boride and carbide ceramics, in *Materials Science and Technology*, Vol. 11, R. W. Cahn, P. Haasen, and E. J. Kramer (Eds.). Structure and Properties of Ceramics. Ed. M. Swain. VCH, Weinheim, Germany, 1994, 175.
3 G. V. D. Goor, P. Sägesser, and K. Berroth. Electrically conductive ceramic composites. Solid State Ionics 101–103 (1997), 1163–1170.
4 P. Ettmayer, H. Kolaska, W. Lengauer, and K. Dreyer. Ti(C,N) cermets—Metallurgy and properties. Int. J. Refract. Met. Hard Mater. 13 (1995), 343–351.
5 G. E. D'Errico, S. Bugliosi, and E. Gugliemi. Tool-life reliability of cermet inserts in milling tests. J. Mater. Proc. Technol. 77 (1998), 337–343.
6 D. Moskowitz and M. Humenik Jr. Cemented titanium carbide cutting tools. Mod. Dev. Powder Metall. 3 (1966), 83–94.
7 S. Put, J. Vleugels, G. Anne, and O. Van Der Biest. Functionally graded ceramic and ceramic–metal composites shaped by electrophoretic deposition. Colloids and Surfaces A: Physicochemical and Engineering Aspects 222(1–3) (2003), 223–232.
8 Y. Liu, X. Liao, F. Tang, and Z. Chen. Observations on the operating of TiB_2-coated cathode reduction cells, in *Light Metals*, E. Cutshall (Ed.). The Minerals, Metals and Materials Society, Warrendale, PA, 1991, 427–429.
9 F. De Mestral and F. Thevenot. Ceramic composites: TiB_2–TiC-SiC—Part I properties and microstructures in the ternary system. J. Mater. Sci. 26 (1991), 5547–5560.
10 (a) B. Basu, J. Vleugels, and O. Van Der Biest. Toughness tailoring of yttria-doped zirconia ceramics. Mater. Sci. Eng. A 380 (2004), 215–221.(b) B. Basu, J.-H. Lee, and D.-Y. Kim. Development of nanocrystalline wear resistant Y-TZP ceramics. J. Am. Ceram. Soc. 87(9) (2004), 1771–1774.
11 (a) B. Basu. Toughening of yttria-stabilised tetragonal zirconia ceramics. Int. Mater. Rev. 50(4) (2005), 239–256.(b) B. Basu, J. Vleugels, and O. Van Der Biest. Processing and mechanical properties of ZrO_2-TiB_2 composites. J. Eur. Ceram. Soc. 25 (2005), 3629–3637.
12 S. Zang. Titanium carbonitride–based cermets: Processes and properties. Mater. Sci. Eng. A 163 (1993), 141–148.
13 W. D. Schubert, A. Bock, and B. Lux. General aspects and limits of conventional ultrafine WC powder manufacture and hard metal production. Int. J. Refract. Met. Hard Mater. 13 (1995), 281–296.
14 T. Yamamoto, A. Jaroenworaluck, Y. Ikuhara, and T. Sakuma. Nanoprobe analysis of core-rim structure of carbides in TiC-20 wt% Mo_2C wt% Ni cermet. J. Mater. Res. 14 (1999), (11)4129–4131.
15 L. S. Siegl and K. A. Schwetz. TiB_2-based cemented borides. A new generation of hardmetals. Powder Metall. Int. 23 (1991), 221.
16 A. Bellosi, G. De Portu, and S. Guicciardi. Preparation and properties of electroconductive Al_2O_3-based composites. J. Eur. Ceram. Soc. 10 (1992), 307–315.
17 A. H. Jones, C. Trueman, R. S. Dobedoe, J. Huddleston, and M. H. Lewis. Production and EDM of Si_3N_4-TiB_2 ceramic composites. Br. Ceram. Trans. 100(2) (2001), 49–54.
18 L. Rao, E. G. Gillan, and R. B. Kaner. Synthesis of refractory ceramics via rapid metathesis reactions between solid state precursors. J. Mater. Res. 10 (1995), 333.

19 S. E. Bates, W. E. Buhro, C. A. Frey, S. M. L. Sastry, and K. F. Kelton. Synthesis of titanium boride (TiB_2) nanocrystallites by solution phase processing. J. Mater. Res. 10 (1995), 2599.

20 R. L. Axelbaum, D. P. DuFaux, C. A. Frey, K. F. Kelton, S. A. Lawton, L. J. Rosen, and S. M. L. Sastry. Gas-phase combustion synthesis of titanium boride [TiB_2] nanocrystallites. J. Mater. Res. 11 (1996), 948–954.

21 A. Y. Hwang and J. K. Lee. Preparation of TiB_2 powders by mechanical alloying. Mater. Lett. 54 (2002), 1–7.

22 A. K. Khanra, L. C. Pathak, S. K. Mishra, and M. M. Godkhindi. Effect of NaCl on the synthesis of TiB_2 powder by a SHS technique. Mater. Lett. 58 (2004), 733–738.

23 Y. Gu, Y. Qian, L. Chen, and F. Zhou. A mild solvothermal route to nanocrystalline titanium diboride. J. Alloys Compd. 352 (2003), 325–327.

24 L. Chen, Y. Gu, Y. Qian, L. Shi, Z. Yang, and J. Ma. A facile one-step route to nanocrystalline TiB_2 powders. Mater. Res. Bull. 39 (2004), 609–613.

25 S. Baik and P. F. Becher. Effect of oxygen contamination on densification of TiB_2. J. Am. Ceram. Soc. 70(8) (1987), 527–530.

26 W. Wang, Z. Fu, H. Wang, and R. Yuan. Influence of hot pressing sintering temperature and time on microstructure and mechanical properties of TiB_2 ceramics. J. Eur. Ceram. Soc. 22 (2002), 1045–1049.

27 H. R. Baumgartner and R. A. Steiger. Effects of the sintering atmosphere and Ni content on the liquid-phase sintering of TiB_2-Ni. J. Am. Ceram. Soc. 67(3) (1984), 207–212.

28 S. H. Kang, D. J. Kim, E. S. Kang, and S. S. Baek. Pressureless sintering and properties of titanium diboride ceramics containing chromium and iron. J. Am. Ceram. Soc. 84(4) (2001), 893–895.

29 E. S. Kang, C. W. Jang, C. H. Lee, C. H. Kim, and D. K. Kim. Effect of iron and boron carbide on the densification and mechanical properties of titanium Di-boride ceramics. J. Am. Ceram. Soc. 72 (1989), (10)1868–1872.

30 M. Einarsrud, E. Hagen, G. Pettersen, and T. Grande. Pressureless sintering of titanium boride (TiB_2) with nickel, nickel boride and iron additives. J. Am. Ceram. Soc. 80 (1997), (12)3013–3020.

31 L. H. Li, H. E. Kim, and E. S. Kang. Sintering and mechanical properties of titanium diboride with aluminum nitride as a sintering aid. J. Eur. Ceram. Soc. 22 (2002), 973–977.

32 S. Torizuka and T. Kishi. Effect of SiC on sinterability and mechanical properties of titanium nitride, titanium carbide and titanium diboride. Mater. Trans. JIM 37 (1996), 782–787.

33 S. Torizuka, K. Sato, J. Harada, H. Yamamoto, and H. Nishio. Microstructure and sintering mechanism of TiB_2-ZrO_2-SiC composite. J. Ceram. Soc. Jpn. 100 (1992), 392–397.

34 R. Telle, S. Meyer, G. Petzow, and E. D. Franz. Sintering behaviour and phase reactions of TiB_2 with ZrO_2 additives. Mater. Sci. Eng. A 105/106 (1988), 125–129.

35 S. Torizuka, J. Harada, and H. Nishio. High strength TiB_2. Ceram. Eng. Sci. 11 (1989), 1454–1457.

36 Y. Murata, H. P. Julien, and E. D. Whitney. Densification and wear resistance of ceramic systems: I. Titanium diboride. Ceram. Bull. 46(7) (1967), 643–648.

37 J. H. Park, Y. H. Koh, H. E. Kim, C. S. Hwang, and E. Kong. Densification and mechanical properties of titanium diboride with silicon nitride as sintering aid. J. Am. Ceram. Soc. 82 (1999), (11)3037–3042.

38 J. H. Park, Y. H. Lee, Y. H. Koh, H. E. Kim, and S. S. Baek. Effect of hot-pressing temperature on densification and mechanical properties of titanium diboride with silicon nitride as a sintering aid. J. Am. Ceram. Soc. 83(6) (2000), 1542–1544.

39 Y. Muraoka, M. Yoshinaka, K. Hirota, and O. Yamaguchi. Hot isostatic pressing of TiB_2-ZrO_2(2 mol% Y_2O_3) composite powders. Mater. Res. Bull. 31(7) (1996), 787–792.

40 C. E. Holcombe and N. L. Dykes. Microwave sintering of titanium diboride. J. Mater. Sci. 26 (1991), 3730–3738.

41 H. Itoh, S. Naka, T. Matsudaira, and H. Hamamoto. Preparation of TiB_2 sintered compacts by hot pressing. J. Mater. Sci. 25 (1990), 533–536.

42 T. Graziani and A. Bellosi. Sintering and characterization of TiB_2-B_4C-ZrO_2 composites. Mater. Manuf. Process. 9(4) (1994), 767–780.

43 S. Torizuka, K. Sato, H. Nishio, and T. Kishi. Effect of SiC on interface reaction and sintering mechanism of TiB_2. J. Am. Ceram. Soc. 78(6) (1995), 1606–1610.

44 E. S. Kang and C. H. Kim. Improvements in mechanical properties of TiB_2 by the dispersion of B_4C particles. J. Mater. Sci. 25 (1990), 580–584.

45 S. K. Bhaumik, C. Diwakar, A. K. Singh, and G. S. Upadhyaya. Synthesis and sintering of TiB_2 and TiB_2–TiC composite under high pressure. Mater. Sci. Eng. A 279 (2000), 275–281.

46 H. J. Kim, H. J. Choi, and J. G. Lee. Mechanochemical synthesis and pressureless sintering of TiB_2–AIN composite. J. Am. Ceram. Soc. 85(4) (2002), 1022–1024.

47 A. S. Nakane, Y. Takano, M. Yoshinaka, K. Hirota, and O. Yamaguchi. Fabrication and mechanical properties of titanium boride ceramics. J. Am. Ceram. Soc. 82(6) (1999), 1627–1628.

48 G. Ivaldi, S. Tuffe, J. Dubois, G. Fantozzi, and G. Barbier. Densification, microstructure and mechanical properties of TiB_2–B_4C based composites. Int. J. Refract. Met. Hard Mater. 14 (1996), 305–310.

49 (a) T. S. R. Ch. Murthy, B. Basu, R. Balasubramaniam, A. K. Suri, C. Subramonian, and R. K. Fotedar. Processing and properties of TiB_2 with $MoSi_2$ sinter-additive: A first report. J. Am. Ceram. Soc. 89(1) (2006), 131–138.(b) K. Biswas, B. Basu, A. K. Suri, and K. Chattopadhyay. A TEM study on TiB_2–20% $MoSi_2$ composite: Microstructure development and densification mechanisms. Scr. Mater. 54 (2006), 1363–1368.

50 G. Kennedy, L. Ferranti, M. Zhou, and N. Thadhani. Dynamic high-strain-rate mechanical behavior of microstructurally biased two-phase TIB_2 + AL_2O_3 ceramics. J. Appl. Phys. 91(4) (2002), 1921–1927.

51 D. P. Dandekar and D. C. Benfanti. Strength of titanium diboride under shock wave loading. J. Appl. Phys. 73(2) (1993), 673–679.

52 A. R. Keller and M. Zhou. Effect of microstructure on dynamic failure resistance of titanium diboride/alumina ceramics. J. Am. Ceram. Soc. 86(3) (2003), 449–457.

53 C. Roberson and P. J. Hazell. Resistance of four different ceramic materials to penetration by a tungsten carbide cored projectile. *Ceramic Armor Materials by Design (Ceramic Transactions, Volume 151) Edited by Eugen Medvedovski*. The American Ceramic Society, Nashville, TN, 2003.

54 A. Krell and E. Strassburger. High-purity submicron α-Al_2O_3 armor ceramics—design, manufacture, and ballistic performance. *Ceramic Armor Materials by Design (Ceramic Transactions, Volume 134) Edited by W. McCauley, A. Crowson, W.A. Gooch Jr., A. M. Rajendra, S. J. Bless, K. V. Logan, M. Normandia, and S. Wax*. The American Ceramic Society, Nashville, TN, 2002.

55 M. Dionne, G. L'esperance, and A. Mirchi. Microscopic characterization of a TiB_2-carbon material composite: Raw materials and composite characterization. Metall. Mater. Trans. A 32 (2001), 2649–2656.

56 E. W. Dewing. The solubility of titanium diboride in aluminum. Metall. Mater. Trans. A 20 (1989), 2185–2187.

57 A. A. Abdel-Hamid and F. Durand. Discussion of "the grain refining of aluminum and phase relationships in the Al-Ti-B system.". Metall. Mater. Trans. A 17 (1986), 349–351.

58 K. D. Watson and J. M. Toguri. The wettability of carbon/TiB_2 composite materials by aluminum in cryolite metals. Metall. Mater. Trans. B 22 (1991), 617–621.

59 Q. Fang and R. Knodler. Porous TiB_2 electrodes for the alkali metal thermoelectric convertor. J. Mater. Sci. 27 (1992), 6725–6729.

60 C. S. Choi, Q. Wang, C. M. Osburn, G. A. Ruggles, and A. S. Shah. Electrical characteristics of TiB_2 ULSI applications. IEEE Trans. Electron Devices 39 (1992), (10)2341–2345.

61 Y. H. Koh, H. W. Kim, and H. E. Kim. Improvement in oxidation resistance of TiB_2 by formation of protective SiO_2 layer on surface. J. Mater. Res. 16(1) (2001), 132–137.

Chapter 14

Processing and Properties of TiB_2 and ZrB_2 with Sinter-Additives

In the case of non-oxide ceramics, such as titanium diboride (TiB_2), the type and amount of sinter-aid influences the microstructural development and, consequently, the mechanical properties. In this chapter, the experimental results obtained with silicide sinter-aid to TiB_2 are summarized. It is illustrated how $MoSi_2$ addition leads to $TiSi_2$ formation via sintering reactions and how this has motivated researchers to assess whether $TiSi_2$ addition can further reduce sintering temperature to obtain dense TiB_2. This chapter also shows how the silicide addition, if not tailored within a narrow window, can lead to degradation in properties of TiB_2-based materials. Toward the end of the chapter, the results of spark plasma sintering (SPS) on optimizing the densification and mechanical properties of ZrB_2–18SiC–$X$$TiSi_2$ ($X = 0$, 2.5, 5) ceramic composites are discussed.

14.1 INTRODUCTION

From the perspective of high-temperature applications, different transition metal borides with melting points higher than 3000°C have attracted wider attention for a range of technological applications.[1–5] Ceramics such as borides, carbides, and nitrides of Group IVB and VB elements are known as ultra-high-temperature ceramics (UHTCs).[6,7] There has been a growing interest in UHTCs due to the increasing demand for hypersonic aerospace vehicles and reusable atmospheric re-entry vehicles for space applications, which requires temperature capabilities of more than 1800°C.[1,6,7] Also, diborides of transition metals, such as zirconium, titanium, and hafnium, are candidate materials for various structural applications, including furnace elements, high-temperature electrodes, refractory linings, microelectronics, and cutting tools, besides the aerospace applications.[1,2,8–11] Among the UHTCs,

Advanced Structural Ceramics, First Edition. Bikramjit Basu, Kantesh Balani.

zirconium diboride (ZrB_2) is a promising material because of the combination of high-temperature strength and good oxidation and corrosion resistance.

Among the borides, titanium diboride (TiB_2) is one of the materials of choice for cathodes in aluminum electrosmelting, armor components, cutting tools, wear-resistant parts, and various other high-temperature applications.[7,12,13] However, such a wider application of TiB_2 is limited because of the difficulties in obtaining full density. TiB_2 is difficult to sinter mainly because of the covalent bonding and low self-diffusion coefficient.[1,6]

Generally, pressure-assisted sintering and high sintering temperatures (≥1800°C) are needed to attain full densification of monolithic borides (TiB_2, ZrB_2, etc.).[6] Hence, the use of sintering aids is necessary to overcome the intrinsically low sinterability of ZrB_2 ceramics. Metallic additives and liquid-phase sintering (LPS) techniques have been utilized to enhance densification. However, the residual secondary phases degrade the properties of the borides. Since 2000, attempts have been made toward using various nonmetallic additives (such as SiC, ZrC, HfN, AlN, Y_2O_3, $TaSi_2$, Si_3N_4, and $MoSi_2$), often in combination with advanced techniques, such as SPS and reactive hot pressing (HP).[14–32]

The first part of this chapter shows how the optimization of HP temperatures and sinter-additive content can lead to development of dense TiB_2. The experimental results published elsewhere are summarized.[27–31] In the second part, the results of SPS experiments on ZrB_2-based materials are presented. A survey of the existing literature indicates that ZrB_2–SiC composites have good high-temperature oxidation properties.[7,11–13] However, the hardness, toughness, and strength of ZrB_2-SiC composites require improvement. In developing TiB_2-based ceramics with better properties,[32] it was demonstrated that $TiSi_2$ could be used as an effective sinter-aid for TiB_2. In view of this, a small amount of $TiSi_2$ (0–5 wt%), in addition to 18 wt% SiC, was added to ZrB_2 to enhance the densification at lower sintering temperatures and shorter sintering time using SPS.

14.2 MATERIALS PROCESSING

All the TiB_2–$TiSi_2$ and TiB_2–$MoSi_2$ composites were hot pressed in the temperature range 1400–1800°C for 1 hour. The TiB_2 powder was characterized by a median particle size (D_{50}) of 1.2 μm and specific surface area 1.49 m^2/g. $MoSi_2$ powder was used as a sintering-aid and the powders had D_{50} of 3.4 μm and surface area of 0.29 m^2/g.

Commercially available powders of ZrB_2, SiC, and $TiSi_2$ were used as starting materials in developing ZrB_2-based composites, which were sintered using SPS at a temperature of 1600°C for 10 minutes, with an applied pressure of 50 MPa, at a vacuum of 6 Pa. The selection of such SPS conditions was based on our earlier work, where we demonstrated good densification of ZrB_2–(6 wt%)Cu cermets processed at 1600°C for 10 minutes.[33] As the melting point of $TiSi_2$ sintering additive is about 1540°C, the selected SPS temperature of 1600°C was expected to cause melting of $TiSi_2$ in ZrB_2 composites.

14.3 TiB_2–$MoSi_2$ SYSTEM

14.3.1 Densification, Microstructure, and Sintering Reactions

The sinter density data of TiB_2 as a function of temperature and $MoSi_2$ content are plotted in Figure 14.1a. At 1700°C, low relative density (~91% ρ_{th}) was recorded with monolithic TiB_2, while ~98% ρ_{th} could be achieved with increasing the hot-pressing temperature to 1800°C. In contrast, almost full densification (>99% ρ_{th}) was attained at a hot-pressing temperature of 1700°C for the TiB_2 composites containing 2.5 wt% $MoSi_2$ sinter-additive. This implies that a small amount of $MoSi_2$ sinter-aid enhances the densification of TiB_2 at lower hot-pressing temperature, compared with binderless TiB_2. However, further addition of $MoSi_2$ (>5 wt%) lowers density due to the increased volume fraction of various secondary phases (Mo_5Si_3 and Ti_5Si_3).

From x-ray diffraction (XRD) observations, the observable increase in relative x-ray peak intensities of Ti_5Si_3 and Mo_5Si_3 phases with an increase in the amount of $MoSi_2$ is indicative of the concomitant increment in volume fraction of the secondary phases (see Fig 14.1b). The bright field scanning transmission electron microscope (STEM) image of hot-pressed TiB_2–(10 wt%)$MoSi_2$ reveals the presence of polygonal TiB_2 grains (darker contrast) and the $MoSi_2$ grains (gray contrast) (see Fig. 14.2a). The grain size of TiB_2 varies in the range 1.0–1.5 μm, while that of $MoSi_2$ varies in the range 3–4 μm. Figure 14.2b illustrates a conventional bright field transmission electron microscopy (TEM) micrograph and the analysis of selected area diffraction (SAD) patterns confirms the presence of Mo_5Si_3 and Ti_5Si_3 at the triple junction. The possible reaction pathways that can explain the sintered reaction products are as follows:

$$5\,TiO_2 + 5.714\,MoSi_2 \rightarrow 1.143\,Mo_5Si_3 + Ti_5Si_3 + 5\,SiO_2: \quad (14.1)$$

$$2.5\,Ti_3O_2 + 5\,MoSi_2 \rightarrow Mo_5Si_3 + 1.5\,Ti_5Si_3 + 2.5\,SiO_2. \quad (14.2)$$

The thermodynamic calculations estimate that the free energy changes (ΔG^o) are −53.67 kcal and −76.17 kcal at 1700°C (hot-pressing temperature) for Reactions 14.1 and 14.2, respectively,[34] thereby indicating their thermodynamic feasibility at or below 1700°C. Another possible reason for densification enhancement can be correlated to the dislocation activity (Fig. 14.2a), since the dislocations can act as short-circuit diffusion paths for mass transport during sintering.

14.3.2 Mechanical Properties

A comparison of the hardness possessed by TiB_2-based ceramics is presented in Figure 14.3a. Importantly, the addition of $MoSi_2$, compared with other ceramic sinter-additives, results in considerably higher hardness of TiB_2-based ceramics.

A plot showing the hardness variation with indent load is presented in Figure 14.3b. It can be noted that monolithic TiB_2 (HP, 1800°C) and $MoSi_2$-reinforced

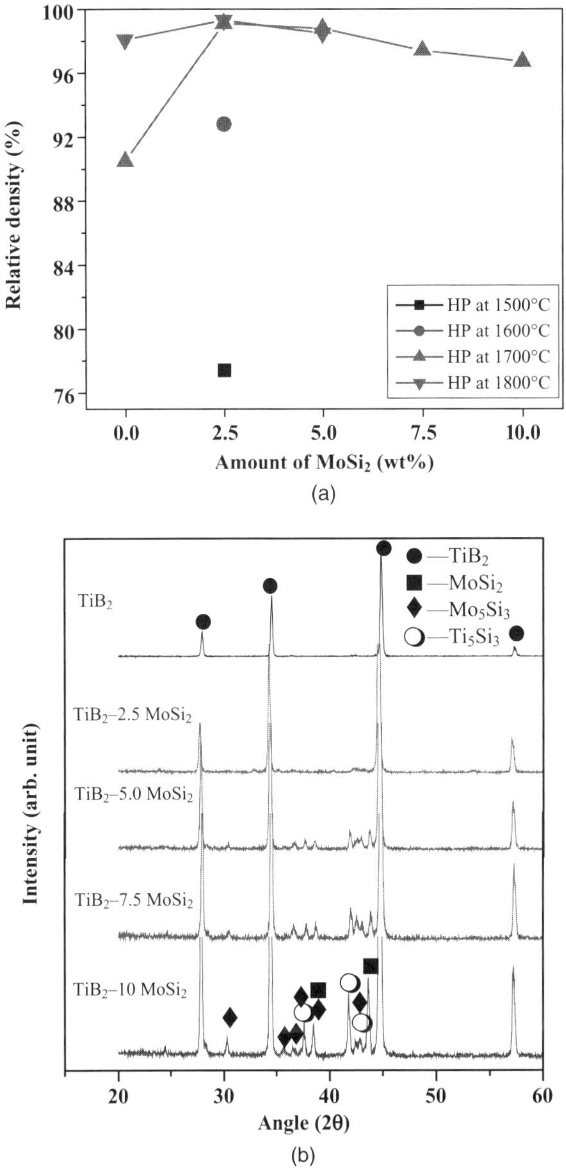

Figure 14.1 (a) Relative density of TiB_2 specimens as function of the $MoSi_2$ content at various temperatures after hot pressing for 1 hour, in vacuum; (b) the XRD patterns of the TiB_2–$MoSi_2$ samples hot pressed at 1700°C for 1 hour.[31]

TiB_2-based composites (HP, 1700°C) exhibit comparatively higher hardness. However, monolithic TiB_2, hot pressed at 1700°C, is measured with low hardness at all indent loads. Such interesting observations along with the instrumented indentation results (Fig. 14.4) are critically analyzed in a later section.

Fracture toughness results, obtained via a short crack method,[24] of the TiB_2-based ceramics are provided in Figure 14.5. A modest increment in fracture

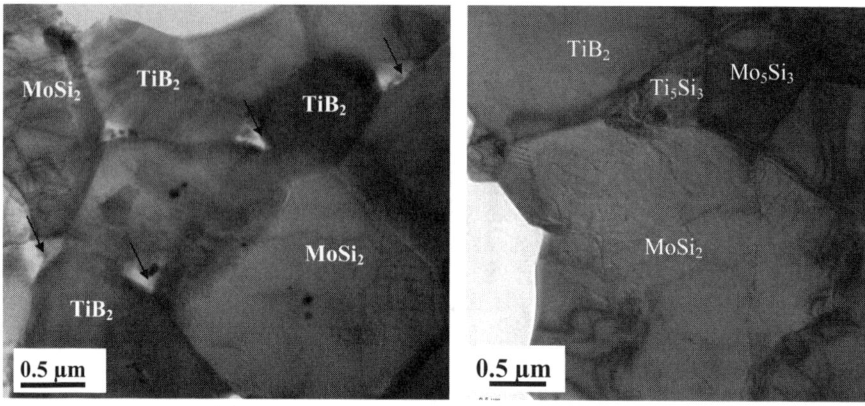

(a) (b)

Figure 14.2 STEM bright field image (a) and the conventional bright field TEM image (b) show the phase assemblage and grain structure of TiB_2–(10 wt%)$MoSi_2$ hot pressed at 1700°C for 1 hour. The single-headed arrows indicate grain boundary phases.[31]

toughness (up to ~6 MPa·m$^{1/2}$) of the nominally brittle TiB_2 ceramics was obtained with a small amount of $MoSi_2$ added (2.5 wt%). However, the fracture toughness reduces with further increase in $MoSi_2$ content. Comparing these fracture toughness values with those previously achieved using other ceramic additives,[7,14–17] it has been noticed that toughness improvement can be achieved at some optimum reinforcement content, beyond which toughness generally deteriorates, irrespective of the sinter-additive.

14.3.3 Depth Sensing Instrumented Indentation Response

A critical analysis of indentation response can be extracted from measurements of load versus penetration depth, using an instrumented Vickers indenter (see Fig. 14.4a). In the case of TiB_2 materials, the variation of residual indentation depths, after completion of the unloading cycle, correlates well with the measured hardness variation with $MoSi_2$ content. Conventionally, hardness is an indication of the amount of plastic deformation in the indented volume at a particular indent load. Hence, using the load-versus-penetration curves (*P–d* curves), the correlation between the experimentally determined Vickers hardness values ($H_{v0.2}$); measurement of indent diagonal lengths and the estimated plastic work done can be made. Following work of Chollacoop and coworkers,[35] the area under the loading curve is defined as the total work done during indentation (W_t), with the area under the unloading curve being the recovered elastic work (W_e). At any given indent load, the net plastic work (W_p) can be obtained as

$$W_p = W_t - W_e. \tag{14.3}$$

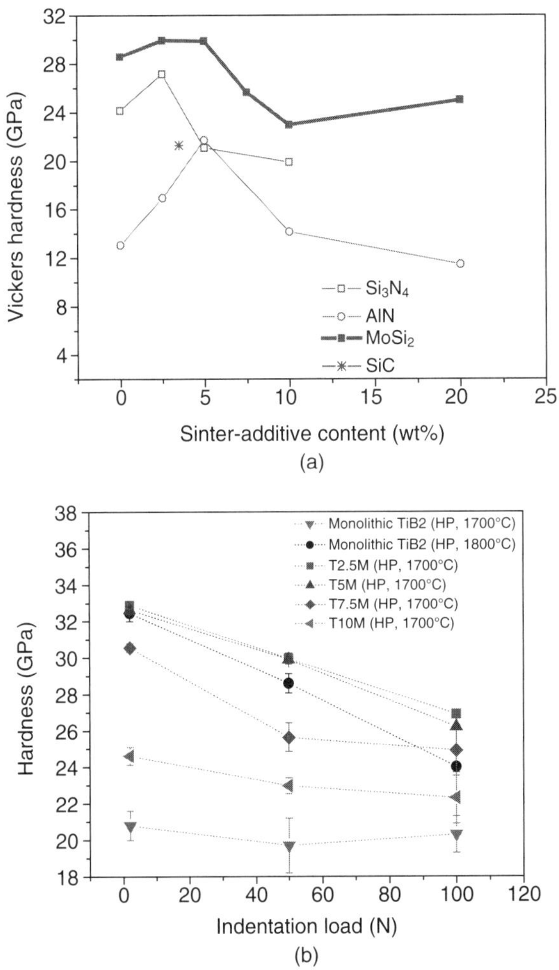

Figure 14.3 (a) Variation of Vickers hardness of TiB_2-based ceramics reinforced with varying amount of different ceramic sinter-additives.[12] (b) Variation of Vickers hardness with indent load for monolithic TiB_2 (HP, 1800°C) as well as monolithic TiB_2 and TiB_2-based ceramics, reinforced with various amounts of $MoSi_2$ (HP, 1700°C).[29]

The characteristic areas corresponding to W_e and W_p are shown in Figure 14.4b, and the corresponding values for the investigated TiB_2-based ceramics are reported in Table 14.1. The elastic moduli (*E*) values, derived from the estimated system modulus by measurement of the slopes of the initial part of the unloading curves (Oliver and Pharr[36,37]) are summarized in Table 14.1, assuming a Poisson's ratio of 0.2 for the TiB_2-based ceramics. The elastic modulus value of ~500 GPa, as estimated from the instrumented indentation data, for monolithic TiB_2 (HP, 1800°C), agrees well with the values previously reported using more conventional techniques.[2,17,26] Furthermore, the *E*-modulus decreases with $MoSi_2$ sinter-additive, which can be expected in light of the lower stiffness of $MoSi_2$ and the other secondary phases, compared with TiB_2.[2]

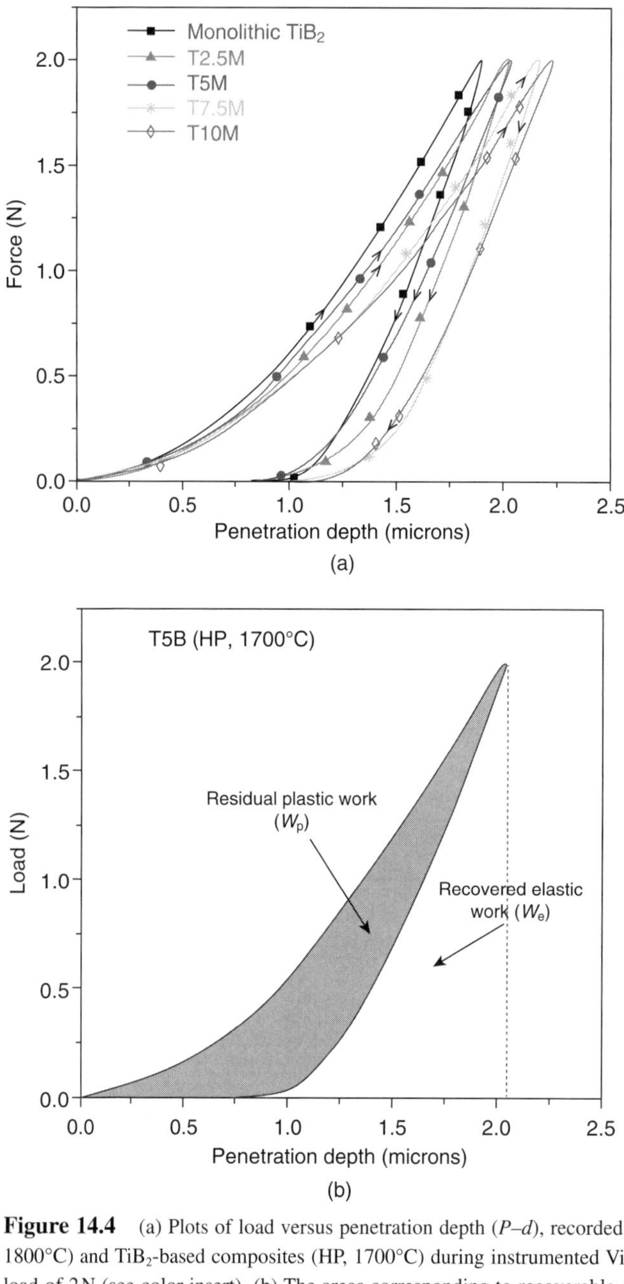

Figure 14.4 (a) Plots of load versus penetration depth (*P–d*), recorded for monolithic TiB_2 (HP, 1800°C) and TiB_2-based composites (HP, 1700°C) during instrumented Vickers indentation at a peak load of 2 N (see color insert). (b) The areas corresponding to recoverable elastic work (W_e) and residual plastic work (W_p) performed during the instrumented indentation are indicated in a typical *P–d* plot.[29]

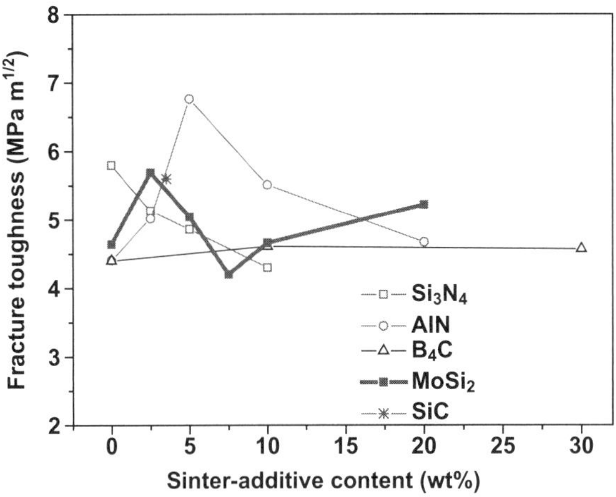

Figure 14.5 Comparison between the fracture toughness achieved on using varying amounts of $MoSi_2$ and other ceramic additives as reinforcements for TiB_2.[12]

14.3.4 Residual Strain-Induced Property Degradation

From Table 14.1, it can be observed that, with $MoSi_2$ addition, the total work done (W_t; estimated from the load-versus-penetration curves) during indentation increases. This can be attributed to lower stiffness and hardness of the secondary phases ($MoSi_2$, Mo_5Si_3, and Ti_5Si_3), which aids increased elastic as well as plastic deformation, respectively. However, Table 14.1 reveals that the amount of permanent plastic work (W_p) is lower for the TiB_2–(2.5%)$MoSi_2$ or TiB_2–(5%)$MoSi_2$ composites than for monolithic TiB_2. Such observations are commensurate with the improvement in densification and hardness on reinforcement with small amounts (2.5 and 5.0 wt%) of $MoSi_2$. On the contrary, a significantly larger amount of plastic work done was recorded during indentation of the composites with higher $MoSi_2$ (>5 wt%) reinforcement. Another reason for the permanent deformation in composites with higher $MoSi_2$ content could be a larger fraction of the brittle reaction product phases (Ti_5Si_3 and Mo_5Si_3), which might have resulted in the release of a considerable amount of elastic strain energy via brittle fracture (microcracking) in the deformation zone. This would restrict elastic recovery on release of the indentation load, which would contribute to the significant increase in the measured residual plastic work, and hence reduction in hardness (H_v), for such materials.

The property degradation with large $MoSi_2$ addition can be explained as follows. The differential strain (ε_{gb}) set up near a grain boundary can be expressed as[38]

$$\varepsilon_{gb} = \Delta\alpha T/(1-\nu), \tag{14.4}$$

where ΔT is the temperature range over which the strain develops and ν is Poisson's ratio (~0.2). In the case of the TiB_2–$MoSi_2$ system, $\Delta\alpha$ between TiB_2 and Ti_5Si_3 is $\sim 12 \times 10^{-6}\,K^{-1}$ and that between TiB_2 and Mo_5Si_3 is $\sim 5 \times 10^{-6}\,K^{-1}$; the elastic

Table 14.1. Values of Work Done ($\boldsymbol{W}_t$, $\boldsymbol{W}_e$, and $\boldsymbol{W}_p$), Hardness ($\boldsymbol{H}_{v0.2}$), and Elastic Modulus ($\boldsymbol{E}$), of the Densified TiB_2-Based Ceramics, as Determined from Instrumented Indentation Experiments, Performed at a Peak Load of 2 N[29]

Composition	Hot-pressing temperature (°C)	Residual plastic work (W_p) ($\times 10^{-6}$ J)	Recovered elastic work (W_e) ($\times 10^{-6}$ J)	Total work done (W_t) ($\times 10^{-6}$ J)	Hardness ($H_{v0.2}$) (GPa)	Elastic modulus (E) (GPa)
Monolithic TiB_2	1800	0.701 ± 0.035	0.668 ± 0.037	1.369 ± 0.039	32.44 ± 0.45	497 ± 15
TiB_2–(2.5)$MoSi_2$	1700	0.664 ± 0.014	0.721 ± 0.021	1.385 ± 0.022	32.89 ± 0.16	486 ± 11
TiB_2–(5)$MoSi_2$	1700	0.688 ± 0.017	0.727 ± 0.019	1.415 ± 0.025	32.63 ± 0.25	479 ± 12
TiB_2–(7.5)$MoSi_2$	1700	0.789 ± 0.021	0.728 ± 0.023	1.516 ± 0.028	30.55 ± 0.24	464 ± 17
TiB_2–(10)$MoSi_2$	1700	0.894 ± 0.027	0.731 ± 0.025	1.625 ± 0.031	24.61 ± 0.49	459 ± 16

strains developed at the respective interfaces are ~24.9×10^{-3} and ~10.4×10^{-3}. Also, the $\Delta\alpha$ between TiB_2 and $MoSi_2$ is ~$2 \times 10^{-6} K^{-1}$, which leads to development of a comparatively lesser strain (~4.2×10^{-3}) at the interface between TiB_2 and $MoSi_2$.

Following the model proposed by Clarke,[39] the critical elastic strain (ε_{gbc}), set up near a grain boundary, resulting in spontaneous fracture is given by

$$\varepsilon_{gbc} = (24\gamma_{gb} / El_b)^{1/2}, \tag{14.5}$$

where γ_{gb} is the grain boundary fracture energy, E is the elastic modulus, and l_b is the length of the common interface between the two phases.

On close observation of the various micrographs showing the phase assemblage in the TiB_2-based composites, the average length scale (l_b) of the common interface between TiB_2 and Ti_5Si_3/Mo_5Si_3 or $MoSi_2$ phases is found to be nearly ~1 μm. Assuming γ_{gb} to be ~1 J/m^2, ε_{gbc} can be roughly estimated to be ~7.2×10^{-3}, which is lower than the elastic strains possibly developed at the TiB_2/Ti_5Si_3 and TiB_2/Mo_5Si_3 interfaces. Hence, relatively larger residual strains along the TiB_2/Ti_5Si_3 and TiB_2/Mo_5Si_3 interfaces can lead to microcracking, especially at the indentation stress field. Such interfacial cracks are bound to have a considerable influence on the mechanical behavior of the TiB_2–$MoSi_2$-based composites.

14.3.5 Relationship between Indentation Work Done and Phase Assemblage

The experimentally measured Vickers hardness values with varying indent loads (2–100 N) revealed considerable variation with indentation load for all the materials (see Fig. 14.3b). The decrease in hardness with increasing indent load is suggestive of an "indentation size effect." Such size-versus-load effects for different ceramic materials were reported earlier,[40–42] and such observations have been attributed to the occurrence of the easier process of multiplication of a sufficient number of pre-existing elements of plasticity (dislocations, twins) at higher loads.

Although not shown here, except for the radial cracks emanating from the corner of such indents, no additional cracking can be observed in the vicinity of the "plastically" deformed region. However, in addition to such radial cracking, a concentric array of cracks were observed within the indents (at the faces) obtained at an indent load of 100 N. Furthermore, lateral cracking leading to chipping of material has been observed in some of the indents obtained at the highest load of 100 N. The preceding observations point toward the increased severity of surface and subsurface cracking with increase in indent load. It must be noted that such considerable crack openings (at the indent faces) result in additional displacements, and this is manifested at the macroscopic level as permanent (plastic) deformation. It must also be mentioned here that Richter and Ruthendorf[42] made a similar observation at higher loads with transition metal carbides.

14.4 TiB_2–$TiSi_2$ SYSTEM

14.4.1 Sintering Reactions and Densification Mechanisms

The densification data of the hot-pressed TiB_2–$TiSi_2$ ceramics are provided in Table 14.2. Critical observation of the data presented in Table 14.2 reveals that sintering density increases with the amount of $TiSi_2$ sintering-additive and the maximum of ~99.6% ρ_{th} can be obtained with $TiSi_2$ addition (5–10 wt%) when all samples are hot pressed at 1650°C. From these observations, it should be clear that by optimizing the "processing and compositional window," high sinter density of 99% ρ_{th} can be obtained by the hot-pressing route.

XRD results (not shown) indicated the formation of $TiSi_3$ when 5% or more $TiSi_2$ was added to TiB_2. In Figure 14.6a, a high-angle annular dark field (HAADF) TEM image reveals the existence of various phases in the TiB_2–(10 wt%) $TiSi_2$. The presence of various phases, that is, TiB_2, $TiSi_2$, and Ti_5Si_3, was confirmed by analyzing the selected area diffraction patterns (SADPs). The average grain size of TiB_2 varied between 2 and 3 μm, while the size of the Ti_5Si_3 grains ranged from 100 to 150 nm. A bright field conventional TEM image, as shown in Figure 14.6b, reveals that Ti_5Si_3 phase is observed to be located at the triple pocket and surrounded by the TiB_2 and $TiSi_2$ grains. The morphology of Ti_5Si_3 grains at the triple points is a clear signature of liquid phase sintering. The attainment of high sinter density in TiB_2 reinforced with $TiSi_2$ at such low hot-pressing temperature can be attributed to LPS, since the hot-pressing experiments were carried out above the melting point (1540°C) of $TiSi_2$.

In addition to the direct reaction between TiB_2 and $TiSi_2$, the reaction pathways that explain the formation of Ti_5Si_3 could be as follows:

$$3TiB_2 + 2TiSi_2 \rightarrow Ti_5Si_3 + SiB_6; \tag{14.6}$$

$$2TiO_2 + TiB_2 + 2TiSi_2 \rightarrow Ti_5Si_3 + SiO_2 + B_2O_2; \tag{14.7}$$

$$7TiO_2 + 8TiSi_2 \rightarrow 3Ti_5Si_3 + 7SiO_2; \tag{14.8}$$

$$5TiO_2 + 5TiSi_2 + 2C \rightarrow 2Ti_5Si_3 + 4SiO_2 + 2CO_{(g)}; \tag{14.9}$$

$$Ti_3O_2 + 2TiSi_2 \rightarrow Ti_5Si_3 + SiO_2; \tag{14.10}$$

$$2.8TiO + 2.2TiSi_2 \rightarrow Ti_5Si_3 + 1.4SiO_2. \tag{14.11}$$

In this set of reactions, TiO_2 and various sub-oxides, TiO_x, are considered to be present, as TiB_2 particles are known to be covered with surface oxides. Importantly, thermodynamic calculations reveal that the net free energy change for the third possible reaction (Eq. 14.8), that is, ΔG_3, is negative over a broad temperature range (up to sintering temperature). For example, ΔG_3 is –74.727 kcal/mole at 1923 K. The Gibbs free energy of the reaction is positive up to 1323 K and becomes negative ($\Delta G_4 = -56.95$ kcal/mole) at 1923 K. ΔG_5 is relatively moderate (–30.24 kcal/mole) and ΔG_6 is very low (–4.49 kcal/mole) at 1923 K. From the detailed thermodynamic analysis, it is realized that Reactions 14.8–14.10 are more probable reaction pathways for Ti_5Si_3 formation.

Table 14.2. Densification Data and Properties of the Developed Ceramic Materials, Which Were Hot Pressed for 1 hour in Argon[29–31]

Material composition (in wt%)	Processing details	Relative density (% ρ_{th})	Elastic modulus (GPa)	Vickers hardness, H_v (GPa)	Indentation toughness (K_{IC}), $MPa{\cdot}m^{1/2}$	Four-Point flexural Strength (MPa)
TiB_2–(2.5)$TiSi_2$	HP 1600°C	95.52	–	22.8 ± 1.5	–	–
TiB_2–(5.0)$TiSi_2$	HP 1600°C	95.58	–	22.1 ± 2.1	–	–
Monolithic TiB_2	HP 1650°C	94.45	434.9 ± 12	18.3 ± 1.2	3.8 ± 0.6	365 ± 88.5
TiB_2–(2.5)$TiSi_2$	HP 1650°C	98.77	509.2 ± 8	24.8 ± 0.9	4.3 ± 0.3	380.9 ± 74.0
TiB_2–(5.0)$TiSi_2$	HP 1650°C	99.61	517.9 ± 11	25.2 ± 0.6	5.8 ± 0.5	425.7 ± 68.8
TiB_2–(10.0)$TiSi_2$	HP 1650°C	99.60	470.3 ± 15	23.5 ± 1.0	4.2 ± 0.8	337.9 ± 67.9
TiB_2	HP, 1800°C	97.50	–	26.0	5.1	–
TiB_2	HP, 1700°C	88.10	–	–	–	–
TiB_2–(10)$MoSi_2$	HP, 1700°C	99.30	–	27.0	4.0	–

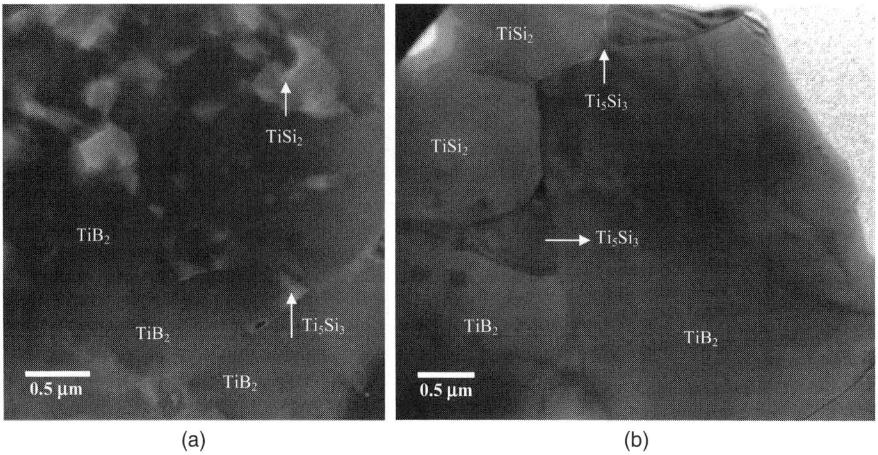

Figure 14.6 (a) Representative HAADF TEM image reveals various contrasting phases in the hot-pressed TiB_2–(10 wt%)$TiSi_2$ ceramic. (b) A conventional TEM bright field image showing the grain morphology of the various constituent phases.[28]

14.4.2 Mechanical Properties

Table 14.2 reveals that the addition of 2.5 wt% $TiSi_2$ did not cause significant improvement in room-temperature strength, whereas 5 wt% $TiSi_2$ addition increased strength to ~426 MPa due to high densification and uniform microstructure with finer TiB_2 grains. The lowest strength (~338 MPa), however, was recorded with TiB_2–(10 wt%)$TiSi_2$, because of the large amount of second phase formation. Similar to flexural strength, maximum E-modulus of ~518 GPa was measured with the TiB_2–(5 wt%)$TiSi_2$ composite. Similarly, TiB_2–(5 wt%)$TiSi_2$ exhibited high hardness of 25 GPa. However, further addition of $TiSi_2$ reduced hardness to 23.5 GPa in TiB_2–(10 wt%)$TiSi_2$ composites. It can be noted here that high hardness values of 19, 22–23, and 28 GPa were reported for TiB_2–(10 vol%)B_4C,[15] TiB_2–(15 wt%)TiC,[17] and TiB_2–(3 wt%)CrB_2[19] composites, respectively. The measured toughness of the investigated materials varies in the range 3.8–5.8 MPa·m$^{1/2}$. The mechanical properties data imply that the $TiSi_2$ addition to TiB_2 needs to be restricted to 5 wt% and any further addition degrades the properties.

14.4.3 Residual Stress or Strain and Property Degradation

From Table 14.2, it is clear that $TiSi_2$ addition has an obvious influence on the mechanical properties of TiB_2 materials. In various noncubic ceramics (such as TiB_2), the grain size dependence of fracture strength is correlated with residual stress, introduced by the anisotropy of the coefficient of thermal expansion (CTE).[16]

The residual stress or strain also influences the fracture toughness of TiB_2–$TiSi_2$. It is known that residual strain can be generated due to elastic modulus and/or thermal expansion mismatch between the matrix and second phase as well as from the anisotropy of thermal expansion in noncubic materials. Therefore, the resultant radial tensile or compressive stresses around a particle can develop by a suitable of choice of low-expansion matrix and a high-expansion additive. It has been observed that crack deflection occurs due to a residual strain field. In calculating residual stress, the thermal expansion coefficient along the crystallographic *a* and *c* axes for TiB_2 are $\alpha_a = 6.4 \times 10^{-6}\,K^{-1}$ and $\alpha_c = 9.2 \times 10^{-6}\,K^{-1}$ and the elastic modulus of TiB_2 is ~565 GPa;[3] for Ti_5Si_3, $\alpha_a = 5.1 \times 10^{-6}\,K^{-1}$, $\alpha_c = 22.2 \times 10^{-6}\,K^{-1}$, and elastic modulus ~156 GPa were considered.[30]

A number of models are available to calculate the residual stresses in particle-reinforced ceramic composites.[43–45] The residual strain in TiB_2–$TiSi_2$ composites is estimated from a model proposed by Taya et al.[44]:

$$\alpha_1^* = \int_{T_P}^{T_R} (\alpha_p - \alpha_m)\delta dT, \tag{14.12}$$

where α_p and α_m are the CTEs of the particle and matrix phases, respectively, δ is the isotropic tensor (Kronecker's delta), and T_p is the sintering temperature from which the ceramic composite is cooled to T_R, the room temperature. The isotropic average stress fields in the particles and matrix are given by $\langle\sigma\rangle_p$ and $\langle\sigma\rangle_m$, for a given volume fraction of particles (f_p):

$$\frac{\langle\sigma\rangle_p}{E_m} = \frac{-2(1-f_p)\beta\alpha_1^*}{A} \tag{14.13}$$

and

$$\frac{\langle\sigma\rangle_m}{E_m} = \frac{2f_p\beta\alpha_1^*}{A}, \tag{14.14}$$

where

$$A = (1-f_p)(\beta+2)(1+\nu_m) + 3\beta f_p(1-\nu_m) \tag{14.15}$$

and

$$\beta = \left(\frac{1+\nu_m}{1+\nu_p}\right)\left(\frac{E_p}{E_m}\right). \tag{14.16}$$

ν_m and ν_p are Poisson's ratio of the matrix and particles, and E_m and E_p are elastic modulus of the matrix and particles, respectively. Based on the preceding set of equations, the misfit strain (α_1^*) in the particles is estimated to be around -5.645×10^{-3}, and $\langle\sigma\rangle_m$ = (−112.2 MPa) and $\langle\sigma\rangle_p$ = 1444.6 MPa. Kang et al.[46] reported the stress at the interface between the B_4C particles and TiB_2 as −260 MPa. Blugan et al.[47] estimated $\langle\sigma\rangle_m$ = −490 MPa and $\langle\sigma\rangle_p$ = 1080 MPa for Si_3N_4–(30 wt%)TiN composite. In the case of TiB_2–$TiSi_2$ composites, if high tensile stress exceeds the bonding strength of the matrix–particulate interface, then the interface will be debonded. The residual compressive stress in the TiB_2 matrix potentially results in an increase in toughness of TiB_2–$TiSi_2$ composites.

14.5 ZrB_2–SiC–$TiSi_2$ COMPOSITES

The density and mechanical properties of SPSed ZrB_2–SiC–$TiSi_2$ samples are presented in Table 14.3. It can be observed that the density of ZrB_2–SiC increases from 97.8% ρ_{th} to ~100% ρ_{th} with the addition of $TiSi_2$ up to 5 wt%, when all are spark plasma sintered at 1600°C, for 10 minutes, at 50 MPa. It can be noticed that high sintering temperatures (above 1800°C) are imperative for the full densification (>99% ρ_{th}) of ZrB_2 despite the use of various processing routes. These results therefore indicate that $TiSi_2$ aids in improving the sinterability of ZrB_2–SiC. This can be attributed to the LPS of ZrB_2 ceramics in the presence of $TiSi_2$. Wang et al.[48] reported that sintering aids play a vital role in enhancing densification of ZrB_2. Guo et al.[49]

Table 14.3. Summary of the Mechanical Properties of ZrB_2 Sintered with Different Sinter-Additives

Material composition (in wt%)	Processing details	Relative density (% ρ_{th})	Vickers hardness, H_v (GPa)	Indentation toughness (K_{IC}), MPa·m$^{1/2}$	Reference
ZrB_2	PS, 2150°C, 9 hours	98	14.5	–	23
ZrB_2–(20)$MoSi_2$	PS, 1850°C, 30 minutes	99.1	16.1	2.3	24
ZrB_2–(30)SiC	RHP, 1800°C	99	27	2.1	25
ZrB_2–(15)$TaSi_2$	HP, 1900°C	99	17.8	3.8	26
ZrB_2–(18.5) SiC– (3.7)Si_3N_4–(1) Al_2O_3–(0.5)Y_2O_3	HP, 1760°C, 10 minutes	98	14.2	4.6	9
ZrB_2–(25.2)SiC	SPS, 1400°C, 30 MPa, 12 minutes	98.3	17.3	5.3	52
ZrB_2–(40)ZrC–(12) SiC	SPS, 1800°C, 20 MPa, 10 minutes	99.5	16.9	5.9	53
ZrB_2–(60)SiC	SPS, 2100°C, 20 MPa, 180 seconds	99	26.8	3.5	54
ZrB_2–(6)Cu	SPS, 1500°C, 15 minutes	~95	19.1	7.4	33
ZrB_2–(18)SiC–(0) $TiSi_2$	–	97.8	21.0 ± 0.8	3.2 ± 0.7	55
ZrB_2–(18)SiC–(2.5) $TiSi_2$	–	98.2	25.6 ± 0.4	4.1 ± 1.2	55
ZrB_2–(18)SiC–(5) $TiSi_2$	–	~100	26.4 ± 0.5	5.1 ± 0.3	55

HP, hot pressing; PS, pressureless sintering; RHP, reactive hot pressing; SPS, spark plasma sintering.

reported that shrinkage rate increases with increasing heating rate of ZrB_2. During heating, the shrinkage rate peak is shifted to a lower temperature and increases with increasing heating rates. In view of such observations, it is possible that the high heating rate in SPS possibly resulted in achieving full densification at lower sintering temperatures.

To assess the potentiality of the newly developed ZrB_2–SiC–$TiSi_2$ materials, the processing details along with sinter density data and mechanical properties of various ZrB_2 materials are summarized in Table 14.3. It can be observed that full densification of ZrB_2-based materials via a hot-pressing route is possible only at temperature ≥1800°C. The hardness of previously developed ZrB_2 materials varied from 14 to 27 GPa and indentation toughness from 2.1 to 7.4 MPa·m$^{1/2}$. An intriguing observation is that the materials exhibiting the highest hardness (27 GPa) were measured to have low fracture toughness (about 2 MPa·m$^{1/2}$). The SPS-processed dense ZrB_2–(18)SiC–(5)$TiSi_2$ composite also exhibited an excellent combination of properties with maximum hardness of 26.4 GPa and indentation fracture toughness of ~5.1 MPa·m$^{1/2}$.

The flexural strength measurements reveal a moderate strength of 497 MPa for ZrB_2–(18)SiC–(5)$TiSi_2$ and 373 MPa for ZrB_2–(18)SiC–(0)$TiSi_2$ composites. Such modest strength improvement can be attributed either to better density, that is, the absence of crack-initiating flaws or pores, or to silicide addition.

The measured toughness values of the investigated ZrB_2 materials varied in the range of 3.2–5.1 MPa·m$^{1/2}$ (see Table 14.3). In a recent review, Guo et al. reported that the fracture toughness (measured by indentation) of ZrB_2–SiC–$MoSi_2$ and ZrB_2–$ZrSi_2$ systems[50,51] varied in the range of 2.6–3.7 MPa·m$^{1/2}$ and 3.8–4.8 MPa·m$^{1/2}$, respectively. From these results, the addition of $TiSi_2$ appears to have beneficial effects in enhancing the toughness properties. A better toughness in ZrB_2–SiC–$TiSi_2$ is mainly due to crack deflection in the residual stress field (see Fig. 14.7). The indentation-induced crack appears to propagate along ZrB_2 grains and crack bridging was also observed. The residual strain in the composite and/or weak matrix–second-phase interfaces, as mentioned earlier, can lead to crack deflection. In the ZrB_2 composites, residual strain can be generated due to elastic modulus and/or thermal expansion mismatch between the ZrB_2 matrix and the secondary phases (SiC and $TiSi_2$).

14.6 CONCLUDING REMARKS

As a concluding note, the experimental results summarized in this chapter strengthen the idea that the sintering reactions play an important role in determining the densification behavior and material properties of the non-oxide ceramics. On the basis of the hot-pressing experiments on TiB_2–$MoSi_2$ and TiB_2–$TiSi_2$ systems, it is realized that even under situations when LPS can be promoted at lower temperature, full densification of TiB_2 still requires a hot-pressing temperature, in excess of 1600°C.

Another interesting message is that the sinter-aid addition needs to be optimized in a close window and even small additions (≤5 wt%) can still promote sintering reactions in the case of TiB_2-based materials. In view of the favorable

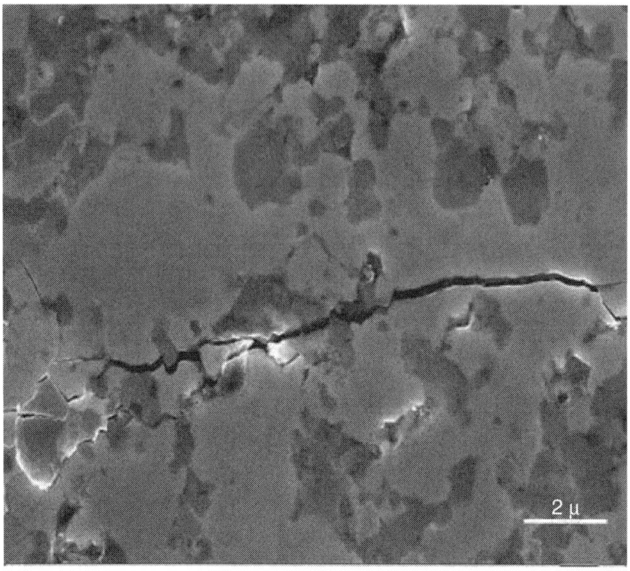

Figure 14.7 Scanning electron microscopy (SEM) image illustrating crack generation from the indent corner (20 N) on ZrB_2–(18)SiC–(5)$TiSi_2$ SPSed for 10 minutes at 1600°C and 50 MPa.[55]

high-temperature properties of Ti_5Si_3, the presence of Ti_5Si_3 in TiB_2–$TiSi_2$ ceramics would be certainly beneficial for high-temperature applications as Ti_5Si_3 has relatively high melting point (2130°C) and better mechanical properties.

The mechanical property combination obtained with ZrB_2–(18)SiC–(5)$TiSi_2$ is observed to be much better than for previously developed ZrB_2-based materials. However, the structure–property correlation in this material system needs to be established.

REFERENCES

1 W. G. Fahrenholtz, G. E. Hilmas, I. G. Talmy, and J. A. Zaykoski. Refractory diborides of zirconium and hafnium. J. Am. Ceram. Soc. 90(5) (2007), 1347–1364.

2 J. R. Ramberg and W. S. Williams. High temperature deformation of titanium diboride. J. Mater. Sci. 22 (1987), 1815–1826.

3 J. J. Melendez-Martinez, A. Dominguez-Rodriguez, F. Monteverde, C. Melandri, and G. de Portu. Characterisation and high temperature mechanical properties of zirconium boride-based materials. J. Eur. Ceram. Soc. 22 (2002), 2543–2549.

4 F. Peng and R. F. Speyer. Oxidation resistance of fully dense ZrB_2 with SiC, TaB_2 and $TaSi_2$ additives. J. Am. Ceram. Soc. 91(5) (2008), 1489–1494.

5 R. A. Cutler. *Engineering Properties of Borides, Engineering Materials Handbook, Ceramic and Glasses*. ASM International, Vol. 4, Metals Park, OH, 1991, 787–803.

6 F. Monteverde and L. Scatteia. Resistance to thermal shock and to oxidation of metal diborides–SiC ceramics for aerospace application. J. Am. Ceram. Soc. 90(4) (2007), 1130–1138.

7 R. G. Munro. Material properties of titanium diboride. J. Res. Natl. Inst. Stand. Technol. 105(5) (2000), 709–720.

8 A. K. Kuriakose and J. L. Margrave. The oxidation kinetics of zirconium diboride and zirconium carbide at high temperatures. J. Electrochem. Soc. 111(7) (1964), 827–831.

9 F. Monteverde, S. Guicciardi, and A. Bellosi. Advances in microstructure and mechanical properties of zirconium diboride based ceramics. Mater. Sci. Eng. A 346 (2003), 310–319.

10 Y. Yan, Z. Huang, S. Dong, and D. Jiang. Pressureless sintering of high-density ZrB_2-SiC ceramic composites. J. Am. Ceram. Soc. 89(11) (2006), 3589–3592.

11 F. Monteverde, C. Melandri, and S. Guicciardi. Microstructure and mechanical properties of an HfB_2 + 30 vol% SiC composite consolidated by spark plasma sintering. Mater. Chem. Phys. 100 (2006), 513–519.

12 B. Basu, G. B. Raju, and A. K. Suri. Processing and properties of monolithic TiB_2 based materials. Int. Mater. Rev. 51 (2006), 352–374.

13 W. Wang, Z. Fu, H. Wang, and R. Yuan. Influence of hot pressing sintering temperature and time on microstructure and mechanical properties of TiB_2 ceramics. J. Eur. Ceram. Soc. 22 (2002), 1045–1049.

14 J. J. Melendez-Martinez, A. Dominguez-Rodriguez, F. Monteverde, C. Melandri, and G. de Portu. Characterisation and high temperature mechanical properties of zirconium diboride-based materials. J. Eur. Ceram. Soc. 22 (2002), 2543–2549.

15 F. Monteverde and A. Bellosi. Effect of the addition of silicon nitride on sintering behavior and microstructure of zirconium diboride. Scr. Mater. 46 (2002), 223–228.

16 F. Monteverde and A. Bellosi. Beneficial effects of AlN as sintering aid on microstructure and mechanical properties of hot pressed ZrB_2. Adv. Eng. Mater. 5 (2003), 508–512.

17 F. Monteverde and A. Bellosi. Efficacy of HfN as sintering aid in the manufacturing of ultra high temperature metal diboride-matrix ceramics. J. Mater. Res. 19 (2004), 3576–3585.

18 F. Monteverde and A. Bellosi. Development and characterization of metal-diboride-based composites toughened with ultra-fine SiC particulates. Solid State Sci. 7 (2005), 622–630.

19 S. S. Hwang, A. L. Vasiliev, and N. P. Padture. Improved processing and oxidation resistance of ZrB_2 ultra-high temperature ceramics containing SiC nanodispersoids. Mater. Sci. Eng. A 464 (2007), 216–224.

20 S. Zhu, W. G. Fahrenholtz, and G. E. Hilmas. Influence of silicon carbide particles size on the microstructure and mechanical properties of zirconium diboride-silicon carbide ceramics. J. Eur. Ceram. Soc. 27 (2007), 2077–2083.

21 L. Rangaraj, C. Divakar, and V. Jayaram. Fabrication and mechanisms of densification of ZrB_2-based ultra high temperature ceramics by reactive hot pressing. J. Eur. Ceram. Soc. 30 (2010), 129–138.

22 Z. Wang, S. Wang, X. Zhang, P. Hu, W. Han, and C. Hong. Effect of graphite flake on microstructure as well as mechanical properties and thermal shock resistance of ZrB_2–SiC matrix ultrahigh temperature ceramics. J. Alloys and Comp. 484 (2009), 390–394.

23 A. L. Chamberlain, W. G. Fahrenholtz, and G. E. Hilmas. Pressureless sintering of zirconium diboride. J. Am. Ceram. Soc. 89(2) (2006), 450–456.

24 D. Sciti, S. Guicciardi, A. Bellosi, and G. Pezzotti. Properties of a pressureless-sintered ZrB_2-$MoSi_2$ cermic composite. J. Am. Ceram. Soc. 89(7) (2006), 2320–2322.

25 A. L. Chamberlain, W. G. Fahrenholtz, and G. E. Hilmas. Low-temperature densification of zirconium diboride ceramics by reactive hot pressing. J. Am. Ceram. Soc. 89(12) (2006), 3638–3645.

26 D. Sciti, L. Silvestroni, G. Celotti, C. Melandri, and S. Guicciardi. Sintering and mechanical properties of ZrB_2-$TaSi_2$ and HfB_2-$TaSi_2$ ceramic composites. J. Am. Ceram. Soc. 91(10) (2008), 3285–3291.

27 G. Brahma Raju and B. Basu. Thermal and electrical properties of TiB_2-$MoSi_2$. Int. J. Refract. Met. Hard Mater. 28 (2010), 174–179.

28 G. Brahma Raju, K. Biswas, and B. Basu. Microstructural characterization and isothermal oxidation behavior of hot-pressed TiB_2-10 wt% $TiSi_2$ composite. Scr. Mater. 61 (2009), 674–677.

29 A. Mukhopadhyay, G. B. Raju, A. K. Suri, and B. Basu. Correlation between phase evolution, mechanical properties and instrumented indentation response of TiB_2-based ceramics. J. Eur. Ceram. Soc. 29 (2009), 505–516.

30 G. Brahma Raju and B. Basu. Densification, sintering reactions, and properties of titanium diboride with titanium disilicide as a sintering aid. J. Am. Ceram. Soc. 90(11) (2007), 3415–3423.

31 G. Brahma Raju, K. Biswas, A. Mukhopadhyay, and B. Basu. Densification and high temperature mechanical properties of hot pressed TiB_2-(0–10 wt. %) $MoSi_2$ composites. Scr. Mater. 61 (2009), 674–677.

32 G. Brahma Raju, B. Basu, N. H. Tak, and S. J. Cho. Temperature dependent hardness and strength properties of TiB_2 with $TiSi_2$ sinter-aid. J. Eur. Ceram. Soc. 29(10) (2009), 2119–2128.

33 T. Venkateswaran, B. Basu, G. B. Raju, and D.-Y. Kim. Densification and properties of transition metal borides-based cermets via spark plasma sintering. J. Eur. Ceram. Soc. 26 (2006), 2431–2440.

34 A. Roine. Chemical reaction and equilibrium software with extensive thermochemical database, Outokumpu HSC Chemistry for Windows (version 5.1). Pori, Finland.

35 N. Chollacoop, M. Dao, and S. Suresh. Depth-sensing instrumented indentation with dual sharp indenters. Acta Mater. 51 (2003), 3713–3729.

36 W. C. Oliver and G. M. Pharr. An improved technique for determining hardness and elastic modulus using load and displacement sensing indentation experiments. J. Mater. Res. 7(6) (1992), 1564–1583.

37 G. M. Pharr, W. C. Oliver, and F. R. Brotzen. On the generality of the relationship among contact stiffness, contact area, and elastic modulus during indentation. J. Mater. Res. 7(3) (1992), 613–617.

38 R. W. Davidge. Mechanical behaviour of ceramics. In *Cambridge Solid State Science Series*, ed. R. W. Cahn, M. W. Thompson and I. M. Ward. University Press, Cambridge, UK, 1979.

39 F. J. P. Clarke. Residual strain and the fracture stress-grain size relationship in brittle solids. Acta Metall. 12 (1964), 139–143.

40 A. Krell. A new look at grain size and load effects in the hardness of ceramics. Mater. Sci. Eng. A. 245 (1998), 277–284.

41 A. Carpinetri, and S. Puzzi. A fractal approach to indentation size effect. Eng. Fract. Mech. 73(15) (2006), 2110–2122.

42 V. Ritcher, and M. V. Ruthendorf. On hardness and toughness of ultrafine and nanocrystalline hard materials. Int. J. Ref. Met. Hard. Mater. 17 (1999), 141–152.

43 A. G. Evans and K. T. Faber. Toughening of ceramics by circumferential microcracking. J. Am. Ceram. Soc. 64 (7) (1981), 394–398.

44 M. Taya, S. Hayashi, A. S. Kobayashi, and H. S. Yoon. Toughening of a particulate-reinforced ceramic–matrix composite by thermal residual stress. J. Am. Ceram. Soc. 73(6) (1990), 1382–1391.

45 K. T. Faber and A. G. Evans. Crack deflection processes I. Theory. Acta Metall. 31(4) (1983), 565–576.

46 S. H. Kang, D. J. Kim, E. S. Kang, and S. S. Baek. Pressureless sintering and properties of titanium diboride ceramics containing chromium and iron. J. Am. Ceram. Soc. 84(4) (2001), 893–895.

47 G. Blugan, M. Hadad, J. Janczak-Rusch, J. Kuebler, and T. Graule. Fractography, mechanical properties, and microstructure of commercial silicon nitride–titanium nitride composites. J. Am. Ceram. Soc. 88(4) (2005), 926–933.

48 H. Wang, C. A. Wang, X. Yao, and D. Fang. Processing and mechanical properties of zirconium diboride-based ceramics prepared by spark plasma sintering. J. Am. Ceram. Soc. 90(7) (2007), 1992–1997.

49 S. Q. Guo, T. Nishimura, Y. Kagawa, and J. M. Yang. Spark plasma sintering of zirconium diborides. J. Am. Ceram. Soc. 91(9) (2008), 2848–2855.

50 S. Q. Guo, Y. Kagawa, and T. Nishimura. Mechanical behavior of two step hot-pressed ZrB_2-based composites with $ZrSi_2$. J. Eur. Ceram. Soc. 29(4) (2009), 787–794.

51 S. Q. Guo, T. Nishimura, T. Mizuguchi, and Y. Kagawa. Mechanical properties of hot-pressed ZrB_2–$MoSi_2$–SiC composites. J. Eur. Ceram. Soc. 28(9) (2008), 1891–1898.

52 Y. Zhao, L. J. Wang, G. J. Zhang, W. Jiang, and L. D. Chen. Effect of holding time and pressure on properties of ZrB_2–SiC composite fabricated by the spark plasma sintering reactive synthesis method. Int. J. Refract. Met. Hard Mater. 27 (2009), 177–180.

53 R. Licheri, R. Orru, C. Musa, and G. Cao. Combination of SHS and SPS techniques for fabrication of fully dense ZrB2-ZrC-SiC composites. Mater. Lett. 62(3) (2008), 432–435.

54 I. Akin, M. Hotta, F. C. Sahin, O. Yucel, G. Goller, and T. Goto. Microstructure and densification of ZrB_2–SiC composites prepared by spark plasma sintering. J. Eur. Ceram. Soc. 29(2009), 2379–2385.

55 K. Pavani, K. Madhav Reddy, and B. Basu. Unpublished work, 2010.

Chapter 15

High-Temperature Mechanical and Oxidation Properties

This chapter discusses the influence of $TiSi_2/MoSi_2$ addition (up to 10 wt%) and temperature on hardness and strength of TiB_2. Another important aspect of the discussion in this chapter is to assess how the presence of sinter-aid influences the oxidation resistance. The results of isothermal oxidation tests on TiB_2–(x wt%) $MoSi_2/TiSi_2$ ($x \leq 10$) composites are also analyzed for this purpose.

15.1 INTRODUCTION

It can be reiterated here that ceramics in last few decades have emerged as high-temperature materials for aerospace and other structural applications. In this context, very few reports are available on high-temperature mechanical properties of TiB_2 materials.[1] Hot hardness measurements are necessary to evaluate the high-temperature mechanical behavior of various ceramics, including borides.[1–5] Elevated-temperature mechanical strength properties of some of the advanced ceramic composites are also investigated.[1,6–19] The high-temperature strength properties are found to be sensitive to microstructural phase assemblage or sinter-additive content. As far as are concerned, Most of the investigations of the high-temperature properties of TiB_2 are related to monolithic TiB_2. In preceding chapters, the development of TiB_2–silicide composites is discussed. This chapter reports on the high-temperature hardness and strength properties of these materials.

Another area of discussion in this chapter is the oxidation properties of high-temperature ceramics. High-temperature oxidation is a form of corrosion that does not require the presence of a liquid electrolyte and therefore is known as *dry corrosion* or *scaling*. Table 15.1 compares the oxidation properties of various potential high-temperature ceramics. Though the oxidation rate constant of TiB_2-based materials is comparable with that of other ceramics, prolonged exposure of TiB_2 above 1000°C in air degrades oxidation résistance, whereas SiC and $MoSi_2$ have good oxidation resistance due to the formation of protective oxide scale. Experimental

Advanced Structural Ceramics, First Edition. Bikramjit Basu, Kantesh Balani.

Table 15.1. Summary of Research Results Illustrating the Effect of Temperature on Flexural Strength of TiB_2–$MoSi_2$/$TiSi_2$ Ceramics

Material composition (wt%)	Sintering conditions (°C, min, MPa)	Relative density (% ρ_{th})	Grain size (μm)	Bend test conditions	Flexural strength (MPa)			Reference
					RT	500°C	1000°C	
TiB_2–(0)$MoSi_2$	HP, 1800, 60, 30	98	1.5	4-P, air, 3 × 4 × 40 mm^3	387 ± 52	422 ± 29	546 ± 33	28
TiB_2–(2.5)$MoSi_2$	HP, 1700, 60, 30	99	1.2	4-P, air, 3 × 4 × 40 mm^3	391 ± 31	442 ± 34	503 ± 27	28
TiB_2–(10.0)$MoSi_2$	HP, 1700, 60, 30	97	1.3	4-P, air, 3 × 4 × 40 mm^3	268 ± 70	312 ± 28	261 ± 30	28
TiB_2–(2.5)$TiSi_2$	HP, 1650, 60, 30	>98	2.3	4-P, air, 3 × 4 × 40 mm^3	381 ± 74	–	433 ± 17	31
TiB_2–(5.0)$TiSi_2$	HP, 1650, 60, 30	>99	3.0	4-P, air, 3 × 4 × 40 mm^3	426 ± 69	479 ± 33	314 ± 17	31
TiB_2–(10.0)$TiSi_2$	HP, 1650, 60, 30	>99	3.5	4-P, air, 3 × 4 × 40 mm^3	345 ± 60	375 ± 50	325 ± 25	31

HP, hot pressing; 4-P, four point flexural configuration.

investigation revealed that the oxidation mechanism of TiB_2 was influenced by partial pressure of oxygen, time of exposure, porosity, and the nature of sintering additives. Kulpa and Trocszynski[20] reported that the oxidation of TiB_2 powder starts below 400°C with the formation of $TiBO_3$. They proposed the following reactions:

$$4TiB_2 + 9O_2 \rightarrow 4TiBO_3 + 2B_2O_3 \text{ (<400°C and 0.05 ppm of } O_2\text{)}; \quad (15.1)$$

$$4TiBO_3 + O_2 \rightarrow 4TiO_2 + 2B_2O_3 \text{ (400–900°C and 10 ppm of } O_2\text{)}. \quad (15.2)$$

It was experimentally observed that both of these oxidation reactions can take place concurrently in the temperature range of 400–900°C. A comparison of oxidation resistance of various TiB_2-based materials indicates that monolithic TiB_2 (without sinter-additive) has poor oxidation resistance, compared with composites of TiB_2 with Si- or Al-based sinter-additives. These additives can facilitate the formation of a SiO_2[21] or Al_2O_3[22] layer on the composite surface. Koh et al.[23] reported the improved oxidation resistance of TiB_2 due to a coating of protective amorphous SiO_2 layer on the surface.

In an effort to evaluate oxidation characteristics, the oxidation rate constants at various temperatures for monolithic TiB_2 and TiB_2-based materials are summarized in Figure 15.1. Up to 900°C, all materials, except TiB_2-cermet, exhibited diffusion-controlled kinetics, that is, a parabolic rate law: $(\Delta W/s)^2 \approx K_P t$, where K_P is the parabolic oxidation rate constant, ΔW is the weight gain after time $t = t$, and s is the

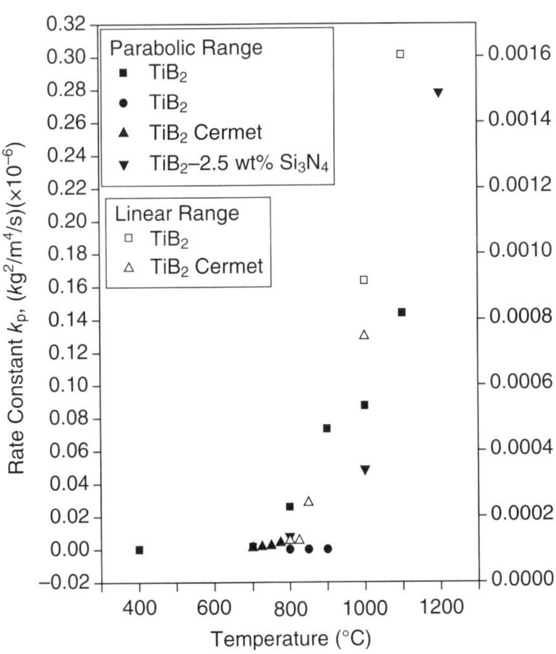

Figure 15.1 Oxidation rate constant in different temperature regions, recorded for monolithic TiB_2 (sintered without any sintering aid), TiB_2 sintered with nonmetallic sinter-additive (Si_3N_4), and TiB_2 sintered with metallic binder (10.54% Fe).[1]

surface area of the material exposed to the oxidizing environment. A noticeable increase in K_P values was experimentally measured at $T > 900°C$. In the temperature range 800–1000°C, the oxidation follows linear behavior ($\Delta W/s \approx K_L t$), where K_L is the linear oxidation rate constant. Similar linear oxidation behavior was also recorded with monolithic TiB_2. Interestingly, parabolic oxidation behavior was also recorded with TiB_2–(2.5 wt%)Si_3N_4 materials up to 1200°C. It has been reported that TiB_2 starts to oxidize in air at 400–500°C and the oxidation process was controlled by a diffusion-controlled mechanism up to 900°C.[24]

Koh et al.[25] studied the oxidation behavior of hot-pressed (HP) TiB_2–(2.5wt%) Si_3N_4 at 800–1200°C for up to 10 hours in air. They reported that TiB_2–(2.5 wt%) Si_3N_4 could exhibit better oxidation resistance at high temperatures (below 1000°C) due to formation of a protective oxide layer on the surface. At temperatures below 1000°C, parabolic weight gains were measured as a result of the formation of TiO_2 and B_2O_3 (*l*) on the surface. At temperatures above 1000°C, crystalline TiO_2 was observed along with volatile B_2O_3 and the surface was covered with only a thick crystalline TiO_2 layer.

Graziani et al. reported the parabolic oxidation kinetics of HP TiB_2–(12.1 wt%) B_4C–(2.1 wt%)Ni material.[26] As regards the kinetics, the formation of the oxide product B_2O_3 was observed initially, because of the small radius of the boron atom.[27] The diffusion of boron to the surface is more intensive than the diffusion of the metal atom of the boride, and this results in the formation of large amounts of B_2O_3. The glassy nature of the B_2O_3 film presents an additional diffusion barrier for atmospheric oxygen during oxidation.

While oxidation studies were conducted on a few TiB_2-based materials, a few attempts were made to improve oxidation resistance. In an innovative approach, TiB_2 materials were coated with a SiO_2 layer by placing them in a bed of SiC powder in flowing H_2 containing 0.1% H_2O at 1450°C for 2 hours.[23] The coating layer was found to be effective in restricting the oxidation of TiB_2. The oxidation rate of coated TiB_2 decreased by about a factor of 10, primarily because of the reduced oxygen transport through the coating layer as well as the consumption of oxygen via reaction with Ti_2O_3 to form TiO_2.

The oxidation of TiB_2 can exert a negative influence on the mechanical performance of components. Flexural strength of TiB_2 specimens without and with the coating layer (after treatment in a bed of SiC powder) was measured,[23] and the data are summarized in Figure 15.2. Without the coating layer, the strength was reduced remarkably after oxidation at 1000°C for 10 hours. It was observed that the formation of a thick oxide layer and the presence of cracks contributed to a reduction in strength. When the TiB_2 specimen was coated with a SiO_2 layer, however, the strength decreased to a lower extent and this was due to blunting of surface cracks with the coating layer. In fact, when the coated specimen was oxidized at 800°C for 10 hours, the reduction in strength was minimal.

In this chapter, high-temperature mechanical properties as well as oxidation properties of TiB_2–$MoSi_2$ and TiB_2–$TiSi_2$ materials are summarized.[28–31] A comparison with the published results is also provided to illustrate the relative performance of these newly developed materials.

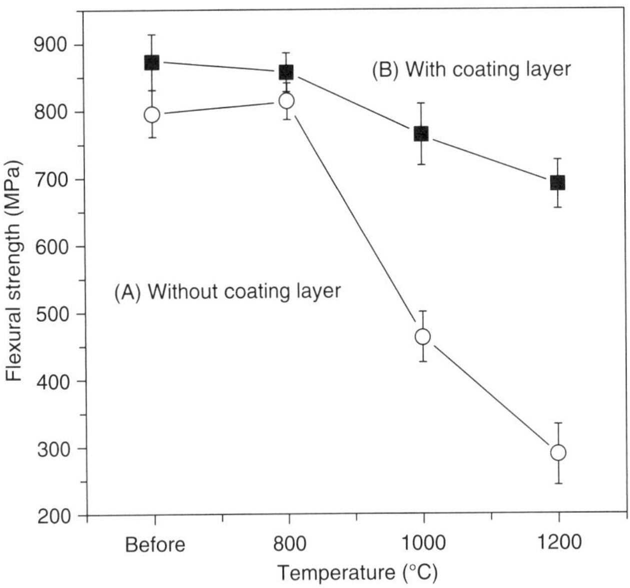

Figure 15.2 Flexural strength of TiB_2 specimens after oxidation in air for 10 hours at various temperatures (A) without and (B) with coating layer.[23] Much lower strength decrease with coated TiB_2 reveals better resistance toward oxidation-induced material property degradation.

Table 15.2. Hot Hardness Values of TiB_2–$MoSi_2$/$TiSi_2$ Ceramics

Material (wt%)	Hot hardness (GPa) at various temperatures (°C)						Reference
	23	200	300	600	800	900	
TiB_2–(0)$MoSi_2$	25.6	–	15.1	11.5	–	8.5	28
TiB_2–(2.5)$MoSi_2$	27.6	–	18.9	13.6	–	10.5	28
TiB_2–(2.5)$TiSi_2$	27.0	–	15.1	11.5	–	8.9	31
TiB_2	25.0	14.7	12.8	–	5.3	–	31
TiB_2–(5)$TiSi_2$	27.0	–	13.3	11.0		7.0	31
TiB_2–(10)$TiSi_2$	24.0	–	13.0	10.0		5.0	31
TiB_2	28.0	24.0	18.0	14.0	8.0	7.0	56
ZrB_2	20.0	–	12.0	9.0	7.0	7.0	56
HfB_2	27.0	26.0	16.0	10.0	9.0	6.0	56

15.2 HIGH-TEMPERATURE PROPERTY MEASUREMENTS

The processing of the TiB_2–$MoSi_2$/TiB_2–$TiSi_2$ is described in Chapter 14 and the densification data of the HP samples is presented in Table 15.2. It is known that the anisotropy of the hexagonal crystal structure results in deleterious internal stresses and the onset of spontaneous microcracking during cooling, if the grain size of TiB_2 exceeds the critical grain size of 15 μm.[9] The average grain size in both TiB_2–$MoSi_2$ and TiB_2–$TiSi_2$ (≤3.6 μm) is well below the critical grain size. The samples (5 mm × 5 mm × 10 mm) for hot hardness measurements were indented at room temperature (RT, 23°C) and at 300, 600, and 900°C with a load of 9.8 N using a high-temperature hardness tester in a vacuum of less than 5×10^{-3} Pa. The specimens for strength measurements were obtained from the HP disks and machined into bar shapes with dimensions of 3 × 4 × 40 mm. The flexural strength was recorded on a four-point bending configuration using a silicon carbide fixture, with a crosshead speed of 0.5 mm/min and inner and outer spans of 10 and 30 mm, respectively. Short-term (12 hours) oxidation tests were performed on ceramic coupons of 3 × 4 × 10 mm at 1200°C in dry air atmosphere with a heating rate of 30°C/min.

15.3 HIGH-TEMPERATURE MECHANICAL PROPERTIES

15.3.1 High-Temperature Flexural Strength

The high-temperature flexural strength of the HP TiB_2–$MoSi_2$ samples is provided in Table 15.1. At RT, the four-point flexural strength values of both monolithic TiB_2 and TiB_2–(2.5 wt%)$MoSi_2$ composite were measured to be ~390 MPa. Typically, the flexural strength of HP TiB_2, as reported by others, varied in the range 300–400 MPa.[5] The addition of 2.5 wt% $MoSi_2$ does not degrade the strength properties of TiB_2. However, the lowest strength (~268 MPa) was recorded with the TiB_2-based composite densified using 10 wt% $MoSi_2$ sinter-additive.

The fracture strength of the TiB_2, irrespective of $MoSi_2$ content, increases with temperature up to 500°C. In an earlier study, Baumgartner and Steiger[32] reported an increase in strength of monolithic TiB_2 with temperature and they attributed it to the relief of residual internal stresses. In fact, the strength properties measured with TiB_2–$(x)MoSi_2$ ($x \leq 2.5$ wt%) are relatively high, compared with other TiB_2 (see Table 15.1). The high strength of TiB_2–$MoSi_2$ ceramics is due to finer grain size of TiB_2.

At 1000°C, the flexural strength of the TiB_2 composite with 10 wt% $MoSi_2$ sinter-additive is reduced (~261 MPa), compared with strength at 500°C (~312 MPa). Composites with higher $MoSi_2$ content could retain the RT strength up to 1000°C.

Among all the TiB_2–$TiSi_2$ compositions, TiB_2–(5 wt%)$TiSi_2$ exhibited the highest room-temperature strength (~426 MPa; see Table 15.1). Up to 500°C, the fracture strength increases for all the TiB_2 compositions, which were densified to more than 97% of the theoretical density (ρ_{th}). Both the baseline monolithic TiB_2 and TiB_2–(2.5 wt%)$TiSi_2$ ceramics could retain flexural strength of more than 400 MPa up to 1000°C, whereas the flexural strength of the other monolithic TiB_2 (HP at 1650°C) and TiB_2 specimens with $TiSi_2$ content of either 5 wt% or more decreased with increasing temperature. However, a minimum of 79% RT strength was measured for all TiB_2–$TiSi_2$ compositions.

At 1000°C, the flexural strength of reference TiB_2 is observed to be higher than all the TiB_2–$TiSi_2$ compositions. It implies that at elevated temperature the grain boundary sliding at TiB_2/Ti_5Si_3 and $TiB_2/TiSi_2$ interfaces results in fracture at low loads for the TiB_2 composites. Since the brittle-to-ductile transition temperature of $TiSi_2$ is 800°C and lies in the range of 1000–1200°C for Ti_5Si_3, these phases could exhibit plasticity at or above 800°C with the application of load.[33–37] Therefore, the plastic deformation of $TiSi_2$ and Ti_5Si_3 at high temperature can lead to strength degradation in the TiB_2 composites.

The preceding discussion implies that it is advantageous to use the $TiSi_2/MoSi_2$ as a sintering aid to retain high-temperature strength and hardness properties. To achieve high density (>97% ρ_{th}) with monolithic TiB_2, a high hot-pressing temperature of 1800°C is needed. One could achieve maximum density of 99% ρ_{th} at a lower hot-pressing temperature of 1650°C with the use of $TiSi_2$ as a sintering aid. Also, the retention of strength at high temperature is only possible with better density and minimal amount of sinter-additive. The TiB_2–(2.5 wt%)$TiSi_2$ composite exhibited a better combination of hardness and strength values at high temperatures.

15.3.2 Hot Hardness Property

For the hot hardness measurements, TiB_2–(2.5 wt%)$MoSi_2$ composite possessing the maximum sinter density among the composites HP at 1700°C was selected, along with monolithic TiB_2 (HP, 1800°C). The measured hot hardness values are provided in Table 15.2. The hardness of TiB_2 decreased from ~26 GPa at RT to ~8.5 GPa at 900°C, while the measured hardness reduced from ~28 GPa at RT to ~10.5 GPa at 900°C for TiB_2–(2.5 wt%)$MoSi_2$ composite. It is known that brittle materials can be plastically deformed even at temperatures below 0.5 T_{mp} (melting temperature).[38] However, the maximum hardness of TiB_2 composite can be attributed to better sinter density (~99% ρ_{th}). Even though $MoSi_2$ is softer (H_v ~ 9 GPa) compared with TiB_2, addition of small amounts of $MoSi_2$ to TiB_2 did not have any negative effect on the hardness.[14]

It has been reported in the literature that the hardness of TiB_2 cermets is much lower than both monolithic TiB_2 and TiB_2 reinforced with the ceramic additives. The hardness of TiB_2-based cermets varied between 7.3 GPa (for TiB_2–(5 vol%)[Fe-Fe_2B]) and 4.8 GPa (for TiB_2–(20 vol%)[Fe-Cr-Ni-Fe_2B]) at 800°C[39]. Also, monolithic TiB_2 could retain maximum hardness of ~5 GPa at 800°C.[7] Compared with the earlier reported values, TiB_2–$MoSi_2$ materials had better hardness values of more than 10 GPa at 900°C.

Among all the TiB_2–$TiSi_2$ samples, the hardness varied from 27 GPa at RT to 8.9 GPa at 900°C for TiB_2–(2.5 wt%)$TiSi_2$ (see Table 15.2). At RT, the monolithic TiB_2 (HP at 1800°C) shows a little lower hardness than the TiB_2–(5 wt%)$TiSi_2$. However, the hardness of the monolithic TiB_2 is comparable with other TiB_2 samples (containing ≥5 wt% $TiSi_2$) at elevated temperatures. For example, the hardness of monolithic TiB_2 and TiB_2–(5 wt%)$TiSi_2$ were recorded as 8.2 and 7.8 GPa, respectively, at 900°C.

The RT hardness of TiB_2–$TiSi_2$ ceramics varies from 21 to 27 GPa (see Table 15.2). Such a hardness variation can be attributed to differences in sinter density and amount of sinter-additive. The hardness increased with $TiSi_2$ addition (up to 5 wt% $TiSi_2$), further increase in the $TiSi_2$ content to 10 wt% lowers hardness (24 GPa) of TiB_2 despite its full densification (99.6% ρ_{th}). Although the TiB_2 composites consist of relatively softer phases such as $TiSi_2$ (8.7 GPa) and Ti_5Si_3 (9.8 GPa),[40] the addition of small amounts of $TiSi_2$ (≤ 5 wt%) does not degrade the hardness of TiB_2.

15.4 OXIDATION BEHAVIOR OF TiB_2–$MoSi_2$

The oxidation test results, that is, the variation of weight gain per unit surface area ($\Delta W/s$) with time t were analyzed to determine the kinetic parameters, namely, oxidation exponent (n), rate constant (k), and the parabolic rate constant (k_p) as per the following expression:

$$(\Delta W/s)n = k_p t. \tag{15.3}$$

The parabolic rate constant (k_p) is generally determined from the slope of the linear regression–fitted line of $(\Delta W/s)^2$-versus-time plots (not shown). Table 15.3 reveals

Table 15.3. Comparison of the Weight Gain, Oxide Layer Thickness, and Oxidation Rate Constant of Various Ultra-High-Temperature TiB_2-, ZrB_2-, and HfB_2-Based Materials (From Reference 30)

Material composition (wt%)	Oxidation conditions (°C, time)	Weight gain (mg/cm²)	Oxide layer thickness (μm)	Parabolic rate constant, k_p (mg²/cm⁴·s)
TiB_2	1100°C, 15 hours	8.8	–	–
TiB_2–(20)B_4C–(1)Ni	1300°C, 30 hours	34.0	–	–
TiB_2–(2.5)Si_3N_4	1200°C, 10 hours	11.0	–	–
ZrB_2–(20)$MoSi_2$	1200°C, 30 hours	0.7	Less than a few micrometers	0.95×10^{-5}
ZrB_2–(20)$MoSi_2$	1300°C, 30 hours	2.5	100	0.67×10^{-5}
ZrB_2–(41)TiB_2–(4)Ni	1000°C, 30 hours	20.0	–	–
ZrB_2–(15)SiC	1450°C, 20 hours	3.4	50	–
ZrB_2	1300°C, 2 hours	10.0	160	–
ZrB_2–(5)Si_3N_4	1300°C, 2 hours	15.0	140	–
ZrB_2–(15)Ta_5Si_3	1400°C, 2 hours	8.0	115	–
HfB_2–(19)SiC–(5.8) Si_3N_4	1250°C, 1 hours	0.5	–	0.69×10^{-4}
HfB_2–(22.1(SiC–(5.9) HfC	1450°C, 20 hours	1.5	20	–
TiB_2–(0)$MoSi_2$	1200°C for 12 hours	45.0	286	4.9×10^{-2}
TiB_2–(2.5)$MoSi_2$	1200°C for 12 hours	44.1	273	4.6×10^{-2}
TiB_2–(10)$MoSi_2$	1200°C for 12 hours	28.8	235	2.1×10^{-2}

that $MoSi_2$ addition (up to 10 wt%) to TiB_2 lowers the oxidation kinetics. The oxidized surface of TiB_2 samples, exposed to 1200°C in air for 12 hours, was analyzed using x-ray diffraction (XRD), scanning electron microscopy energy-dispersive x-ray spectrometry (SEM-EDS), and x-ray mapping. Figure 15.3a shows the characteristics of oxide scale on TiB_2–(0%)$MoSi_2$ and its thickness is measured to be about 286 μm. Measurements of the oxide layer thickness of other TiB_2 samples are

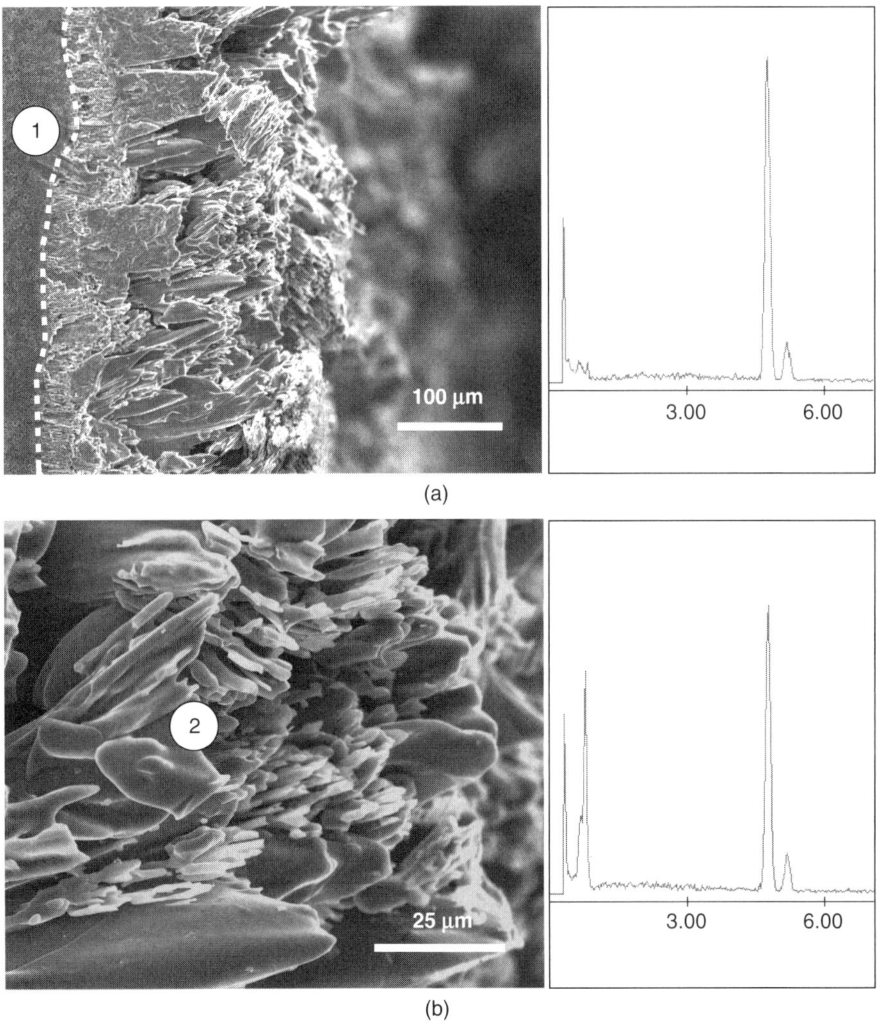

Figure 15.3 (a) SEM image from the cross section of the TM0 sample (TiB_2–(0%)$MoSi_2$) after oxidation at 1200°C for 12 hours. (b) Magnified image of oxide layer shows the highly textured TiO_2 crystals. Energy-dispersive x-ray spectrometry (EDS) patterns recorded from the different microstructural phases correspond to (1) unoxidized base material and (2) oxide scale (reproduced from Reference 30).

presented in Table 15.3. Highly textured TiO_2 crystals can be seen in the oxide scale (Fig. 15.3b).

Next, the oxidation behavior is discussed in reference to the literature results. Tampieri and Bellosi reported the oxidation kinetic of monolithic TiB_2, when tested at $T \geq 1100°C$.[24] When TiB_2 is oxidized in air at elevated temperatures, the following reactions are expected to occur:

$$TiB_2 + (5/2)O_2 \rightarrow TiO_2 + B_2O_3; \quad (15.4)$$

$$B_2O_3\ (l) \rightarrow B_2O_3(g). \quad (15.5)$$

At 800°C, the oxide scale of TiB_2 is reported to consist of both the TiO_2 and B_2O_3 phases.[24] The diffusion coefficient of oxygen in B_2O_3 at 900°C is about $10^{-11}\,m^2/s$; whereas for the TiO_2, the oxygen diffusivity at 1000°C is $10^{-18}\,m^2/s$.[23–25] Hence, at or above 1000°C, the protective role of TiO_2 increases with temperature with the formation of rutile scale. At high temperatures (1000–1200°C), B_2O_3 evaporates due to high volatility.[23–25] At above 1000°C, the oxidation of TiB_2 is controlled by the following reaction:

$$TiB_2\ (s) + (5/2)O_{2\,(g)} \rightarrow TiO_{2\,(s)} + B_2O_{3\,(g)}\ (\Delta G = -288.12\ \text{kcal/mol at } 1200°C). \quad (15.6)$$

From the cross-sectional scanning electron microscopy (SEM) image, it is found that a thick dense oxide scale of ~286 μm is formed on the surface (see Fig. 15.3a). Koh et al. also observed a thick oxide layer (~170 μm) on the surface of TiB_2 after oxidation in air at 1200°C for 10 hours.[23,25] When a TiB_2–(2.5 wt%)Si_3N_4 specimen was tested at 1200°C for 2 hours, a single TiO_2 oxide layer of 100 μm was recorded.[25] The morphology of oxide scale on TiB_2 consists of highly textured rutile crystals (Fig. 15.3b). Similar characteristics of textured rutile crystals were also observed by Tampieri et al., and it is attributed to the epitaxial growth of rutile crystals in the [2 1 1] and [1 0 1] directions.[24] TiO_2 is an n-type oxide and contains both titanium interstitial ions and oxygen vacancies. TiO_2 is also prone to oxidation by diffusion of both cations and anions.[25] Hence, a thick TiO_2 oxide scale is observed along with the parabolic oxidation kinetics.

The following oxidation reactions may also take place in the case of TiB_2–$TiSi_2$/TiB_2–$MoSi_2$:

$$5MoSi_2 + 7O_2 \rightarrow Mo_5Si_3 + 7SiO_2\ (\Delta G = -1029.43\ \text{kcal/mol at } 1200°C); \quad (15.7)$$

$$2Mo_5Si_3 + 21O_2 \rightarrow 10MoO_3 + 6SiO_2\ (\Delta G = -1728.88\ \text{kcal/mol at } 1200°C); \quad (15.8)$$

$$Ti_5Si_3 + 8O_2 \rightarrow 5TiO_2 + 3SiO_2\ (\Delta G = -1135.96\ \text{kcal/mol at } 1200°C). \quad (15.9)$$

At 1200°C, Reactions 15.6–15.9 are thermodynamically feasible according to the free energies calculations by HSC software.[41] Also, TiO_2, B_2O_3, MoO_3, and SiO_2 are the expected oxidation products; however, the B_2O_3 and MoO_3 phases can evaporate at temperatures above 1000°C.

From Table 15.3, the weight gain, oxide layer thickness, and parabolic rate constant values all decrease with an increase in the amount of $MoSi_2$ sinter-additive. Also, a decrease in oxidation rate of TiB_2–(10%)$MoSi_2$ can be attributed to the formation of a protective SiO_2 layer in addition to TiO_2, which acts as a barrier for the diffusion of oxygen. It appears that at least 10 wt% $MoSi_2$ is needed to form protective SiO_2 for improving the oxidation resistance of TiB_2.

The oxidation properties of various ultra-high-temperature ceramics, particularly transition metal diborides, are summarized in Table 15.3 with a view to comparing the performance of TiB_2–$MoSi_2$ materials with other competing materials. It can be noticed that TiB_2-based materials have poor oxidation resistance (orders of magnitude higher k_p value, see Table 15.3), compared with ZrB_2 and HfB_2 at high temperatures. Overall, the TiB_2 materials experience large mass gains, oxide layer thickness, and high parabolic rate constant, thus reflecting their poor oxidation properties. Kaufman and Clougherty reported that oxidation resistance increases in the order $TiB_2 > ZrB_2 > HfB_2$.[42] The formation of oxidation products (ZrO_2, HfO_2, and B_2O_3) can provide good oxidation resistance to pure ZrB_2 and HfB_2 up to 1200°C.[43] Much more research effort was also directed toward the study of high-temperature oxidation properties of these borides.[1,2,7,44–50] In general, sintering additives modify chemical composition of the oxide layer and decrease the inward diffusion of oxygen at very high temperatures. From Table 15.3, it is clear that higher amounts (15–25 vol%) of silicon-based sintering additives should be used with ZrB_2 and HfB_2 ceramics. Such high amounts of sintering additives form good protective silica and or borosilicate oxide glass layers to ensure oxidation resistance. For example, the presence of SiC particles results in formation of a protective borosilicate glassy coating in the HfB_2–(19 vol%)SiC–(5.8 vol%)Si_3N_4 composite at 1400°C.[51] For pressureless sintered ZrB_2–(20 vol%)$MoSi_2$ composite, Sciti et al. commented that the silica resulting from $MoSi_2$ facilitated the oxidation resistance at 1200°C.[51] These ZrB_2 composites could retain four-point flexural strength of 500 MPa up to 1500°C.[51]

15.5 OXIDATION BEHAVIOR OF TiB_2–$TiSi_2$

15.5.1 Oxidation Kinetics

In this section, the oxidation results are discussed to illustrate the influence of $TiSi_2$ addition on oxidation resistance of TiB_2. Figure 15.4a provides a typical plot showing the variation of weight gain (ΔW) per unit surface area (S) as a function of temperature (T). For comparison, the weight gain data of monolithic TiB_2 (HP, 1800°C for 1 hour) are also plotted in Figure 15.4a,b. The analysis of weight change data yields $n = 1.8$ for monolithic TiB_2 and $n = 1.9$ for the TiB_2–(10 wt%)$TiSi_2$. Hence, the parabolic rate constant ($k_p \sim 4.9 \times 10^{-2}\,mg^2/cm^4{\cdot}s$ for monolithic TiB_2 and $k_p \sim 2.9 \times 10^{-2}\,mg^2/cm^4{\cdot}s$ for the TiB_2–(10 wt%)$TiSi_2$ composite) can be determined from the slope of the linear regression–fitted line of the $(\Delta W/s)^2$-versus-time plot. At 1200°C, it is interesting to note that the oxidation kinetics of the TiB_2–(10%) $TiSi_2$ is relatively lower than that of the monolithic TiB_2. The weight gain of the TiB_2–(10 wt%)$TiSi_2$ composite is relatively lower than monolithic TiB_2. In general, the weight gain shows an increasing trend with the temperature. Slow oxidation above 750°C is mainly due to the formation of molten B_2O_3, which inhibits the diffusion of oxygen. However, a sharp increase in oxidation above 1000°C is attributed

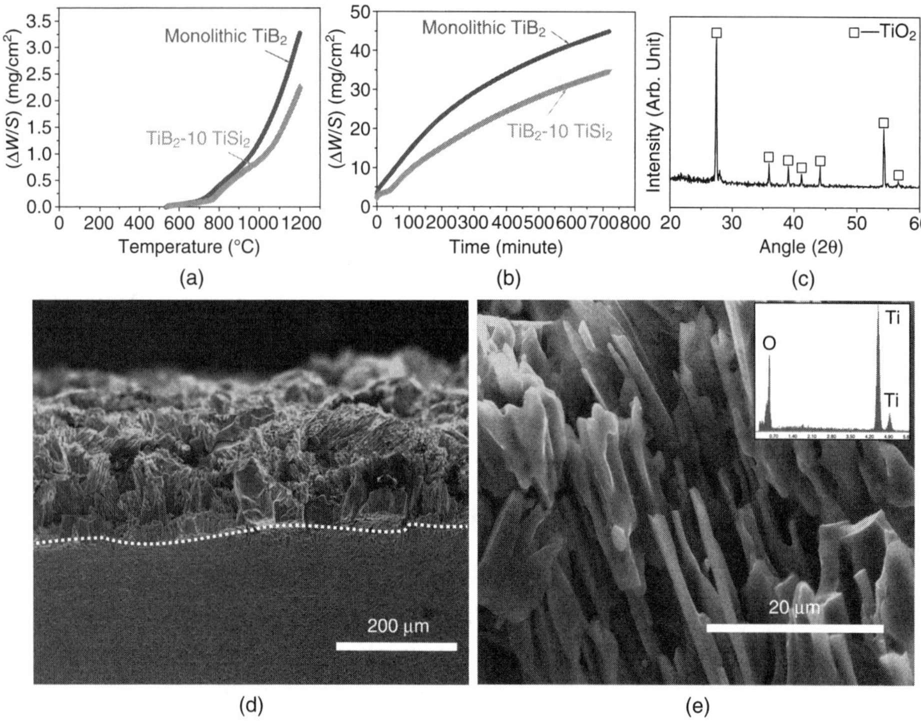

Figure 15.4 Thermogravimetric data of the monolithic TiB_2 (hot pressed at 1800°C for 1 hour) and TiB_2–(10 wt%)$TiSi_2$ composite (hot pressed at 1650°C for 1 hour): (a) change of weight per unit surface area ($\Delta W/s$) versus temperature and (b) change of weight per unit surface area ($\Delta W/s$) versus time. (c) The XRD analysis of the oxidized surface of the TiB_2–(10 wt%)$TiSi_2$ composite sample shows the presence of crystalline TiO_2 (rutile) phase. The SEM image taken from the cross section of the TiB_2–(10 wt%)$TiSi_2$ composite after oxidation at 1200°C for 12 hours (d) and a magnified image of oxide layer shows the highly textured rodlike TiO_2 crystals (e). The EDS pattern (e, inset) recorded from the oxide scale shows evidence of the presence of Ti and O (reproduced from Ref. 29).

to a rapid increase in evaporation rate of B_2O_3. It has been widely reported that the oxidation of TiB_2 is mainly governed by a diffusion mechanism up to 900°C.[24,52,53] Mechanistically, the oxidation of TiB_2 depends on either the inward diffusion of O^{2-} ions or the outward diffusion of M^{n+} ions.

Tampieri et al. reported that oxidation kinetics of TiB_2 was controlled by diffusion of oxygen at 1100°C up to about 500 minutes and by a linear law at higher temperature.[22] The parabolic rate constant k_p of monolithic TiB_2 during oxidation at 1100°C was recorded as ~0.14×10^{-2} mg^2/cm^4·s. The TiB_2 cermets containing Fe-rich binder exhibited parabolic oxidation behavior (k_p ~ 0.02×10^{-2} mg^2/cm^4·s) below 750°C and a linear oxidation kinetics above 750°C.[54,55] The parabolic rate constant k_p ~ 0.51×10^{-2} mg^2/cm^4·s was measured for TiB_2–(2.5 wt%)Si_3N_4 composite after isothermal oxidation at 1200°C for 2 hours.[25] From the preceding observations, the oxidation rate constant measured for TiB_2–(10 wt%)$TiSi_2$ is found to be one order of magnitude higher.

15.5.2 Morphological Characteristics of Oxidized Surfaces

XRD analysis of oxidized TiB_2–(10%)$TiSi_2$ surface reveals the presence of only crystalline rutile (TiO_2) phase, as shown in Figure 15.4c. Figure 15.4d shows a secondary electron image of thick oxide scale. The oxide scale is characterized by highly textured elongated rodlike rutile crystals (Fig. 15.4e). X-ray mapping of the oxidized surface of the TiB_2–(10 wt%)$TiSi_2$ surface reveals the distribution of the Ti, B, O, and Si, respectively (see Fig. 15.4). From the microstructural analysis, it is therefore obvious that the oxide layer consists of SiO_2 and TiO_2, which must have resulted from the oxidation of TiB_2, $TiSi_2$, and Ti_5Si_3. The following reactions are most likely to occur thermodynamically as the change in standard Gibbs free energy at the oxidation temperature is measured to be negative, using HSC chemistry thermochemical database software[41]:

$$TiB_2 + (5/2)O_2 \rightarrow TiO_2(s) + B_2O_{3\,(g)} \; (\Delta G^{o} = -288.12 \text{ kcal at } 1200°C); \quad (15.10)$$

$$5TiSi_2 + 7O_2 \rightarrow TiO_2 + 7SiO_2 \; (\Delta G^{o} = -1029.43 \text{ kcal at } 1200°C); \quad (15.11)$$

$$Ti_5Si_3 + 8O_2 \rightarrow 5TiO_2 + 3SiO_2 \; (\Delta G^{o} = -1135.96 \text{ kcal at } 1200°C). \quad (15.12)$$

At the oxidation temperature of 1200°C, Equations 15.10–15.12 are found to be thermodynamically feasible and, therefore, these reactions explain the formation of oxides on the ceramic surfaces.

15.6 CONCLUDING REMARKS

On the basis of the experimental results presented in this chapter, it can be stated that it is essential to optimize the sinter-aid addition within a narrow window (0–10 wt%); in fact, the results also demonstrate that a considerable reduction in strength at 1000°C can take place if the silicide addition is increased from 2.5 to 5.0 or 10.0 wt%. High flexural strength of 550 MPa at 1000°C can be achieved for monolithic TiB_2 as well as TiB_2–(2.5 wt%)$MoSi_2$. Among all the compositions, TiB_2–(2.5 wt%)$TiSi_2$ was measured to have better hardness properties (27 GPa at RT and 9 GPa at 900°C) and TiB_2–(5 wt%)$TiSi_2$ exhibited better strength (~479 MPa) properties due to its high sinter density and intergranular mode of fracture at 500°C.

The experimental results clearly reveal the advantages of $MoSi_2$ addition in terms of oxidation resistance. The parabolic rate constant of TiB_2 decreases with an increase $MoSi_2$ addition, and this indicates the importance of $MoSi_2$ addition in imparting enhanced oxidation resistance to TiB_2. The morphology of oxide scale is characterized by highly textured elongated rodlike rutile crystals. At higher temperatures, the growth of TiO_2 oxide scale results from the diffusion of both cations and anions. As far as the oxidation kinetics of TiB_2–(10 wt%)$TiSi_2$ is concerned, the observation of a parabolic rate law during oxidation for 12 hours at 1200°C is a promising result. Also, the addition of $TiSi_2$ sinter-additive enhances both the sinterability and the oxidation resistance of TiB_2. From the perspective of ultra-high-temperature applications, the TiB_2-based materials are, however, found to be inferior to ZrB_2 and HfB_2 in view of inferior oxidation resistance.

REFERENCES

1 B. Basu, G. B. Raju, and A. K. Suri. Processing and properties of TiB_2-based materials: A review. Int. Mater. Rev. 51(6) (2006), 352–374.

2 H. L. Wang and M. H. Hon. Temperature dependence of ceramics hardness. Ceram. Int. 25 (1999), 267–271.

3 R. D. Koester and D. P. Moak. Hot hardness of selected borides, oxides and carbides to 1900°C. J. Am. Ceram. Soc. 50(6) (1967), 290–296.

4 W. A. Sanders and H. B. Probst. Hardness of five borides at 1625°C. J. Am. Ceram. Soc. 49(4) (1966), 231–232.

5 K. Nakano, H. Matsubara, and T. Imura. High temperature hardness of titanium diboride single crystal. Jpn. J. Appl. Phys. 13(6) (1974), 1005–1006.

6 S. Maloy, A. H. Heuer, J. Lewandowski, and J. Petrovic. Carbon additions to molybdenum disilicide: Improved high-temperature mechanical properties. J. Am. Ceram. Soc. 74(10) (1991), 2704–2706.

7 X. Zhong and H. Zao. High temperature properties of refractory composites. Am. Ceram. Soc. Bull. 60 (1999), 98–101.

8 L. A. Pierce, D. M. Mieskowski, and W. Sanders. Effect of grain-boundary crystallization on the high-temperature strength of silicon nitride. J. Mater. Sci. 21 (1986), 1345–1348.

9 M. Keppeler, H. G. Reichert, J. M. Broadley, G. Thurn, I. Wiedmann, and F. Aldinger. High-temperature mechanical behavior of liquid-phase-sintered silicon carbide. J. Eur. Ceram. Soc. 18 (1998), 521–526.

10 D. Chen, M. E. Sixta, X. F. Zhang, L. C. De Jonghe, and R. O. Ritchie. Role of the grain-boundary phase on the elevated temperature strength, toughness, fatigue, and creep resistance of silicon carbide sintered with Al, B, and C. Acta Mater. 48 (2000), 4599–4608.

11 G. Rixecker, I. Wiedmann, A. Rosinus, and F. Aldinger. High-temperature effects in the fracture mechanical behavior of silicon carbide liquid-phase sintered with AlN-Y_2O_3. Addit. J. Eur. Ceram. Soc. 21 (2001), 1013–1019.

12 J. J. Melendez-Martinez, A. Dominguez-Rodriguez, F. Monteverde, C. Melandri, and G. de Portu. Characterization and high temperature mechanical properties of zirconium boride-based materials. J. Eur. Ceram. Soc. 22 (2002), 2543–2549.

13 S. Guo, N. Hirosaki, Y. Yamamoto, T. Nishimura, and M. Mitomo. Improvement of high-temperature strength of hot-press sintering silicon nitride with Lu_2O_3 addition. Scr. Mater. 45 (2001), 74–86.

14 D. S. Park, B. D. Hahn, B. C. Bae, and C. Park. Improved high-temperature strength of silicon nitride toughened with aligned whisker seeds. J. Am. Ceram. Soc. 88(2) (2005), 383–389.

15 Q. Zhu and K. Shobu. High-temperature mechanical properties of SiC-$Mo_5(Si,Al)_3C$ composites. J. Am. Ceram. Soc. 84(2) (2001), 413–419.

16 Y. W. Kim, M. Mitomo, and T. Nishimura. High-temperature strength of liquid-phase-sintered SiC with AlN and RE_2O_3 (RE=Y, Yb). J. Am. Ceram. Soc. 85(4) (2002), 1007–1009.

17 H. Shimizu, M. Yoshinaka, K. Hirota, and O. Yamaguchi. Fabrication and mechanical properties of monolithic $MoSi_2$ by spark plasma sintering. Mater. Res. Bull. 37 (2002), 1557–1563.

18 G. M. Song, Y. J. Wang, and Y. Zhou. Thermomechanical properties of TiC particle-reinforced tungsten composites for high temperature applications. Int. J. Refract. Met. Hard Mater. 21 (2003), 1–12.

19 G. W. Wen and X. X. Huang. Increased high temperature strength and oxidation resistance of Al_4SiC_4 ceramics. J. Eur. Ceram. Soc. 26 (2006), 1281–1286.

20 A. Kulpa and T. Troczynski. Oxidation of TiB_2 powders below 900°C. J. Am. Ceram. Soc. 79(2) (1996), 518–520.

21 S. Torizuka and T. Kishi. Effect of SiC and ZrO_2 on sinterability and mechanical properties of titanium nitride, titanium carbonitride and titanium diboride. Mater. Trans. JIM 37(4) (1996), 782–787.

22 A. K. Khanra, L. C. Pathak, S. K. Mishra, and M. M. Godkhindi. Effect of NaCl on the synthesis of TiB_2 powder by a self-propagating high-temperature synthesis. Mater. Lett. 58 (2004), 733–738.

23 Y. H. Koh, H. W. Kim, and H. E. Kim. Improvement in oxidation resistance of TiB_2 by formation of protective SiO_2 layer on surface. J. Mater. Res. 16(1) (2001), 132–137.

24 A. Tampieri and A. Bellosi. Oxidation of monolithic TiB_2 and of Al_2O_3-TiB_2 composite. J. Mater. Sci. 28 (1993), 649–653.

25 Y. H. Koh, S. Y. Lee, and H. E. Kim. Oxidation behavior of titanium boride at elevated temperatures. J. Am. Ceram. Soc. 84(1) (2001), 239–241.

26 T. Graziani, E. Landi, and A. Bellosi. Oxidation of TiB_2–20 vol.% B_4C composite. J. Mater. Sci. Lett. 12 (1993), 691–694.

27 V. A. Lavrenko, S. S. Chuprov, A. P. Umanskii, T. G. Protsenko, and E. S. Lugovskaya. High-temperature oxidation of composite materials based on titanium diboride. Powder Metall. Met. Ceram. (Engl. transl.), 26(9) (1987), 761–762.

28 G. Brahma Raju, K. Biswas, A. Mukhopadhyay, and B. Basu. Densification and high temperature mechanical properties of hot pressed TiB_2-(0–10 wt. %) $MoSi_2$ composites. Scr. Mater. 61 (2009), 674–677.

29 G. Brahma Raju, K. Biswas, and B. Basu. Microstructural characterization and isothermal oxidation behavior of hot-pressed TiB_2-10 wt% $TiSi_2$ composite. Scr. Mater. 61 (2009), 674–677.

30 G. Brahma Raju, B. Basu, and A. K. Suri. Oxidation kinetics and mechanisms of hot pressed TiB_2-$MoSi_2$ composites. J. Am. Ceram. Soc. 91(10) (2008), 3320–3327.

31 G. Brahma Raju, B. Basu, N. H. Tak, and S. J. Cho. Temperature dependent hardness and strength properties of TiB_2 with $TiSi_2$ sinter-aid. J. Eur. Ceram. Soc. 29(10) (2009), 2119–2128.

32 H. R. Baumgartner and R. A. Steiger. Sintering and properties of TiB_2 made from powder synthesized in a plasma-arc heater. J. Am. Ceram. Soc. 67(3) (1984), 207–212.

33 R. Mitra. Mechanical behavior and oxidation resistance of structural silicides. Int. Mater. Rev. 51(1) (2006), 13–64.

34 H. Inui, M. Moriwaki, N. Okamoto, and M. Yamaguchi. Plastic deformation of single crystals of $TiSi_2$ with the C54 structure. Acta Mater. 51 (2003), 1409–1420.

35 J. Li, D. Jiang, and S. Tan. Microstructure and mechanical properties of in situ produced SiC/$TiSi_2$ nanocomposites. J. Eur. Ceram. Soc. 20 (2000), 227–233.

36 J. Li, D. Jiang, and S. Tan. Microstructure and mechanical properties of in situ produced Ti_5Si_3/TiC nanocomposites. J. Eur. Ceram. Soc. 22 (2002), 551–558.

37 R. Rosenkranz and G. Frommeyer. Microstructures and properties of high melting point intermetallic Ti_5Si_3 and $TiSi_2$ compounds. Mater. Sci. Eng. A 152 (1992), 288–294.

38 H. L. Wang and M. H. Hon. Temperature dependence of ceramics hardness. Ceram. Int. 25 (1999), 267–271.

39 T. Jungling, L. S. Sigl, R. Oberacker, F. Thummler, and K. A. Schwetz. New hardmetals based on TiB_2. Int. J. Refract. Met. Hard Mater. 12 (1993), 71–88.

40 G. Berg, C. Friedrich, E. Broszeit, and C. Berger. Data collection of properties of hard materials, in *Handbook of Ceramic Hard Materials*, Vol. 2, R. Riedel (Ed.). Wiley-VCH Verlag GmbH, Weinheim, Germany, 2000, 965–990.

41 A. Roine. Chemical reaction and equilibrium software with extensive thermochemical database, Outokumpu HSC Chemistry for Windows (version 5.1). 2002.

42 L. Kaufman and E. V. Clougherty. Investigation of boride compounds for very high temperature applications. ManLabs Report RTD-EDR-63-4096, Part I USA, December, 1963.

43 I. G. Talmy, J. A. Zaykoski, and M. M. Opeka. High-temperature chemistry and oxidation of ZrB_2 ceramics containing SiC, Si_3N_4, Ta_5Si_3, and $TaSi_2$. J. Am. Ceram. Soc. 91(7) (2008), 2250–2257.

44 J. B. Berkowitz-Mattuck. High-temperature oxidation III. Zirconium and hafnium diborides. J. Electrochem. Soc. 113(9) (1966), 908–914.
45 W. G. Fahrenholtz. Thermodynamic analysis of ZrB_2-SiC oxidation: Formation of a SiC-depleted region. J. Am. Ceram. Soc. 90(1) (2007), 143–148.
46 F. Peng and R. F. Speyer. Oxidation resistance of fully dense ZrB_2 with SiC, TaB_2, and $TaSi_2$ additives. J. Am. Ceram. Soc. 91(5) (2008), 1489–1494.
47 F. Monteverde, A. Bellosi, and S. Guicciardi. Processing and properties of zirconium diboride-based composites. J. Eur. Ceram. Soc. 22 (2002), 279–288.
48 F. Monteverde and L. Scatteia. Resistance to thermal shock and to oxidation of metal diborides-SiC ceramics for aerospace application. J. Am. Ceram. Soc. 90(4) (2007), 1130–1138.
49 F. Monteverde and A. Bellosi. The resistance to oxidation of an HfB_2-SiC composite. J. Eur. Ceram. Soc. 25 (2005), 1025–1031.
50 M. M. Opeka, I. G. Talmy, and J. A. Zaykoski. Oxidation-based materials selection for 2000°C hypersonic aerosurfaces: Theoretical considerations and historical experience. J. Mater. Sci. 39 (2004), 5887–5904.
51 D. Sciti, S. Guicciardi, A. Bellosi, and G. Pezzotti. Properties of a pressureless-sintered ZrB_2-$MoSi_2$ ceramic composite. J. Am. Ceram. Soc. 89(7) (2006), 2320–2322.
52 V. B. Voitovich, V. A. Lavrenko, and V. M. Adejev. High-temperature oxidation of titanium diboride of different purity. Oxidation Metals 42(1/2) (1994), 145–161.
53 R. J. Irving and I. G. Worsley. The oxidation of titanium diboride and zirconium diboride at high temperatures. J. Less-Common Metals 16 (1968), 103–112.
54 M. G. Barandika, J. J. Echeberria, and F. Castro. Oxidation resistance of two TiB2-based cermets. Mater. Res. Bull. 34 (1999), 1001–1011.
55 M. G. Barandika, J. J. Echeberria, J. M. Sanchez, and F. Castro. Oxidation resistance and microstructure of the oxide layers for TiB2-based cermets. J. Mater. Chem. 8(8) (1998), 1851–1857.
56 R. D. Koester and D. P. Moak. Hot hardness of selected borides, oxides, and carbides to 1900°C. J. Am. Ceram. Soc. 50(6) (1967), 290–296.

Section Six

Nanoceramic Composites

Chapter 16

Overview: Relevance, Characteristics, and Applications of Nanostructured Ceramics

In the last few decades, bulk nanoceramic materials, characterized by grain sizes smaller than 100 nm and with some appealing mechanical, physical, and tribological properties, have attracted wider attention in the ceramics community. One of the major focuses in the research on bulk nanostructured ceramics encompasses processing-related challenges, in particular. This chapter shed light on some of the outstanding issues involved in the processing of nanoceramics and ceramic nanocomposites, and it critically analyzes the property modifications resulting from microstructural refinement. After mentioning the potential fields of application for ceramic nanomaterials, this chapter concludes with some of the unresolved issues related to bulk nanoceramics, along with mentioning the scope for future research.

16.1 INTRODUCTION

Conventionally nanostructured materials are generally defined as materials composed of structural units with a size scale of less than 100 nm in any dimension.[1] The characteristic length scale refers to particle diameter, grain size, layer thickness, or even the width of a conducting line on an electronic chip. Based on dimensions, nanostructured materials can be classified into different categories, including zero-dimensional (nanosized powders), one-dimensional (nanocrystalline multilayer), two-dimensional (filamentary rods of nanoscaled thickness), and three-dimensional (bulk materials with at least one nanocrystalline phase).[2] A wide variety of applications are projected for bulk nanoceramics and nanoceramic composites, such as durable ceramic parts for automotive engines, cutting tools, heat engine components,

Advanced Structural Ceramics, First Edition. Bikramjit Basu, Kantesh Balani.

wear-resistant parts, aerospace-related industrial applications, ultrafine filters, flexible superconducting wire, and fiber-optic connector components. Despite such an appreciable range of projected applications of bulk nanoceramics and ceramic nanocomposites, they have not yet penetrated the commercial market in a big way. The major challenge lies in the restriction of grain growth during processing, which is difficult to achieve using conventional sintering techniques. Against this backdrop, the adoption of advanced processing techniques has been the subject of extensive research in the last few decades. The advanced sintering techniques—in particular, spark plasma sintering (SPS), sinter-forging, and sinter–hot isostatic pressing (sinter-HIPing)—are some of the successful laboratory-scale processes for synthesizing bulk nanomaterials. Among them, SPS is currently one of the most widely tested processing routes for developing bulk nanostructured ceramics or ceramic matrix composites (CMCs). To provide some examples of nanostructured ceramics and composites, a summary of sintered grain size data vis-à-vis starting powder particle size data is presented in Table 16.1. It should be evident from Table 16.1 that, with the exception of one or two ceramic systems, it is possible to retain grain size below 100 nm.

At the microscopic level, ultrafine grain sizes and the corresponding increase in interfacial area result in the presence of a significant fraction of atoms at or near the grain boundary region in bulk nanomaterials. It has been reported that around 14–27% of all atoms reside in a region within 0.5–1 nm of a grain boundary for a grain size of ~10 nm.[3] In addition to the advantages obtained due to superior strength, another unique property of bulk nanomaterials is that they can be superplastically deformed at relatively lower temperatures, primarily by grain boundary sliding. This property can be exploited for the ease of processing of brittle materials, such as ceramics and intermetallics.

Table 16.1. Retention of Nanocrystalline Grain Sizes via Spark Plasma Sintering of Nanocrystalline Monolith and Composite Powders (Taken from Reference 1)

Ceramic (monolith/composite)	Initial powder particle sizes (nm)	Final matrix grain sizes (nm)
ZrO_2 (3 mol% yttria)	27	70–80
ZrO_2 (3 mol% yttria)	60	100
TiN	70	90–100
ZnO	20	100
TiO_2	20	200
WC	7	25
γ-Al_2O_3–(20 vol%)SiC_w	32	118
WC–(12 wt%)Co (process: PPS)	60	50
ZrO_2–(10 mol%)Al_2O_3	<10	<100
3Y-TZP–(40 vol%)HAp	10	50
Si_3N_4–(30 vol%)TiN	5–20	50
Mullite–(10 vol%)SiC	100	240

Among other ceramics, yttria-stabilized tetragonal zirconia polycrystals (Y-TZPs) exhibit considerable superplasticity at ultrafine and nanocrystalline grain sizes.[4] Additionally, ZrO_2/Al_2O_3-based nanocomposites are reported to behave superplastically at relatively higher strain rates ($0.4\,s^{-1}$).[5]

As pioneered by Niihara,[6] significant improvement in strength, hardness, and fracture toughness can be achieved in ceramics by nanocomposite design. Ceramic nanocomposites can also be classified, depending on whether the matrix or the reinforcement or both are nanocrystalline and on the distribution of the nanocrystalline reinforcement (Fig. 16.1). Accordingly, specific improvements in mechanical properties can be predicted, as mentioned in Figure 16.1. Apart from enhancement in room-temperature mechanical properties such as hardness and strength, some

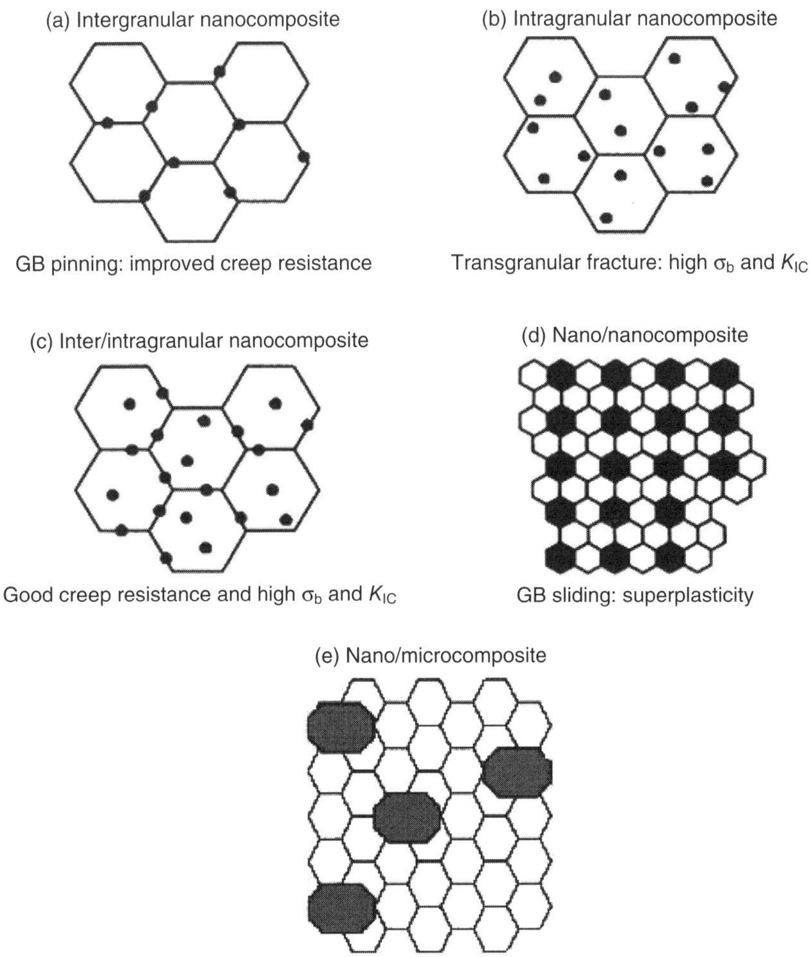

Figure 16.1 Schematic representation of the microstructural features of the various nanocomposites as well as nano/nanocomposites and nano/microcomposites.[1] GB, grain boundary.

nanoceramic composites can exhibit better high-temperature mechanical properties such as hot hardness, high-temperature strength retention, creep resistance, and fatigue fracture resistance.[7]

16.2 PROBLEMS ASSOCIATED WITH SYNTHESIS OF NANOSIZED POWDERS

Broadly, powder-based processing approaches are mostly employed for the processing of bulk nanoceramics. The three basic steps involved in such processing include (1) obtaining unagglomerated nanosized powders with uniform size distribution, (2) cold compaction to obtain crack-free green bodies, and (3) sintering to near-theoretical density without grain growth. In reviewing the existing literature base, copious research efforts are noted to have produced nanocrystalline ceramic powders in gram quantities.[8]

16.2.1 Methods of Synthesis of Nanoscaled Ceramic Powders

The synthesis of nanocrystalline powders is a first step in the processing of bulk ceramic nanomaterials. Various synthesis routes, adopted for synthesizing nanophase ceramic powders, are classified as chemical methods and physical methods.

The main advantages of chemical synthesis methods include the ability to produce a large variety of compositions and homogeneous (atomic level) mixing of the constituent particles.[9] Chemical vapor deposition (CVD), in which a precursor is converted to gas, is one of the conventional techniques of synthesizing ceramic nanopowders. Apart from the commonly synthesized nanoceramic powders, CVD is also utilized for synthesis of a technologically important material—carbon nanotubes.[10] A similar process, the inert gas condensation technique developed for metals by Gleiter,[11] has been modified to produce nanoscale oxides.[12] This technique involves the synthesis of powders from supersaturated vapor phase by either physical cooling of the vapor or by gas-phase chemical reactions. For example, nanosized TiO_2 powders have been synthesized via a two-step inert gas condensation process.[11] Also, nanosized TiCN powders are synthesized via rapid condensation of precursors from gas phase using high-frequency plasma.[13] It should be mentioned here that the vapor condensation techniques have distinct advantages in terms of having good control over particle sizes, producing a narrow size distribution, being free of agglomeration, and having lower levels of contamination.

Chemical pyrolysis, involving decomposition of chemical precursors to obtain a new compound under suitable thermal conditions, has been used to synthesize nanoceramic powders. This process also enables good control over production of high purity powders. However, the major challenge remains particle agglomeration and broadening of the particle size distribution due to high synthesis temperatures.[14] To overcome these difficulties, dispersants and atomization of precursor solutions are used. Moreover, lasers, plasma, and microwaves are used to generate a favorable temperature profile during powder synthesis. The polymer pyrolysis technique enables the development of amorphous Si–C–N nanopowders, which are subsequently used to produce Si_3N_4–SiC nanocomposites.[15]

In chemical precipitation processes, a solution containing hydroxide or oxalic acid is added to the solution containing cations of the desired oxide; this is followed by firing the precipitates (oxalates or hydroxides). Using a coprecipitation route, nanocrystalline (<30 nm) tetragonal zitconia (t-ZrO_2) has been produced by hydrothermal treatment of amorphous zirconia.[16] In fact, nanosized ZrO_2 powders, synthesized via coprecipitation techniques, are quite regularly used in the development of bulk nanoceramics[17] and nanocomposites.[18] Chemical reaction between precursors in the gaseous state is also used for the synthesis of nanopowders. For example, nanosized SiC powders (particle size ~ 15 nm) are synthesized by inducing reaction between gaseous SiH_4 and C_2H_4 using a CO_2 laser.[19]

Among the various synthesis routes, sol–gel processing involves gelation of organic or inorganic precursors to form an interconnected three-dimensional network after hydrolysis and polycondensation in organic solvents, which is followed by dehydration and calcination of the gel to form nanopowders. Simultaneous control over particle size, morphology, and surface chemistry is the most important advantage of using sol–gel processes for nanocluster formation.[20] However, high costs of alkoxide precursors limits the wider use of sol–gel processes.

Among the physical methods of nanopowder preparation, mechanical attrition or high-energy ball milling (HEBM) has gained major acceptance, due to the high-energy collisions among the hard balls and the precursor powders during vibrational motion of the milling vial. Even though nanopowders synthesized via HEBM suffer from contamination and broad particle size distributions, the method's high yield has led to its wide acceptability. To date, a large variety of ceramic nanocomposites have been produced using nanopowders produced via HEBM.[21–24]

HEBM has an important limitation: there is a lower limit, below which particle size or crystallite size cannot be further reduced. This lower limit corresponds to the point at which the probability of finding any internal defect or surface notches in the milled powders becomes negligible. The experimental work, as well as theoretical analysis, has led to the prediction that HEBM cannot reduce the particle sizes below 25–50 nm.[25,26]

16.2.2 Challenges Posed by the Typical Properties of Nanoscaled Powders

The ultrafine ceramic starting powders possess very high surface-to-volume ratio. The large surface area provides a strong driving force for the powder particles to form agglomerates or lumps. Due to such problems, appropriate control against agglomeration is needed during the stages of particle synthesis, drying as well as sintering. The use of surfactants and electrostatic repulsion, which induce surface charges, are generally applied to prevent particles from agglomerating during the synthesis stage. The introduction of large surface charges on zirconia particles at low pH has been reported to lower the propensity for agglomeration.[27] An important aspect that should be kept in mind is that the organic molecules have to be much shorter than those used conventionally.[28] The method of drying precursor powders during synthesis is also important with respect to control of agglomeration. Freeze drying,[29] as well as replacing water by organic solvents prior to drying,[30] is generally used to avoid agglomerate formation. Centrifugation, ultrasonication, and ball milling are generally used as postsynthesis deagglomeration techniques. Another essential characteristic is narrow size distribution. A wider particle size distribution results in abnormal grain growth during sintering, which is extremely detrimental to successful consolidation into bulk nanoceramics.

In addition to the size-related effects, nanocrystalline powders suffer from contamination by atmospheric adsorbates. This causes difficulties during compaction and densification. For example, room-temperature oxidation tests on 20-nm SiC powders reveal a significantly high rate of oxygen pickup.[7] Hence, it is recommended that complete processing as well as intermediate handling be carried out under inert gas conditions. Similarly, it can be noted that the application of special chemical techniques, such as washing oxygen-rich SiC powder (50 nm) with concentrated hydrofluoric acid, can reduce the oxygen content from 10 wt% to less than 1 wt%.[31]

16.3 CHALLENGES FACED DURING PROCESSING

16.3.1 Problems Arising due to Fine Powders

From the consolidation aspect, it is difficult to obtain a green compact of ultrafine precursor powders at the applied pressures (~200 MPa), commonly used for conventional powders. The total number of interparticle contacts becomes increasingly larger as the particle sizes become progressively finer. The enhanced number of contacts results in an augmentation of the frictional resistance to the applied pressure during compaction.[32] Hence, higher pressure is required for compaction of nanocrystalline ceramic powders. In the case of 3Y-TZP nanopowders, a compaction pressure of around 480 MPa in uniaxial pressing resulted in green density of around 45%.[33] Also, the application of pressure in the gigapascal range (~1 GPa) can lead to 51 and 62% green densification for ZrO_2 particles (~10 nm)[34] and TiO_2 particles (~15 nm), respectively.[35] It has been reported that "superhigh" isostatic pressure of 3 GPa was necessary to obtain 60% green densification of 3Y-TZP nanopowders (~10 nm).[36]

The processing of nanocrystalline materials requires minimization of grain growth during the final stage of densification. It is recognized that the rate of grain growth for a polycrystalline material is given by

$$dg/dt = 2M_b\gamma(3/g), \tag{16.1}$$

where g is the grain size, M_b is the grain boundary mobility, and γ is the interfacial energy. It can be easily conceived, from Equation 16.1, that the smaller the grain size, the faster the grain growth. Hence, owing to faster grain growth during sintering of nanocrystalline powders, the retention of nanocrystalline grains in sintered microstructures becomes highly difficult. For this reason, lower sintering temperatures and shorter sintering times are necessary for production of bulk nanocrystalline ceramic materials. The problem of grain growth can also be handled by some innovative techniques. For example, the formation of Si_3N_4 layer on nano-SiC particles by heat treatment in nitrogen can hinder grain growth of nanosized SiC.[37]

Since powders with smaller (nanocrystalline) particles can be densified at a faster rate than microcrystalline powders, inhomogeneous densification becomes apparent when sintering at faster heating rates.[38] This can be more significant in nanoceramics, especially for nanoceramic materials with very low thermal conductivity (ZrO_2, Al_2O_3). The bulk compact experiences a thermal gradient because of the faster heating rate, with the outer surface being densified at a faster rate than the control region. Thus, a hard shell forms at the outside surface, which restricts the core of the sample from shrinking to the maximum extent. Hence, cracks and pores are produced due to strain incompatibility. Also, the grain size changes take place simultaneously with the density gradient.

16.3.2 Challenges Faced due to Agglomerated Powders

In addition to the problems arising from finer powders, another major problem arises from the enhanced tendency of the powders to form agglomerates. A schematic representation, along with typical transmission electron microscopy (TEM) micrographs, of nanocrystalline agglomerated powders is illustrated in Figure 16.2. In the case of agglomerated powders, agglomerate sizes and not the crystallite sizes largely influence the densification behavior. The larger the agglomerate sizes, the coarser would be the interagglomerate pore sizes. Larger pore sizes increase the diffusion distance, resulting in lowering the densification rate (since $d\rho/dt \propto 1/r$; where r is the pore size). To compensate for this, higher sintering temperature becomes necessary. However, higher sintering temperature leads to enhanced grain growth, which poses difficulties in maintaining nanocrystalline grain sizes in the as-sintered bulk material. Another reason for enhanced grain growth stems from the bimodal pore size distribution, characteristic of agglomerated powders. The intercrystallite–intra-agglomerate pore sizes are, however, much smaller than the interagglomerate pore sizes. During sintering, the smaller intercrystallite/intra-agglomerate pores are removed at a much faster rate than the larger interagglomerate pores due to higher driving force for annihilation. Here, it must be noted that an early annihilation of the intercrystallite/intra-agglomerate pores will reduce the constraint for grain growth. Hence, during the final stage of sintering the grains can grow in the absence of pore drag, until they meet the more widely spaced interagglomerate pores. This results in grain coarsening in the dense ceramics. It can be mentioned here that agglomeration is, however, not such a great problem in the consolidation of bulk nanocrystalline metals, due to the unique ability of metal powders to squeeze shut the larger interagglomerate pores via plastic deformation.

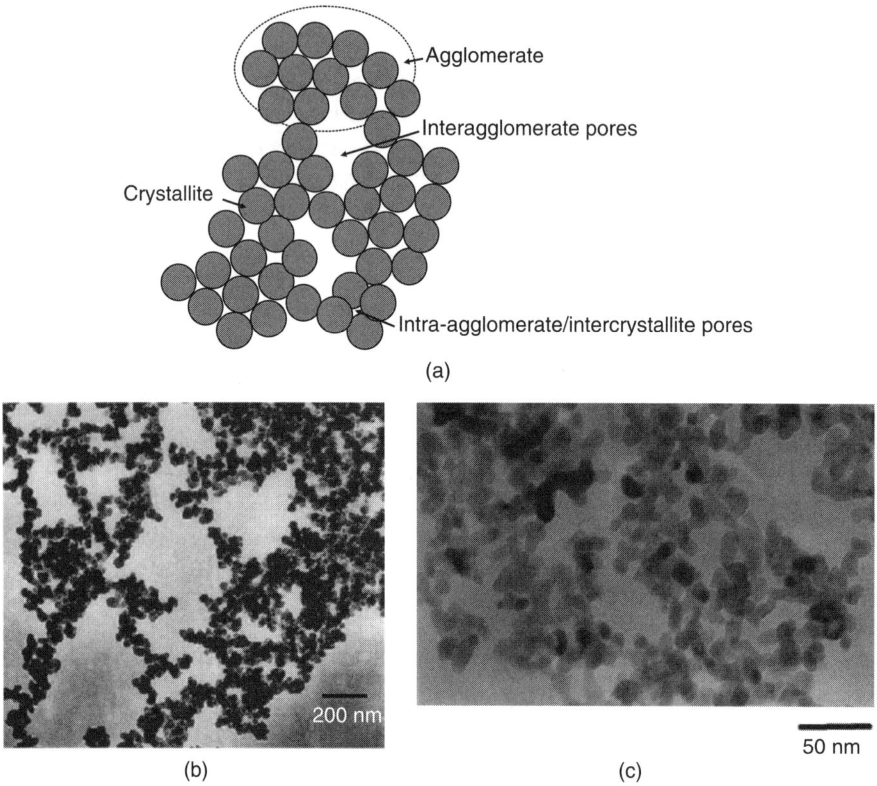

Figure 16.2 (a) A schematic illustration of particle agglomeration, an inherent problem related to nanopowders, (b) TEM image of SiC nanopowders,[7] and (c) TEM image of ZrO_2(3Y) nanopowders[35] showing chainlike powder network as a result of agglomeration.

16.4 PROCESSING OF BULK NANOCRYSTALLINE CERAMICS

16.4.1 Processes Used for Developing Bulk Nanocrystalline Ceramics

As mentioned in the previous section, the consolidation of nanopowders to obtain bulk nanocomposites requires precise control of coarsening–densification competition during the sintering process, that is, enhancing coarsening while suppressing densification (see also Chapter 5, Section 5.5). A combination of much lower sintering temperature with shorter sintering time, in short "activated sintering," has been used in the past decade in order to realize the dream of successfully developing nanocomposites. Fundamentally, activated sintering implies either enhancing the driving force for sintering or enabling the process of sintering to become kinetically faster by physical or chemical treatment.

In the case of conventional pressureless sintering, due to the absence of any external driving force, higher temperature and longer time are required to obtain near-theoretical density. It is known that pressureless sintering results in significant grain growth, especially in the last stage of sintering.[39] However, the application of a two-step sintering process,[40] which involves lowering the temperature after reaching the peak sintering temperature, demonstrates the feasibility of processing nanoceramics via pressureless sintering. According to the explanation provided by Chen and Wang[40] the kinetics of grain boundary mobility is reduced at the lower holding temperature, while maintaining sufficient grain boundary diffusion kinetics to ensure elimination of residual pores. Here, it can be noted that pressureless sintering, when used in conjunction with application of "superhigh" pressure (~3 GPa) during green compaction, can also result in development of bulk nanoceramics in some ceramic systems, such as 3Y-TZP (~80-nm crystallite size).[36]

The enhancement of the densification rate at lower sintering temperature and time can, however, be achieved by application of external pressure during sintering (hot pressing, sinter-forging, sinter-HIPing, and SPS). Near-theoretical densification was recorded via uniaxial hot pressing for γ-Al_2O_3 powders, while maintaining a grain size of ~50 nm.[41] Uchic et al.[42] sintered nanocrystalline TiO_2 powders to near-theoretical density (97% ρ_{th}) at 650°C (<$0.5T_m$) via sinter-forging and, importantly, retained nanocrystalline grains (50–60 nm). HIPing can also result in more than 98% densification of Si_3N_4–SiC nanocomposites, with the SiC grain size in the sintered compact being less than 30 nm.[43] Nanocrystalline SiC (grain size ~ 150 nm) was obtained after HIPing at 1600–1650°C and 350-MPa pressure.[44]

The success of pressure-assisted sintering in processing of bulk nanocrystalline ceramics, in some ceramic systems, has been attributed to pressure-assisted phase transformations. The consolidation of metastable ceramic powers at high pressures and lower temperatures can favor faster nucleation of the product phase, while suppressing the growth rate. Careful optimization of the consolidation pressure and temperature enables retention of nanocrystalline grains with sizes even smaller than those of the starting powders. This process is known as transformation-assisted consolidation (TAC).[45]

Another variant of activated sintering is known as the field-assisted sintering technique (FAST), which enhances densification with simultaneous application of an electric field and pressure. FAST is known under different names, such as SPS, plasma activated sintering (PAS), pulse electric current sintering (PECS),[22] and plasma pressure compaction (PPC).[46] In Chapter 6, a brief description of the SPS process is provided and the next two chapters (Chapters 17 and 18) show how to optimize SPS process parameters to densify nanoceramics and nanoceramic composites. Using the SPS process, it has been demonstrated (in 2010) that one can obtain oxide (Al_2O_3, ZrO_2) and non-oxide ceramics (TiB_2) with uniform mechanical properties by carefully tailoring the multistage heating schedule.[47–49]

16.4.2 Mechanisms Leading to Enhanced Sintering Kinetics on Pressure Application

All the aforementioned processes are capable of resulting in near-theoretical densification of ceramic powder compacts with minimum undesirable microstructural changes. It is known that applied pressure increases the vacancy flux from the pores (vacancy sources) to the grain boundaries (vacancy sinks). In the case of nanocrystalline ceramics, the intrinsic sintering stress is higher due to the very small initial particle sizes and extremely small pore sizes:

$$\sigma = -2\gamma/r, \tag{16.2}$$

where σ is the sintering stress, γ is the surface tension, and r is the pore radius. The estimated sintering stress is generally on the order of 500 MPa, compared with that of ~5 MPa for a conventional micron-sized powder compact. Hence, externally applied pressure (30–100 MPa) during a pressure-assisted densification process fails to enhance further the sintering kinetics of nanocrystalline powder compact, to any appreciable degree.

However, one plausible reason for the observed increase in densification kinetics on application of external pressure lies in the presence of larger interagglomerate pores, which actually reduces the intrinsic sintering stress. The pressure application facilitates the breaking of agglomerates and hence accelerates the densification process. The pore shrinkage mechanism during pressure-assisted sintering of nanocrystalline ceramics is due to the ability of these ultra-fine-grained materials to undergo plastic deformation at $T/T_m \approx 0.5$. This kind of deformation has been reported in nanocrystalline ceramics.[3,50–53] Therefore, the plastic deformation allows pores to be squeezed shut, leading to elimination of even very large pores.[26] Hence, the application of external pressure during sintering results in faster elimination of pores, even the deleterious larger interagglomerate pores, which results in enhanced densification kinetics with minimal grain growth.

16.5 MECHANICAL PROPERTIES OF BULK CERAMIC NANOMATERIALS

16.5.1 Mechanical Properties

In this subsection, the existing literature is reviewed to illustrate how different is the mechanical behavior of nanophase ceramics, compared with conventional ceramics. Figure 16.3 presents a summary of experimental results for SiC ceramics and WC–Co cemented carbides; an analysis of Figure 16.3 reveals that, apart from increase in strength and hardness, Weibull modulus is also improved due to grain refinement, which points toward the improved reliability of fine-grained ceramics compared with their coarser grained counterparts. This can be attributed to improved homogeneity of microstructure and reduction of critical flaw size and density.

16.5.1.1 Hardness and Yield Strength

It can be recalled here that an increase in yield strength and hardness follows the Hall–Petch relation:

$$H_d = H_o + \text{const}/d^{0.5}, \tag{16.3}$$

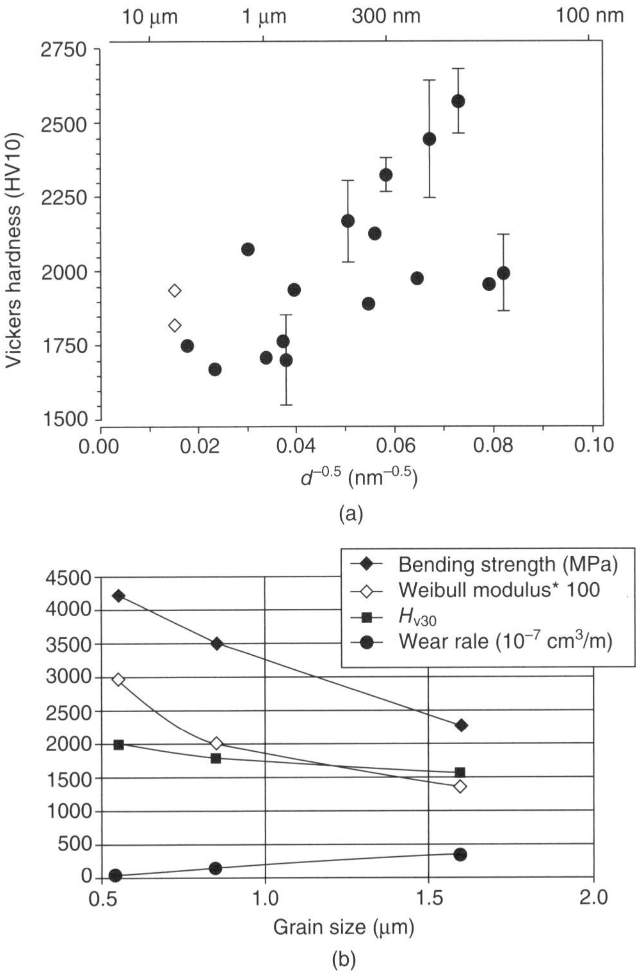

Figure 16.3 Literature result, showing (a) the experimentally measured variation of hardness with decreasing grain in fine-grained SiC and (b) hardness, bending strength, Weibull modulus, and abrasive wear resistance variation with grain sizes for WC–Co cemented carbides.[7]

where H_d is the hardness at grain size d, H_0 is the hardness of single crystal of the same material, and d is the grain diameter. In the Hall–Petch relationship, hardness (H_v) and yield strength (σ) are used interchangeably. Following this relationship, hardness and strength are expected theoretically to keep on increasing with decreasing grain size and reach an extremely high level. However, the increment in strength for nanocrystalline materials falls below the theoretical prediction based on the Hall–Petch equation. In fact, the grain size refinement, beyond a critical grain size, often results in an inverse Hall–Petch relationship. Besides TiAl,[54,55] a decrease in hardness and strength with grain refinement below a certain nanocrystalline level has also been reported for certain ceramic systems, such as TiO_2.[56] As has been extensively reviewed by Ovidko,[57] several mechanisms, based on dislocation motion and interface diffusion, can possibly explain this typical behavior of nanocrystalline materials.

It has been proposed that the deformation mechanism in nanocrystalline ceramics can be correlated with enhanced diffusion along grain boundaries and grain boundary triple junctions. For example, triple junction diffusion coefficient (D_{tj}) is higher (by three or more orders) than the grain boundary diffusion coefficient. Considering the fact that nanocrystalline materials are characterized by the presence of a very high volume fraction of triple junctions, the contribution of triple junction diffusion to deformation of nanocrystalline bulk ceramics should have been considerable. Chokshi et al.[58] proposed that in the lower regime of grain size (<20 nm), room-temperature Coble creep (resulting from grain boundary diffusion) is the dominant deformation mechanism. According to a model[59] based on two deformation mechanisms (lattice dislocation slip vs. grain boundary–triple junction diffusion), deformation by lattice dislocation slip is dominant at larger grain sizes (20–200 nm). Grain boundary sliding is predominant at finer grain sizes, particularly in nanocrystalline materials.[60–62] Such phenomena can expectedly result in strength and hardness reduction of the nanoceramics. Also, grain boundary sliding, accommodated by diffusion (Ashby–Verrall creep) has been attributed to the inverse Hall–Petch relationship for nanocrystalline TiAl.[63] However, this phenomenon requires confirmation in the case of ceramic systems.

It has been experimentally reported that brittle materials undergo considerable grain boundary microcracking under indentation load. According to Ovidko,[57] the Hall–Petch relation needs to be refined to incorporate a negative term (σ_{GB}) related to the grain boundary strength,

$$\sigma_f(\mathrm{d}) = \sigma_f(d = \alpha) - \sigma_{GB}(D/d) + A^*\Phi(D/d)^*d^{-1/2}, \tag{16.4}$$

where $\sigma_f(d)$ is the flow stress of a polycrystal with grain size d, $\sigma_f(d=\alpha)$ is the flow stress of a polycrystal with very large grains, D is a typical dimension of the indented zone, and $A^*\Phi(D/d)^*d^{-1/2}$ is a term related to dislocation pileup at grain boundaries. It can be assumed that, since at the finest grain sizes in the nanocrystalline regime the contribution of dislocation motion to plastic deformation is still negligible, the impact of the term $A^*\Phi(D/d)^*d^{-1/2}$ to the overall $\sigma_f(d)$ can therefore be neglected. Hence, with further reduction in grain size, the dislocation pileup-induced hardening becomes insignificant. Hence, at the finest grain sizes, when the volume fraction of grain boundaries becomes very large, the amount of deformation in ceramics due to cracking increases. Moreover, from Equation 16.4, it can be observed that, when $D \gg \mathrm{d}$, the negative term $\sigma_{GB}(D/d)$ has a significant contribution to the overall $\sigma_f(d)$. Thus, this model also provides an insight into the negative Hall–Petch relationship, as observed at the finest grain sizes.

However, the consideration of a similar model coupled with Griffith's theory of crack propagation in brittle materials[64] leads to a prediction of completely opposite behavior for nanocrystalline materials. From Griffith's criterion, the stress required for crack propagation is proportional to the inverse square root of critical flaw size. Since the flaw size is reduced with reduction in grain size, a significantly higher stress is required for crack propagation at the finest grain sizes. Furthermore, as has been stated previously, at the smallest grain sizes, plastic deformation is not possible by dislocation motion, and only microcracking can lead to deformation. These imply that the amount of deformation, possible at a given stress should decrease with grain refinement and, hence, hardness should continue to increase at the finest grain sizes. In fact, Veprek et al.[65] experimentally observed a hardness increment with grain refinement up to 3.5 nm in $TiN–Si_3N_4$ and $W_2N–Si_3N_4$ nanocomposites. This proves the validity of the preceding hypothesis.

16.5.1.2 Fracture Strength and Fracture Toughness

The considerable increase in fracture strength due to microstructural refinement can be correlated with the synergistic effects of the reduction in flaw sizes with reduced microstructural scale[66,67] and to the lowering of thermal residual stresses arising from anisotropic crystal structure and coefficient of thermal expansion mismatch among the microstructural phase assemblage.[68] It is an established fact that fracture strength of brittle materials is determined by the size and distribution of the critical flaws. Hence, due to alleviation of the flaw sizes by maintaining nanocrystalline microstructure, the fracture strength can be significantly enhanced. Moreover, as predicted by Gao et al.[69] below a certain critical length scale (h^*), the fracture strength of a cracked crystal can be compared with that of a perfect crystal. The fracture strength ($\sigma_{\mathrm{m}}^{\mathrm{f}}$) of a crystal containing the flaws can be described by Griffith's criterion:

$$\sigma_m^f = \alpha E_m \psi, \tag{16.5}$$

where $\psi = \sqrt{(\gamma/E_m\, h)}$, E_m is the elastic modulus, γ is the surface energy, h is the crystal dimension, and α is a parameter that depends on crack geometry. Using Equation 16.5, a comparison of the strength increment on reduction of h predicts that there exists a critical dimension (h^*) (given by Eq. 16.6) at which the actual strength (from Eq. 16.5) becomes equal to that of the theoretical strength of the perfect crystal:

$$h^* \approx \alpha^2 \frac{\gamma E_m}{\sigma_{th}^2}. \tag{16.6}$$

Above this critical size, the crystal strength is controlled by the presence of flaws. However, below this critical size, strength would be insensitive to flaws; that is, the strength would be limited by the theoretical strength of the flawless "perfect" crystal. An empirical plot illustrating such an outstanding prediction is presented in Figure 16.4.

Another reason cited for the improved fracture strength is the presence of lower residual stress.[70] Such residual stress is generated during cooling from the sintering temperature and is due to anisotropic crystal structure as well as mismatch of coefficient of thermal expansion between different microstructural phases. However, the sintering temperatures are generally lower for nanoceramics than for their conventional counterparts. It is also reported that the residual stress can be relieved to a greater extent in nanoceramics.[71]

Despite significant improvement of fracture strength, it has been experimentally observed that fracture toughness of monolithic nanoceramics is not improved by the reduction of grain sizes to nanocrystalline levels.[71,72] Fracture toughness is, however, improved to a moderate extent by nanocomposite design.[5] Stating this concept in a broad way, the incorporation of reinforcements in a conventional or nanostructured

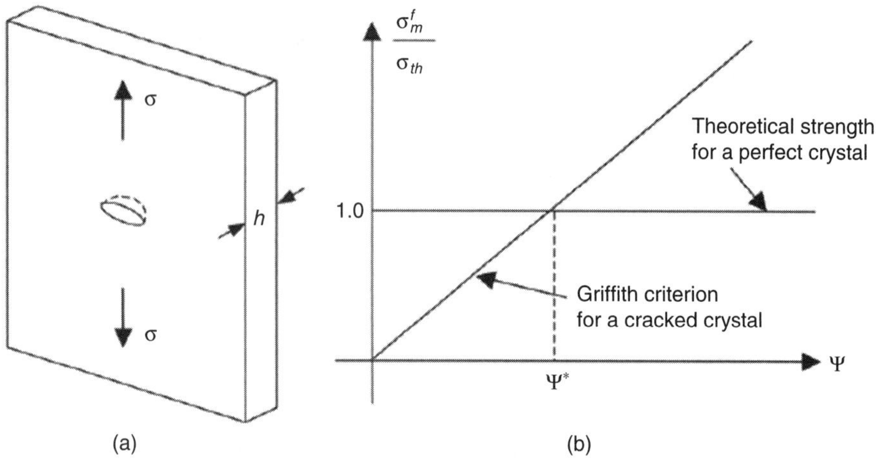

Figure 16.4 (a) A schematic representation of platelet/crystal with a surface crack. (b) Comparison of the fracture strength of a cracked platelet calculated from the Griffith criterion with the strength of a perfect crystal.[1] One can critically observe that above a critical ψ^* (below a critical h^*) the strength of the cracked crystal becomes equal to that of a perfect crystal.

ceramic matrix (ceramic nanocomposites) could potentially result in increase of strength and fracture toughness. Niihara and Nakahira reported an increase in flexural strength of brittle ceramics by almost three times via nanocomposite design.[73,74] The fracture toughness has been attributed to the development of localized residual stresses within and around reinforcements due to mismatch of coefficient of thermal expansion between the constituent phases. The compressive residual stresses within the nanocrystalline reinforcements is reported to cause crack-deflection toughening and change in fracture mode from intergranular to transgranular.[63,75] On the other hand, stresses generated near the reinforcement–matrix interfaces, open up microcracks or nanocracks at the crack tip of a propagating crack. Microcracking or nanocracking expands the size of the fractal process zone (FPZ), resulting in enhancement of fracture toughness.[76] Furthermore, crack bridging by the nanodispersoids has also resulted in modest improvement of fracture toughness.[77] Such enhancement of the resistance to crack propagation leads to significant improvement of the fracture strength on nanocomposite formation.

Here it must be noted that, although the beneficial effects of residual stresses are also present in conventional ceramic composites, yet the mechanical property improvement resulting from such effects will be greater in ceramic nanocomposites. According to Niihara and coworkers,[77] three effects of nanosized reinforcements need to be considered.

First, residual stresses result in a competition between toughening from particle bridging and fracture toughness reduction due to residual traction in the matrix. Both these effects depend on the average reinforcement dimension, which is represented graphically in Figure 16.5a. Figure 16.5a shows that fracture toughness increases steeply up to particle size of ~100 nm and then flattens out considerably. In contrast,

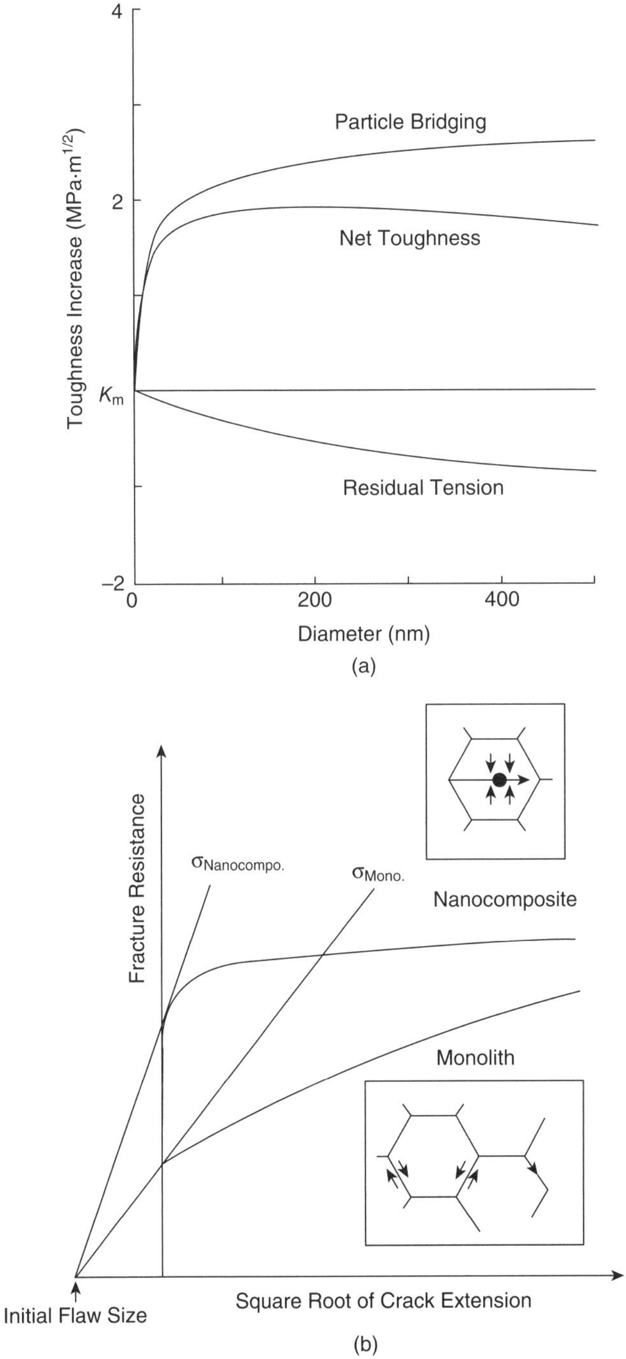

Figure 16.5 (a) Plot depicting how reinforcement size influences the bridging-induced fracture toughness increase, the residual-tension-induced fracture toughness reduction, and (b) the combined net fracture toughness and a comparison of R-curve behavior of monolithic ceramics and ceramic nanocomposites.[1]

the fracture toughness reduction due to matrix residual traction increases monotonically with increase in reinforcement size. In accordance with the preceding observation by Niihara, Levin et al.[78] used computational modeling to demonstrate how smaller reinforcement size would reduce the negative effect of tensile matrix residual stress on the net fracture toughness.

Second, it has been observed that crack propagation in many of the ceramic nanocomposites (e.g., Al_2O_3–SiC nanocomposite) is via fracture of the matrix–reinforcement interface. Therefore, an increase in interfacial strength can result in impeding crack propagation. Also, a reduction in reinforcement dimension to nanoscale leads to an increase in the interfacial strength due to better lattice matches observed in ceramic nanocomposites, compared with those in the conventional microcomposites. A better lattice match is a result of nanosize effect, since atomic diffusion through much smaller distances can lead to near-equilibrium structure.

Third, for a given reinforcement volume fraction, the interparticle distance (λ) reduces with particle size. Considering crack bridging as the major toughening mechanism, the frequency of interaction of cracks with the nanosized reinforcement increases considerably due to a reduction in λ. The tortuous nature of cracks and frequent obstruction to crack propagation results in a steeper slope of the R curve (crack growth resistance curve) in the nanocomposites, compared with that in microcomposites or monolithic ceramics (Fig. 16.5b). Another effect that can lead to enhanced fracture resistance is that, at a given reinforcement volume fraction, the number of particles increase on reduction of particle size. This leads to more uniform distribution of stress among the particles, and such stress distribution would reduce the propensity for fracture at a given applied load and hence would increase the strength.

16.5.1.3 Superplasticity

An outstanding characteristic of nanoceramics is their ability to undergo superplastic deformation at relatively lower temperatures. Though superplasticity at extremely low temperature (<500°C) for various ceramic systems is still awaited, yet a number of nanoceramics successfully exhibit superplasticity at moderately lower temperatures.[49–52] The present consensus is that grain boundary sliding and augmented Coble creep rate, arising from considerable increase in interfacial area (grain boundary and triple junctions) are the underlying mechanisms of superplasticity. It is reported that Coble creep involves high-temperature deformation at relatively lower stress ($\sigma/G < 10^{-4}$) via grain boundary diffusion of atoms or vacancies. The kinetics of Coble creep strain (ε_{Co}) may be described by the following expression:

$$\varepsilon_{Co} = A_{Co}\frac{D_{gb}Gb}{kT}\left(\frac{b}{d}\right)^3\frac{\sigma}{G}, \qquad (16.7)$$

where A_{Co} is a dimensionless coefficient (~66), σ is the applied stress, G is the shear modulus, b is the Burgers vector of the mobile dislocations at temperature T (in Kelvin), D_{gb} is the grain boundary diffusion coefficient, and d is the grain size. From Equation 16.7, one can emphasize that reduction of grain size can result in appreciable creep rate. Extrapolation of this concept to materials with nanocrystalline grains can lead to the realization of significant enhancement of Coble creep rate ($\propto d^{-3}$) and, hence, to a prediction of the possible occurrence of superplasticity. With this possibility in mind, researchers have made efforts to enhance the deformability of traditionally brittle materials, such as ceramics and intermetallics, by grain refinement down to the nanoscale level. Karch et al.[79] observed the possibility of brittle materials exhibiting room-temperature deformability on grain refinement down to 100 nm. However, subsequent research could not establish this expectation in nanocrystalline ceramics. This has been tentatively attributed to the fact that the room-temperature grain boundary diffusion coefficient (D_{gb}) is as low as $10^{-30}\,cm^2/s$ for ceramics with melting point above 1500°C.[80] Also, nanocrystalline 5Y-TZP[81] and TiO_2[82] exhibited extensive superplastic deformation at temperatures of around 1000°C. Superplasticity at strain rates of the order of $10^{-4}\,s^{-1}$ is experimentally observed for nanoceramics, based on TiO_2,[50] TiAl,[52] Si_3N_4,[83] Y-TZP,[3,4,84] and SiC.[85] Also, the strain rate of $10^{-4}\,s^{-1}$ observed for nanocrystalline (~80 nm) 3Y-TZP is about 34 times higher than the strain rates observed during superplastic deformation of submicron (0.3 μm) zirconia. Importantly, superplastic deformation, resulting in a tensile elongation of up to 1050% at a relatively higher strain rate of $0.4\,s^{-1}$, has also been observed for nanocomposites of zirconia, alumina, and spinels.[4] It must be emphasized that superplastic deformation of nanoceramics at lower temperatures and higher strain rates can be advantageous in lowering of sintering temperatures and in near-net-shaping processes.

16.6 APPLICATIONS OF NANOCERAMICS

The improvement in mechanical properties, in particular strength and hardness, can lead to better performance of nanocrystalline ceramics in engineering applications, currently served by conventional and microcrystalline ceramics. Furthermore, the uniformity of microstructure and the reduction of processing flaw sizes lead to enhanced mechanical reliability, thus improving feasibility in critical engineering applications. Owing to higher wear resistance,[16,22] these novel materials are also candidates for heavy-duty wear applications. In particular, better tribological performance bestows longer lifetime on nanostructured tool components. For example, Al_2O_3–SiC nanocomposites can find application as abrasive grits, with performance and cost intermediate between the commonly used Al_2O_3 and superabrasives BN-diamond.[28] It must also be mentioned that incorporation of nanosized ceramic fillers such as nano-Al_2O_3 (even in very small amounts ~1–5 wt%) significantly enhances (over 3000 times) the wear resistance of polytetrafluoroethylene (PTFE).[86]

One of the most significant benefits of nanocrystalline ceramics is their ability to undergo superplastic deformation at lower temperatures and higher strain rates.[3,4,51,59,83–86] Superplasticity enables near-net-shaping of the nanocrystalline ceramics and, in future, it might be even possible to commercially shape them via conventional forming techniques. For example, superplastic behavior of nanocrystalline (~150 nm) 3 mol% yttria-stabilized tetragonal zirconia tubes, at 1550°C, was successfully exploited for net shaping applications (Fig. 16.6a).[51] Even though

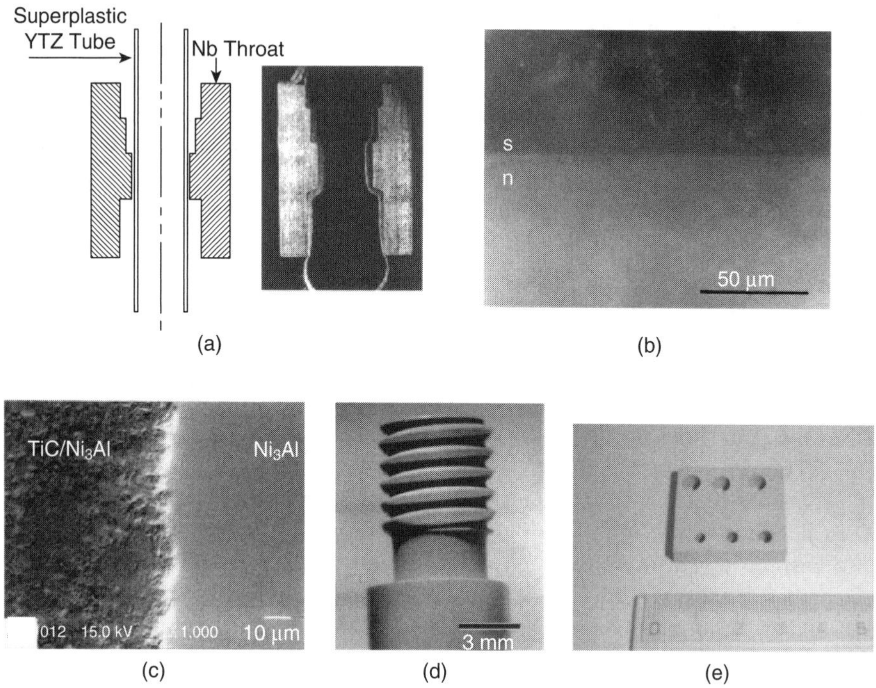

Figure 16.6 (a) Schematic diagram of an experiment to expand superplastic Y-TZP tube within a rigid Nb shell (left); the cross section after forming at 1550°C (right);[51]; (b) flawless interface produced by joining nanocrystalline and microcrystalline 3Y-TZP at lower temperature of 1150°C (s denotes submicron and n denotes nanocrystalline); (c) interface between nanocrystalline TiC–Ni_3Al and Ni_3Al formed *in situ* via SPS; (d) and (e) machinable Si_3N_4–BN nanocomposite.[1]

higher strain rates and forming rates can be achieved by nanoceramics, yet the highest strain rates possible to date are much slower than desired for commercial-scale economical forming of nanoceramics.

The successful commercial application of large or complex-shaped nanoceramic composites also demands the adaptation of joining techniques. Superplastic flow and enhanced diffusivity of nanoceramics are beneficial for easier joining. Moreover, when used as interlayers, the grain size difference between conventional ceramic components and nanocrystalline interlayer may lead to grain growth. This can facilitate chemical bonding, along with mechanical interlocking across the interfaces. Mayo et al. were able to join two nanocrystalline 3Y-TZP ceramics directly at 1090°C at 10-MPa pressure.[39] Also, a flawless and strong joint could be obtained by sandwiching a nanocrystalline Y-TZP layer between two submicron 3Y-TZPs at 1150°C, which was 200°C lower than that required to produce a similar joint sans the nanocrystalline Y-TZP layer.[87] To provide such evidence, Figure 16.6b illustrates a joint interface between the submicron component and the nanocrystalline zirconia interlayer. Liu and Naka[88] successfully attempted *in situ* joining of nanocrystalline Ni_3Al and Ni_3Al–TiC in nanocomposites via SPS at 1100°C. Even after completion of the joining process, monolithic Ni_3Al had a grain size of 96 nm, while the Ni_3Al

and TiC had grain sizes of ~60 nm. Most importantly, cracks or voids were not detected on the joint surface (Fig. 16.6c) and shear strength was around 765 MPa.

As mentioned in the previous section, machinable Si_3N_4–BN nanocomposites, in addition to superior mechanical properties, also possess better machinability. Illustrative examples of drilled parts are shown in Figure 16.6d,e. As for the quality of machined surfaces, the drilled surfaces of the nanocomposites possessed better surface finish (lower roughness; R_{max} ~ 2 µm for nanocomposite vs. 4 µm for microcomposite).

It must be further noted that the combination of lower thermal conductivity, higher thermal shock resistance,[7] and superior mechanical properties enables nanoceramics to be used in applications involving thermal insulation (thermal barrier coatings).

The potential of nanostructured ceramics for biomedical applications needs to be investigated more in future. In fact, improved mechanical properties and biocompatibility should ideally render nanostructured ceramics as potential candidates for load-bearing biomedical applications. For example, nanoceramics and nanoceramic composites based on bioactive materials, such as hydroxyapatite (HAp) are deemed to be superior for dental and orthopedic implants owing to the better possibility of controlling the compositions, surface, and mechanical properties, to make them similar to those of physiological bone. Importantly, Webster et al. reported that some nanostructured ceramics support better bone cell (e.g., osteoblast) functionality, and therefore the nanophase ceramics can have better efficacy in orthopedic applications.[89,90]

16.7 CONCLUSION AND OUTLOOK

As mentioned in this overview chapter, extensive research on the processing, microstructure, and properties of nanoceramics and nanoceramic composites has provided considerable information and understanding of these novel materials. Such insight can be helpful in modulating the properties by careful selection of the processing techniques and parameters, with concomitant control of the microstructures and compositional design.

It is useful to reiterate here that agglomeration and contamination during nanopowder handling and considerable grain growth via conventional sintering often cause problems in maintaining nanocrystalline grains. Though still challenging, various techniques are being developed for minimizing the problem of powder agglomeration and contamination. It must be mentioned that non-agglomerated nanosized powders are being produced at a commercial level (Nanophase Technologies, Tosoh). The adaptation of advanced processing techniques, in particular "activated sintering techniques," has been successful for development of bulk ceramic nanomaterials. Among them, SPS has been a major success. By providing high heating rate, lower sintering temperatures, and short holding times, SPS is capable of restricting grain growth during densification to near-theoretical density. Nevertheless, SPS processing of nanoceramics has been restricted to laboratory-scale synthesis. It is an intriguing issue whether processing of larger components can suffer from densification gradient. Moreover, SPS is limited only to the production of simple shapes (cylindrical, square). Furthermore, future research should answer some key questions regarding the underlying mechanisms of SPS processes. With better understanding of the underlying densification mechanisms of SPS, it would be possible to closely tune the SPS parameters to develop dense nanostructured materials in a large number of ceramic systems.

It is an established fact that reduction of microstructural scale leads to enhancement in hardness and strength. However, it is intriguing the extent to which such increment is possible for ceramics characterized by the finest possible grain sizes (<20 nm). It has also been observed that, below a critical grain size, softening takes place.

One of the drawbacks in the developmental research on nanoceramics appears to be the inability to enhance fracture toughness to a satisfactory level with grain refinement in single-phase ceramic materials. It can be noted that fracture toughness is an important criterion for various applications (including tribological applications) of ceramics. However, the toughness issue could be addressed to a certain extent by judicious incorporation of nanocrystalline second-phase reinforcements. The possibility of fracture toughness improvement in ceramics by a nanocomposite approach, incorporation of tougher reinforcements, piezoelectric or ferroelectric nanoparticles, and proper tuning of the compositions is already being explored. One notable advantage is that the incorporation of a considerable amount of a softer phase for toughening is possible without compromising the hardness to any considerable extent. Thus, bulk nanoceramics and ceramic nanocomposites possess a superior combination of mechanical properties. Also, the theoretical studies concerning the mechanical behavior of nanoceramic composites have been conducted based on experimental results of limited ceramic systems such as Al_2O_3–SiC[75–79] and Si_3N_4–SiC.[28]

Improved mechanical behavior due to nanocrystallinity has opened up new vistas for applications, including near-net-shape forming, demanding tribological applications, thermal barrier coatings, and applications demanding better machinability and enhanced biocompatibility. Further research into the beneficial effects of nanocrystallinity on commercial aspects of such applications is needed to ensure more applications of nanoceramics. Last, in spite of the current state of development of the processing and properties of bulk nanoceramics, this is still an emerging field. More research is definitely required to optimize processing parameters, develop processing techniques for large-scale production, improve understanding of their mechanical behavior, improve fracture toughness, and render them useful in commercial applications.

As a concluding note, the major issues related to the development of nanoceramics and nanoceramic composites are summarized in Figure 16.7. The major processing-related challenges include the synthesis of non-agglomerated nanosized ceramic powders as well as adopting an appropriate densification route with the capability of enabling sintering at a much faster rate, while inhibiting grain growth at the final stage of sintering. As far as the mechanical properties are concerned, while better hardness and strength can be achieved in nanostructured ceramics compared with conventional ceramics, the enhancement of fracture toughness remains to be achieved in many technologically important ceramic systems. This issue requires further attention.

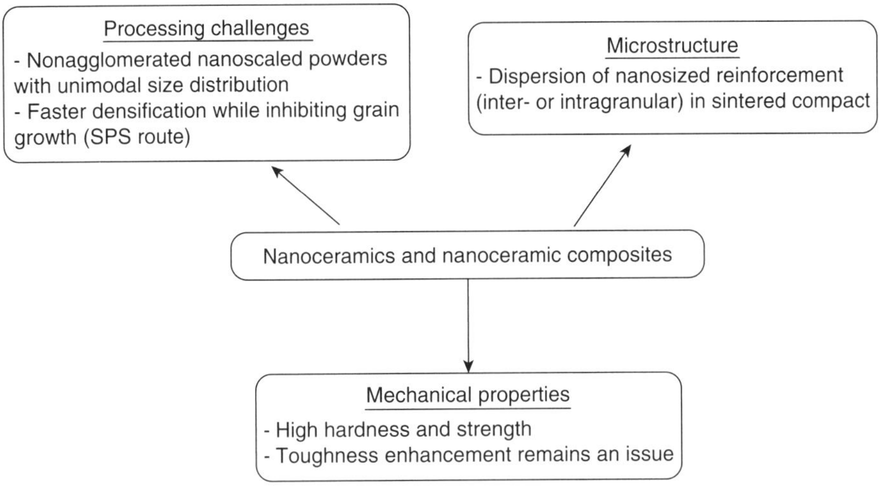

Figure 16.7 Schematic illustrating the major issues with the development of nanostructured ceramics.

REFERENCES

1 A. Mukhopadhyay and B. Basu. Consolidation-microstructure-property relationships in bulk nanoceramics and ceramic nanocomposites: A review. Int. Mater. Rev. 52(5) (2007), 257–288.

2 H. Gleiter. Nanocrystalline materials. Prog. Mat. Sci. 33 (1989), 223–315.

3 K. S. Kumar, H. V. Swygenhoven, and S. Suresh. Mechanical behavior of nanocrystalline metals and alloys. Acta Mater. 51 (2003), 5743–5774.

4 T. G. Nieh and J. Wadsworth. Superelastic behaviour of a fine-grained, yttria-stabilized, tetragonal zirconia polycrystal (Y-TZP). Acta Metal. Mater. 38 (1990), 1121–1133.

5 B. N. Kim, K. Hiraga, K. Morita, and Y. Sakka. A high-strain-rate superplastic ceramic. Nature 413 (2001), 288.

6 K. Niihara. New design concept for structural ceramics—Ceramic nanocomposites. J. Ceram. Soc. Jpn 99(10) (1991), 974–982. The Centennial Memorial Issue.

7 K. Niihara and Y. Suzuki. Strong monolithic and composite $MoSi_2$ materials by nanostructured design. Mater. Sci. Eng. A 261 (1999), 6–15.

8 R. Vaben and D. Stover. Processing and properties of nanophase ceramics. J. Mater. Process. Tech. 92–93 (1999), 77–84.

9 H. S. Nalwa. *Handbook of Nanostructured Materials and Nanotechnology*, Vol. 1. Synthesis and Processing, Academic Press, San Diego, CA, 2002.

10 S. Lijima. Helical microtubules of graphitic carbon. Nature 354 (1991), 56–58.

11 H. Gleiter. Nanostructured materials: basic concepts and microstructure. Acta Materialia 48(1) (2000), 1–29.

12 R. W. Siegel, S. Ramasamy, H. Hahn, L. Zongquan, L. Ting, and R. Gronsky. Synthesis, characterization, and properties of nanophase TiO_2. J. Mater. Res. 3 (1988), 1367–1372.

13 I. Zalite, S. Ordanyan, and G. Korb. Synthesis of transitionmetal nitride/carbonitride nanopowders and their application for modification of structure of hardmetals. Powder Metall. 46 (2003), 143–147.

14 Z. L. Wang (Ed.). *Nanophase and Nanostructured Materials*. Kluwer Academic Press, Beijing, 2002, 37.

15 J. Wan, R. G. Duan, and A. K. Mukherjee. Spark plasma sintering of silicon nitride/silicon carbide nanocomposites with reduced additive amounts. Scr. Mater. 53 (2005), 663–667.

16 E. Tani, M. Yoshimura, and S. Somiya. Formation of ultrafine tetragonal ZrO_2 powder under hydrothermal conditions., J. Am. Ceram. Soc. 66 (1983), 11–14.

17 B. Basu, J.-H. Lee, and D.-Y. Kim. Development of nanocrystalline wear resistant Y-TZP ceramics. J. Am. Ceram. Soc. 87(9) (2004), 1771–1774.

18 B. Basu, T. Venkateswaran, and D.-Y. Kim. Microstructure and properties of spark plasma sintered ZrO_2-ZrB_2 nanoceramic composites. J. Am. Ceram. Soc. 89(8) (2006), 2405–2412.

19 J. S. Haggerty and W. R. Cannon. Chapter 3, in *Laser Induced Chemical Processes*, J. I. Steinfeld (Ed.). Plenum Press, New York, 1981, 165–241.

20 E. A. Barringer and H. K. Bowen. Formation, packing, and sintering of monodisperse TiO_2 powders. J. Am. Ceram. Soc. 65 (1982), C199–C201.

21 G. D. Zhan, J. Kuntz, J. Wan, J. Garay, and A. K. Mukherjee. A novel processing route to develop a dense nanocrystalline alumina matrix (<100 nm) nanocomposite material. J. Am. Ceram. Soc. 86(1) (2003), 200–202.

22 G. D. Zhan, J. D. Kuntz, R. G. Duan, and A. K. Mukherjee. Spark plasma sintering of silicon carbide whiskers (SiC_w) reinforced nanocrystalline alumina. J. Am. Ceram. Soc. 87(12) (2004), 2297–2300.

23 M. Yoshimura, O. Komura, and A. Yamakawa. Microstructure and tribological properties of nano-sized Si_3N_4. Scr. Mater. 44 (2001), 1517–1521.

24 Y. Zhang, L. Wang, W. Jiang, L. Chen, and G. Bai. Microstructure and properties of Al_2O_3-TiC nanocomposites fabricated by spark plasma sintering from high-energy ball milled reactants. J. Eur. Ceram. Soc. 26 (2006), 3393–3397.

25 D. L. Zang, J. Liang, and J. Wu. Processing Ti3Al0SiC nanocomposites using high energy mechanical milling. Mater. Sci. Eng. A 375–377 (2004), 911–916.

26 Y. D. Kim, S. T. Oh, K. H. Min, H. Jeon, and I. H. Moon. Synthesis of Cu dispersed Al_2O_3 nanocomposites by high energy ball milling and pulse electric current sintering. Scr. Mater. 44 (2001), 293–297.

27 M. J. Mayo. Grain growth and the processing of nanocrystalline ceramics. Mater. Sci. Forum 204–206 (1996), 389–398.

28 R. Nass, S. Albayrak, M. Aslan, and H. Schmidt. Proceedings of the Topical Symposium VII on Advanced Materials in Optics, Electro-Optics and Communications Technologies of the Eighth CIMTEC-World Ceramic Congress and Forum on New Materials, TECHNA, Faenza, Italy, P. Vincenzini and G. C. Righini (Eds.). 1995, 47.

29 M. Sternitzke. Structural ceramic nanocomposites. J. Eur. Ceram. Soc. 17 (1997), 1061–1082.

30 D. L. Chen and M. J. Mayo. Densification and grain growth of ultrafine 3 mol % Y_2O_3-ZrO_2 ceramics. J. Nanostructured Mater. 2 (1993), 469–478.

31 K. Wetzel, G. Rixecker, G. Kaiser, and F. Aldinger. Preparation of dense nanocrystalline silicon carbide ceramics by sinter forging in the presence of a liquid phase. Adv. Eng. Mater. 7 (2005), 520–524.

32 A. O. Boschi and E. Gilbart. Wet forming processes as a potential solution to agglomeration problems, in *Advanced Ceramic Processing and Technology*, Vol. 1, J. G. P. Binner (Ed.). Noyes Publications, Park Ridge, NJ, 1990, 73–93.

33 W. H. Rhodes. Agglomeration and particle size effects on mixing yttria stabilized zirconia. J. Am. Ceram. Soc. 64 (1981), 19–22.

34 G. Skandan. Processing of nanostructured zirconia ceramics. Nanostructured Mater. 5 (1995), 111–126.

35 W. Wagner, R. S. Averbach, H. Hahn, W. Petry, and A. Wiedenmann. Sintering characteristics of nanocrystalline TiO_2—A study combining small angle neutron scattering and nitrogen absorption–BET. J. Mater. Res. 6 (1991), 2193–2198.

36 L. Gao, W. Li, H. Z. Wang, J. X. Zhou, Z. J. Chao, and Q. Z. Zai. Fabrication of nano Y–TZP materials by superhigh pressure compaction. J. Eur. Ceram. Soc. 21 (2001), 135–138.

37 A. Kaiser, R. Vassen, and D. Stover. Grain growth of dense Si-C-N/additive and SiC at high temperatures. J. Mater. Sci. Lett. 15 (1996), 1960–1962.

38 W. Li and L. Gao. Rapid sintering of nanocrystalline ZrO_2(3Y) by spark plasma sintering [J]. J. Eur. Ceram. Soc. 20 (2000), 2441–2445.

39 M. J. Mayo, D. C. Hague, and D. J. Chen. Processing nanocrystalline ceramics for applications in superplasticity. Mater. Sci. Eng. A 166 (1993), 145–159.

40 I. W. Chen and X. H. Wang. Sintering dense nanocrystalline oxide without final stage grain growth. Nature 404 (2000), 168–171.

41 S. C. LIAO, Y. J. CHEN, B. H. KEAR, and W. E. MAYO. High pressure low temperature sintering of nanocrystalline Al_2O_3. Nanostructured Mater. 10 (1998), 1063–1079.
42 M. UCHIC, H. J. HOÈFLER, W. J. FLICK, R. TAO, P. KURATH, and R. S. AVERBACK. Sinter-forging of nanophase TiO_2. Scr. Metall. Mater. 26 (1992), 791–796.
43 R. VAßEN, A. KAISER, J. FÖRSTER, H. P. BUCHKREMER, and D. STÖVER. Densification of ultrafine SiC powders. J. Mater. Sci. 31 (1996), 3623–3637.
44 R. VASSEN and D. STOVER. Processing and properties of nanograin silicon carbide. J. Am. Ceram. Soc. 82 (1999), 2585–2593.
45 B. H. KEAR, J. COLAIZZI, W. E. MAYO, and S. C. LIAO. On the processing of nanocrystalline and nanocomposite ceramics. Scr. Mater. 44 (2001), 2065–2068.
46 S. H. YOO, T. S. SUNDARSHAN, K. SETHURAM, G. SUBHASH, and R. J. DOWDING. Consolidation and high strain rate mechanical behavior of nanocrystalline tantalum powder. Nanostructured Mater. 12 (1999), 23–28.
47 K. MADHAV REDDY, A. MUKHOPADHYAY, and B. BASU. Microstructure-mechanical-tribological property correlation of multistage spark plasma sintered tetragonal ZrO_2. J. Eur. Ceram. Soc. 30 (2010), 3363–3375.
48 K. MADHAV REDDY, N. KUMAR, and B. BASU. Innovative multi-stage spark plasma sintering to obtain strong and tough ultrafine grained ceramics. Scr. Mater. 62 (2010), 435–438.
49 D. JAIN, K. MADHAV REDDY, A. MUKHOPADHYAY, and B. BASU. Achieving uniform microstructure and superior mechanical properties in ultrafine grained TiB_2-$TiSi_2$ composites using innovative multi stage spark plasma sintering. Mater. Sci. Eng. A 528 (2010), 200–207.
50 M. MITOMO, H. HIROTSURU, and Y.-W. KIM. Superplastic silicon carbide sintered body. U.S. Patent 5,591,685, January 7, 1997. Japan Patent 2,671,945, July 11, 1997.
51 H. HAHN, J. LOGAS, H. J. HOÈFLER, P. KURATH, and R. S. AVERBACK. Mater. Res. Soc. Symp. Proc. 196 (1990), 71.
52 J. WITTENAUER. In *Processing and Fabrication of Advanced Materials III*, V. A. Ravi, T. S. Srivatsan, and J. J. Moore, (Ed.). The Minerals, Metals & Materials Society, Warrendale, PA, 1994, 197.
53 K. AMEYAMA, A. MIYAZAKI, and M. TOKIZANE. Superplasticity in advanced materials, in S. Hori, M. Tokizane, and N. Furushiro (Eds.). Japanese Society of Research on Superplasticity, 1991, 317.
54 H. CHANG, C. J. ALSTETTER, and R. S. AVERBACK. Characteristics of nanophase TiAl produced by inert gas condensation. J. Am. Ceram. Soc. 74 (1991), 2672.
55 R. S. AVERBACK, H. J. HOFLER, and R. TAO. Processing of nano-grained materials. Mater. Sci. Eng. A 166 (1993), 169–177.
56 M. GUERMAZI, H. J. HOFLER, R. HAHN, and R. S. AVERBACK. Temperature dependence of the hardness of nanocrystalline titanium dioxide. J. Am. Ceram. Soc. 74 (1991), 2672.
57 I. A. OVIDKO. Deformation and diffusion modes in nanocrystalline materials. Int. Mater. Rev. 50(2) (2005), 65–82.
58 A. H. CHOKSHI, A. ROSEN, J. KARCH, and H. GLEITER. Influence of grain size on the mechanical behaviour of some high strength materials. Scr. Metall. 23 (1989), 1679–1683.
59 R. A. MASUMURA, P. M. HAZZLEDINE, and C. S. PANDE. Yield stress of fine grained materials. Acta Mater. 46 (1998), 4527–4534.
60 F. A. MOHAMED and Y. LI. Creep and superplasticity in noanocrystalline materials: Current understanding and future prospects. Mater. Sci. Eng. A 298 (2001), 1–15.
61 R. Z. VAILEV and I. V. ALEXANDROV. *Nanostructured Materials Prepared by Severe Plastic Deformation*. Logos, Moscow, 2000.
62 M. Y. GUTKIN, I. A. OVID'KO, and C. S. PANDE. Yield stress of nanocrystalline materials: Role of grain-boundary dislocations, triple junctions and coble creep. Phil. Mag. 84 (2004), 847.
63 Y. XU, A. ZANGVIL, and A. KERBERB. SiC nanoparticle-reinforced Al_2O_3 matrix composites: Role of intra- and intergranular particles. J. Eur. Ceram. Soc. 17 (1997), 921–928.
64 R. W. HERTZBERG. *Deformation and Fracture Mechanics of Engineering Materials*. Wiley, New York, 1898.
65 S. VEPREK, M. HAUSSMANN, S. REIPRICH, L. SHIZHI, and J. DIAN. Novel thermodynamically stable and oxidation resistant superhard coating materials. Surf. Coating Technol. 86/87 (1996), 394–401.
66 R. W. RICE. Proceedings: Strength and fracture of hot-pressed MgO. Br. Ceram. Soc. 20 (1972), 329–363.

67 R. W. Rice. Effect of hot extrusion, other constituents and temperature on the strength and fracture of polycrystalline MgO. J. Am. Ceram. Soc. 76(12) (1993), 3009–3018.
68 R. W. Rice, S. W. Freiman, and P. F. Becher. Grain-size dependence of fracture energy in ceramics: II, A model for noncubic materials. J. Am. Ceram. Soc. 64(6) (1981), 350–354.
69 H. Gao, B. Ji, I. L. Jäger, E. Arzt, and P. Fratzl. Materials become insensitive to flaws at nanoscale: Lessons from nature. Proc. Natl. Acad. Sci. U.S.A. 100 (2003), 5597–5600.
70 M. J. Mayo. Synthesis and applications of nanocrystalline ceramics. Mater. Des. 14(6) (1993), 323–329.
71 R. S. Mishra, C. E. Leshier, and A. K. Mukherjee. High pressure sintering of nanocrystalline Y-Al_2O_3. J. Am. Ceram. Soc. 79(11) (1996), 2989–2992.
72 R. Vaben and D. Stover. Properties of silicon-based ceramics produced by hot isostatic pressing ultrafine powders. Phil. Mag. B76(4) (1997), 585–591.
73 K. Niihara and A. Nakahira. Particulate strengthened oxide ceramics—nanocomposites, in *Advanced Structural Inorganic Composites*, P. Vincenzini (Ed.). Elsevier, Amsterdam, 1991, 637–644.
74 K. Niihara and A. Nakahira. Strengthening of oxide ceramics by SiC and Si_3N_4 dispersions. Proceedings of the Third International Symposium on Ceramic Materials and Components for Engines, American Ceramic Society, Westerville, OH, 1988, 919–926.
75 D. Galusek, J. Sedlacek, P. Svancarek, R. Riedel, R. Satet, and M. Hoffmann. The influence of post sintering HIP on the microstructure hardness, and indentation fracture toughness of polymer derived Al_2O_3–SiC nanocomposites. J. Eur. Ceram. Soc. 27 (2007), 1237–1245.
76 H. Awaji, S. M. Choi, and E. Yagi. Mechanisms of toughening and straightening in ceramic-based nanocomposites. Mech. Mater. 34 (2002), 411–422.
77 T. Ohji, Y. K. Jeong, Y. H. Choa, and K. Niihara. Strengthening and toughening mechanisms of ceramic nanocomposites. J. Am. Ceram. Soc. 81(6) (1998), 1453–1460.
78 I. Levin, W. D. Kaplan, and D. G. Brandon. Effect of SiC submicron particle size and content on fracture toughness of alumina-SiC nanocomposites. J. Am. Ceram. Soc. 78(1) (1995), 254–256.
79 J. Karch, R. Birringer, and H. Gleiter. Ceramics ductile at low temperature. Nature 330 (1987), 556–558.
80 W. D. Kingery, H. K. Bowen, and D. R. Uhlmann. *Introduction to Ceramics*. Wiley, 2nd ed. New York, 1976, 231–232.
81 U. Betz, G. Scipione, E. Bonetti, and H. Hahn. Low-temperature deformation behavior of nanocrystalline 5 mol% yttria stabilized zirconia under tensile stresses. Nanostructured Mater. 8(7) (1997), 845–853.
82 H. Hahn and R. S. Averback. Low-temperature creep of nanocrystalline titanium(IV) oxide. J. Am. Ceram. Soc. 74 (1991), 2918.
83 X. Xu, T. Nishimura, N. Hirosaki, R. J. Xie, Y. Yamamoto, and H. Tanaka. Superplastic deformation of nano-sized silicon nitride ceramics. Acta Mater. 54 (2006), 255–262.
84 M. J. Mayo. In *Superplasticity in Advanced Materials*, S. Hori, M. Tokizane, and N. Furushiro (Eds.). Japan Society for Research on Superplasticity, Osaka, 1991, 541.
85 M. Mitomo, H. Hirotsuru, and Y. Kim. Superplastic silicon carbide sintered body. U.S. Patent No. 5,591,685, 1995, 1997.
86 D. L. Burris and W. G. Sawyer. Improved wear resistance in alumina-PTFE nanocomposites with regular shaped nanoparticles. Wear 260 (2006), 915–918.
87 F. G. Mora, A. D. Rodriguez, J. L. Routbort, R. Chaim, and F. Guiberteau. Joining of yttria-tetragonal stabilized zirconia polycrystals using nanocrystals. Scr. Mater. 41(5) (1999), 455–460.
88 W. Liu and M. Naka. In situ joining of dissimilar nanocrystalline materials by spark plasma sintering. Scr. Mater. 48 (2003), 1225–1230.
89 T. J. Webster, C. Ergun, R. H. Doremus, R. W. Siegel, and R. Bizios. Enhanced functions of osteoblasts on nanophase ceramics. Biomaterials 21(17) (2000), 1803–1810.
90 L. Zhang, S. Sirivisoot, G. Balasundaram, and T. J. Webster. Nanomaterials for improved orthopedic and bone tissue engineering applications. In *Advanced Biomaterials: Fundamentals, Processing, and Applications*, ed. B. Basu, D. S. Katti, and A. Kumar. John Wiley & Sons, Hoboken, NJ, 2009, 205–244.

Chapter 17

Oxide Nanoceramic Composites

This chapter first reviews the recent developments in the area of Al_2O_3- and ZrO_2-based ceramic nanocomposites. As a case study, the optimization of processing and properties of ZrO_2 nanoceramics and ZrO_2–ZrB_2 nanoceramic composites are further discussed.

17.1 OVERVIEW

In efforts to confirm postulated superiority of ceramic nanomaterials in terms of better material properties, considerable efforts have been invested toward the processing and characterization of these materials. In particular, spark plasma sintering (SPS) has enabled successful synthesis of many bulk nanocrystalline ceramics and ceramic matrix composites (CMCs). In discussions that follow, the materials with grain sizes, of at least one of the phases, less than about 100 nm are referred to as nanocrystalline materials.

Among various oxide materials, 3 mol% yttria-doped ZrO_2 [ZrO_2(3Y)/3Y-TZP] can be sintered to near-theoretical density (ρ_{th}) by SPS at 1200°C under 30-MPa pressure from 27-nm-sized powders.[1] The grain sizes of the final compact were less than 90 nm (Fig. 17.1a) and the nanoceramics exhibited higher hardness (~14.5 GPa), in comparison with that of conventionally sintered monolithic ZrO_2 (~11 GPa). Importantly, the hardness of nanocrystalline zirconia is superior to those obtained by conventional composites of zirconia with 30 vol% harder reinforcements, such as ZrB_2,[2,3] TiB_2,[4] and WC.[5] In SPS experiments at 1100°C under 50 MPa pressure, Nygren and Shen[6] densified nanocrystalline powders (20 nm) of yttria-stabilized tetragonal zirconia to near-theoretical densification, while maintaining the final grain size below 100 nm. In another work, pressureless sintering at a lower temperature of 1150°C of yttria-stabilized tetragonal zirconia polycrystal (Y-TZP) powders, prepared via a colloidal technique, enabled retention of nanocrystalline grain size (~110 nm).[7] A maximum hardness of ~13 GPa and fracture toughness of ~14 MPa $m^{1/2}$ were obtained for 2Y-TZP. For comparison, it can be noted that conventional

Advanced Structural Ceramics, First Edition. Bikramjit Basu, Kantesh Balani.

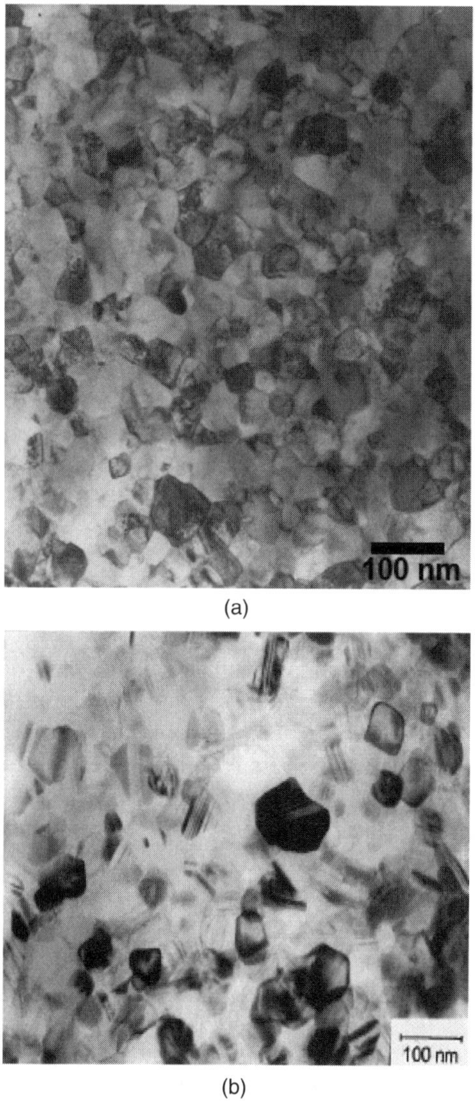

Figure 17.1 Bright field TEM images representative of various dense nanoceramics: (a) 3Y-TZP nanoceramic, spark plasma sintered at 1200°C for 5 minutes[1]; (b) nanoscaled grain size of SiC (~70 nm) after HIPing of SiC.[11]

submicron-grained (<1 μm) Y-TZP ceramics (hot pressed, 1450°C) exhibit a hardness of ~11 GPa and a maximum fracture toughness of ~10 GPa.[8]

Nanocrystalline α-Al_2O_3 with grain sizes of ~150 nm, developed via high-pressure sintering (~1 GPa, ~1000°C), exhibited excellent hardness of ~25 GPa, and this value is higher than that of conventional monolithic alumina (~20 GPa).[9] SPS was also used to develop nanocrystalline β-Si_3N_4 ceramics at 1600°C.[10] The grain size of the nanoceramic was ~68 nm, which resulted in high hardness of 18 GPa, better than conventional β-Si_3N_4 ceramics (~16 GPa). Sinter-forging has also been used to develop dense (~97% ρ_{th}) nanocrystalline SiC, with grain size of ~60 nm.[11] A bright field transmission electron microscopy (TEM) image revealing the finer-scale microstructure of the developed SiC nanoceramic is shown in Figure 17.1b.

From the preceding discussion, it is evident that development of bulk nanocrystalline ceramics, without second-phase reinforcement, can be successfully achieved via the advanced sintering techniques such as sinter–hot isostatic pressing (sinter-HIPing) and SPS. Also, judicious combination of novel powder preparation with sintering techniques helps in maintaining nanocrystalline grains in the densified products. As mentioned earlier, nanoceramics have distinct advantages in terms of superior hardness and strength. However, no significant improvement of fracture toughness has been reported in the nanocrystalline monoliths. To couple fracture toughness improvement with other benefits, nanocomposite design is recommended.[12] The following chapter (Chapter 18) shows how the careful optimization of processing and sinter-aid can lead to the development of WC-based ceramic nanocomposites with reasonably high fracture toughness.

17.2 Al_2O_3-BASED NANOCOMPOSITES

Compared with other oxide ceramics, the densification of monolithic alumina to near-theoretical density, with nanoscaled microstructure is difficult, even using some of the advanced sintering techniques. This is because alumina (in particular, γ-alumina) requires comparatively higher sintering temperature, owing to the sluggish densification of γ-alumina—and this can be attributed to the nucleation-controlled phase transformation from γ- to α-alumina. Such a phase transition also results in the formation of interlocking vermicular structure, which is detrimental to densification.[13,14]

With the advent of SPS, the densification of γ-alumina with better hardness has been possible at temperatures lower than that of the γ-to-α transformation temperature (1200°C).[15] However, any noticeable improvement of fracture toughness and fracture strength is not measured with nanocrystalline monolithic alumina. A summary of processing conditions is presented in Table 17.1. Zhan et al.[16] reported a modest fracture toughness of ~3.30 MPa m$^{1/2}$ for monolithic alumina (grain size ~350 nm). Still lower fracture toughness of ~3.03 MPa m$^{1/2}$, in nanocrystalline (~150 nm) alumina, has been measured by Mishra and Mukherjee.[13]

To obtain alumina-based ceramics with higher fracture strength and fracture toughness, the nanocomposite approach has been reported to be quite successful. Among various alumina nanocomposites, Al_2O_3–SiC nanocomposites, characterized by uniform dispersions of nano-SiC within a submicron Al_2O_3 matrix have been widely investigated.[12,17–21] Ever since the pioneering work of Niihara,[12] nano-SiC reinforcements in Al_2O_3 matrix have resulted in a significant increase in fracture strength (even >1 GPa, compared with ~350 MPa for monolithic Al_2O_3) along with a modest improvement in fracture toughness (~4 MPa m$^{1/2}$, compared with ~3 MPa m$^{1/2}$ for monolithic alumina). Crack deflection and crack bridging by nano-SiC reinforcements have been attributed to the improved fracture properties of the

Table 17.1. Mechanical Properties of Al_2O_3 Ceramic Nanocomposites Processed via SPS and Conventional Sintering Techniques[55]

System (material)	Processing conditions	Heating rate (K/min)	Grain sizes (nm)	Hardness (GPa)	Fracture toughness (K_{IC}) (MPa m$^{1/2}$)	Flexural strength (three-point bending) (MPa)	Remarks
Monolithic alumina	SPS at 1150–1300°C for 3 minutes	370	600	21–22	3.5 ± 0.5	–	Microstructure contained submicron grains and gave higher hardness than microcrystalline alumina.
Monolithic alumina	HPS	–	152	25	3.0	–	Nanocrystalline monolithic alumina showing lower fracture toughness.
γ-Al_2O_3 – (20 vol%) 3Y-TZP	SPS at 1100°C for 3 minutes	500	α-Al_2O_3: 96 ZrO_2: 265	15.2	8.9	–	Incorporation of near nanocrystalline zirconia results in high fracture toughness increment.
γ-Al_2O_3 – (20 vol%) SiC_w	SPS at 1125°C for 3 minutes	500	γ-Al_2O_3: 500	26.1	6.2	–	Nanocrystalline grains of alumina increases hardness, while whisker reinforcement improves fracture toughness.
Al_2O_3 – (5 vol%)SiC	SPS at 1450°C with no holding time	600	Nano-SiC	19.0	4.0	980	Intragranular nano-SiC dispersions led to tremendous increase in strength.

Al_2O_3 – (15 vol%) ZrO_2(3Y)–(5 vol%)SiC	SPS at 1450°C with no holding time	600	Nano-SiC	18.0	5.0	1200	Intragranular nano-SiC dispersions led to tremendous increase in strength.
Al_2O_3 – (9 vol%) $Nd_2Ti_2O_7$	SPS at 1100°C for 3 minutes	200	α-Al_2O_3: 281 and nanocrystalline $Nd_2Ti_2O_7$	–	5.7	–	Toughening due to nanoscaled piezoelectric reinforcements.
γ-Al_2O_3 – (7.5 vol%) $BaTiO_3$	SPS at 1150°C for 3 minutes	200	α-Al_2O_3: 190 and nanocrystalline $BaTiO_3$	–	5.3	–	Toughening due to nanoscaled ferroelectric reinforcements.
Al_2O_3 – (6 vol%) diamond	HPS at 900°C	–	Matrix grain sizes of ~105 nm	32	3.5	–	High hardness due to nanocrystalline grain size and harder reinforcements
Al_2O_3 – (10 vol%)Nb		–	40–50 nm matrix grain size	20–23	8		High fracture toughness due to incorporation of ductile metallic phase while maintaining high hardness due to nanocrystalline grain size

SPS, spark plasma sintering; HPS, high-pressure sintering.[53]

nanocomposites. The reduction in flaw size and density due to microstructural refinement were identified as another major reason for the enhanced resistance to fracture. Several research publications and review articles are available, demonstrating the improvement in mechanical properties of nano-SiC–reinforced alumina.[12,18,19,21,22] It should be noted that second-phase reinforcements can also lead to enhanced fracture toughness due to other toughening mechanisms, such as transformation toughening as for ZrO_2,[22] ferroelastic domain switching as for $BaTiO_3$,[23] and ZrO_2[24] and piezoelectric toughening for $Nd_2Ti_2O_7$.[16]

Zhan et al.[24] developed ZrO_2-toughened Al_2O_3 nanocomposite via an SPS route. The nanocomposite, densified at 1100°C (heating rate of 500°C/min), had a finer microstructure with alumina matrix grain size of 96 nm and the zirconia reinforcements having a size of 265 nm. A combination of mechanical properties, in particular, hardness of approximately 15.2 GPa and fracture toughness of around 8.9 MPa m$^{1/2}$ were recorded with the developed Al_2O_3–(20 vol%)ZrO_2 nanocomposite. The toughness increment was found to be due to ferroelastic domain switching of the ultrafine tetragonal zirconia (t-ZrO_2). It should be noted that, despite the presence of a significant amount of softer ZrO_2 phase, high hardness was measured as a direct consequence of the microstructural refinement down to nanocrystalline regime.

With respect to Al_2O_3–SiC nanocomposites, SPS of Al_2O_3–(5 vol%)SiC powders at 1450°C resulted in fully densified intragranular ceramic nanocomposite.[20,21] The nanocomposites exhibited superior strength of around 980 MPa, which is much higher than the 350 MPa normally measured with monolithic Al_2O_3. The property improvement is in consensus with earlier findings of Niihara, for hot pressed Al_2O_3–(5–10 vol%)SiC nanocomposites.[12] Gao et al. also fabricated Al_2O_3–SiC–ZrO_2(3Y) nanocomposites via SPS under similar conditions and obtained an even superior strength of ~1.2 GPa.[20] Despite the incorporation of ZrO_2, the nanocomposite exhibited a modest fracture toughness of around 4 MPa m$^{1/2}$ (slightly superior to that of monolithic alumina ~3.5 MPa m$^{1/2}$). However, high hardness of around 19 GPa similar to that of pure alumina was measured. A novel method for preparation of Al_2O_3–SiC nanocomposites utilizes infiltration of porous alumina matrix by polycarbosilane precursor, followed by subsequent sinter-HIPing.[25] Sintering at 1800°C, followed by HIPing at 1700°C, enabled near-theoretical densification of Al_2O_3–(x vol%)SiC nanocomposites (x ~ 3–8). The nanocomposites possessed a toughness of ~5 MPa m$^{1/2}$, irrespective of the reinforcement content. An alumina-based nanocomposite, containing nanocrystalline $Ti(C_{0.7}N_{0.3})$ (30 wt%) and nano-SiC (5 wt%) as second phases, was also obtained by hot pressing at 1650°C (30 MPa).[26] The obtained nanocomposite exhibited high fracture toughness of ~8 MPa m$^{1/2}$, along with a considerably improved strength of ~800 MPa. In accordance with the available toughening models,[17,27] the change in fracture mode and obstruction of crack propagation by the nanoreinforcements were identified as the possible reasons for the improved fracture properties.

Another processing approach to development of Al_2O_3–(18 vol%)SiC nanocomposites was reactive hot pressing of a powder mixture, containing mullite, aluminum, and carbon. TEM investigation found the presence of nanocrystalline SiC particles within the matrix alumina grains, while submicron SiC reinforcements were located in the intergranular regions (Fig. 17.2a). The nanocomposite displayed good strength of ~800 MPa with moderate fracture toughness of ~3.1 MPa m$^{1/2}$.[28]

Among various Al_2O_3-based nanocomposites, combination of high-energy ball milling (HEBM) and SPS (1480°C, 4 minutes) of powder mixture of Ti, graphite, and Al_2O_3 enabled to obtain Al_2O_3–(35 vol%)TiC nanocomposite (grain sizes: Al_2O_3,

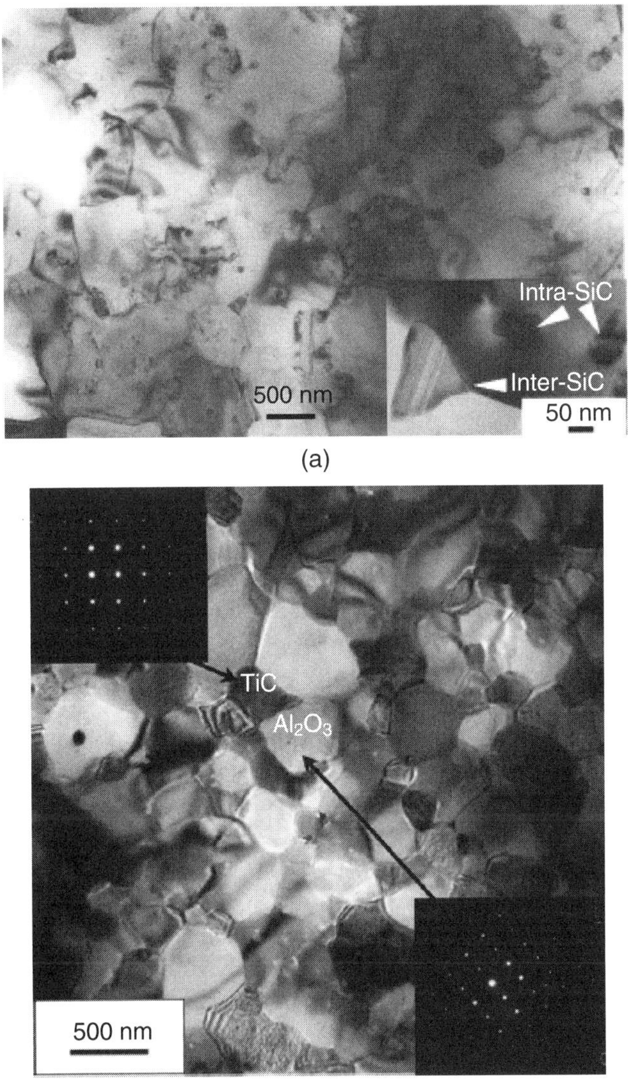

(b)

Figure 17.2 Microstructures of ceramic nanocomposites showing (a) TEM micrograph of alumina–silicon carbide nanocomposites consolidated via reactive hot pressing[20]; (b) TEM image and selected area diffraction patterns (SADPs) of Al_2O_3–TiC nanocomposite, developed *in situ* via a combination of high-energy ball milling (HEBM) and SPS.[29]

400 nm; and TiC, 200 nm; see Fig. 17.2b). It was reported that Al_2O_3–TiC *in situ* nanocomposites had a better fracture strength of ~950 MPa, compared with that of hot pressed Al_2O_3–TiC micron composites (~800 MPa). However, the hardness (~20 GPa) and fracture toughness (~4 MPa m$^{1/2}$) were comparable, irrespective of processing route.[29] The microstructural refinement resulted in enhancement of strength.

In a different approach the incorporation of nanoscaled piezoelectric ceramic phase in the alumina matrix has been attempted. Zhan et al.[16] developed Al_2O_3–$Nd_2Ti_2O_7$ nanocomposite via SPS at 1000–1150°C. The Al_2O_3–(9 vol%)$Nd_2Ti_2O_7$ nanocomposites possessed better fracture toughness of 5.7 ± 0.4 MPa m$^{1/2}$. The fracture toughness increment, vis-à-vis low fracture toughness of pure α-Al_2O_3 (3.30 ± 0.14 MPa m$^{1/2}$), was attributed to the conversion of mechanical energy to electrical energy during localized deformation in the presence of the piezoelectric second phase ($Nd_2Ti_2O_7$). In their follow-up work dense nanocomposite of γ-Al_2O_3–(7.5 vol%)$BaTiO_3$ was densified via SPS under similar conditions. The mean grain size of the alumina matrix was maintained at 190 nm, with the $BaTiO_3$ size below 100 nm. The fracture toughness was appreciably increased to 5.3 ± 0.3 MPa m$^{1/2}$ due to a domain switching effect.[30,31]

The Al_2O_3–(10 vol%)Nb nanocomposite was densified via high-pressure (2 GPa) sintering at 900°C.[32] Incorporation of a metallic phase (Nb) resulted in attaining higher fracture toughness (~8 MPa m$^{1/2}$). Interestingly, the incorporation of a metallic phase did not reduce the excellent hardness of alumina ceramics and the composite possessed a hardness of 20–23 GPa. This superior mechanical property was attributed to the nanocrystalline (~40 nm) matrix grains. Mishra and Mukherjee[32] also reported the preparation of Al_2O_3–(6 vol%)diamond nanocomposite (~100-nm matrix grains), which possessed high hardness of 32 GPa.

It can be concluded from the literature on processing–microstructure–property relationships of alumina-based nanocomposites that the advent SPS has triggered and assisted the development of alumina-based ultra-fine-grained nanocomposites. The most significant advance has been a threefold increase in fracture toughness. Furthermore, the incorporation of a softer phase to enhance fracture toughness has been achieved without compromising on hardness. A comparison of the mechanical properties of the various alumina-based nanoceramics and nanocomposites with conventional alumina is summarized in Figure 17.3. On critical observation of Figure 17.3, the zenith in strength can be seen at 5 vol% SiC reinforcement. The nanocomposites always exhibit higher strength compared with that of conventional alumina ceramics. A similar observation, with respect to only SiC reinforcement, was originally made by Niihara.[12] Furthermore, Sun et al. reported 5 vol% SiC nanoparticles as the optimum reinforcement to attain maximum strengthening of the Al_2O_3–SiC nanocomposite.[21] Similar behavior, with respect to fracture toughness, was predicted on the basis of a theoretical modeling study by Levin and co-workers.[33] Additionally, the strength reduction with increased additive content has been attributed to agglomeration effects. However, it must be emphasized here that no universally accepted reason has emerged to explain this variation of fracture property with reinforcement content.

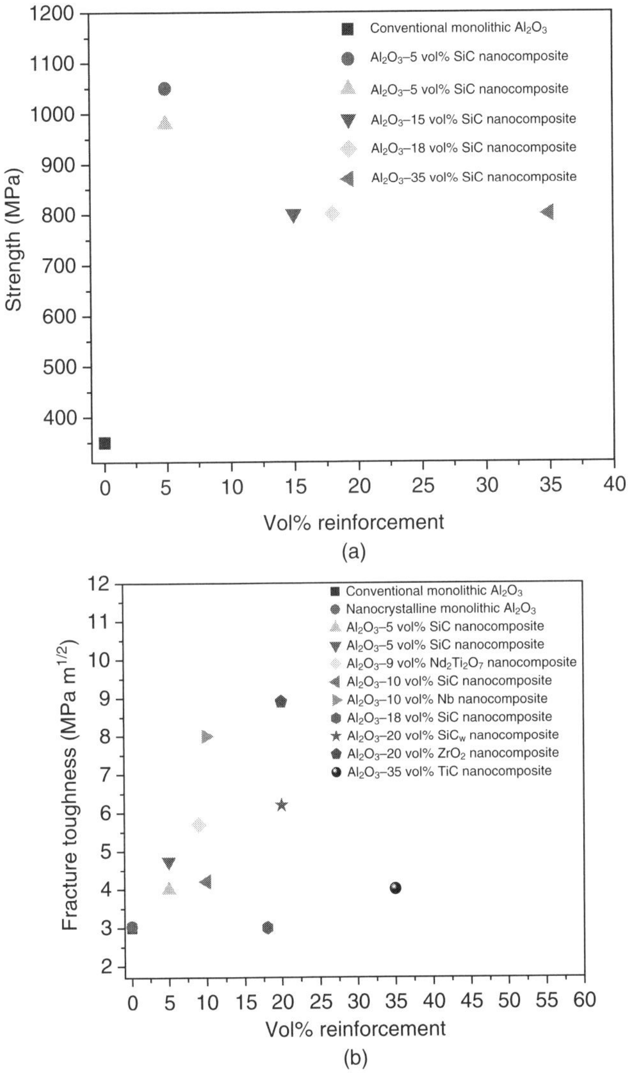

Figure 17.3 Comparison of mechanical properties of various nanoceramics and nanocomposites based on alumina.[55]

17.3 ZrO_2-BASED NANOCOMPOSITES

Among various structural ceramics, tetragonal zirconia polycrystals (TZPs), in view of high fracture toughness, have been widely investigated. The high fracture toughness of Y-TZP (2–20 MPa $m^{1/2}$) stems from the crack tip shielding by volume increments due to stress-induced transformation of metastable t-ZrO_2 to stable monoclinic zirconia (m-ZrO_2) in the crack tip stress field, a phenomenon known as transformation toughening.[22] In spite of having high fracture toughness and excellent strength (700–1200 MPa), conventional TZP monoliths possess lower hardness (10–11 GPa), compared with other structural ceramics, such as Si_3N_4 or Al_2O_3. In light of this limitation, the incorporation of hard reinforcements (such as ZrB_2, TiB_2) and, in particular, development of nanoceramics and nanocomposites based on TZP have been attempted.

Yoshimura et al.[34] sintered ZrO_2–Al_2O_3 nanocomposites using SPS route and the microstructure is characterized by nanosized (~50 nm) matrix (ZrO_2) grains. Jiang and co-workers[35] fabricated ZrO_2 (2Y)–(5–40 vol%)WC nanocomposites via hot pressing at 1450°C (1 hour). The nanocomposites (with 40 vol% nano-WC) possess an excellent combination of mechanical properties, such as strength of ~2 GPa, hardness of ~15 GPa and fracture toughness of ~9 MPa m$^{1/2}$. The strength and hardness values were better than conventional monolithic ZrO_2/ZrO_2–WC composites, containing similar volume percent of micron-sized WC.[5]

Hirvonen et al.[36] obtained zirconia–zircon ($ZrSiO_4$) nanocomposite by adopting the reaction between zirconia and cordierite via conventional pressureless sintering. The microstructure was characterized by submicron ZrO_2 grains (~200 nm) and an intergranular zircon glassy phase. An interesting observation was the significant lowering of thermal conductivity, when 15 vol% cordierite was used in the precursor powders. With 20 vol% cordierite, the composites sintered at 1500°C exhibited thermal conductivity even lower than 2 W/m·K. These results somehow indicate that the observed low thermal conductivity can be correlated with the ultrafine microstructure.

In attempts to develop machinable ZrO_2-based materials, Li et al.[37] fabricated ZrO_2 nanocomposites by dispersing BN particulates within ZrO_2 matrix. Their experiments demonstrated that ZrO_2(3Y)–BN nanocomposites can be sintered via hot pressing of nanocrystalline BN-coated ZrO_2 particles at 1400°C. Intergranular ceramic nanocomposites with dispersion of nanocrystalline h-BN at the grain boundaries of submicron ZrO_2 matrix was developed. It was observed that though the mechanical properties degraded to some extent with addition of BN (up to 30 vol%), nevertheless the properties were far superior to those of the microcomposites of corresponding compositions. For example, upon addition of 30 vol% BN, the nanocomposites exhibited a high strength of ~800 MPa and fracture toughness of ~8 MPa m$^{1/2}$, while the microcomposites of similar composition had moderate strength of ~500 MPa and lower fracture toughness of ~5 MPa m$^{1/2}$.

17.4 CASE STUDY

17.4.1 Yttria-Stabilized Tetragonal Zirconia Polycrystal Nanoceramics

This section summarizes the experimental results to illustrate how yttria-stabilized tetragonal zirconia polycrystal (Y-TZP) nanoceramics with high hardness (~14.5 GPa) can be obtained at lower sintering temperature of 1200°C using a SPS route. To optimize processing conditions, the commercially available 3 mol% yttria co-precipitated ZrO_2 (Tosoh grade TZ-3Y) powders were sintered at a high heating rate of around 600°C/min at different temperatures in the range of 1100–1300°C.

Looking at Table 17.2, the superiority of the SPS route over the conventional hot pressing (HP) route can be noticed. Li and Gao could achieve full density after SPS treatment (heating rate 600°C/min) at 1400°C for 3 minutes, while the same authors obtained full density at 1180°C with holding time of 9 minutes.[38] It should be mentioned that the specific advantage of the SPS process is not the general shift to a lower sintering temperature (1200°C), but shorter sintering time (5 minutes) due to faster densification mechanisms.

Bright field TEM image (Fig. 17.1) confirms the presence of finer tetragonal zirconia grains (70–80 nm). However, the presence of coarser t-ZrO_2 grains (110–130 nm) is also observed. Critically assessing the data presented in Reference 1 it is observed that the grain sizes of the pressureless sintered 3Y-TZP, densified at 1200°C for 2 hours, lie around 240–280 nm. The observation of much finer grain sizes (~100 nm) can be attributed to extremely high heating rate (650°C/min) and lower holding time (5 minutes). Furthermore, considering the crystal size of the starting powder (27 nm), Figure 17.1 clearly indicates that full densification without promoting significant grain growth is feasible in the SPS route.

Observing the data presented in Table 17.1, considerably higher hardness of around 14.5 GPa is achieved in SPS-processed TZP nanoceramic in contrast to lower hardness (12 GPa) in hot pressed Y-TZP. It can be noted that the hardness of Y-TZP ceramics in the range of 12–13 GPa is commonly reported in the literature.[8] Figure 17.4 plots the hardness versus grain size for monolithic ZrO_2 (3Y) as well as for ZrO_2–ZrB_2 composites. Based on the preceding discussion and Figure 17.4, it can be concluded that the drawback of ZrO_2, in terms of lower hardness, can be circumvented by grain refinement down to nanocrystalline level. The fracture toughness data are summarized in Table 17.2 and plotted in Figure 17.5. The difference in toughness data can be interpreted in terms of the difference in ZrO_2 matrix stabilization. A decrease in overall yttria content from 3 to 2 mol% considerably enhances the fracture toughness from 6.9 to 10.0 MPa m$^{1/2}$.

17.4.2 ZrO_2–ZrB_2 Nanoceramic Composites

The experimental results obtained with spark plasma sintered ZrO_2–ZrB_2 nanocomposites are presented in this section, to address some specific issues, including (1) whether ZrO_2-based composites with ZrB_2 reinforcements can be densified under the optimal SPS conditions for TZP, (2) whether improved hardness can be obtained in the composites when 30 vol% ZrB_2 is incorporated, and (3) whether the toughness can be tailored by varying the ZrO_2-matrix stabilization as well as retaining finer ZrO_2 grains. To develop ZrO_2–ZrB_2 nanoceramic composites, the SPS experiments were performed under optimal densification conditions for 3Y-ZrO_2 (1200°C, 5 minutes); 30 vol% 5 ZrB_2 was used as reinforcement and all ZrO-based composites

Table 17.2. Mechanical Properties of ZrO_2 Ceramic Nanocomposites Processed via SPS and Conventional Sintering Techniques (Taken from Reference 55)

System (material)	Processing conditions	Heating rate (K/min)	Grain sizes (nm)	Hardness (GPa)	Fracture toughness (K_{IC}) ($MPa\,m^{1/2}$)	Flexural strength (three-point bending) (MPa)	Remarks
ZrO_2 (3Y)	SPS at 1200°C for 5 minutes	600	70–80	14.5	–	–	High hardness due to nanocrystalline microstructure
ZrO_2–(30 vol%) ZrB_2	SPS at 1200°C for 5 minutes	600	ZrO_2: 100–300 ZrB_2: 2000–3000	14	9.9	–	Presence of nanosized matrix grains and hard reinforcements improve hardness
ZrO_2–(40 vol%) WC			Nanocrystalline WC, in matrix of microcrystalline ZrO_2	14.8	9.4	2	Improved properties with respect to similar compositions containing microcrystalline WC

SPS, spark plasma sintering.

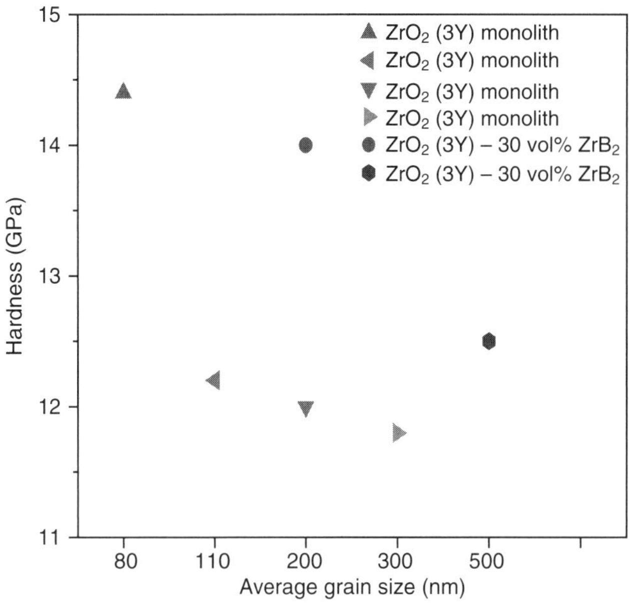

Figure 17.4 Variation of hardness of monolithic ZrO_2 (3Y) as well as ZrO_2–(30 vol%)ZrB_2 composites as a function of grain size. Notice the significant hardening with grain size reduction below 100 nm.[55]

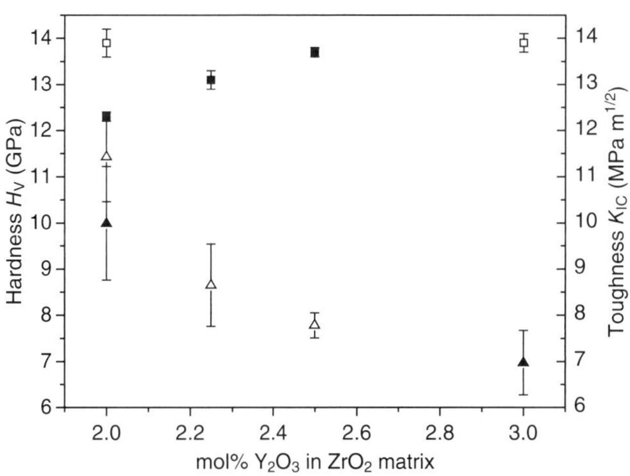

Figure 17.5 Variation of hardness and indentation toughness vs. yttria stabilization ZrO_2 matrix for spark plasma sintered ZrO_2–(30%)ZrB_2 nanocomposites. Different symbols represent the mechanical property measured with various composites based on ZrO_2 matrix processed from either co-precipitated or mixed grade ZrO_2 powders: ■, H_{v10} of the mixed grades; □, H_{v10} of the co-precipitated grades; ▲, K_{IC} of the co-precipitated grades (using Palmqvist formulae); △, K_{IC} of the mixed grades (using Palmqvist formulae).[39]

Figure 17.6 Scanning electron microscopy (SEM) fractograph of ZrO_2–(30 vol%)ZrB_2 nanocomposites developed via SPS.[39]

were sintered to more than 97% ρ_{th} after SPS at 1200°C for 5 minutes. The results are presented in detail elsewhere.[39]

The microstructures of the SPS-processed composites were characterized by coarser tabular–elongated ZrB_2 particles (~2–3 μm) and the equiaxed nano-ZrO_2 matrix grains (~100–200 nm). In view of such a phase assemblage, these composites have been referred to as nano/microcrystalline composites (Fig. 17.6). The ZrO_2–ZrB_2 nanocomposites dispalyed superior hardness of ~14 GPa, with the fracture toughness varying between 4 and 10 MPa m$^{1/2}$, depending on Y_2O_3 stabilization. Transformation toughening was identified as the main toughening mechanism, along with limited contribution from crack deflection by hard ZrB_2 reinforcements.

The mechanical property data, presented in Figure 17.4 and Table 17.2, reveal some interesting facts concerning the mechanical behavior of the developed zirconia nanoceramic composites. The *E*-modulus data, as presented in Table 17.2, reveal that the addition of ZrB_2 increased modulus values up to ~266 GPa. As far as hardness is concerned (see Table 17.2), SPS-processed nanoceramic composites are characterized by moderate hardness varying in the range of 12–14 GPa. Importantly, no observable hardness improvement was recorded, despite incorporating 30 vol% of harder ZrB_2 particulates. Such observations can be rationalized on the basis of similar observations made in other ceramic systems.[40–42] In an earlier study, Green measured lower hardness of 10–12 GPa in Al_2O_3–(10 vol%)ZrO_2[40] composites. Such an observation was attributed to the presence of microcracks, formed due to post-fabrication cooling. In a different work, Chamberlain et al.[41] obtained lower hardness of 14.5 GPa in a pressureless sintered (2150°C, 9 hours, helium atmosphere) monolithic ZrB_2. In contrast, much higher hardness of 23 GPa was recorded with hot pressed monolithic ZrB_2 (1900°C, 45 minutes, vacuum). Such a significant difference in hardness was attributed to differences in ZrB_2 sizes, which are larger in pressureless sintered material (average size ~9.1 μm). Similar observation was also made by Lee and Speyer,[42] who recorded a decrease in hardness from 18.3 GPa to 16.5 GPa, as a result of increase in grain size from 2.2 μm to 3.1 μm, respectively, in the case of pressureless sintered (90% theoretical density) monolithic B_4C. From the mechanistic point of view, the frequency with which the movement of dislocations is impeded will decrease with increase in grain size, and this can lower the stress required for deformation.[41,42]

The absence of any observable hardness increment in SPSed composites can additionally be attributed to both the microcracking effect and coarser particle size of ZrB_2. In the developed ZrO_2–ZrB_2 materials, the thermal residual stress is developed due to two factors, as discussed in earlier chapters.[43,44] Because of the tensile residual stress in ZrO_2, microcracking can occur either during postfabrication cooling or in the indented region. Besides the possible contribution from microcracking, the coarser ZrB_2 particles, like in other ceramic systems,[45–47] can cause lowering in hardness of ZrO_2–ZrB_2 nanoceramic composites. As far as the toughness measurement is concerned, the toughness of brittle materials is reported to be dependent on testing techniques,[45–51] which are widely classified into long crack and short crack methods. In estimating the toughness values of ZrO_2–ZrB_2 nanocomposites, various indentation fracture mechanics models are used:

a) Anstis's model[48] for radial–median cracks ($l/a > 2.5$),

$$K_{IC} = \eta(E/H)^{1/2} P/c^{3/2}, \tag{17.1}$$

where E is the Elastic modulus, H is the Vickers hardness, P is the indent load, and c is half of the crack length;

b) the toughness equation for Palmqvist-type cracks ($0.25 < l/a < 2.5$),[47]

$$K_{IC} = \eta(E/H)^{2/5} P/(al^{1/2}), \tag{17.2}$$

where $2a$ is the average indent diagonal length, $2c$ is the crack length, and $l = c - a$.

Kaliszewski and co-workers[50] have reported that an appropriate value of $\eta = 0.019$ can be used. In an earlier work, Niihara et al.[51] experimentally determined the value of η to be 0.0089 and 0.0122 for l/a ratios varying in the range of 0.25–2.5 and 1–2.5, respectively. Using $\eta = 0.0089$, the toughness of ZrO_2–ZrB_2 nanocomposites was estimated. A decrease in overall Y-content from 3 to 2 mol% increases the toughness from 6.9 to 11.4 MPa m$^{1/2}$. Because of a greater increase in toughness by decreasing Y-content from 2.5 to 2.0 mol% rather than by decreasing in Y-content from 3.0 to 2.5 mol%, one composite with ZrO_2 matrix having 2.25 mol% Y-stabilization was also developed. The toughness estimation for the TM2.25B composite indicated a moderate toughness improvement as revealed by K_{IC} data: 8.7 MPa m$^{1/2}$ for TM2.25B grade compared with 7.8 MPa m$^{1/2}$ for TM2.5B grade. Overall, the toughness data indicate that the toughness can be tailored by carefully optimizing the Y-stabilization of the ZrO_2 matrix.

Regarding the microstructure–toughness relationship, the effect of grain size on toughness should be considered. In SPSed ZrO_2–ZrB_2 nanocomposites, finer t-ZrO_2

grain sizes (100–300 nm) are retained and such finer grains are also able to transform in the crack tip stress field, leading to higher transformation toughness in the nanocomposites. Such a finer zirconia microstructure (in the case of Y-TZP monoliths) should be extremely stable and would not undergo transformation at the crack tip, according to the earlier literature reports.[8,13,52,53]

Another important factor, apart from the yttria content and grain size, is the residual stress. Because of the mismatch of the coefficient of thermal expansion, as mentioned earlier, ZrO_2 grains are subjected to residual tension and, therefore, the critical stress needed to induce the transformation of ZrO_2 grains in the crack tip would be lower according to the established literature.[51–53] This increases the transformability of t-ZrO_2.

The crack path–microstructure interaction in the SPSed nanocomposites is shown in Figure 17.7. Figure 17.7 reveals the increase in crack path tortuosity (more like a sinusoidal crack path) due to crack deflection by ZrB_2 particles. Closer observation of Figure 17.7 also indicates the crack wake debonding of the ZrB_2, around the ZrO_2/ZrB_2 interface. All these observations indicate that both crack deflection and crack wake debonding also contribute to achieving high indentation toughness in SPS-processed ZrO_2–ZrB_2 nanocomposites.

Summarizing, near-theoretical density of the yttria-stabilized tetragonal zirconia nanocomposites, reinforced with 30 vol% of ZrB_2 via an SPS route can be achieved at 1200°C with a holding time of 5 minutes and heating rate of 600 K/min. The sintering temperature is 200–250°C less and total processing times (~20–25 minutes) are considerably less, compared with established densification processes such as hot

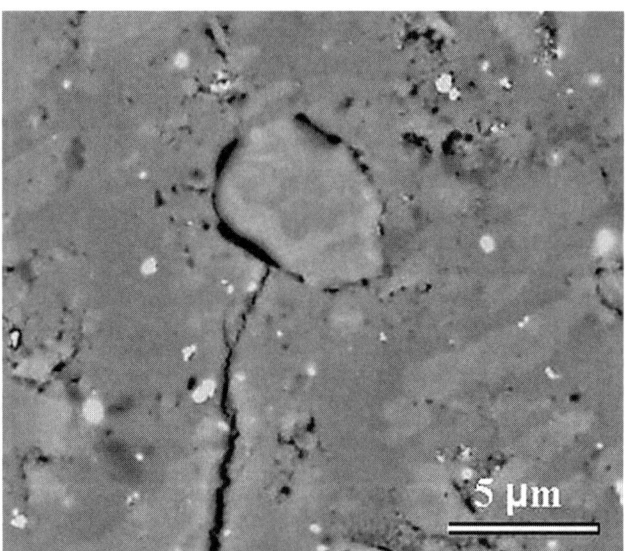

Figure 17.7 SEM image revealing crack-wake debonding of the coarser ZrB_2 particles at the interface in T2B nanocomposite, SPSed at 1200°C for 5 minutes (heating rate 600 K/min).[39]

pressing or sinter-HIPing. The nanocomposite materials exhibit finer microstructure, with ZrB_2 sizes of 2–3 μm and ZrO_2 in the range of 100–300 nm. The hardness of the developed composites remains moderate, varying in the range of 12–14 GPa. Careful use of indentation data provides an estimate of toughness in the range of 6–11 MPa $m^{1/2}$, depending on the Y_2O_3 stabilization level of ZrO_2 matrix. An important observation has been that transformation toughening with a ZrO_2 matrix having finer tetragonal grains (100–300 nm) occurs in ZrB_2-containing nanocomposites.

REFERENCES

1 B. Basu, J.-H. Lee, and D.-Y. Kim. Development of nanocrystalline wear resistant Y-TZP ceramics. J. Am. Ceram. Soc. 87(9) (2004), 1771–1774.

2 B. Basu, J. Vleugels, and O. Van Der Biest. Development of ZrO_2-ZrB_2 composites. J. Alloys Comp. 334(1–2) (2002), 200–204.

3 A. Mukhopadhyay, B. Basu, S. Das Bakshi, and S. K. Mishra. Pressureless sintering of ZrO_2-ZrB_2 composites: Microstructure and properties. Int. J. Refract. Metals Hard Mater. 25 (2007), 179.

4 B. Basu, J. Vleugels, and O. Van Der Biest. Development of ZrO_2-TiB_2 composites: Role of residual stress and starting powders. J. Alloys Comp. 365(1–2) (2004), 266–270.

5 G. Anne, S. Put, K. Vanmeensel, D. Jiang, J. Vleugels, and O. VanderBiest. Hard, tough and strong ZrO_2-WC composites from nanosized powders. J. Eur. Ceram. Soc. 25 (2005), 55–63.

6 M. Nygren and Z. Shen. On the preparation of bio-, nano- and structural ceramics and composites by spark plasma sintering. Solid State Sci. 5 (2003), 125–131.

7 O. Vasylkiv, Y. Sakka, and V. V. Skorokhod. Low temperature processing and mechanical properties of zirconia and zirconia alumina nanoceramics. J. Am. Ceram. Soc. 86(2) (2003), 299–304.

8 B. Basu, J. Vleugels, and O. Van Der Biest. Microstructure-toughness-wear relationship of tetragonal zirconia ceramics. J. Eur. Cer. Soc. 24(7) (2004), 2031–2040.

9 R. S. Mishra, C. E. Leshier, and A. K. Mukherjee. High pressure sintering of nanocrystalline y-Al_2O_3. J. Am. Ceram. Soc. 79(11) (1996), 2989–2992.

10 X. Xu, T. Nishimura, N. Hirosaki, R. J. Xie, Y. Zhu, Y. Yamamoto, and H. Tanaka. New strategies for preparing nanosized silicon nitride ceramics. J. Am. Ceram. Soc. 88(4) (2005), 934–937.

11 R. Vaben and D. Stover. Processing and properties of nanophase non-oxide ceramics. J. Mater. Process. Tech. 92–93 (1999), 77–84.

12 K. Niihara. New design concept for structural ceramics–ceramic nanocomposites. J. Ceram. Soc. Jpn. 99(10) (1991), 974–982. The Centennial Memorial Issue.

13 R. S. Mishra and A. K. Mukherjee. Electric pulse assisted rapid consolidation of ultrafine grained alumina matrix composites. Mater. Sci. Eng. A 287 (2000), 178–182.

14 F. W. Dynys and J. W. Halloran. Alpha alumina formation in alum-derived gamma alumina. J. Am. Ceram. Soc. 65 (1982), 442–448.

15 G. D. Zhan, J. D. Kuntz, R. G. Duan, and A. K. Mukherjee. Spark-plasma sintering of silicon carbide whiskers (SiCw) reinforced nanocrystalline alumina. J. Am. Ceram. Soc. 87(12) (2004), 2297–2300.

16 G. D. Zhan, J. Kuntz, J. Wan, J. Garay, and A. K. Mukherjee. Alumina-based nanocomposites consolidated by spark plasma sintering [J]. Scr. Mater. 47 (2002), 737–741.

17 M. Sternitzke. ChemInform abstract: Structural ceramic nanocomposites. J. Eur. Ceram. Soc. 17 (1997), 1061–1082.

18 T. Ohji, Y. K. Jeong, Y. H. Choa, and K. Niihara. Strengthening and toughening mechanisms of ceramic nanocomposites. J. Am. Ceram. Soc. 81(6) (1998), 1453–1460.

19 J. L. O. Merino and R. I. Todd. Relationship between wear rate, surface pullout and microstructure during abrasive wear of alumina and alumina/SiC nanocomposites. Acta Mater. 53(12) (2005), 3345–3357.

20 L. Gao, H. Z. Wang, J. S. Hong, H. Miyamoto, K. Miyamoto, Y. Nishikawa, and S. D. D. L. Torre. Mechanical properties and microstructure of nano-SiC-Al_2O_3 composites densified by spark plasma sintering. J. Eur. Ceram. Soc. 19 (1999), 609–613.
21 X. Sun, J. G. Li, S. Guo, Z. Xiu, K. Duan, and X. Z. Hu. Intragranular particle residual stress strengthening of Al_2O_3–SiC nanocomposites. J. Am. Ceram. Soc. 88(6) (2005), 1536–1543.
22 B. Basu. Toughening of Y-stabilized tetragonal zirconia ceramics. Int. Mater. Rev. 50(4) (2005), 239–256.
23 G. D. Zhan, J. Kuntz, J. Wan, J. Garay, and A. K. Mukherjee. Spark plasma sintered $BaTiO_3$/ $Al_2O_3Al_2O_3$ nanocomposites. Mater. Sci. Eng. A 356 (2003), 443–446.
24 G. Zhan, J. Kuntz, J. Wan, J. Garay, and A. K. Mukherjee. A novel processing route to develop a dense nanocrystalline alumina matrix (<100 nm) nanocomposite material. J. Am.Ceram.Soc. 86(1) (2003), 200–202.
25 D. Galuseka, J. Sedláčeka, P. Švančáreka, R. Riedel, R. Satet, and M. Hoffmann. The influence of post-sintering HIP on the microstructure, hardness, and indentation fracture toughness of polymer-derived Al_2O_3–SiC nanocomposites. J. Eur. Ceram. Soc. 27(2–3) (2007), 1237–1245.
26 H. Liu, C. Huang, J. Wang, and X. Teng. Fabrication and mechanical properties of Al_2O_3/Ti($C_{0.7}N_{0.3}$) nanocomposites. Mater. Res. Bull. 41 (2006), 1215–1224.
27 W. Zhang, M. Sui, Y. Zhou, Y. Zhong, and D. Li. Orientated nanometer-sized fragmentation of TiC particles by electropulsing. Adv. Eng. Mater. 4 (2002), 697–700.
28 G. J. Zhang, J. F. Yang, M. Ando, and T. Ohji. Reactive hot pressing of alumina-silicon carbide nanocomposites. J. Am. Ceram. Soc. 87 (2004), 299–301.
29 Y. Zhang, L. Wang, W. Jiang, L. Chen, and G. Bai. Microstructure and properties of Al_2O_3–TiC nanocomposites fabricated by spark plasma sintering from high-energy ball milled reactants. J. Eur. Ceram. Soc. 26 (2006), 3393–3397.
30 G. Winfield, F. Azough, and R. Freer. DiP224: Neodymium titanate ($Nd_2Ti_2O_7$)ceramics. Ferroelectrics 133 (1992), 181–186.
31 G. H. Haertling. Ferroelectric ceramics: History and technology. J. Am. Ceram. Soc. 82 (1999), 797–818.
32 R. S. Mishra and A. K. Mukherjee. Processing of high hardness-high toughness alumina matrix nanocomposites. Mater. Sci. Eng. A 301 (2001), 97–101.
33 I. Levin, W. D. Kaplan, and D. G. Brandon. Effect of SiC submicrometer particle size and content on fracture toughness of alumina-SiC nanocomposites. J. Am. Ceram. Soc. 78(1) (1995), 254–256.
34 M. Yoshimura, M. Sando, Y. H. Choa, T. Sekino, and K. Niihara. Fabrication of dense ZrO_2-based nano/nano type composites by new powder preparation method and controlled sintering process. Key Eng. Mater. 161–163 (1999), 423–426.
35 D. Jiang, O. VanderBiest, and J. Vleugels. ZrO_2–WC nanocomposites with superior properties. J. Eur. Ceram. Soc. 27(2–3) (2007), 1247–1251.
36 A. Hirvonen, R. Nowak, Y. Yamamoto, T. Sekino, and K. Niihara. Fabrication, structure, mechanical and thermal properties of zirconia-based ceramic nanocomposites. J. Eur. Ceram. Soc. 26(8) (2005), 1497–1505.
37 Y. Li, J. Zhang, G. Qiao, and Z. Jin. Fabrication and properties of machinable 3Y–ZrO_2/BN nanocomposites. Mater. Sci. Eng. A 397 (2005), 35–40.
38 W. Li and L. Gao. Rapid sintering of nanocrystalline ZrO_2 (3Y) by spark plasma sintering. J. Eur. Ceram. Soc. 20 (2000), 2441.
39 B. Basu, T. Venkateswaran, and D.-Y. Kim. Microstructure and properties of spark plasma sintered ZrO_2-ZrB_2 nanoceramic composites. J. Am. Ceram. Soc. 89(8) (2006), 2405–2412.
40 D. J. Green. Critical Microstructures for Microcracking in Al_2O_3-ZrO_2 composites. J. Am. Ceram. Soc. 65(12) (1982), 610–614.
41 L. Chamberlain, W. C. Fahrenholtz, and G. E. Hilmans. Pressureless sintering of zirconium diboride. J. Am. Ceram. Soc. 89(2) (2006), 450–456.
42 H. Lee and R. F. Speyer. Hardness and fracture toughness of pressureless-sintered boron carbide (B_4C). J. Am. Ceram. Soc. 85(5) (2002), 1291–1293.

43 S. Schmauder and H. Schubert. Significance of internal stress on the martensite transformation in yttria-stabilised tetragonal zirconia polycrystals during degradation. J. Am. Ceram. Soc. 69(7) (1986), 534–540.

44 R. N. L. Okamoto, M. Kusakaari, K. Tanaka, H. Inui, and S. Otani. Temperature dependence of thermal expansion and elastic constants of single crystals of ZrB_2 and the suitability of ZrB_2 as a substrate for GaN film. J. Appl. Phys. 93(1) (2003), 88–93.

45 D. K. Shetty, I. G. Wright, P. N. Mincer, and A. H. Clauer. Indentation fracture of WC-Co cermets. J. Mater. Sci. 20 (1985), 1873–1882.

46 P. Chantikul, G. R. Anstis, B. R. Lawn, and D. B. Marshall. A critical evaluation of indentation techniques for measuring fracture toughness. J. Am. Ceram. Soc. 64(9) (1981), 539–543.

47 S. Palmqvist. Occurrence of crack formation during Vickers indentation as a measure of the toughness of hard materials. Arch. Eisenhuettenwes. 33 (1962), 629–333.

48 G. R. Anstis, P. Chantikul, B. R. Lawn, and D. B. Marshall. A critical evaluation of indentation techniques for measuring fracture toughness. J. Am. Ceram. Soc. 64 (1981), 553–557.

49 B. R. Lawn. *Fracture of Brittle Solids*, 2nd ed. Cambridge University Press, Cambridge, UK, 1992.

50 M. S. Kaliszewski, G. Behrens, A. H. Heuer, M. C. Shaw, D. B. Marshall, G. W. Dransmann, and R. W. Steinbrech. Indentation studies on Y_2O_3-stabilized ZrO_2: I, development of indentation-induced cracks. J. Am. Ceram. Soc. 77 (1994), 1185–1193.

51 K. Niihara, R. Morena, and D. P. H. Hasselman. Evaluation of K_{IC} of brittle solids by the indentation method with low crack-to-indent ratios. J. Mater. Sci. Lett. 1 (1982), 13–16.

52 B. Budiansky, J. W. Hutchinson, and J. C. Lambropoulos. Continuum theory of dilatant transformation toughening in ceramics. Int. J. Solids Struct. 19 (1983), 337.

53 A. G. Evans. Perspective on the development of high-toughness ceramics. J. Am. Ceram. Soc. 73(2) (1990), 187–206.

54 A. G. Evans and R. M. Cannon. Toughening of brittle solids by martensitic transformations. Acta Mater. 34(5) (1986), 761–800.

55 A. Mukhopadhyay and B. Basu. Consolidation-microstructure-property relationships in bulk nanoceramics and ceramic nanocomposites: A review. Int. Mater. Rev. 52(4) (2007), 1–31.

Chapter 18

Microstructure Development and Properties of Non-Oxide Ceramic Nanocomposites

The improved as well as novel properties achievable with ceramic nanocomposites in reference to conventional materials have inspired significant research activities on the synthesis of nanoceramic composites in technologically important ceramic systems.[1–8] These nanocomposites can be broadly classified into two categories[9]: (1) nanocomposites, fabricated by dispersion of nanosized particles within micron-sized matrix grains or at the grain boundaries of the matrix, and (2) nano/nanocomposites, in which matrix grains are also in the nanosize scale. After a review of the development of Si_3N_4-based nanocomposites, reports on the development of other composites are briefly summarized. Importantly, this chapter discusses the development of the tungsten carbide (WC)-based nanocomposites with ZrO_2 and Co sinter-aids and shows how the composition of the WC–ZrO_2–Co system can be optimized to obtain a superior combination of strength and toughness.

18.1 NANOCOMPOSITES BASED ON Si_3N_4

In view of their excellent thermomechanical properties, Si_3N_4-based materials are potential candidates for high-temperature applications. However, it has been widely recognized that the presence of amorphous grain boundary phases, formed due to reaction with additives used for facilitating liquid-phase sintering, limit their high-temperature applicability to some extent. This limitation can be surpassed either by strength increments due to reduction of matrix grain size or by dispersion of ultrafine crystalline phase along the grain boundaries. In Table 18.1, the mechanical properties of various Si_3N_4-based nanocomposites are summarized.

Park et al.[10] reported the microstructure and properties of Si_3N_4–SiC nanocomposites by hot pressing a mixture of Si_3N_4 and β-SiC at 1800°C for 2 hours. They

Advanced Structural Ceramics, First Edition. Bikramjit Basu, Kantesh Balani.

Table 18.1. Mechanical Properties of Nanoceramics and Nanocomposites Processed via SPS and Conventional Sintering Techniques (Taken from Reference 79)

System (material)	Processing conditions	Heating rate (K/min)	Grain sizes (nm)	Hardness (GPa)	Fracture toughness, K_{IC} ($MPa\,m^{1/2}$)	Flexural strength (three-point bending) (MPa)	Remarks
Binderless WC	SPS at 1900°C		1×10^6	24		1000	Contains very large grains and exhibits extremely poor fracture toughness
Binderless WC	PAS at 1963 K for 0.18 Ks		25	23	4	–	Tremendous increase in hardness due to the presence of nanocrystalline grains
WC–(6 wt%) Co	PS	–	Microcrystalline (1–2 μm) WC and submicron (<1 μm) ZrO_2	16	14	–	Presence of metallic binder enhances fracture toughness but reduces hardness
WC–(10 wt%) Co	SPS at 1000°C for 510 minutes	100	Matrix grain size of <300 nm	18	12	–	Nanocrystalline grain size results in hardness enhancement
WC–(18 wt%) MgO	PAS at 1963 K for 0.18 Ks		MgO: 50 nm	15	14	–	Combination of high hardness (due to nanocrystalline microstucture) and fracture toughness (due to MgO)
Si_3N_4 – (20 vol%)SiC	HP at 1800°C for 2 hours	–	–	–	–	1150, maintained up to 1200°C	Intrergranular SiC dispersions inhibited grain growth and also led to enhanced strength

(Continued)

Table 18.1. *(Continued)*

System (material)	Processing conditions	Heating rate (K/min)	Grain sizes (nm)	Hardness (GPa)	Fracture toughness, K_{IC} (MPa m$^{1/2}$)	Flexural strength (three-point bending) (MPa)	Remarks
Si_3N_4–SiC	HP at 1800°C for 2 hours	–	–	–	–	1100 at RT and 800 at 1400°C	Intergranular SiC dispersions inhibited grain growth and this, along with crystallization of the grain boundary phase, led to higher high-temperature strength.
Si_3N_4 – (30 vol%)SiC	HP at 1800°C for 1 hours					1080 at 1400°C	Inter-/intragranular nanocrystalline (~100 nm) SiC dispersed in nearly equiaxed matrix (Si_3N_4) grains
Mullite–(5 vol%)SiC	SPS at 1500°C	200	Submicron mullite and nanocrystalline SiC	–	2.2–2.3	466	Presence of nano-SiC reinforcements led to strength increment from that of monolithic mullite (200–300 MPa)
YAG–(5 vol%) SiC	SPS at 1550°C for 1 minute	200	YAG matrix grains of ~1000 nm and nanocrystalline SiC	–	–	565	Nano-SiC reinforcements increased strength from that of monolithic YAG (348 MPa)
SiC–(30 vol%) TiC	SPS at 1800°C for 20 minutes	200–600°C, 150 from 600°C to 1700°C and 10 from 1700°C to 1800°C	Nanocrystalline TiC	–	6.2	646	Dispersions of nano-TiC led to transgranular fracture SiC grains, resulting in improvement of bend strength, while intergranular fracture of TiC resulted in the improvement of fracture toughness.

SPS, spark plasma sintering; HP, hot pressing; PS, pressureless sintering.

reported that heat treatment at 1500°C prior to the densification resulted in the development of a microstructure characterized by the presence of nanocrystalline SiC particles at the grain boundaries of the fine-grained Si_3N_4 matrix. The developed nanocomposite, with 20 vol% nano-SiC reinforcements, had high flexural strength (~1150 MPa) at room temperature, which was maintained up to 1200°C. Hot-pressed Si_3N_4–SiC nanocomposites, as developed by Oh et al.,[11] are characterized by intergranular and intragranular nanocrystalline SiC. Such nanocomposites also exhibited superior room-temperature strength of ~1100 MPa, and ~800 MPa at 1400°C. Hirano and Niihara[12] also reported that Si_3N_4–(30 vol%)SiC nanocomposites (hot pressed at 1850°C, 1 hour, 30 MPa) maintained a strength of ~1100 MPa at 1400°C, whereas for monolithic Si_3N_4, strength decreased with increasing temperature (~900 MPa at 1400°C). In a different work with hot-pressed monolithic Si_3N_4 and Si_3N_4–SiC nanocomposites, it was confirmed that the nanocomposites exhibit much improved strength and fracture toughness at room temperature as well as at elevated temperature.[13] Experimental results also indicate that the limitations imposed on the high-temperature strength by the amorphous grain boundary phase can be alleviated by reducing the additive content.[14,15] SPS enabled the densification of amorphous Si–C–N, with reduced additive content (~1 wt% yttria), at a nominal temperature of 1600°C.[14] Predominantly, solid-state sintering restricted grain growth of Si_3N_4 and SiC and resulted in a nano/nanoceramic composite (Fig. 18.1a).

To develop Si_3N_4-based ceramics with better machinability, h-BN is added as the second phase.[16,17] However, fracture strength decreases considerably with the addition of micron-sized BN. Thus, in order to attain a balance between mechanical properties and machinability, Si_3N_4–BN nanocomposites are being developed.[18–20] The surface roughness of the drilled surface of the nanocomposites (R_a ~ 0.37 μm) was smoother than that of the microcomposites (R_a ~ 0.65 μm). The strength of these nanocomposites (~800 MPa for 15 wt% BN) was found to be superior to that of conventional Si_3N_4–BN composites. Figure 18.1b presents a transmission electron microscopy (TEM) image of a machinable Si_3N_4–BN nanocomposite, with a high-resolution transmission electron microscopy (HRTEM) image of a Si_3N_4–BN grain boundary as an inset.

The preceding discussion signifies that the mechanical properties, in particular the strength, of ceramics based on Si_3N_4 can undergo potential improvement via nanocomposite design. In fact, Niihara,[9,12] while reviewing the property improvement of ceramics by nanocomposite design, reported that the strength of Si_3N_4-based materials can be improved from ~850 MPa, for monolithic Si_3N_4, to a high value of ~1500 MPa for Si_3N_4–SiC nanocomposites. Also, the machinability characteristics and wear resistance are enhanced for the nanocomposites. Nanocomposite formation is useful in terms of reduced requirement for sinter-additive and, concurrently, lower amount of amorphous grain boundary phase formation. Multiple research groups reported that the nanocomposites based on Si_3N_4 exhibit high strength retention at elevated temperature.[10–12,14,15,17–23] In view of the better high-temperature strength, the application temperature can be raised from a maximum of 1200°C for conventional Si_3N_4 ceramics to around 1400°C for the Si_3N_4-based nanocomposites.

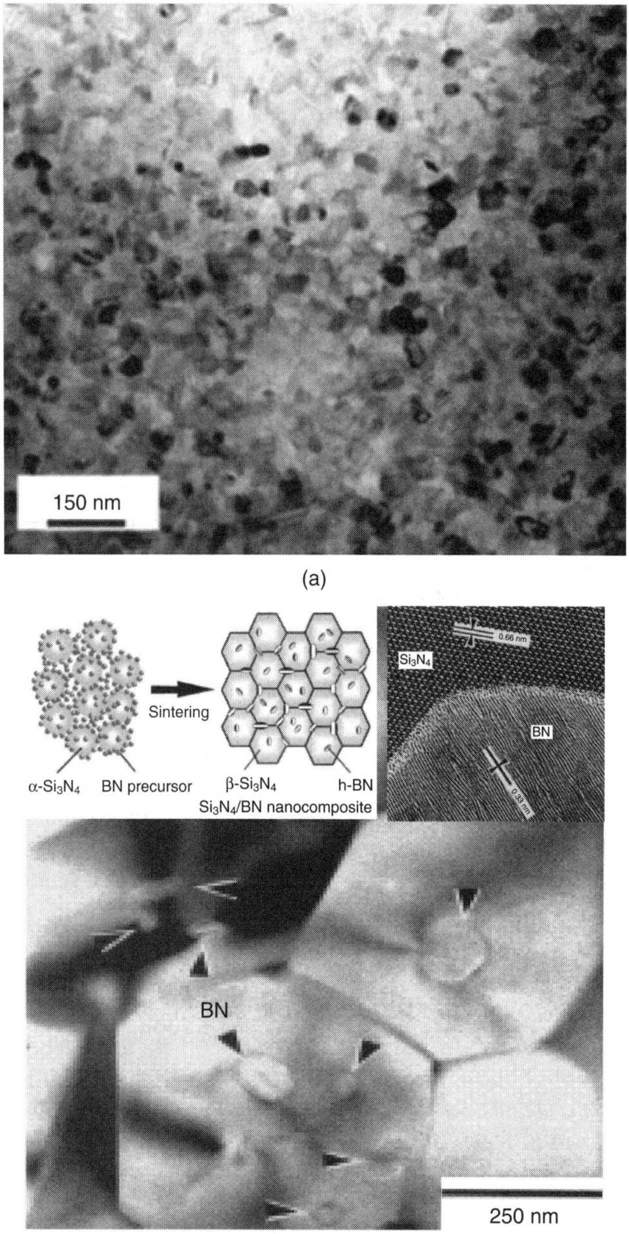

Figure 18.1 Microstructures of ceramic nanocomposites showing (a) TEM image of Si_3N_4/SiC nano/nanocomposite developed by SPS of amorphous Si–C–N powders (1 wt% yttria additive);[86]; (b) TEM micrograph (with insets showing schematic of starting powder distribution and HRTEM of the nanocomposite) of hot-pressed machinable Si_3N_4–(15 vol%)BN nanocomposite processed through the chemical route.[8]

18.2 OTHER ADVANCED NANOCOMPOSITES

Apart from the aforementioned nanocomposites, SPS and related techniques have been used to develop a variety of technically important ceramic materials having nanoscaled microstructures. A few such examples are cited in the following subsections.

18.2.1 Mullite–SiC

Gao et al.[21] obtained fully dense mullite–(5 and 10 vol%)SiC nanocomposites, after SPS at 1500°C, while conventional sintering of mullite requires consolidation temperatures of 1600–1700°C. The nano-SiC particles were found to be located within the matrix grain. The flexural strengths (466 and 454 MPa for 5 and 10 vol% SiC reinforcements, respectively) were better than those of the mullite monoliths (200–300 MPa). It was proposed that the residual stresses in nanocomposites can account for the strength increment. However, the fracture toughness (2.2–2.3 MPa $m^{1/2}$) remained low (see Table 18.1).

18.2.2 Yttrium Aluminum Garnet–SiC

The addition of nanosized SiC particles in yttrium aluminum garnet (YAG) has been reported to improve mechanical properties. For example, YAG–(5 vol%)SiC nanocomposites,[22] fully densified by SPS (1550°C), had high bending strength of 565 MPa, which is considerably higher than that of SPS-processed pure YAG (348 MPa) (see Table 18.1). The nanosized SiC dispersions were mostly located at the intragranular regions. The transgranular fracture mode was attributed to residual-stress strengthening of the grain boundaries.

18.2.3 SiC–TiC

The SiC-based nanocomposites, containing dispersions of nanosized 30 vol% TiC, were obtained with near-theoretical density via SPS at 1800°C.[23] The nanocomposites exhibited a good combination of fracture toughness (6.2 MPa $m^{1/2}$) and bend strength (~650 MPa) (see Table 18.1). Transgranular fracture has been reported to result in improvement of bend strength.

18.2.4 Hydroxyapatite–ZrO_2 Nanobiocomposites

In an attempt to develop bioactive nanocomposites, which potentially exhibit better biocompatibility than conventional biomaterials,[24–29] an yttria-stabilized tetragonal zirconia polycrystalline (Y-TZP) nanocomposite, 3Y-TZP–(40 vol%) hydroxyapatite (HAp), was sintered via SPS (1150°C, 3 minutes).[30] The dense nanocomposite microstructure was characterized by the presence of nanocrystalline ZrO_2 (~50 nm) and

HAp (~100 nm). The absence of tricalcium phosphate was attributed to the short sintering time in SPS.

18.2.5 Stress-Sensing Nanocomposites

The stress-sensing ability in structural ceramics has been demonstrated in materials having good magnetic and magnetomechanical properties. Nakayama et al.[31] developed MgO-based nanocomposites containing nanocrystalline Fe–FeCo using pulsed electric current sintering (1473°C, 5 minutes). Such nanocomposites displayed ferromagnetism and high coercive force (H_c), approximately two orders of magnitude higher than that for pure iron. Importantly, H_c was found to be dependent on grain size and dislocation density.[32,33] From a phenomenological point of view, with reduction in particle size, the characteristic magnetic structure varies from a multidomain to a single-domain state, thus reducing the total energy loss during a magnetization–demagnetization cycle. Thus, the nanocrystalline dispersoids in the MgO–Fe or MgO–FeCo resulted in extremely high values of H_c. Moreover, the inverse magnetostrictive response to the applied uniaxial stress can enable their use as stress sensors.

18.3 WC-BASED NANOCOMPOSITES

18.3.1 Background

Due to the attractive combination of high hardness, high elastic modulus, superior wear resistance, good thermal conductivity, and low thermal expansion coefficient, WC-based cermets are conventionally used as cutting tools, wear parts, tools, and dies. Bulk monolithic WC, despite possessing the aforementioned attractive properties, are not useful in technological applications, due to processing difficulties and inherent brittleness. The addition of cobalt has traditionally been used to improve sinterability and fracture resistance. Liquid-phase sintered cemented carbides, which are characterized by WC particles bonded in a Co/Ni matrix, derive their properties from the constituent phases—the brittle carbide and the ductile binder. It can be mentioned here that cemented carbides are commercially one of the oldest and most successful powder metallurgy products.

Conventional cemented carbides having composition WC–(6 wt%)Co exhibit a high fracture toughness of around 14 MPa m$^{1/2}$, which is significantly higher than the fracture toughness of monolithic WC (~4 MPa m$^{1/2}$). However, the hardness (~16 GPa) is lowered and this is attributed to the presence of the softer metal binder.[34] To obtain cemented carbides with a better combination of properties, attempts to develop nanocrystalline cemented carbides were made using a combination of spray conversion processing for nanoscaled powder synthesis[35] and subsequent consolidation via advanced sintering techniques. Grain growth was restricted by addition of grain growth inhibitors, such as VC–TaC, while densifying via conventional sintering technique. It was reported that a reduction in WC grain size to the ultrafine-nanoscale range enhanced the hardness and strength drastically. Additionally, the solid solution strengthening of the binder phase due to dissolution of W and C in Co binder resulted in further improvement of mechanical properties. Sivaprahasam et al.[36] developed WC–(12 wt%)Co cemented carbides with higher hardness (~15.2 GPa) via SPS, compared with conventional pressureless sintering (PS) (~13.8 GPa). Microstructural refinement was also reported by Cha et al.[37] in SPS (1000°C, 10 minutes, 50–100 MPa) processed WC–(10 wt%)Co nanomaterial, which possessed a hardness of ~18 GPa, while conventionally sintered (1100°C) similar compositions had a maximum hardness of ~16 GPa. A better hardness of more than 20 GPa for bulk WC–Co nanomaterials has also been recorded by Jia et al.[38] Interestingly, Richter and Ruthendorf[39] observed that the hardness of nanocrystalline WC-based ceramics can reach a value close to, or even exceeding, that of WC single crystal.

Despite excellent hardness, the fracture toughness of the nanocrystalline cemented carbides is inferior to that of cemented carbides with micron-sized grains. According to Richter and Ruthendorf,[39] the fracture toughness of ultrafine and nanocrystalline WC–Co remained as low as 5–6 MPa $m^{1/2}$. The inferior fracture toughness has been attributed to an increase in constraint to plastic deformation ahead of the propagating crack tip. It has even been hypothesized that the metallic Co phase actually behaves in a brittle manner below a critical mean free path, as mean free path in the binder phase is reduced with decrease in WC size. Another possible reason for lower fracture toughness is the reduction in the hcp/fcc (hexagonal close packed/face centered close packed) ratio of the binder phase due to increased concentration of solute (W,C).[40] This limits the toughening, which arises from hcp-to-fcc transformation of the Co binder phase. Compared with conventional cemented carbides, fracture toughness of nanocrystalline WC–Co is not affected to any observable extent by increasing hardness (decreasing grain size) (see Fig. 18.2). Thus, the increasing constraint to plastic deformation of the binder phase, with reduction of grain size in the nanocrystalline domain, does not influence the bulk fracture toughness. Jia et al.[38] observed that fracture toughness of nanocrystalline WC–Co could increase moderately with increase in bulk hardness (with grain size reduction). This variation of fracture toughness with nanocrystalline grain size indicates that ductile metal bridging, and not plastic deformation of the binder, is the dominant toughening nmechanism in nanostructured cermets.[28] In contrast to the findings of Jia et al.[38] and Ritcher and Ruthendorf,[39] a study by Kim et al.[41] revealed that the fracture toughness increases significantly with decreasing hardness, even for SPS-processed nanocrystalline/ultrafine WC–Co. In the same work,[41] a fracture toughness of ~12 MPa $m^{1/2}$ was measured for WC–(10 wt%)Co (WC grain size ~ 350 nm), which is superior to that of 6 MPa $m^{1/2}$ reported by Ritcher and Ruthendorf[39] for the same composition with similar (~400 nm) grain sizes. Additionally, in the work of Michalski et al.,[42] pulse plasma sintering (PPS)-processed WC–(12 wt%)Co nanomaterials (~50 nm) exhibited a superior fracture toughness of ~15 MPa $m^{1/2}$. At this stage, no accepted theory exists to explain these discrepancies in mechanical behaviour of bulk WC–Co nanomaterials. It has also been stated that, by increasing the binder content in nanocrystalline WC–Co, fracture toughness can be achieved near to that of conventional cemented carbides, without compromising on hardness owing to the grain size refinement.

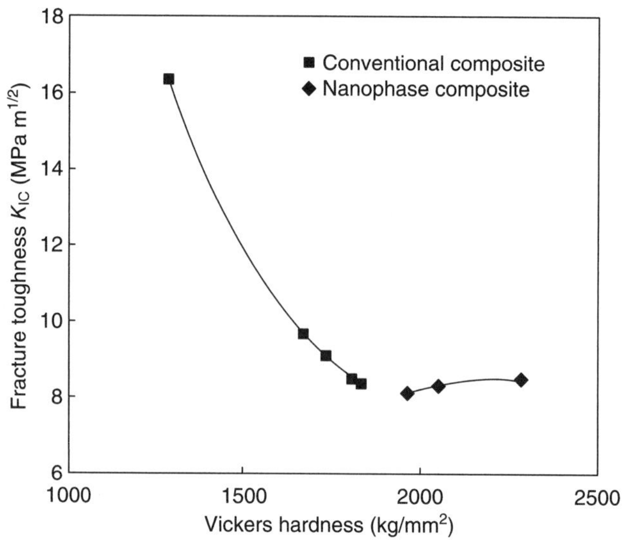

Figure 18.2 Variation of indentation fracture toughness with Vickers hardness for conventional and nanocrystalline WC–Co cemented carbides.[79]

Eskandarany[43] sintered WC–(18 wt%)MgO nanocomposites using plasma-assisted sintering (PAS) at 1690°C. The microstructure was characterized by dispersion of MgO (<50 nm) in nanocrystalline WC matrix. This nanocomposite exhibited a high fracture toughness of around 14 MPa m$^{1/2}$, along with moderate hardness (15 GPa) and elastic modulus (413 GPa). The fracture toughness is much better than that of binderless nanocrystalline WC (4 MPa m$^{1/2}$), synthesized using the same technique (see Table 18.1). The hardness achieved with various WC-based nanoceramics and nanocomposites is presented in Figure 18.3. Importantly, the presence of relatively softer MgO does not degrade the hardness property. Thus, the developed WC–MgO nanocomposite can potentially replace conventional WC-based cermets in many existing as well as futuristic applications.

Summarizing the benefits of WC-based nanocomposites, the following points emerge: (1) nanocrystalline WC–Co cemented carbides exhibit much improved hardness over their conventional counterparts, which should impact better efficiency in cutting and wear-resistant applications; (2) the replacement of metallic additives by ceramic additives has been possible without any significant reduction in the hardness compared with conventional WC–Co cemented carbides.

It can be reiterated here that cermets have been extensively used for decades in various engineering applications, such as cutting tools, rock drill tips, tools, and die, as well as general wear parts.[35–38,40,44–54] For further performance improvement in structural and tribological applications, the last few decades have witnessed an increasing surge toward the development of WC–Co cermets with nanoscale microstructure. Furthermore, with the advent of spark plasma sintering (SPS), the development of dense WC-based cermets possessing submicron–nanosized WC grains has been pursued in various research laboratories.[8–13]

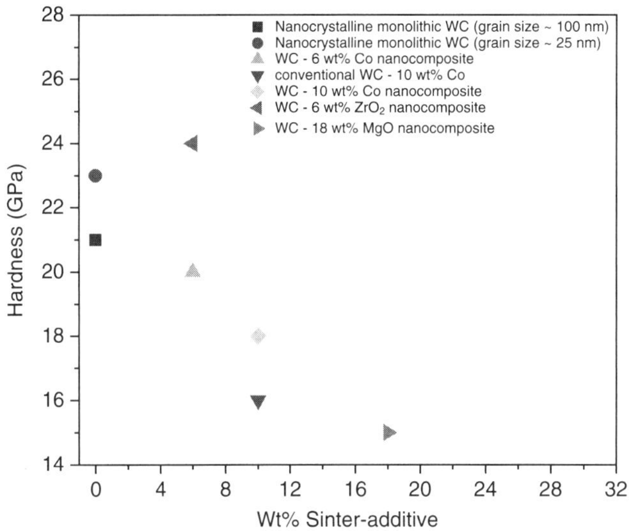

Figure 18.3 Comparison of hardness possessed by various nanoceramics and nanocomposites based on WC.[79]

One of the major industrial applications of WC–Co cermets is cutting tool inserts.[55] However, WC–Co has inherent limitations because of the presence of the binder phase, which leads to failure at high temperature due to softening and also failure under sudden change in loading. To overcome this lacuna, researchers have attempted different combinations of metallic binder to achieve improved physicomechanical properties.[56–58] Furthermore, such metallic phases with inferior corrosion and oxidation resistance are considered as the preferential sites for the initiation of unwanted corrosion- and oxidation-induced failures,[14–16,59–62] thus limiting the lifetime and performance of the cermets in corrosion-prone environments. To overcome such drawbacks, a few researchers have pursued the development of binderless monolithic WC using SPS.[11,18–20] However, extremely high sintering temperatures (1700–1900°C) are required to obtain dense (~97–98% ρ_{th}) bulk WC even with the use of SPS. Furthermore, monolithic WC ceramics exihibit significantly inferior fracture toughness (~5–6 MPa m$^{1/2}$) compared with the Co-containing cermets (~12 MPa m$^{1/2}$).[11,18,19]

18.3.2 WC–ZrO_2 Nanoceramic Composites

To address the aforementioned issues, the possibility of replacing metallic binders by ceramic sinter-additives has been explored to obtain dense WC-based nanocomposites using SPS techniques.[1,63,64] The commercially available high purity WC (primary crystallite size 200 nm, H.C. Strack, Germany) and Y-ZrO_2 powders (primary crystallite size 27 nm, Tosoh grade) were used in processing WC–ZrO_2 composites. Different grades of yttria-stabilized ZrO_2, that is, 2Y-ZrO_2 (2 mol%

Y_2O_3-stabilized), 3Y-ZrO_2 (3 mol% Y_2O_3-stabilized), and undoped ZrO_2, were used as reinforcement. Following a traditional powder metallurgical processing route, dried powders were compacted using cold isostatic pressing at 275 MPa for 5 minutes to obtain green compacts (10mm diameter, 2-mm height) with a green density of ~55% theoretical density. The sintering of the powder compacts were performed via a PS route with temperature varying in the range 1500–1700°C for varying time periods (1–3 hours) at high vacuum (6×10^{-2} torr). In another set of experiments, WC–(6 wt%)ZrO_2 powder mixtures were spark plasma sintered at 1300°C for 5 minutes with heating rate of 600°C/min.

For comparison, fully dense WC–(6 wt%)Co materials were obtained at 1500°C with 1-hour sintering. In the case of WC–(6 wt%)ZrO_2(2Y), full densification requires slightly higher sintering temperature (1600°C) and longer sintering time (3 hours). When using 3Y-ZrO_2 sinter-additive (6 wt%), a higher sintering temperature of 1700°C (1-hour soaking time) is required to obtain dense WC materials (~99.5% ρ_{th}). The effectiveness of ZrO_2 additive in achieving densification of WC composites is therefore lower than when Co binder is used. It should be realized that Co binder facilitates liquid-phase sintering at lower temperature, while the addition of ZrO_2 leads to solid-state sintering. The density data of SPS samples are compared with the conventionally sintered material in Table 18.2. In normalizing the sintered density with respect to the composite density (14.26 g/cm^3), the theoretical density (ρ_{th}) of WC and ZrO_2 is taken as 15.26 and 6.1 g/cm^3 respectively. The data, as shown in Table 18.2, reveal that fully dense (98–99% theoretical density) WC composites reinforced with 6 wt% ZrO_2 are achieved after sintering at 1300°C for 5 minutes via SPS route. The attainment of such high density via SPS processing is found to be independent of the Y_2O_3-dopant content (0, 2, and 3 mol%) in the ZrO_2 particles. Looking at Table 18.2, it is evident that lower SPS temperature (1300°C) is required

Table 18.2. Mechanical Properties of WC–(6 wt%)ZrO_2 Composites and WC–(6 wt%)Co Cermets Sintered via Spark Plasma Sintering (SPS) for 5 Minutes and Pressureless Sintering (PS) in Vacuum[1]

Material composition	Sintering conditions	ρ (g/cm^3)	E (GPa)	H_{v10} (GPa)	K_{IC} (MPa m$^{1/2}$)
WC–ZrO_2 (3Y)	SPS, 1200°C	13.19	479	21.3 ± 0.9	5.6 ± 0.1
WC–ZrO_2 (3Y)	SPS, 1250°C	13.89	527	21.6 ± 0.6	6.3 ± 0.3
WC–ZrO_2 (3Y)	SPS, 1300°C	14.25	575	23.6 ± 0.3	5.9 ± 0.1
WC–ZrO_2 (2Y)	SPS, 1300°C	14.24	540	23.9 ± 0.2	5.8 ± 0.2
WC–ZrO_2 (0Y)	SPS, 1300°C	14.24	562	21.5 ± 0.3	5.7 ± 0.2
WC–ZrO_2 (3Y)	PS, 1700°C, 1h	14.22	515	23.3 ± 0.2	4.5 ± 0.1
WC–Co	SPS, 1300°C	13.6	–	–	–
WC–Co	SPS, 1350°C	14.4	–	–	–
WC–Co	SPS, 1400°C	15.1	649	15.3 ± 0.1	13.2 ± 0.4
WC–Co	PS, 1500°C, 1h	15.1	584	16.8 ± 0.1	14.0 ± 0.9

to achieve full densification of WC materials with ZrO_2 additive compared with metallic (Co) binder (1400°C). This also shows the effectiveness of ZrO_2 sinter-aid in achieving faster densification of WC composites than with the use of Co binder. Considering that conventional sintering (heating rate 5–10°C/min) requires the rather higher densification temperature of 1500°C (1 hour) for WC–6 wt% Co cermets (see Table 18.1), SPS processing (heating rate 600°C/min) drives faster densification of similar compositions at the lower temperature of 1400°C (5 minutes). Probably, the incipient melting of Co phase around WC particles starts at around 1400°C during SPS and drives the faster densification rate. Furthermore, PS of WC–ZrO_2(3Y) requires a sintering temperature of 1700°C (heating rate 5–10°C/min) with a dwell time of 1 hour. Hence, SPS processing offers a rather moderate advantage in the case of liquid-phase sintering (around 100°C reduction), compared with solid-state-sintered materials (around 300–400°C lower sintering temperature).

An opposite trend is observed with SPS-processed samples.[1] A lower SPS temperature of 1300°C is necessary to obtain fully dense WC–(6 wt%)ZrO_2(3Y) composite, while WC–(6 wt%)Co cermets can be fully densified at 1400°C in SPS processing. The observed difference in densification behavior of SPS and PS samples can be attributed to the fact that the fast heating rate and the influence of electric field significantly contribute to faster neck growth kinetics in SPS processing.

X-ray diffraction (XRD) investigation of the conventional sintered samples (1700°C) indicated the presence of t-ZrO_2 and WC. XRD results also confirmed the presence of tetragonal zirconia along with WC (major phase) in the SPSed samples. The average grain size of WC was ~1–2 μm and that of ZrO_2 was <1 μm. Considering the starting WC particle size (average ~ 200 nm), the microstructural observation indicates that the presence of ZrO_2 restricts the grain growth of WC.

The mechanical properties of the WC–ZrO_2 composites, densified under different sintering conditions, are summarized in Table 18.2. The data in Table 18.2 reveal that the newly developed WC–(6 wt%)ZrO_2 composite has a high elastic modulus (~500 GPa). This property can be useful in imparting higher resistance to Hertzian contact damage. From Table 18.2, it can be noted that higher hardness (~23 GPa) is achievable with WC–ZrO_2(3Y) composite, sintered at 1700°C for 1 hour. Despite high hardness, the fracture toughness was as low as 4 MPa $m^{1/2}$ in the pressureless sintered WC–(6%)ZrO_2. The higher hardness can be correlated with the observation of finer grain size.

Looking at Table 18.1, the hardness considerably increases from 21.5 GPa for the WC–unstabilized-ZrO_2 composite to around 24 GPa for the WC composites with 2 and 3 mol% Y-stabilized ZrO_2 reinforcements. While the hardness increase is considerable, the variation in the indentation toughness is surprisingly negligible. Little improvement in toughness from 5.8 MPa $m^{1/2}$ (unstabilized) to around 6 MPa $m^{1/2}$ was measured with Y-stabilization of ZrO_2 reinforcement. A representative bright field TEM image (Fig. 18.4a) of the spark plasma sintered WC–ZrO_2 nanocomposite reveals an overall fine microstructure with WC grains varying in the range 0.3–0.5 μm along with ZrO_2 particles. Higher magnification TEM images, presented in Figure 18.4, reveal the distribution of ZrO_2 particles in WC matrix. The

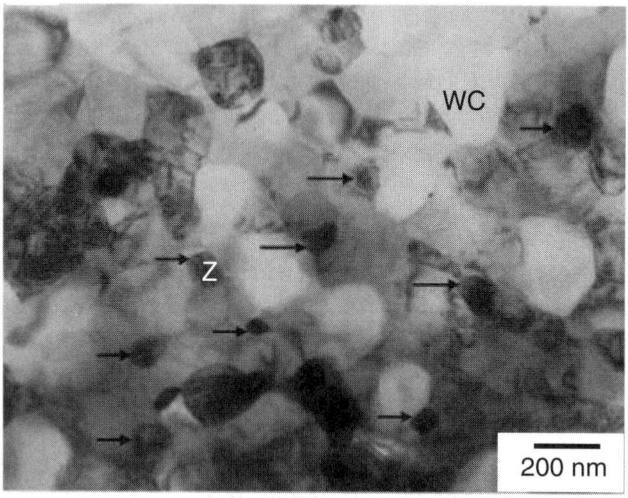

(a)

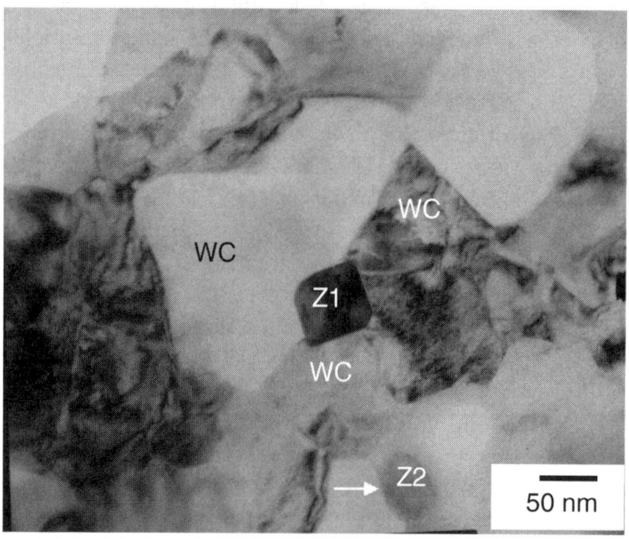

(b)

Figure 18.4 Bright field TEM image of WC–ZrO_2 nanocomposite (spark plasma sintered for 5 minutes at 1300°C) showing distribution of nanocrystalline ZrO_2 particles (indicated by arrows) in submicron WC matrix (a). ZrO_2 particles are distributed in the WC matrix at the triple junction (Z1) and at the grain boundary (Z2) (b) and interior of a WC grain (d). The insets in (b) and (c) show microdiffraction patterns obtained from WC grains (Z stands for ZrO_2).[63]

presence of ZrO_2 particulates, 60–90 nm in size, is also observed at WC triple junctions and grain boundaries (Fig. 18.4). One of the characteristic features is the faceted morphology of these nanosized ZrO_2 particles.

A representative bright field micrograph of WC–ZrO_2 nanocomposite, SPSed at 1300°C for 20 minutes, is shown in Figure 18.5. Most WC grains are 0.4–0.5 μm.

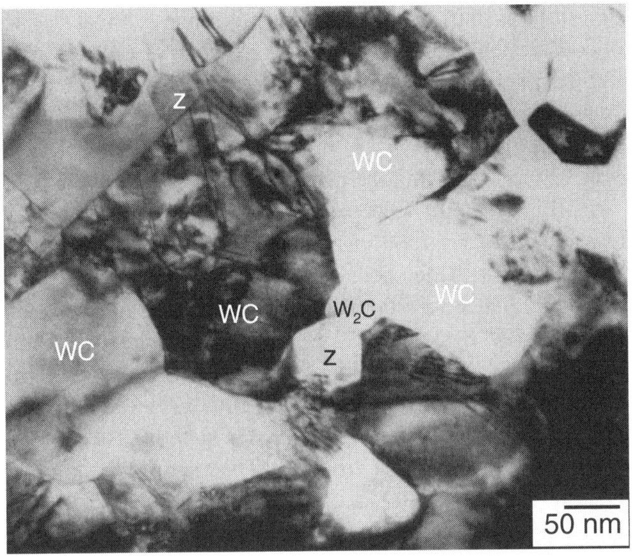

Figure 18.5 Bright field TEM micrograph of WC–(6 wt%)ZrO_2 nanocomposite, spark plasma sintered sample for 20 minutes. The microstructural regions (second-phase particles) corresponding to ZrO_2 are marked as Z. Arrows indicate clustering of the ZrO_2 particles.[63]

This observation indicates that the grain growth is restricted even on sintering for longer duration of 20 minutes. This is in contrast to Omori,[65] who reported abnormal grain growth (about 1-mm grain size) of binderless WC on holding for just over 1 minute, although at a higher spark sintering temperature of 1900°C. An observation of spark sintered WC–ZrO_2 nanocomposite samples also indicates that abnormal grain growth of WC is restricted considerably in contrast to that in binderless WC reported by Cha and Hong.[62] Therefore, ZrO_2 particles are effective in inhibiting grain growth even after longer holding times. Additionally, the presence of α-W_2C at the boundary between WC and ZrO_2 is observed (Fig. 18.5).

The formation of W_2C results from a possible reaction between WC and ZrO_2 at the interface[63]:

$$ZrO_{2-x} + 2y\mathrm{WC} = y\mathrm{W_2C} + ZrO_{2-x-y} + y\mathrm{CO}, \tag{18.1}$$

where x is the oxygen vacancy concentration in the ZrO_2 as a result of the dopant concentration, and y is the additional vacancy concentration created in the ZrO_2 due to reaction with WC.

The preceding reaction creates additional oxygen vacancies and forms a layer of W_2C. The availability of additional oxygen vacancies may enhance the diffusivity to a certain extent, resulting in faster mass transport and hence increased sintering kinetics.

Following critical observations of TEM images (see Fig. 18.4), a significant fraction of the nanocrystalline ZrO_2 particles are found to be present along the WC grain boundaries. These intergranular zirconia particles can potentially restrict the

WC matrix grain boundary mobility during sintering. Such a pinning effect due to ZrO_2 particles can explain considerable inhibition of grain growth of the WC during spark sintering at 1300°C. Since the grain boundary provides a faster diffusion path, the increase of grain boundary area due to nano-ZrO_2 can lead to faster mass transport.

The volume fraction of ZrO_2 is ~14% and the nanosized ZrO_2 particles are well dispersed in the WC matrix (Fig. 18.4). The presence of zirconia can effectively result in a local increase in temperature around a number of WC/ZrO_2 interfaces. Such a phenomenon can result in an enhancement of mass transport with concomitant augmentation of necking and hence resultant sintering kinetics.

The mechanical properties of SPS-processed WC–ZrO_2 composites are compared with the pressureless sintered materials in Table 18.2. The hardness and toughness values are comparable, independent of the processing route. Additionally, the hardness values of WC–ZrO_2 materials (21–24 GPa) are far superior to WC–Co materials (15.5 GPa) with the same binder content (6 wt%). The measured toughness is not suitable for tribological applications, and this has motivated the development of a second generation of WC-based nanocomposites.

18.3.3 WC–ZrO_2–Co Nanocomposites

It is shown in this section how partial or full replacement of Co (metallic phase) with ZrO_2 (ceramic phase) in the WC–Co system, along with SPS, enables the development of high-performance WC-based ceramic nanocomposites.[66] Five different nanocomposite compositions were investigated: WC–(6 wt%)Co (W6Co), WC–(6 wt%)ZrO_2 (W6Zr), WC–(10 wt%)ZrO_2 (W10Zr), WC–(5 wt%)ZrO_2–(1 wt%) Co (W5Zr1Co), and WC–(4 wt%)ZrO_2–(1 wt%)Co (W4Zr2Co). The SPS experiments were carried out at 1300°C for 0 minute with the heating rate of ~175 K/min.

All the nanocomposites could be densified to greater than 97% ρ_{th} even in the absence of Co (see Table 18.3). In particular, the replacement of Co by ZrO_2 did not seem to have any significant influence on densification of the WC–ZrO_2–Co system. XRD analysis did not detect the presence of W_2C etc. in the SPS-processed WC–ZrO_2–Co nanocomposites. In 2009 work by Malek et al.,[67] near-theoretical density for WC–(5/10 vol%)ZrO_2 composites via SPS was achieved after holding for 2–5 minutes at a much higher temperature (1700°C), which resulted in the formation of reaction-product phases, such as ZrC and W_2C. Another important observation is the presence of t-ZrO_2 in all the WC–ZrO_2–Co composites. Most WC grains in the W6Co cermets were observed to be faceted, with the presence of Co phase along the grain boundaries (Fig. 18.6). The development of triangular prism-shaped WC grains, as seen in Figure 18.6a, is commonly observed in WC-based cermets and is considered to be detrimental to the mechanical properties.[9–13,68–71] The average matrix grain size in W6Co cermet was estimated to be 0.4 μm, In contrast, many WC grains appear to be equiaxed in the SPS-processed W6Zr (Fig. 18.6). Hence, the addition of ZrO_2 suppresses the formation of the deleterious "truncated trigonal prism"–shaped WC grains. It can be more clearly observed from Fig. 18.6 that the ZrO_2

Table 18.3. Mechanical Properties of the WC-Based Nanocomposites Developed by Spark Plasma Sintering at 1300°C for 0 Minute[66]

Materials designation	Material composition	Sinter densities (% ρ_{th})	Vickers indentation $H_{v0.1}$ (GPa)	Vickers indentation K_{IC} (10 kg) (MPa m$^{1/2}$)	Flexural strength (four-point bending) (GPa)	Fracture toughness (SEVNB) (MPa m$^{1/2}$)
W6Co	WC–(6 wt%) Co	98.8	20.7 ± 0.3	12.4 ± 0.3	1.1 ± 0. 6	12.7 ± 0.6
W6Zr	WC–(6 wt%) ZrO_2	97.1	20.5 ± 0.8	10.9 ± 0.4	1.3 ± 0. 4	9.8 ± 0.7
W10Zr	WC–(10 wt%) ZrO_2	98.4	23.2 ± 0.2	10.1 ± 0.3	0.9 ± 0.8	9.3 ± 0.8
W5Zr1Co	WC–(5 wt%) ZrO_2–(1 wt%)Co	98.2	22.5 ± 0.6	6.8 ± 0.3	1.2 ± 0.5	7.1 ± 0.6
W4Zr2Co	WC–(4 wt%) ZrO_2–(2 wt%)Co	98.7	22.5 ± 0.3	10.7 ± 0.4	1.2 ± 0.6	10.1 ± 0.5

particles present along the matrix grain boundaries are slightly coarser than those within the grains. More precisely, the intragranular ZrO_2 particles have sizes between 10 and 40 nm (average size, 28 nm), while the sizes of most of the intergranular particles lie between 60 and 80 nm (average size, 71 nm). Based on TEM observations, SPS-processed WC–ZrO_2–Co can be described as an inter/intragranular nanocomposite. Figure 18.6c presents the ultrafine microstructure of the WC–(10 wt%) ZrO_2 nanocomposites. Although the ZrO_2 size distribution is similar to that of W6Zr nanocomposite, clustering of ZrO_2 particles can be observed at many locations throughout the microstructure. Such agglomeration of the ZrO_2 nanoparticles was not observed in the case of WC–(6 wt%) nanocomposites (Figs. 18.4 and 18.6). In addition to intergranular as well as intragranular nanosized ZrO_2 particles, the presence of a Co phase along with a few faceted WC grains was observed in the TEM images, corresponding to W5Zr1Co and W4Zr2Co nanocomposites (see Fig. 18.6).

The hardness, indentation fracture toughness, bulk fracture toughness (measured via single edge V-notched beam [SEVNB]), and flexural strength (measured via four-point bending) of WC–ZrO_2–Co nanocomposites are presented in Table 18.3. The toughness properties were carefully estimated using both an indentation technique (short crack toughness) and the SEVNB route (long crack toughness). It is known that the SEVNB route provides more reliable values of fracture toughness of ceramics, as opposed to the indentation technique, which often overestimates the toughness values.[72] Nevertheless, the indentation toughness values will enable one to compare the investigated materials with other competing materials. Since *c/a*

Figure 18.6 (a) Bright field TEM image of SPS-processed WC–(6 wt%)Co. The microstructural regions (secondary phase) corresponding to cobalt are marked as Co. (b) Higher magnification bright field TEM image of SPS-processed WC–(4 wt%)ZrO_2–(2 wt%)Co nanocomposite. The microstructural regions corresponding to the ZrO_2 particles and cobalt are marked as Z and Co, respectively. Convergent beam electron diffraction (CBED) patterns obtained from the different phases (WC, ZrO_2, and Co) are also presented. (c) Lower magnification bright field TEM image of SPS-processed WC–(10 wt%)ZrO_2 nanocomposite. The microstructural regions (second-phase particles) corresponding to ZrO_2 are marked as Z. Arrows indicate clustering of the ZrO_2 particles.[66]

ratios (where c is half of the crack length of a radial crack and a is half of the diagonal length of an indent) are less than 3, the developed crack system can be classified as Palmqvist cracks.[17,18,73,74] In fact, Palmqvist cracking has been observed during Vickers indentation on WC–Co cermets.[19] The following relationship was used to estimate the indentation toughness from the measured crack lengths:

$$K_{IC} = 0.018\, Ha^{1/2}(E/H)^{0.4}(c/a-1)^{-0.5}, \tag{18.2}$$

where E is the composite elastic modulus (estimated using law of mixture), H is the Vickers hardness, c is half the average crack length, and a is half the average indent diagonal length.

For the flexural strength and fracture toughness measurements (SEVNB), the specimen dimensions were $13.0 \times 2.0 \times 2.5$ mm. For SEVNB, the ratios of the notch depths to specimen thickness lie around 0.2–0.3. The notches were sharpened by razor blade, using diamond paste, up to 1 μm and the tip radii of the notches were ~10 μm. The flexural strengths and fracture toughness were both measured using a four-point bending configuration, with a crosshead loading rate of 5 N/min and inner and outer spans of 5.8 and 10.0 mm, respectively. To estimate the fracture toughness, the following relations were used:[75]:

$$K_{1C} = \frac{P_f(L_o - L_i)}{BW^{3/2}} \frac{3\alpha^{1/2}}{2(1-\alpha)^{3/2}} f(\alpha), \tag{18.3}$$

$$f(\alpha) = 1.9887 - 1.326\alpha - \frac{\alpha(1-\alpha)(3.49 - 0.68\alpha + 1.35\alpha^2)}{(1+\alpha)^2}, \tag{18.4}$$

where P_f, L_o, L_i, B, and W are the experimentally measured fracture load, outer span, inner span, specimen breadth, and specimen width, respectively. The precrack size (notch depth, a) enters Equation 18.4 through the parameter α, where $\alpha = a/W$.

It can be observed that both indentation toughness and the SEVNB fracture toughness values are similar for the investigated nanocomposites. The WC–(6 wt%) Co cermet exhibited higher fracture toughness (~12.5 MPa m$^{1/2}$) among all the investigated materials. Similar fracture toughness (~11–12 MPa m$^{1/2}$) values for SPS-processed WC–Co cermets were also reported earlier.[9–13] Also, a modest decrease (~16%) in fracture toughness occurred on replacement of the entire Co binder with ZrO_2 nanoparticles (6 wt%). However, toughness improvement was not observed on either replacing a part of the ZrO_2 content by Co (up to 2 wt%) or increasing the ZrO_2 content to 10 wt%.

In contrast to the modest reduction in fracture toughness, a strength improvement of ~18% has been measured for the WC–(6 wt%)ZrO_2 ceramic nanocomposite (~1.3 GPa), with respect to the similarly processed WC–(6 wt%)Co cermet (~1.1 GPa). Among other ZrO_2-containing nanocomposites, W5Zr1Co and W4Zr2Co also possess higher flexural strength (~1.2 GPa) with respect to the reference W6Co cermet. Such improvement can be expected due to refinement in microstructural scale and the absence of truncated trigonal prism-shaped matrix grains in the presence of nanosized ZrO_2 particles. However, considerable reduction in the flexural strength (~0.9 GPa) was measured on incorporation of 10 wt% ZrO_2 (W10Zr). The presence of agglomerates and possibly processing defects (see Fig. 18.6) is believed to account for such inferior strength of the W10Zr nanocomposite. Even though the reported values are representative of the nanocomposite properties, the variable nature of strength properties of ceramic-based brittle materials and the errors in the measurements (see Table 18.3) may be taken into account when using the strength values. Most of the ZrO_2-containing nanocomposites (W5Zr1Co, W4Zr2Co, W10Zr) exhibited superior hardness (~23 GPa), with the exception of W6Zr. Although the nanoscaled microstructure is believed to be the reason for the superior hardness and strength, a systematic investigation of the effect of variation in ZrO_2 and WC grain size on the mechanical properties is expected to further elucidate the microstructure–mechanical property relationships for this novel class of materials.

18.3.4 Toughness of $WC–ZrO_2$-Based Nanoceramic Composites

In the $WC–ZrO_2$ nanocomposites, ZrO_2 has a higher thermal expansion coefficient ($\alpha_p \sim 11 \times 10^{-6}\,K^{-1}$; see References [10 and 15]) with respect to that of the matrix (WC, $\alpha_m \sim 5 \times 10^{-6}\,K^{-1}$; see References [10, 76, and 77]). The mismatch in coefficients of thermal expansion (CTEs) is likely to result in the development of residual compressive stresses in the matrix during cooling from the sintering temperatures. According to a model proposed by Taya et al.,[78] misfit strains (ε) that are developed in the particles due to such CTE mismatch can be expressed as

$$\varepsilon = \int_{T_P}^{T_R} (\alpha_p - \alpha_m)\,dT, \tag{18.5}$$

where T_p is the temperature (~1000°C) below which plasticity is considered to be negligible, and T_R is room temperature. The isotropic average stress fields in the particles and matrix are designated as $\langle\sigma\rangle_p$ and $\langle\sigma\rangle_m$, respectively, which are defined for a given volume fraction of second-phase particles (f_p) according to the following equations:

$$\frac{\langle\sigma\rangle_p}{E_m} = \frac{-2(1-f_p)\beta\varepsilon}{A} \tag{18.6}$$

and

$$\frac{\langle\sigma\rangle_m}{E_m} = \frac{2f_p\beta\varepsilon}{A}, \tag{18.7}$$

where

$$A = (1-f_p)(\beta+2)(1+\nu_m) + 3\beta f_p(1-\nu_m) \tag{18.8}$$

$$\beta = \left(\frac{1+\nu_m}{1+\nu_p}\right)\left(\frac{E_p}{E_m}\right), \tag{18.9}$$

where ν_m and E_m are Poisson's ratio (~0.2) and elastic modulus (~700 GPa) of the matrix (WC),[10,24,25] and ν_p and E_p are Poisson's ratio (~0.2) and elastic modulus (~210 GPa) of the second-phase particles (ZrO_2).[10,15] For WC–(6 wt%)ZrO_2 nanocomposite, the volume fraction of ZrO_2 nanoparticles (f_p) is ~0.14. Hence, the preceding equations predict a residual compressive stress of $\langle\sigma\rangle_m \sim -0.2$ GPa in the matrix, while the t-ZrO_2 particles are subjected to a residual tensile stress of $\langle\sigma\rangle_p \sim +1.2$ GPa.

The toughness increment in particulate-reinforced ceramic composites is attributed to CTE-mismatch-induced residual stresses.[26,65,79–81] According to Taya et al.,[78] the existence of such compressive residual stresses in the matrix results in lowering the stress intensity factor, which is given by

$$\Delta K_1 = 2\langle\sigma\rangle_m \sqrt{\frac{2(\lambda - d)}{\pi}} \tag{18.10}$$

where λ is the interparticle distance and d is the average diameter of the second-phase particles. The interparticle distance (λ) for equiaxed particles of the second phase with diameter d and volume fraction f_p can be determined by the following relationship:

$$\lambda = d \times f_p^{-1/2}. \tag{18.11}$$

Considering d to be 50 nm (see Fig. 18.6), the interparticle distance in W6Zr nanocomposite is estimated to be ~150 nm. Hence, Equation 18.10 predicts a toughness increment of $\Delta K_1 \sim 0.1\,\mathrm{MPa\,m^{1/2}}$, over that of unreinforced WC. Such a negligible improvement in fracture toughness due to CTE-mismatch-induced residual stress cannot explain the high toughness values (see Table 18.3). Therefore, other toughening mechanisms must be responsible for the overall fracture toughness of these newly developed ZrO_2-nanoparticle-reinforced WC-based nanocomposites.

The presence of ZrO_2 provides a strong possibility of the contribution of transformation toughening in the WC–ZrO_2 nanocomposites. The presence of a strong monoclinic zirconia (m-ZrO_2) peak, along with the t-ZrO_2 peak with reduced intensity was identified in the XRD pattern recorded from the fracture surface. Such an observation indicates that t-ZrO_2 partially transformed to m-ZrO_2 during the fracture of WC–ZrO_2 nanocomposites. As estimated using the model proposed by Taya et al.,[78] a lower CTE of WC, compared with that of ZrO_2, results in the development of tensile residual stress (~1.2 GPa) in ZrO_2. The presence of such tensile residual stress (σ_r) is believed to aid in increasing t-ZrO_2 transformability by lowering the critical stress (σ_c) needed for transformation. Transformation toughening in the presence of ZrO_2 contributes significantly toward the increment in the fracture toughness of the nanocomposites. For other brittle ceramics that have lower thermal expansion coefficients than ZrO_2, the presence of t-ZrO_2 results in improvements in fracture toughness.[82,83]

In addition to transformation toughening, deflection and bridging of cracks by the second-phase particles (ZrO_2), as can be observed from Figure 18.7, also contribute to high fracture toughness of the investigated WC–ZrO_2 nanocomposites. A model proposed by Faber and Evans[84] estimates that crack deflection leads to an increment in fracture toughness by a factor of ~1.12 over the matrix toughness for a second-phase particle volume fraction between 0.1 and 0.3.

18.3.5 Comparison with Other Ceramic Nanocomposites

In an effort to compare the mechanical properties of WC–ZrO_2–Co with various other ceramic nanocomposites, Table 18.4 summarizes the best results reported in the literature for different ceramic nanocomposite systems. Broadly, six different ceramic nanocomposite systems, that is WC–ZrO_2, ZrO_2–WC, Al_2O_3–SiC,

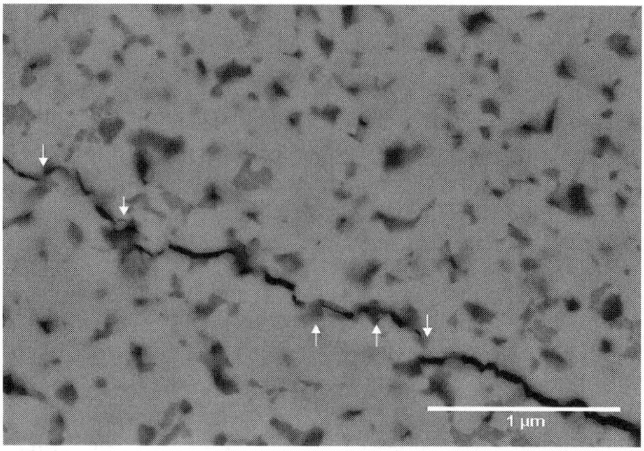

Figure 18.7 SEM images of cracks around Vickers indentations on WC–(4 wt%)ZrO_2–(2 wt%)Co nanocomposites. Arrows indicate the bridging and deflection of the crack by the second-phase particles.[66]

Table 18.4. Summary of the Best Reports in the Literature for Mechanical Properties (Hardness, Fracture Toughness, and Flexural Strength) of Different Ceramic Nanocomposites, Along with Those Obtained for the WC–ZrO_2–Co Nanocomposites

Ceramic nanocomposite system	Hardness (GPa)	Fracture toughness ($MPa\,m^{1/2}$)	Flexural strength (GPa)
WC–(6 wt%)ZrO_2 (3 mol% Y_2O_3)	21	10.0	1.3
WC–(4 wt%)ZrO_2(3 mol% Y_2O_3)–(2 wt%)Co	23	10.0	1.2
WC–(10 vol%)ZrO_2 (2 mol% Y_2O_3)	25	6.0	1.5
WC–(10 wt%)Co	18	12.0	–
ZrO_2–(40 vol%)WC	15	10.0	2.0
Al_2O_3–(5 vol%)SiC	–	5.0	1.5
	20	3.0	1.0
	19	4.0	1.0
Al_2O_3–(20 vol%)ZrO_2	15	9.0	–
Al_2O_3–(15 vol%) ZrO_2–(5 vol%)SiC	18	5.0	1.2
Si_3N_4–(25 vol%)SiC	–	8.0	1.5
MgO–(10 vol%)SiC	–	4.5	0.7

Note that the values obtained for the fracture toughness and strength also depend on the test technique used in the corresponding work.[66]

Si_3N_4–SiC, Al_2O_3–ZrO_2, and MgO–SiC, are being developed.[8,22,27,36–41,85] It needs to be considered that the difference in mechanical properties among various nanocomposite systems depends not only on the microstructural scale of the nanosized reinforcement and the mechanical response of individual phases, but also on the measurement technique and parameters. With respect to the WC–ZrO_2 system, excellent hardness (~25 GPa) and flexural strength (~1.5 GPa) were measured with WC–(10 vol%)ZrO_2 nanocomposites in work by Malek and coworkers.[22] Among other factors, the presence of additional ZrC and W_2C phase in those composites could be the reason for lower fracture toughness, which sheds light on the fact that SPS processing conditions can considerably influence the properties of the as-densified materials. It is also observed that variations in SiC content and particle size result in various combinations of mechanical properties for Al_2O_3–SiC nanocomposites. However, the fracture toughness remains modest and less than 5 MPa $m^{1/2}$ in the Al_2O_3–SiC nanocomposite system. Although high toughness of 10 MPa $m^{1/2}$ and strength of 2 GPa were recorded in the ZrO_2–WC system, the hardness remained considerably lower (~ 15 GPa). Hence, it is clear from Table 18.4 that the WC–ZrO_2–Co nanocomposites demonstrate a superior combination of SEVNB fracture toughness, flexural strength, and hardness, than has ever been reported earlier.[27]

In summary, the partial replacement of metallic Co with ceramic ZrO_2, at the same time as attaining similar densification and maintaining similar fracture toughness, is a significant achievement in the WC–ZrO_2–Co nanocomposite system, and it solves the long-standing problem pertaining to the softening and corrosion of the metallic phase during demanding applications of WC-based cermets.

REFERENCES

1 B. Basu, J. H. Lee, and D. Y. Kim. Development of WC-ZrO_2 nanocomposites by spark plasma sintering. J. Am. Ceram. Soc. 87(2) (2004), 317.

2 M. Sternitzke. Review: Structural ceramic nanocomposites. J. Eur. Ceram. Soc. 17 (1997), 1061.

3 R. D. Shull. View point: Nanocrystalline and nano phase materials. Nano Struct. Mater. 2 (1993), 213.

4 R. S. Averbek, H. J. Holfer, and R. Tao. Processing of nano-grained materials. Mater. Sci. Eng. A 66 (1993), 169.

5 S. Komarneni. Nanocomposites. J. Mater. Chem. 2(12) (1992), 1219–1230.

6 C. Suryanarayana. Nanocrystalline materials. Int. Mater. Rev. 40 (1995), 41–64.

7 H. Gleiter. Nanostructured materials: State of the art and perspectives. Z. Metallkd. 86 (1995), 78.

8 T. Kusunose, T. Sekino, Y. H. Choa, and K. Niihara. Machinability of silicon nitride/boron nitride nanocomposites. J. Am. Ceram. Soc. 85(11) (2002), 2689–2695.

9 K. Niihara. New design concept of structural ceramics—Ceramic nano composites. J. Ceram. Soc. Jpn. 99(10) (1991), 974. The Centennial Memorial Issue.

10 H. Park, H. E. Kim, and K. Niihara. Microstructure and high-temperature strength of Si_3N_4-SiC nanocomposite. J. Eur. Ceram. Soc. 18 (1998), 907–914.

11 Y. S. Oh, C. S. Kim, D. S. Lim, and D. S. Cheong. Fracture strengths and microstructures of Si_3N_4/SiC nanocomposites fabricated by in-situ process. Scr. Mater. 44 (2001), 2079–2081.

12 T. Hirano and K. Niihara. Microstructure and mechanical properties of Si_3N_4/SiC composites. Mater. Lett. 22 (1995), 249–255.

13 Y. H. Koh, H. W. Kim, and H. E. Kim. Mechanical properties and oxidation resistance of Si_3N_4-SiC nanocomposites. Scr. Mater. 44 (2001), 2069–2073.

14 A. Tsuje, K. Nishida, and M. Komatsu. Effect of crystallizing the grain-boundary glass phase on the high-temperature strength of hot-pressed Si_3N_4 containing Y_2O_3. J. Am. Ceram. Soc. 58 (1975), 323–326.
15 M. Mitomo. Pressure sintering of Si_3N_4. J. Mater. Sci. 11 (1976), 1103–1107.
16 T. Funahishi, K. Isomura, A. Harita, and R. Uchimura. In *Ceramic Materials and Components for Engines*. V. J. Tenney (Ed.). American Ceramic Society, Las Vegas, NV, 1989, 968–976.
17 K. Isomura, T. Fukuda, K. Ogasahara, T. Funahashi, and R. Uchimura. Machinable Si_3N_4-BN composite ceramics with high thermal shock resistance, high erosion resistance. In *UNITECR'89 Proceedings*. L. J. Trostel Jr. (Ed.). American Ceramic Society, Westerville, OH, 1989, 624–634.
18 T. Kusunose, T. Sekino, Y. H. Choa, and K. Niihara. Fabrication and microstructure of silicon nitride/boron nitride nanocomposites. J. Am. Ceram. Soc. 85 (2002), 2678–2688.
19 T. Kusunose, T. Sekino, Y. H. Choa, and K. Niihara. Machinability of silicon nitride/boron nitride nanocomposites. J. Am. Ceram. Soc. 85 (2002), 2689–2695.
20 L. Gao, X. Jin, J. Li, Y. Li, and J. Sun. BN/Si_3N_4 nanocomposite with high strength and good machinability. Mater. Sci. Eng. A 415 (2006), 145–148.
21 L. Gao, X. Jin, H. Kawaoka, T. Sekino, and K. Niihara. Microstructure and mechanical properties of SiC-mullite nanocomposite prepared by spark plasma sintering. Mater. Sci. Eng. A 334 (2002), 262–266.
22 L. Gao, H. Wang, H. Kawaoka, T. Sekino, and K. Niihara. Fabrication of YAG–SiC nanocomposites by spark plasma sintering. J. Eur. Ceram. Soc. 22 (2002), 785–789.
23 Y. Luo, S. Li, W. Pan, and L. Li. Fabrication and mechanical evaluation of SiC–TiC nanocomposites by SPS. Mater. Lett. 58 (2003), 150–153.
24 S. J. Kalita, A. Bhardwaj, and H. A. Bhatt. Nanocrystalline calcium phosphate ceramics in biomedical engineering. Mater. Sci. Eng. C 27 (2007), 441–449.
25 J. L. Xu, K. A. Khor, Z. L. Dong, Y. W. Gu, R. Kumar, and P. Cheang. Preparation and characterization of nanosized hydroxypatite powders product in a radio frequency(rf) thermal plasma. Mater. Sci. Eng. A 374 (2004), 101.
26 G. Wei and P. X. Ma. Structure and properties of nano-hydroxyapatite/polymer composite scaffolds for bone tissue engineering. Biomater 25 (2004), 4749–4757.
27 C. Du, F. Z. Cui, X. D. Zhu, and K. D. Groot. Three-dimensional nanoHAp/collagen matrix loading with osteogenic cells in organ culture. J. Biomed. Mater. Res. 44 (1999), 407–415.
28 T. J. Webster, R. W. Siegel, and R. Bizios. Nanoceramic surface roughness enhance osteoblast and osteoclast functions for improved orthopedic/dental implant efficacy. Scr. Mater. 44 (2001), 1639–1642.
29 S. Best and W. Bonfield. Processing behaviour of hydroxyapatite powders with contrasting morphology. J. Mater. Sci. Mater. Med. 5 (1994), 516–521.
30 W. Li and L. Gao. Fabrication of HAp ZrO_2(3Y) nanocomposites by SPS. Biomaterials 24 (2003), 937–940.
31 T. Nakayama, B. S. Kim, H. Kondo, Y. H. Choa, T. Sekino, M. Nagashima, T. Kusunose, Y. Hayashi, and K. Niihara. Fabrication of MgO based nanocomposites with multifunctionality. J. Eur. Ceram. Soc. 24(2) (2004), 259–264.
32 F. E. Luborsky, L. I. Mendelsohn, and T. O. Paine. Reproducing the properties of alnico permanent magnet alloys with elongated single-domain cobalt-iron particles. J. Appl. Phys. 28 (1957), 344–351.
33 F. E. Luborsky and T. O. Paine. Angular variation of the magnetic properties of elongated single-domain iron particles. J. Appl. Phys. 31, (1960), S66–S68.
34 T. Venkateswaran, D. Sarkar, and B. Basu. WC-ZrO_2 composites: Processing and unlubricated tribological properties. Wear 260(1–2) (2006), 1–9.
35 B. K. Kim, G. H. Ha, and D. W. Lee. Sintering and microstructure of nanophase WC/Co hard metals. J. Mater. Process. Technol. 63 (1997), 317–321.
36 D. Sivaprahasam, S. B. Chandrasekar, and R. Sundaresan. Microstructure and mechanical properties of nanocrystalline WC–12Co consolidated by spark plasma sintering. Int. J. Ref. Met. Hard Mater. 25(2) (2007), 144–152.

37 S. I. CHA, S. H. HONG, and B. K. KIM. Spark plasma sintering behavior of nanocrystalline WC–10Co cemented carbide powders. Mater. Sci. Eng. A 351 (2003), 31–38.

38 K. JIA, T. E. FISCHER, and B. GALLOIS. Microstructure, hardness and toughness of nanostructured and conventional WC-Co composites. Nanostructured Mater. 10(5) (1998), 875–891.

39 V. RICHTER and M. V. RUTHENDORF. On hardness and toughness of ultrafine and nanocrystalline hard materials. Int. J. Ref. Met. Hard Mater. 17 (1999), 141–152.

40 S. I. CHA, S. H. HONG, G. K. HA, and B. K. KIM. Microstructure and mechanical properties of nanocrystalline WC-10Co cemented carbides. Scr. Mater. 44 (2001), 1535–1539.

41 H. C. KIM, I. J. SHON, J. K. YOON, and J. M. DOH. Consolidation of ultrafine WC and WC–Co hard materials by pulsed current activated sintering and its mechanical properties. Int. J. Ref. Met. Hard Mater. 25(1) (2007), 46–52. in press.

42 A. MICHALSKI and D. SIEMIASZKO. Nanocrystalline cemented carbides sintered by the pulse plasma method. Int. J. Ref. Met. Hard Mater. 25(2) (2007), 153–158.

43 M. S. E. ESKANDARANY. Fabrication of nanocrystalline WC and nanocomposite WC–MgO refractory materials at room temperature. J. Alloys Comp. 296 (2000), 175–182.

44 V. K. SARIN. Cemented carbide cutting tools, in *Advances in Powder Technology*. D. Y. Chin (Ed.). ASM, Metals Park, OH, 1981, 253–287.

45 H. KOLASKA. The dawn of the hard metal age. Powder Metal. Int. 24(5) (1992), 311–314.

46 A. BOCK, W. D. SCHUBERT, and B. LUX. Inhibition of grain growth on submicron cemented carbides. Powder Metall. Ind. 24(1) (1992), 20–26.

47 F. V. LENEL. *Powder Metallurgy Principles and Applications*. Metal Powder Industries Federation, Princeton, NJ, 1980, 383.

48 S. BERGER, R. PORAT, and R. ROSEN. Nanocrystalline materials: A study of WC-based hard metals. Prog. Mater. Sci. 42(1–4) (1997), 311–320.

49 X. L. SHI, G. Q. SHAO, X. L. DUAN, R. Z. YUAN, and H. H. LIN. Mechanical properties, phases and microstructure of ultrafine hard metals prepared by WC-6.29Co nanocrystalline composite powder. Mater. Sci. Eng. A 392 (2005), 335–339.

50 H. C. KIM, I. J. SHON, J. K. YOON, and J. M. DOH. Consolidation of ultra fine WC and WC-Co hard materials by pulsed current activated sintering and its mechanical properties. Int. J. Ref. Met. Hard Mater. 25(1) (2007), 46–52.

51 H. C. KIM, I. K. JEONG, I. J. SHON, I. Y. KO, and J. M. DOH. Fabrication of WC–8 wt.% Co hard materials by two rapid sintering processes. Int. J. Ref. Met. Hard Mater. 25(4) (2007), 336–340.

52 H. SUZUKI. *Cemented Carbide and Sintered Hard Materials*. Maruzen, Tokyo, Japan, 1986, 262.

53 (a) H. SUZUKI, K. HAYASHI, T. YAMAMOTO, and N. J. NAKAJO. Relations between some properties of sintered WC-10%Ni alloy and its binder phase composition. Journal of the Japan Society of Powder and Powder Metallurgy 13 (1966), 290–295. (b) ASM International Handbook Committee. *Metals Handbook*, 9th ed., Vol. 13. Corrosion, Metals Park, OH, 1987, 855–857.

54 Y. MASUMOTO, K. TAKECHI, and S. IMASATO, Corrosion resistance of cemented carbide. Nippon Tungsten Rev. 19 (1986), 26–33.

55 L. J. PRAKASH. Application of fine grained tungsten carbide based cemented carbides. Int. J. Ref. Met. Hard Mater. 13 (1995), 257–264.

56 A. PARASIRIS, K. T. HARTWIG, and M. N. SRINIVASAN. Formation/consolidation of WC-Co cermets by simple shear. Scr. Mater. 42 (2000), 875–880.

57 H. K. TONSHOFF and C. BLAWIT. Development and evaluation of PACVD coated cermet tools. Surf. Coat. Technol. 99 (1997), 119–127.

58 P. ETTMAYER, H. KOLASKA, W. LENGAUER, and K. DREVER. Ti (C,N)—Metallurgy and properties. Int. J Ref. Mat. Hard Meter. 13 (1995), 343–351.

59 S. IMASATO, K. TOKUMOTO, T. KITADA, and S. SAKAGUCHI. Properties of ultra-fine grain binderless cemented carbide “RCCFN”. Int. J. Ref. Met. Hard Mater. 13 (1995), 305–312.

60 H. C. KIM, I. J. SHON, J. E. GARAY, and Z. A. MUNIR. Consolidation and properties of binderless sub-micron WC by field activated sintering. Int. J. Ref. Metal. Hard Mater. 22 (2004), 257–264.

61 H. C. KIM, I. J. SHON, J. K. YOON, S. K. LEE, and Z. A. MUNIR. One step synthesis and densification of ultra-fine WC by high-frequency induction combustion. Int. J. Ref. Metal. Hard Mater. 24 (2006), 202–209.

62 S. I. Cha and S. H. Hong. Microstructures of binderless tungsten carbides sintered by spark plasma sintering process. Mater. Sci. Eng. A 356 (2003), 381–389.
63 K. Biswas, A. Mukhopadhyay, B. Basu, and K. Chattopadhyay. Densification and microstructure development in spark plasma sintered WC–6 wt.% ZrO_2 nanocomposites. J. Mater. Res. 22(6) (2007), 1491–1501.
64 T. Venkateswaran, D. Sarkar, and B. Basu. Tribological properties of WC-ZrO_2 nanocomposites. J. Am. Ceram. Soc. 88(3) (2005), 691–697.
65 M. Omori. Sintering, consolidation, reaction and crystal growth by the spark plasma system (SPS). Mater. Sci. Eng. A 287 (2000), 183–188.
66 A. Mukhopadhyay, D. Chakrabarty, and B. Basu. Spark plasma sintered WC-ZrO_2-Co nanocomposites with high fracture toughness and strength. J. Am. Ceram. Soc. 93(6) (2010), 1754–1763.
67 O. Malek, B. Lauwers, Y. Perez, P. D. Baets, and J. Vleugels. Processing of ultrafine ZrO_2 toughened WC composites. J. Eur. Ceram. Soc. 29 (2009), 3371–3378.
68 A. V. Shatov, S. A. Firstov, and I. V. Shatova. The shape of WC crystals in cemented carbides. Mater. Sci. Eng. A 242 (1998), 7–14.
69 R. P. Herber, W. D. Schubert, and B. Lux. Hard metals with "rounded" WC grains. Int. J. Ref. Met. Hard Mater. 24(5) (2006), 360–364.
70 P. Klaer, F. Kiefer, K. Stjernberg, and J. J. Oakes. The influence of binder constitution on the shape of WC grains. Adv. Powder Metall. Part. Mater. 3 (1999), 10/51–10/61.
71 H. E. Exner. Physical and chemical nature of cemented carbides. Int. Mater. Rev. 4 (1979), 149–173.
72 G. D. Quinn and R. C. Bradt. On the vickers indentation fracture toughness test. J. Am. Ceram. Soc. 90(3) (2007), 673–680.
73 M. T. Laugier. Palmqvist cracking in WC-Co composites. J. Mater. Sci. Lett. 4 (1985), 207–210.
74 K. Niihara, R. Morena, and D. P. H. Hasselmann. Evaluation of K_{Ic} of brittle solids by the indentation method with low crack-to-indent ratios. J. Mater. Sci. Lett. 1(1) (1982), 13–16.
75 M. Mizuno and H. Okuda. VAMAS round robin on fracture toughness of silicon nitride. J. Am. Ceram. Soc. 78 (1995), 1793–1801.
76 *ASM Engineered Materials Reference Book*. ASM International, Materials Park, OH, 1989, 182.
77 H. J. Scussel. *Friction and Wear of Cemented Carbides*. ASM Handbook 18. ASM Int., Metals Park, OH, 1992, 795.
78 M. Taya, S. Hayashi, A. S. Kabayashi, and H. S. Yoon. Toughening of a particulate-reinforced ceramic-matrix composite by thermal residual stresses. J. Am. Ceram. Soc. 73 (1990), 1382–1391.
79 A. Mukhopadhyay and B. Basu. Consolidation-microstructure-property relationships in bulk nanoceramics and ceramic nanocomposites: A review. Int. Mater. Rev. 52(5) (2007), 257–288.
80 A. Mukhopadhyay. Fabrication and properties of oxide nanocomposites containing uniformly dispersed second phases. PhD thesis, Oxford University, Oxford, UK, 2009.
81 T. Ohji, Y. K. Jeong, Y. H. Choa, and K. Niihara. Strengthening and toughening mechanisms of ceramic nanocomposites. J. Am. Ceram. Soc. 81(6) (1998), 1453–1460.
82 J. Wang and R. Stevens. Zirconia-toughened alumina (ZTA) ceramics. J. Mater. Sci. 24(10) (1989), 3421–3440.
83 T. Watanabe and K. Shoubu. Mechanical properties of hot-pressed TiB_2–ZrO_2 composites. J. Am. Ceram. Soc. 68(2) (1985), C34–C36.
84 K. T. Faber and A. G. Evans. Crack deflection processes—I. Theory. Acta Metall. 31(4) (1983), 565–576.
85 D. Jiang, O. VanderBiest, and J. Vleugels. ZrO_2-WC nanocomposites with superior properties. J. Eur. Ceram. Soc. 27(2–3) (2006), 1247–1251.
86 J. Wan, R. G. Duan, and A. K. Mukherjee. Spark plasma sintering of silicon nitride/silicon carbide nanocomposites with reduced additive amounts. Scr. Mater. 53 (2005), 663–667.

Section Seven

Bioceramics and Biocomposites

Chapter 19

Overview: Introduction to Biomaterials

In last few decades, the development of new materials to obtain better performance in biomedical applications has attracted wider attention. The success of such widespread efforts requires better understanding of various concepts, for example, biocompatibility, host response, and cell–biomaterial interaction. In this overview chapter, we review the fundamental understanding that is required for biomaterials development in the context of hard-tissue replacement. As an illustrative example, the possibility of using SiO_2–MgO–Al_2O_3–K_2O–B_2O_3–F glass-ceramic materials for orthopedic and dental applications is reviewed along with research results.

19.1 INTRODUCTION

One of the major emerging research areas in materials science relates to the application of materials to health care and, in particular, to reconstructive surgery. In the United States, the total health care expenditure in the year 2000 was around \$14 billion, while the U.S. market for biomaterials in 2000 was \$9 billion. It has been projected to touch \$904 billion by the year 2015. It can be further noted here that the respective annual expenses in other developing countries can be around two to three times the U.S. expenses.[1] Therefore, the development of biomaterials and related devices is important.

It has been widely recognized that the field of biomaterials is multidisciplinary and the design of biomaterials therefore requires the synergistic interaction of materials science, biological science, chemical science, medical science, and mechanical science. Such interaction is schematically illustrated in Figure 19.1. Also, Figure 19.1 reveals the necessity of adopting a cross-disciplinary approach, when designing new biomaterials. Among various biomaterials, metals and alloys are used in orthopedics, dentistry, and other load-bearing applications; ceramics are used for their chemically inert nature or for their high bioactivity; polymers are useful for soft-tissue replacement applications. Broadly, all biomaterials are being developed to attain a balance between the physical properties of the replaced tissues and the

Advanced Structural Ceramics, First Edition. Bikramjit Basu, Kantesh Balani.

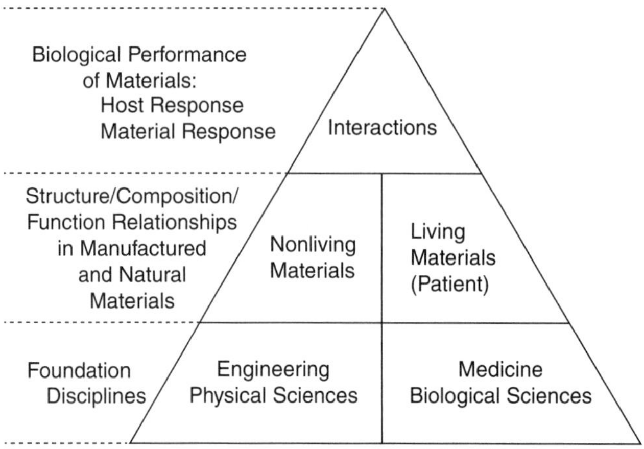

Figure 19.1 Concept triangle illustrating the synergistic interaction of engineering and biological science disciplines involved in designing biomaterials. The schematic also demonstrates the multidisciplinary approach of the science and technology of biomaterials.[1]

biochemical effects of the material on the tissue. However, for most biomedical applications, a range of properties is required, for example, biological activity, mechanical strength, and chemical durability. Therefore, a clinical need can only be met by a designed material that exhibits a tailored combination of several properties such as those mentioned here. Figure 19.2 displays the different organs of a living human body that can be replaced by various biomaterials. Various issues that drive continuous research on biomaterials—and bioceramic implant materials in particular—are summarized in Figure 19.3. Despite significant research on biomaterials, it has been realized that synthetic materials cannot mimic the extremely complex structure of bone in all aspects and the major drawback of synthetic materials is that they cannot repair themselves as living bone does. In a living human, articulating-joint replacements and dental restorations demand the use of hard-tissue and cortical-bone analogue materials, such as high-strength metals and high-hardness ceramics, and such bone replacement materials must have the desired combination of *in vitro* and *in vivo* biocompatibility properties. The major part of this overview chapter discusses this aspect.

19.2 HARD TISSUES

Hard tissues, characterized by their high hardness and elastic modulus (E), include the bones and teeth of human and animal bodies. Bone can be classified as cancellous bone and cortical bone. Cancellous bone (also called trabecular or spongy bone) has a characteristic porous structure; in contrast, cortical bone has a highly anisotropic microstructure that shows higher strength in the loading direction.[2] The mechanical properties of hard tissues are summarized in Table 19.1. The cortical bone has the best combination of strength and modulus, followed by cancellus bone, dentine, and enamel. However, it needs to be remembered that the bone properties are sensitive to anatomical location. All hard tissues are generally formed from four phases:

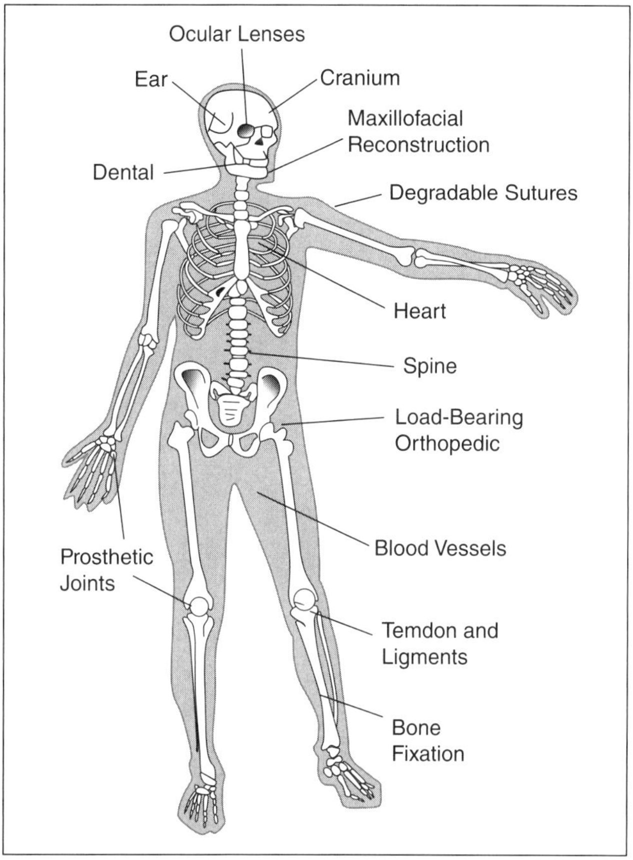

Figure 19.2 A schematic of the various human body parts that can potentially be replaced by synthetic biomaterials.[1]

collagen fibers, mineral (hydroxyapatite [HAp]), organic substances, and water. The relative fraction of each phase varies among bone types and teeth; typical composition of bone is provided in Table 19.2. In the case of tooth enamel, the mineral (i.e., HAp) content is around 95%. Excluding the organic mass and water, natural bone can be described as a natural nanocomposite containing HAp nanoparticles and collagen fiber. The collagen fibers provide strength to the bone, with the HAp particles located between the fibers. Proteins, polysaccharides, and mucopolysaccharides, in combination, act as cement.

19.3 SOME USEFUL DEFINITIONS AND THEIR IMPLICATIONS

19.3.1 Biomaterial

Broadly, *biomaterials* are defined as synthetic materials that are designed to induce a specific biological activity.[3] The major difference between biomaterials and other

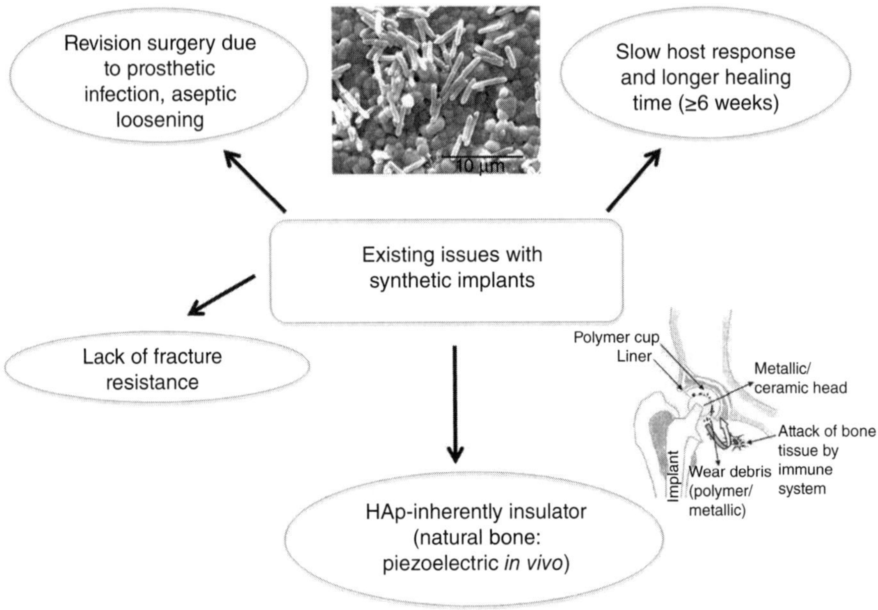

Figure 19.3 Schematic illustration of various issues concerned with biomaterials and, in particular, bioceramic implants.

Table 19.1. Mechanical Properties of Different Hard Tissues of Human System

Tissues	Elastic modulus (GPa)	Tensile strength (MPa)
Cortical bone	17.7	133
Cancellous bone	12.8	52
Enamel	0.4	7.4
Dentin	11.0	39.3

Table 19.2. Composition of Bone

Organic—collagen fibers (type 1)	16%
Mineral—hydroxyapatite [HAp, $Ca_{10}(PO_4)_6(OH)_2$]	60%
Ground substance	2%
Water	23%

materials classes is their ability to function in a biological environment without damaging their surroundings and without getting damaged in that process.[4] It must be emphasized here that the biological properties and response of a material in physiological environment are the determining factors for selecting and defining biomaterials. The most important aspect is, therefore, how a biomaterial interacts when it is implanted in a human or animal body.

19.3.2 Biocompatibility

The fundamental requirement of any biomaterial concerns the ability of the material to perform effectively with an appropriate host response for a targeted application; that is, the material and the tissue environment of the body should coexist without having any undesirable effect on each other. This is mentioned in Figure 19.1. Such a requirement is broadly known as biocompatibility.[5] Broadly, *biocompatibility* is defined as "the ability of a material to perform with an appropriate host response in a specific application." A more recent definition was proposed by D. F. Williams[6]: "Biocompatibility refers to the ability of a biomaterial to perform its desired function with respect to a medical therapy, without eliciting any undesirable local or systemic effect in the recipient or beneficiary of that therapy, but generating the most appropriate beneficial cellular or tissue response in that specific situation, and optimizing the clinically relevant performance of that therapy." From a biological point of view, biocompatibility originates from the acceptability of a nonliving synthetic biomaterial in living mammals and humans. Three important aspects of biocompatibility that a candidate biomaterial needs to achieve in diverse environments, such as bone, blood vessels, and the eye, are as follows: (1) It should be biochemically compatible, nontoxic, nonirritative, nonallergenic, and noncarcinogenic; (2) it should be biomechanically compatible with surrounding tissues; and (3) a bioadhesive contact must be established between the material and the living tissues. It needs to be emphasized here that the biocompatibility of a material depends on the place of application; for example, a specific bone replacement material may not be biocompatible in an application involving direct blood contact. As is discussed later, a range of *in vitro* and *in vivo* tests are suggested to completely describe the biocompatibility properties of a material.

19.3.3 Host Response

When developing new biomaterials, it is desirable to understand the *in vivo* host response of various biomaterials. An ideal scenario would be the formation of a structural and biological bond between the implant material and host tissues. Often materials cause tissue reactions, which may be systemic or local. Depending on the persistence of the systemic effects of a biomaterial in an osseous system, the systemic effect can be classified into four categories according to ISO 10993-11 standards: (1) acute, if observed within 24 hours of the implantation; (2) subacute, if observed within 14–28 days of the implantation; (3) subchronic, if observed within up to 90 days of implantation or within 10% of the animal's life span; and (4) chronic, if observed after more than 90 days postimplantation or more than 10% of the animal's life span.

Depending on the biocompatibility and host response, biomaterials can be broadly classified into three main categories[7]:

(a) ***Bioinert and Biotolerant.*** Bioinert materials cannot induce any interfacial biological bond between implants and bone, for example, Al_2O_3 and ZrO_2.

(b) ***Bioactive.*** Bioactive materials can attach directly to body tissues and form a chemical–biological bond during the early stage of implantation, for example, 45S5 bioglass and calcium phosphates (HAp).

(c) ***Bioresorbable.*** Bioresorbable materials gradually are resorbed and finally replaced by new tissues *in vivo*, for example, tricalcium phosphate (TCP) and bone cement.

19.4 CELL–MATERIAL INTERACTION

It should be evident that *cytocompatibilty*, that is, compatibility of materials with biological–animal cells, is one of the defining criteria in the development of biomaterials. Biologically, a cell can be described as a self-duplicating unit, given the proper nutrients and environment. In Figure 19.4a, the anatomy of a eukaryotic cell has been provided. Various important organelles, as identified in Figure 19.4a, include the mitochondrion (energy warehouse), Golgi apparatus, and endoplasmic reticulum (ER). The structure of the cytoskeleton is also visible; the cytoskeleton is composed of three proteinaceous structures: actin filament, microtubule, and intermediate filaments. In understanding the interaction of biomaterials in a human body, it is important to mention the physicochemical conditions of the human body environment. For example, nominal pH values lie in a wide range from 1.0 (gastric contents) to 7.4 (blood).[8] Additionally, pH values can change depending on health conditions. The normal temperature of the core of the human body is around 37.4°C; however, deviations over a range of temperatures, 20.0–42.5°C, are also reported for diseased patients.[8] Also, the total body burden of Ca, Na, and Cl ions is much higher and trace amounts of Mg, Fe, Zn, Cu, Al, and so on, are found in cytoplasm.

It is critical that any implant material generally, even though to a minimum extent, does not elicit a toxic response that kills cells in the surrounding tissues or release chemicals that can migrate within tissue fluids to cause systemic changes in the physiological environment in vivo. Therefore, it is important to understand the biomaterial–cell interaction. A schematic illustration of the phenomenology of biomaterial–cell interaction is shown in Figures 19.4b and 19.5.

It can be recalled here that, upon implantation of a material, a large number of protein molecules are adsorbed on the biomaterial's surface. This is because the number of protein molecules per eukaryotic cell is estimated to be around 10^9, and a simple calculation shows on the order of 10^{14} eukaryotic cells in a healthy human.

The protein adsorption acts as precursor to the cell–material interaction. A schematic of the protein adsorption phenomenon is provided in Figure 19.6. Importantly, a cell does not adhere directly to a material surface and the initial interaction is established through the interaction of cell-surface receptors with adsorbed protein ligands. Such protein-to-protein binding facilitates the spreading of a cell on the material's surface. Subsequent spreading to cover the biomaterial surface is promoted by cytoskeletal reorganization, as shown schematically in Figure 19.5.

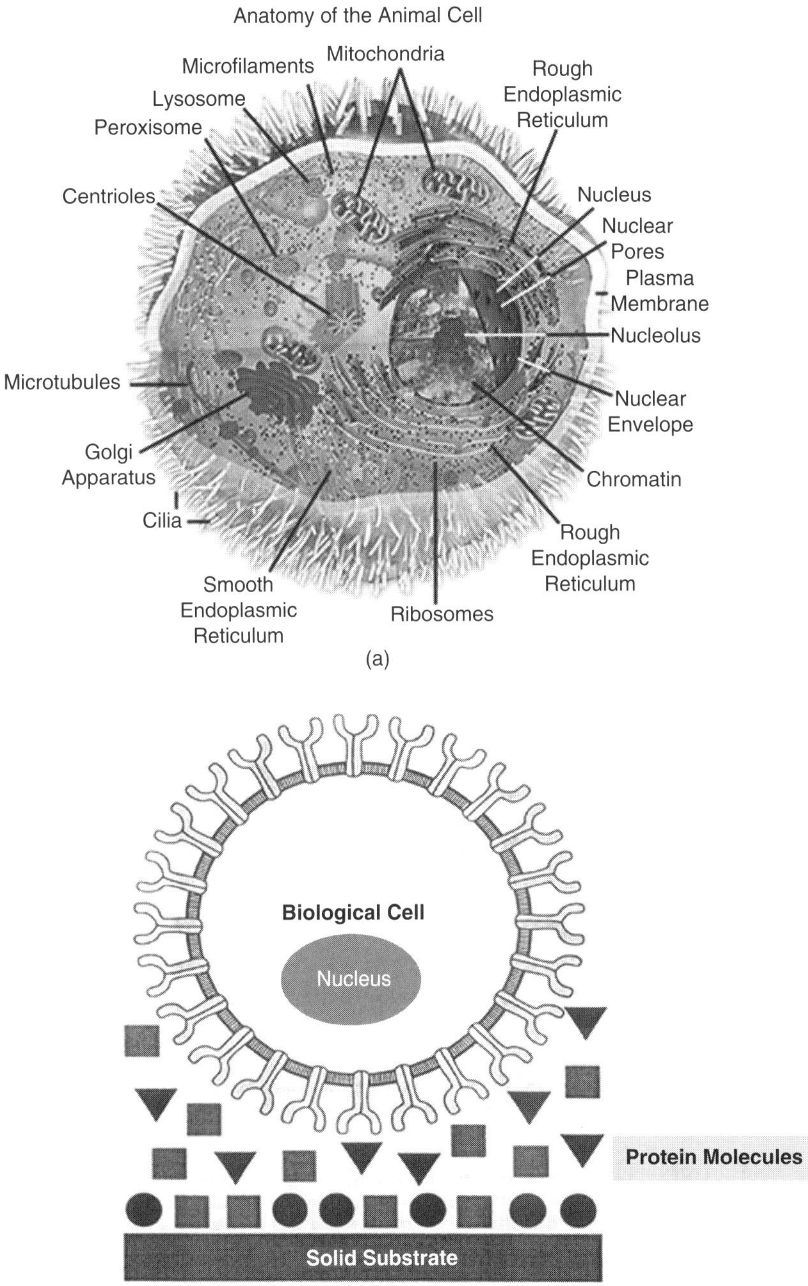

Figure 19.4 (a) Schematic illustration showing the anatomy of a eukaryotic animal cell (see color insert); (b) the fundamental mechanisms involved in biomaterial–cell interaction, established by the adsorbed proteins (circles, boxes, and triangles) with the integrin proteins of a biological cell.[1]

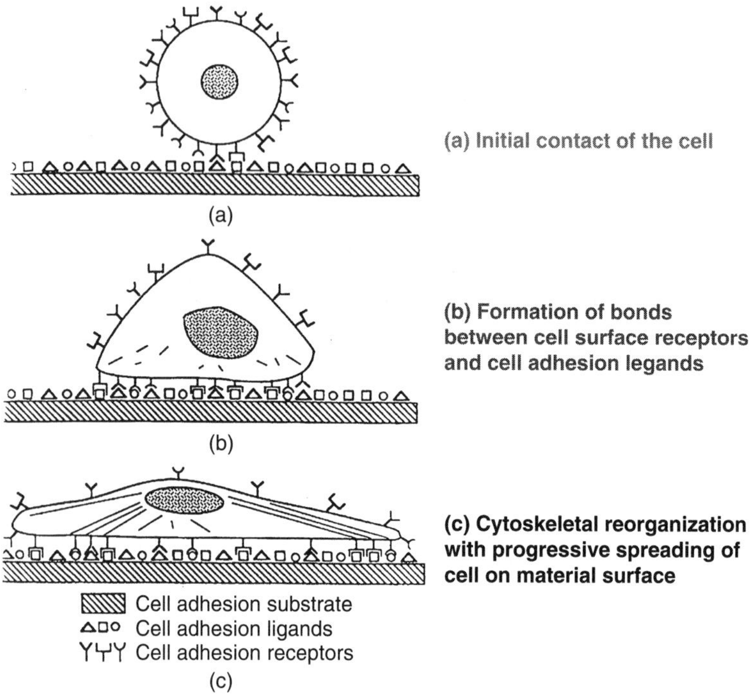

Figure 19.5 The physics of cell attachment process on a biomaterial surface: (A) initial approach of cell to an implanted material; (B) formation of chemical–physical bond between cell surface and adhered protein; and (C) spreading of cell on the surface of the material.[1]

Subsequently, the transport of various cell types toward the biomaterial surface occurs via cell signaling processes, and the interaction of the integrin proteins with the absorbed protein of the biomaterial's surface is established. The secretion of cell enzymes forms an extracellular matrix (ECM). Various cell types adhering in a self-organized manner form a tissue. The formation of small blood vessels (angiogenesis) as well as of large blood vessels (vasculogenesis) eventually takes place within the newly formed tissue layer, and this is necessary for the supply of nutrients locally to various cell types as well as for the removal of waste from the ECM.

19.5 BACTERIAL INFECTION AND BIOFILM FORMATION

Prosthetic infection remains a major challenge for the long-term use of many implanted or intravascular devices, such as joint prostheses, heart valves, vascular catheters, contact lenses, and dentures.[9] Bacterial adhesion to biomaterial surfaces is an essential step in the pathogenesis of these infections. Frequently, the failure of medical implants is reported to be from bacterial biofilm buildup. *Biofilm* consists of a community of microorganisms held together by a matrix, in which the micro-

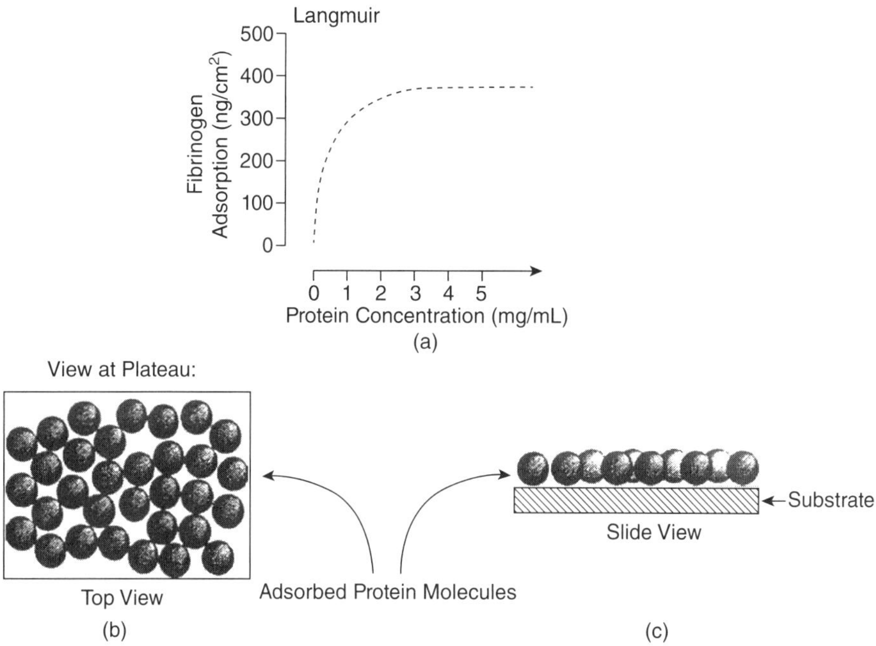

Figure 19.6 (A) The kinetics of protein absorption on biomaterial surface with protein concentration. (B) The adsorbed protein on surface (top view). (C) Protein on surface (side view).[1]

organisms cooperate and interact with one another. Biofilm may contain only one organism or a variety of different microorganisms. The main unit of the biofilm is the *microcolony*, which contains clusters of microorganisms. Microcolonies are located throughout the matrix, and they contain channels for the transport of oxygen, nutrients, waste, and other particles. In some situations, microscale layers within a biofilm contain cells of the same species that exhibit dissimilar phenotypes. Oxygen concentration gradients, pH differences, and other environmental variations are created by these microscale layers.

A *microorganism* or *microbe* is a living organism that is microscopic (too small to be seen by the naked human eye). Microorganisms are very diverse and they include bacteria, fungi, archaea, and protists. *Bacteria* are unicellular prokaryotic microorganisms or simple associations of similar cells. Typically, they are a few micrometers in length and have a wide range of shapes, ranging from spheres to rods and spirals. Cell multiplication is usually accomplished in a binary fission. The bacterial cell is surrounded by a lipid membrane, or cell membrane, which encloses the contents of the cell and acts as a barrier to hold nutrients, proteins, and other essential components of the cytoplasm within the cell. As they are prokaryotes, bacteria do not tend to have membrane-bound organelles in their cytoplasm and thus contain few large intracellular structures. They consequently lack a nucleus, mitochondria, chloroplasts, and the other organelles present in eukaryotic cells, such as the Golgi apparatus and ER. Bacteria do not have a membrane-bound nucleus, and their genetic material is typically a single circular chromosome located in the cytoplasm in an irregularly shaped body called the *nucleoid*. The nucleoid contains the chromosome with associated proteins and RNA. *Flagella* are rigid protein structures, about 20 nm in diameter and up to 20 μm in length, that are used for motility. Flagella are driven by the energy released by the transfer of ions down an electrochemical gradient across the cell membrane. *Fimbriae* are fine filaments of protein, just 2–10 nm in diameter and up to several micrometers in length (see Fig. 19.7a). They are distributed over the surface of the cell and resemble fine hairs when seen under the electron microscope. Fimbriae are believed to be involved in attachment to solid surfaces or to other cells and are essential for the virulence of some bacterial pathogens. *Pili* (singular, pilus) are cellular appendages, slightly larger than fimbriae, that can transfer genetic material between bacterial cells in a process called conjugation.

Bacterial adhesion can be described as physicochemical interactions by which bacteria adhere firmly to a material substrate or to a biological surface (cell or tissues). It has been described as the balance of attractive and repulsive physicochemical interactions between bacteria and surfaces. Bacterial adhesion and subsequent cell growth on a surface have important roles in a variety of systems, including biomaterial development and bacterial delivery systems used for bioremediation.[10,11]

The precursor to bacterial adhesion is slime formation, and the consequence of a large number of bacterial cells is biofilm formation. *Slime* is defined as an extracellular substance (exopolymers composed of mainly polysaccharides) produced by the bacteria that may be partially free from the bacteria after dispersion in a liquid medium (water soluble) and can be removed from bacterial cells by washing. Biofilm is an accumulated biomass of bacteria and extracellular materials (basically slime) on a solid surface. *Substratum* is a solid surface to which a microorganism may adhere. Any structure responsible for adhesive activities can be called *adhesins*. Bacteria may have multiple adhesins for different surfaces (different receptors).[12] A *receptor* is a component (both known and putative) on the surfaces of biomaterials or host tissue that is bound by the active site or adhesin during the process of *specific adhesion*.

Biofilm formation begins with adhesion of a microorganism to a surface. The initial interaction between the microbial cells and the surface is tenuous. It is during this time that biofilms are the most fragile, with cells frequently attaching and detaching from the biofilm surface. Once the cells have attached, they produce an extracellular polysaccharide (EPS) matrix, which provides stability to the biofilm by enabling cell–surface and cell–cell interactions. After stable attachment, the biofilm develops into a more complex environment; additional planktonic cells adhere, microcolonies develop, and complex biofilm architecture forms (see Fig. 19.7b,c).

19.6 DIFFERENT FACTORS INFLUENCING BACTERIAL ADHESION

Bacterial adhesion seems to be the initial step for bacterial infection and it depends on the various factors as discussed in the following subsections.

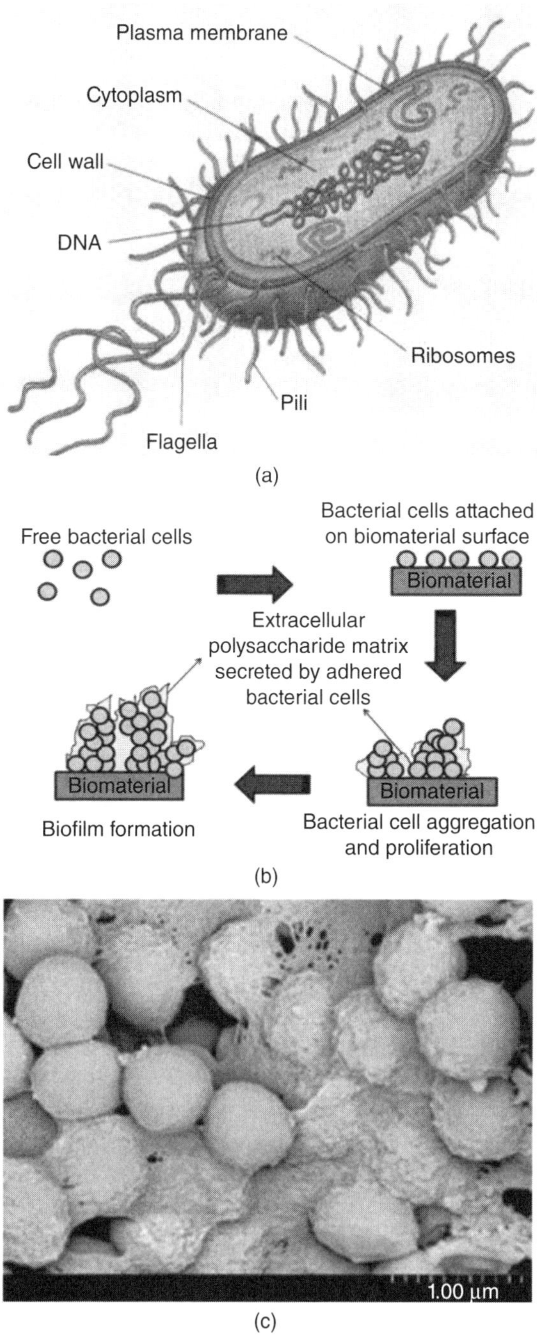

Figure 19.7 (a) Characteristic stricture of a gram-negative bacteria cell, *E. coli*. (b) Schematic representation of biofilm formation on material substrate (see color insert). (c) SEM image illustrating the typical formation of a staphylococcal biofilm. Note that the multiple layers of bacteria are covered with a polysaccharide matrix.[54]

19.6.1 Material Factors

(a) ***Material Surface Composition.*** Material surface composition largely governs bacterial adhesion.[13] In the group of absorbable sutures, polydioxanone sutures exhibit the smallest affinity toward the adherence of both *Escherichia coli* and *Staphylococcus aureus*. A study by Sugarman and Musher[14] reported that adherence of bacteria to the gut was up to 100 times greater than to nylon, and adherence to polyglycolic acid or silk was intermediate. According to Gristina et al.,[15] *Staphylococcus epidermidis* preferentially adheres to polymers and *S. aureus* to metals. If the surface chemistry is changed or modified, such as with an antimicrobial peptide coating, bacterial adhesion to these surfaces is discouraged.[16]

(b) ***Material Surface Properties.*** The physical configuration of the material surface is another underlying factor that influences the adhesion of bacteria on a biomaterial surface. It is basically a morphological description of the pattern of a material surface, such as a monofilament surface, a braided surface, a porous surface, or a gridlike surface, and it is a three-dimensional parameter. Merritt et al.[17] found that implant site infection rates are obviously different between porous and dense dental materials, where porous materials have a much higher rate. This implies that bacteria preferentially adhere to and colonize the porous surface.

(c) ***Surface Roughness.*** Surface roughness is a two-dimensional parameter of a material surface and this parameter is supposed to influence bacterial adhesion. It is a distance measurement between the peak and valley parts on a material surface and does not represent the morphological configurations of the surface. McAllister et al.[18] found that the irregularities of polymeric surfaces promote bacterial adhesion and biofilm deposition. Baker and Greenham[19] found that roughening the surface of either glass or polystyrene with a grindstone greatly increased the rate of bacterial colonization in a river environment. The causes for this phenomenon may include that a rough surface has a greater surface area and the depressions in the roughened surfaces provide more favorable sites for colonization. Clinically, different prostheses or implant devices have different surface roughnesses, which may play a role in bacterial adhesion and implant infection.

(d) ***Surface Hydrophobicity or Hydrophilicity.*** Metal surfaces have a high surface energy and are negatively charged. Therefore, metallic surfaces are hydrophilic, as shown by water contact angles. In contrast, polymers such as ultra-high-molecular-weight polyethylene (UHMWPE) or polytetrafluoroethylene (PTFE; Teflon) have low surface energy. The polymeric surfaces are less electrostatically charged and are hydrophobic. Depending on the hydrophobicity of both bacteria and material surfaces, bacteria adhere differently to materials with different hydrophobicities.[20] Hydrophilic materials are more resistant to bacterial adhesion than hydrophobic materials. Fletcher and Loeb[21] investigated the attachment of a marine *Pseudomonas* sp. to a variety of surfaces. Large numbers of bacteria were able to attach to hydrophobic plastics with little or no surface charge (Teflon, PE, polystyrene, and PE terephthalate); moderate numbers attached to hydrophilic metals with a positive or neutral surface charge; and very few attached to hydrophilic, negatively charged substrata (glass, mica, oxidized plastics). Satou et al.[22] studied the adhesion of two *Streptococcus sanguis* strains and two *Streptococcus mutans* strains to four surface-modified glass slides with different hydrophobicity. The *Streptococcus sanguis* strains (with more hydrophobic surfaces) adhered more to hydrophobic glass slides than others.

19.6.2 Bacteria-Related Factors

(a) ***Characteristics of Bacteria.*** For a given material surface, different bacterial species and strains adhere differently. This can be explained physicochemically, because physicochemical characteristics of bacteria are different between species and strains.

(b) ***Bacterial Hydrophobicity.*** Surface hydrophobicity of bacteria is an important physical factor for adhesion, especially when the substratum surfaces are either hydrophilic or hydrophobic. Hydrophobicity of bacteria can be determined by contact angle measurements, such as the sessile drop method.[23] The hydrophobicity of bacteria varies according to bacterial species and is influenced by growth medium and bacterial surface structure. Krekeler et al.[24] reviewed these factors. Generally, bacteria with hydrophobic properties prefer hydrophobic material surfaces; the ones with hydrophilic characteristics prefer hydrophilic surfaces; also, hydrophobic bacteria adhere to a greater extent than hydrophilic bacteria. Hogt et al.[20] found that one strain of *S. epidermidis* with hydrophobic characteristics showed a significantly higher adhesion to hydrophobic fluorinated ethylene propylene (FEP) than *S. saprophyticus*. Satou et al.[22] also found that *Streptococcus sanguis* strains with hydrophobic surfaces adhered more to hydrophobic glass slides than others with a less hydrophobic character.

(c) ***Bacterial Surface Charge.*** The surface charge of bacteria may be another important physical factor for bacterial adhesion.[20,25] Most particles acquire an electric charge in aqueous suspension due to the ionization of their surface groups. The surface charge attracts ions of opposite charge in the medium and results in the formation of an electric double layer. The surface charge is usually characterized by the isoelectric point,[26] the electrokinetic potential (or zeta potential). Bacteria in aqueous suspension are always negatively charged.[19] A high surface charge is accompanied by a hydrophilic character of the bacteria, but a hydrophobic bacterium may still have a rather high surface charge. The surface charge of bacteria varies according to bacterial species and is influenced by growth medium, the bacteria's age, and bacterial surface structure.[24] Long-range electrostatic forces may influence the initial phase of bacterial adhesion onto solid surfaces.

19.6.3 External Factors

(a) ***Surface Proteins.*** The role of protein adsorption on the material surface or outer membrane of bacteria is crucial to the bacterial adhesion. In the microenvironment, the synthesis of proteins and related enzymes may influence the electrochemical balance of the environment, which can affect its adhesion behavior. Many proteins (serum or tissue proteins) have been studied for their effects on bacterial adhesion to material surfaces, including albumin, fibronectin, fibrinogen, laminin, denatured collagen, and more. They promote or inhibit bacterial adhesion in either binding to substrata surfaces, binding to the bacterial surface, or being present in the liquid medium during the adhesion period. In the following paragraphs, the role of some typical proteins is discussed.

(b) ***Fibronectin.*** Fibronectin (Fn), which is recognized for its ability to mediate surface adhesion of eukaryotic cells, has also been shown to bind to *S. aureus*.[27] Fibronectin clearly promotes *S. aureus* adhesion to the substratum surface. Kuusela et al.[28] demonstrated a time-dependent and Fn-concentration-dependent adhesion of *S. aureus* to Fn-coated cover slips. Staphylococci may be saturated with Fn at a level that suggests the presence of specific receptors (staphylococcal Fn-binding molecules) on bacterial cells, and this Fn-binding molecule has been cloned in *E. coli* and purified.[29] The *S. aureus* binding domain of Fn was also found in the Fn molecule.[30]

(c) ***Albumin.*** Albumin, adsorbed on material surfaces, has shown obvious inhibitory effects on bacterial adhesion to ceramic surfaces.[31] The mechanism of the inhibiting effect of albumin is not clear. Albumin may reduce bacterial adhesion by changing substratum surface hydrophobicity, because in the presence of dissolved and adsorbed bovine serum albumin (BSA), substrata surfaces have been shown to become much less hydrophobic.[32]

(d) ***Fibrinogen.*** Fibrinogen is another important serum protein that mediates bacterial adhesion to biomaterials and host tissues. Most studies have shown that adsorbed fibrinogen promotes adherence of bacteria, especially staphylococci, to biomaterials. In the study by Herrmann et al.,[33] fibrinogen markedly promoted adherence of all *S. aureus* strains, but only a few coagulase-negative strains. The latter finding was supported by the study of Muller et al.[34] Fibrinogen bound to cover slips also increased streptococcal adhesion.[27] In another *in vitro* study, pretreatment of bacteria or both bacteria and PE catheter surfaces with fibrinogen enhanced bacterial adherence, suggesting the presence of ligands for fibrinogen on the staphylococcal cell surface.[35]

19.7 EXPERIMENTAL EVALUATION OF BIOCOMPATIBILITY

The international standard of biocompatibility testing, ISO-10993 draft, categorizes devices on the basis of the nature of their contact with the body:

(a) Surface-containing devices, for example, electrodes, compression bandages, contact lenses, and urinary catheters

(b) External communicating devices, for example, dental cements, arthroscopes, intravascular catheters, and dialysis tubing

(c) Implant devices, for example, hip and knee prostheses, pacemakers, artificial tendons, and heart valves

Any research program on new biomaterials must include a range of *in vitro* and *in vivo* tests, as stated by ISO-10993. ISO guidelines are followed to select the tests used for the biological evaluation of materials and of medical as well as dental devices. It is worth mentioning the difference between *in vitro* and *in vivo* tests. *In vitro* tests are laboratory simulated experiments, which are a must as initial screening tests. However, no information on inflammation and immune response of the materials can be obtained from *in vitro* tests. Also, most of the *in vitro* experiments use a single cell line, which does not simulate the real tissue–material interactions that occur *in vivo*. Nevertheless, the *in vitro* assays are effective as the first step of biocompatibility evaluations. Importantly, *in vivo* tests provide interactions of materials with ECM, blood-borne cells, protein, and molecules. These experiments are regarded as the second step prior to clinical use. Since both cell culture to assess cytotoxicity and bacteria culture to assess antimicrobial properties are significantly used in research on biomaterials for hard-tissue replacement, both the culture protocols and the related biochemical assays are described in detail in the following discussion.

(a) ***Cell Culture.*** This is an *in vitro* test for assessment of cell toxicity. Cytotoxicity experiments are performed in the laboratory, using relevant cell lines, and the cells are seeded on the materials. The first stage involves sterilization of the samples in order to remove other microorganisms from the surface. Sterilization is generally carried out in an autoclave at 15 psi and 121°C, and for some materials sterilization is performed using ultraviolet ray exposure or ethylene oxide (EtO). The culture medium is Dulbecco's Modified Eagle's Medium (DMEM), containing 10% serum, 1% antibiotic cocktail. The samples are incubated for 24 hours at 37.4°C (human body temperature) for the desired time scale (24 hours or longer). Following this, the cells are fixed in glutaraldehyde/formaldehyde and the cell growth–proliferation–adhesion is studied using a fluorescence microscope or scanning electron microscopy (SEM). In Figure 19.8, the experimental steps followed in culturing mammalian cells on material substrates are schematically shown.

(b) ***Cytotoxicity Assay.*** The MTT assay is a widely used colorimetric assay (an assay that measures changes in color) for quantitatively measuring mitochondrially active biological cells. MTT (i.e., 3-(4,5-dimethylthiazol-2-yl)-2,5-diphenyltetrazolium bromide, a tetrazole) forms the dark blue formazan product formed by the reduction of the tetrazolium ring of MTT by the mitochondrial enzyme succinate dehydrogenase. Therefore, the reduction of MTT will be greater if a greater number of viable cells is present. During an MTT assay, a solubilization solution (usually either dimethyl sulfoxide

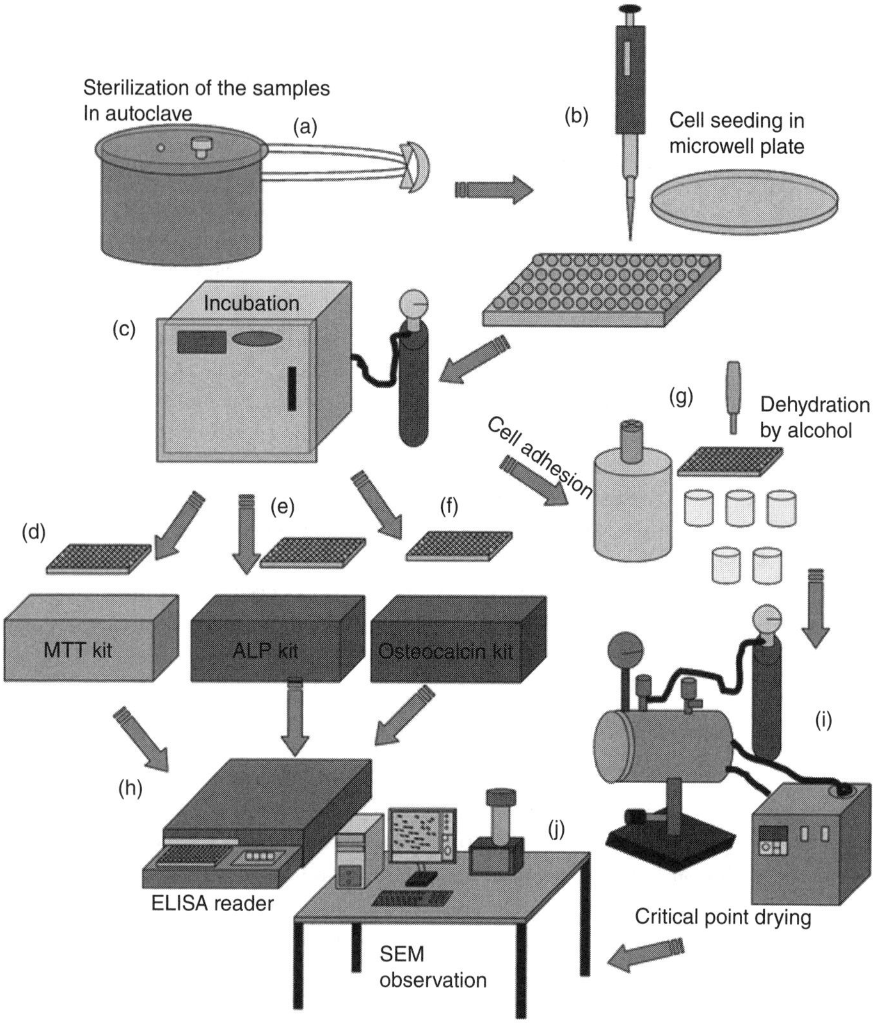

Figure 19.8 Summary of experimental steps and biological protocol involved in cell culture protocol to observe cellular adhesion on biomaterials as well as various biomineralization assays to investigate cell viability and early- versus later-stage osteoblast differentiation in contact with biomaterial surface.[55] Enzyme-linked immunosorbent assay (ELISA) provides quantified values of optical density (OD). ALP, alkaline phosphatase.

or a solution of the detergent sodium dodecyl sulfate in dilute hydrochloric acid) is added to the culture medium to dissolve the insoluble purple formazan product into a colored solution. The absorbance of the treated solution can be quantified using measurements at a certain wavelength (usually between 500 and 600 nm) by a spectrophotometer. This reduction is associated with the activity of mitochondrial reductase enzymes and, therefore, the conversion is directly related to the number of viable (living) cells. The

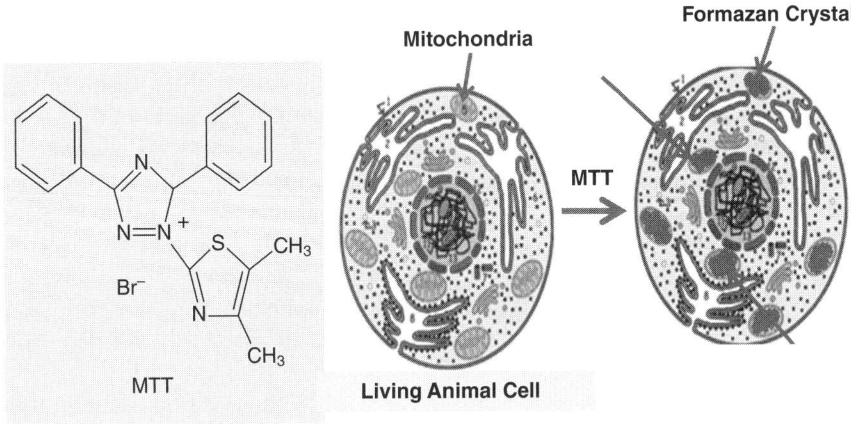

(a)

Antiboby-coated MAB 1 + MAb 2 – HRP + tetramethylbenzydine $\xrightarrow{\text{OC enzyme}}$ Color change of MAb

MAb: monoclonal antibody, HRP: horseradish peroxide

(b)

Figure 19.9 (a) Chemical structure of MTT (left, taken from http://en.wikipedia.org/wiki) as well as fundamental principle involved in MTT assay to investigate cell viability (see color insert) and (b) biochemical reaction involved in osteocalcin (OC) assay.

principle involved in the MTT assay for investigating cell viability is illustrated in Figure 19.9.

Another biochemical assay, the alkaline phosphatase (ALP) assay, is considered as an early-stage differentiation marker for bone cell differentiation, such as osteogenesis, which is used to examine osteogenesis associated with increased expression of ALP. ALP enzyme is bound to the cell membrane of osteoblasts and functions to promote osteogenesis by degrading pyrophosphates.

The osteocalcin (OC) assay is useful in assessing the late-stage osteoblast differentiation ability of a biomaterial. OC is known to be a noncollagenous protein. The OC gene encodes a 6-kDa polypeptide, one of the most abundant noncollagenous bone proteins. In principle, a microwell plate precoated with a biotin-conjugated polyclonal antibody specific for OC reacts with a biomaterial substrate. Avidin conjugated to horseradish peroxidase (HRP) is added to each microplate well and incubated. A solution of OC and 3,3′,5,5′-tetramethyl-benzidine (TMB) is added to each well; for a well containing OC enzyme, the HRP antibody shows a change in color (see Fig. 19.9b). The color change is measured spectrophotometrically at a wavelength of 450 nm.

(c) ***Antimicrobial Tests.*** In the following, the protocol for the culture of a model bacterium, *E. coli*, is presented. Prior to seeding on the sample surface, *E. coli* bacteria are incubated overnight in a nutrient broth supplemented with yeast and beef extract, at 37°C in an incubator. The pure suspension culture having bacterial density of 5×10^7 is normally recommended to be seeded on a biomaterial and incubated for 4 hours at 37°C. Following the standard protocol, the bacteria cells are dehydrated and then dried by 100% hexamethyldisilazane (HMDS) for 10 minutes. To assess the antimicrobial property of biomaterials, both qualitatively and quantitatively, a direct counting method on SEM images and an indirect counting method such as a colony-forming unit (CFU) plate count can be followed. To study the bacterial morphology and its adhesion over the surface, SEM analysis is normally used. A similar protocol can be followed for gram-positive *S. aureus* and *S. epidermidis* bacteria using Luria–Bertani broth. The entire bacteria culture protocol is schematically shown in Figure 19.10a.

A *CFU plate count* is one of the basic methods for the quantification of bacterial adhesion on biomaterials. The test is performed by taking a powder concentration of 50 mg/mL of the test sample. A bacterial suspension of 0.1 absorbance is prepared using an ultraviolet–visible spectrophotometer; the fundamental operating principle is shown in Figure 19.10b. After that, the bacterial suspension is added to vials and incubated at 37°C for 4 hours along with continuous shaking. Agar plates are normally prepared by spreading agar solutions in sterile petri dishes following the incubation. Subsequently, each incubated solution can be spread on the agar plates in an equal number of lines. Finally, those plates are incubated for 24 hours at 37°C and the bacterial colonies on each plate are counted.

(d) ***Genotoxicity.*** In this *in vitro* experiment, the DNA-damaging capability of a biomaterial eluate is assessed using single-cell gel electrophoresis (SCGE; also known as the comet assay) and a micronucleus assay.

(e) ***Hemocompatibility.*** Hemocompatibility signifies the compatibility of a material with red blood cells. In particular, the thrombogenic property of a material is assessed in a blood stream flowing over the biomaterial; a better material should ideally show limited thrombus formation. Such an evaluation is necessary for cardiovascular implant materials. Examples of hemocompatibile materials include PTFE (Teflon) and diamondlike carbon (DLC).

(f) ***Sensitization.*** This is an *in vivo* test in which materials are kept in the subcutaneous region of an animal and the change in skin color, allergic effect, or some other irritations are observed periodically.

(g) ***Carcinogenicity.*** Carcinogenicity is a long-term *in vivo* experiment that determines any cancerous effect of the biomaterial eluate on cells. Examples of carcinogenic materials include Pb and Sn.

(h) ***Implantation and Histopathology Investigation.*** Implantation is an important *in vivo* experiment in which a sample of a predefined shape is placed in the bone defect of a mammal (rabbit, rat, or mouse); after the desired time period, the samples and surrounding tissues are examined histopathologically to investigate the *in vivo* response of the materials. In general, short-term implantation tests are performed for up to 12 weeks and long-term tests for up to 78 weeks. Since the animals are sacrificed at the end of

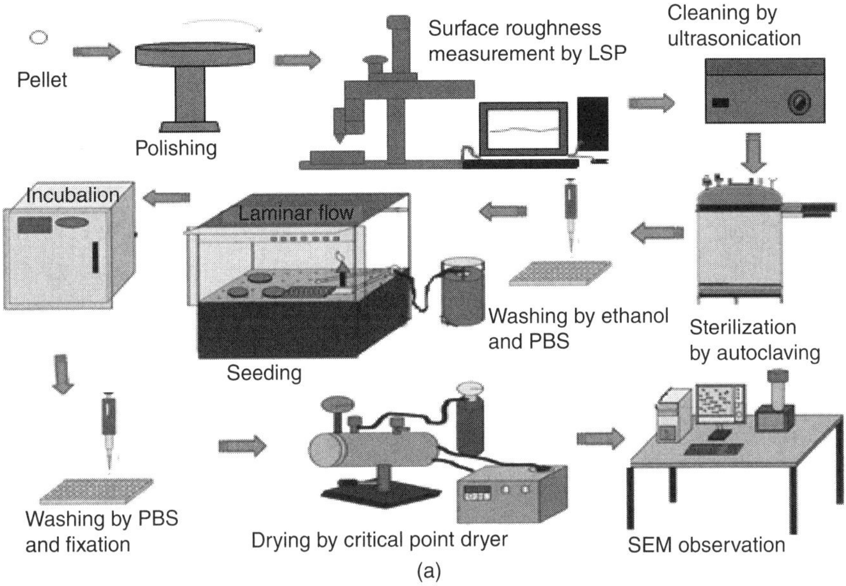

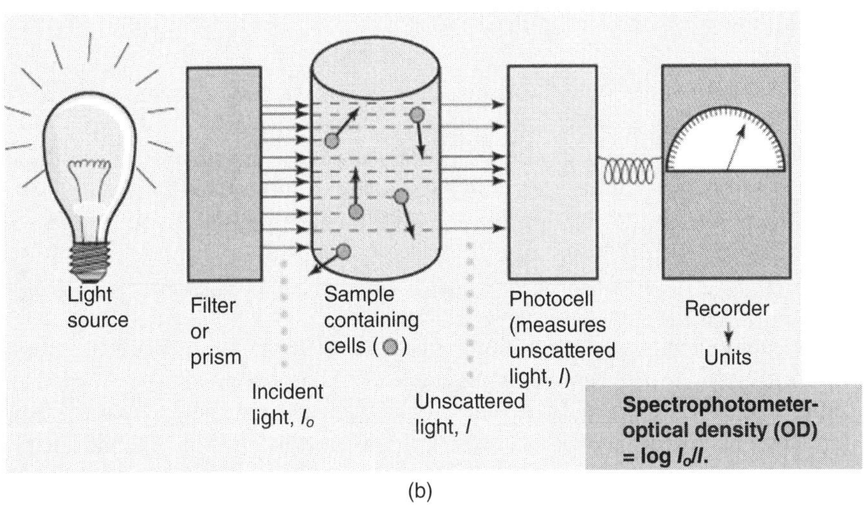

Figure 19.10 (a) Summary of experimental steps and biological protocol involved in bacteria culture protocol to assess antimicrobial properties of biomaterial surface. (b) Fundamental principle, as applied to ultraviolet–visible spectrophotometer, to quantify the number of bacterial cells.[54] LSP, laser surface profilometer; PBS, phosphate buffered saline.

implantation tests, the number of animals is limited to a minimum number from an animal welfare point of view.

As far as the ISO-approved procedure is concerned, all the biomaterial samples are cleaned ultrasonically and sterilized by ethylene oxide (ETO). Control samples, for example, HAp or UHMWPE, are used for comparison. An example of the implantation

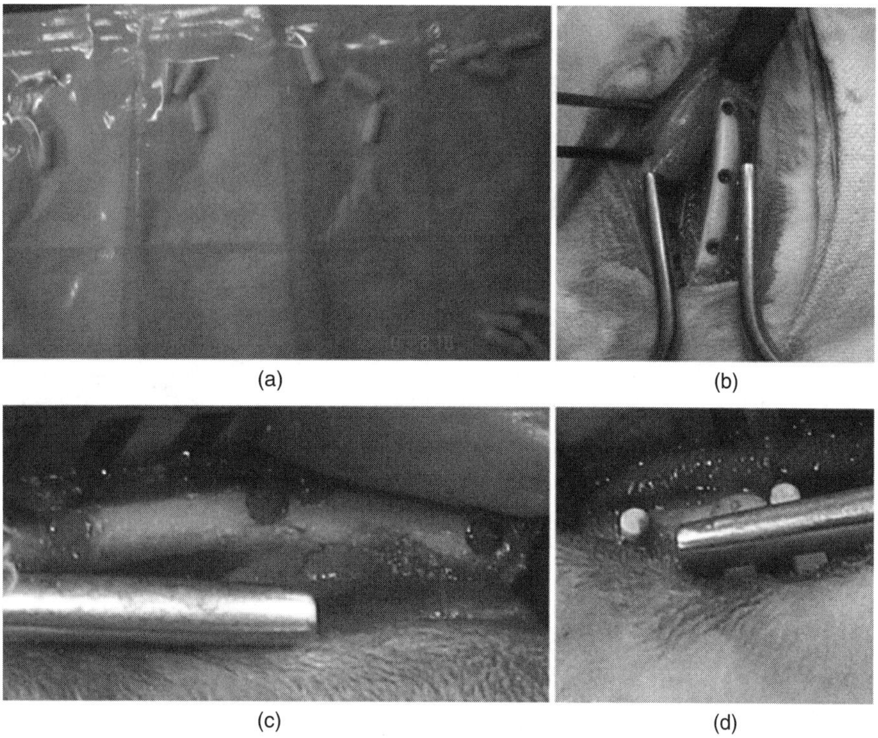

Figure 19.11 Digital camera images showing (a) the pin-shaped (2-mm diameter and 6-mm length) bioceramic implant, prior to implantation, (b) rabbit's femur before implantation, showing the implantation sites (holes), (c) control samples inserted inside the holes, and (d) HAp–(20 wt%) mullite samples, implanted inside the holes.[55] See color insert.

of bioceramic samples in a rabbit's femur is given in Figure 19.11. The implantation procedure is necessarily carried out under clean and aseptic conditions. The experimental animals are premedicated with atropine (0.15 mg/kg) and diazepam (3.0 mg/kg). Subsequently the animals are anesthetized with Xylaxin (5 mg/kg) and ketamin (90 mg/kg) by intramuscular injection. The skins of the anaesthetized rabbits are treated using 70% alcohol followed by Betadine (povidone–iodine) solution. During the operation, cylindrical holes are drilled on the femur and test samples are implanted in the holes; subsequently, the wound is closed using stitches. X-ray radiographs are regularly taken to monitor the implants *in situ*. The implantation sites are macroscopically examined for any evidence of host response and the femur bones with the test materials are removed and fixed in 10% buffered formalin.

Following standard protocols, thin sections of implant with surrounding bone are stained with Stevenel's blue and counterstained with Van Gieson's picrofuchsin. Subsequently, the dried samples can be investigated using SEM and atomic force microscopy (AFM). The entire procedure of preparing samples for histopathological analysis is schematically illustrated in Figure 19.12.

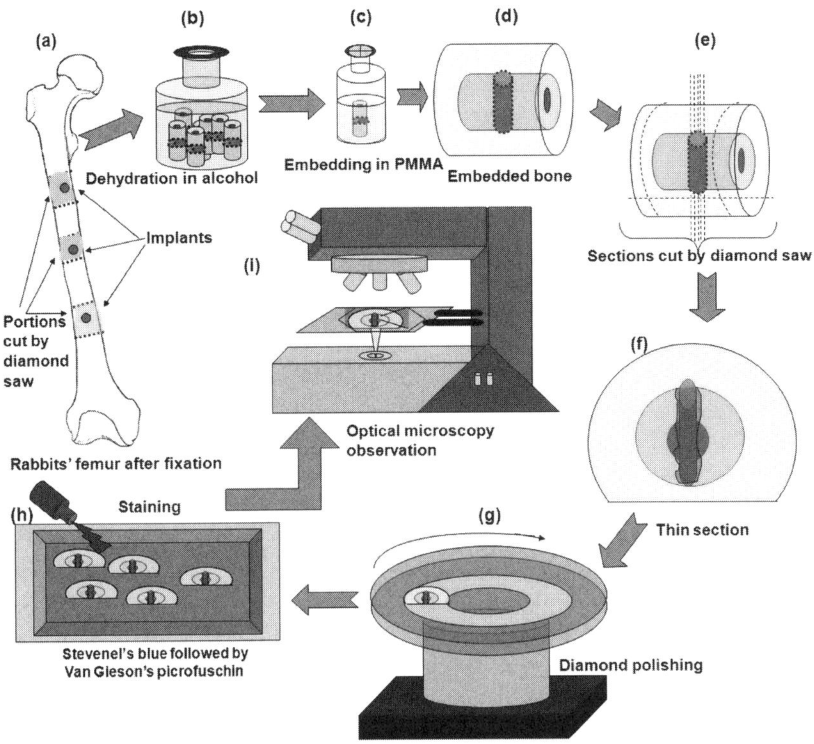

Figure 19.12 Major steps involved during sample preparation for histopathology analysis: (a) the femur of the rabbit after fixation, where the red circles indicate implants; (b) dehydration of bone pieces along with implants; (c) embedding in poly(methyl methacrylate) (PMMA) polymer; (d) embedded bone piece in PMMA, after removal from the bottle; (e) cutting of thin sections by diamond saw, where the thin dotted lines indicate the cutting path; (f) thin section showing the bone (yellow) and bone marrow (pink) embedded in PMMA matrix; (g) polishing of thin section using diamond paste; (h) staining in Stevenel's blue; (i) optical microscopy observation.[55] See color insert.

Furthermore, the ISO-10993 document categorizes implants according to the duration of their interaction with the body: limited exposure (<24 hours), prolonged exposure (>24 hours and <30 days), or permanent contact (>30 days). The duration of interaction and the type of contact between the device and tissues also influence the selection of the test used to assess the device's compatibility.

19.8 OVERVIEW OF PROPERTIES OF SOME BIOMATERIALS

19.8.1 Coating on Metals

As mentioned earlier, an alternative approach to the use of bulk biomaterials is coatings. The driving force for developing various coatings is that the properties of coated bioimplants will combine advantageous properties of both coating and substrate materials. For example, bioactive ceramic coatings on metallic implants can display good strength (due to the metal) as well as good bioactivity (due to the ceramic coating). The coating–substrate adhesion influences the physical properties of coatings. For example, a HAp-containing glass coating on a titanium dental implant can exhibit better adhesion than flame sprayed HAp coatings. HAp-containing glass coatings have advantageous properties, which include increased abrasion resistance, improved aesthetics (color etc.), and enhanced bioactivity. An *in vitro* study was used to investigate the biological response of HAp/Ti–6Al–4V composite coatings in simulated body fluid (SBF) solutions.[36] The coatings were reported to undergo two biointegration processes, that is, dissolution during the initial 4 weeks of soaking in SBF and subsequent bonelike apatite crystal precipitation. The coatings exhibited mechanical stability superior to the pure HAp coatings, indicating a much better long-term stability of the composite coatings in a physiological environment. With the interposition of a composite bond coat (50 vol% HAp and 50 vol% TiO_2), a composite coating on titanium substrate was synthesized using plasma spraying; no chemical reaction was observed between HAp and TiO_2 and it was found that toughness increased with addition of TiO_2[37].

Godley and coworkers[38] proposed that the essential condition for a biomaterial to bond with living bone is the formation of a biologically active bonelike apatite on its surface. In their work, it was demonstrated that chemical treatment is useful in creating a calcium phosphate (CaP) surface layer, which might provide the alkali-treated Nb metal with bone-bonding capability. The formation of a similar CaP layer upon implantation of alkali-treated Nb into the human body is expected to enhance the bonding of the implant to the surrounding bone.

Pajamaki et al.[39] investigated the effect of a glass-ceramic coating on titanium implants. They reported the results with uncoated Ti and showed that, after 52 weeks, the coated metal showed 78% bone ingrowth, whereas in the case of uncoated metal, the bone coverage was only 37%. Munting[40] discussed the merits and demerits of HAp coating on metal implants. The results of using HAp coatings for implant fixation are discussed following a 5-year histomorphological study of the bony incorporation of macroporous stemless hemiarthroplasties in dogs. Munting showed that, importantly, HAp coatings exhibited limited strength and poor fatigue strength.[40] Also, HAp is dissolved *in vivo* in an acidic environment, created by macrophages. Bone ingrowths, however, are observed in contact with the resection surface. Also, coating thickness is a key parameter as a thicker coating has chances to delaminate and a thinner coating has shorter life. A coating thickness of 40–60 μm is reported to be ideal. Wang et al.[41] compared the *in vivo* histocompatibility of plasma-sprayed and electrochemically deposited HAp coatings on Ti–6Al–4V alloy with uncoated Ti–6Al–4V alloy. It was reported that plasma-sprayed HAp coatings had higher bone apposition ratio than those exhibited by bare Ti–6Al–4V and electrochemically deposited HAp coatings after 7 days. However, after 14 days of implantation, both the coated materials exhibited similar bone apposition ratios, much higher than that for uncoated Ti–6Al–4V. Figure 19.13 shows the interaction of tissue with plasma-sprayed and electrochemically deposited coatings.

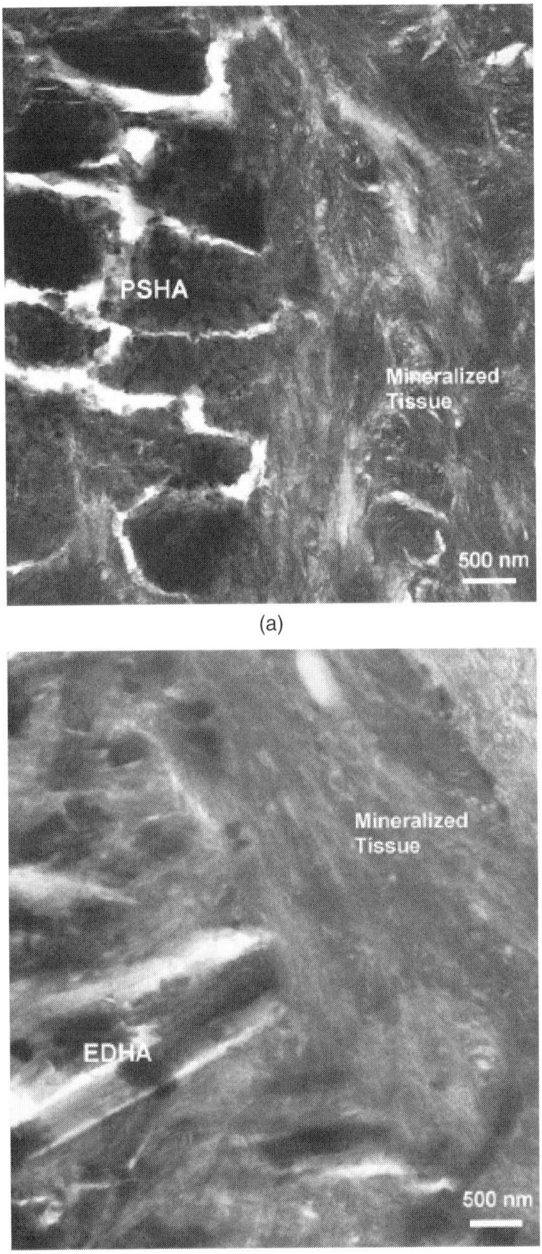

Figure 19.13 Cross-section TEM micrographs of samples of different materials implanted for 14 days in rabbits: (a) plasma-sprayed hydroxyapatite (PSHA) and (b) electrochemically deposited hydroxyapatite (EDHA) coatings; thin sections were stained with uranyl acetate and lead citrate.[41]

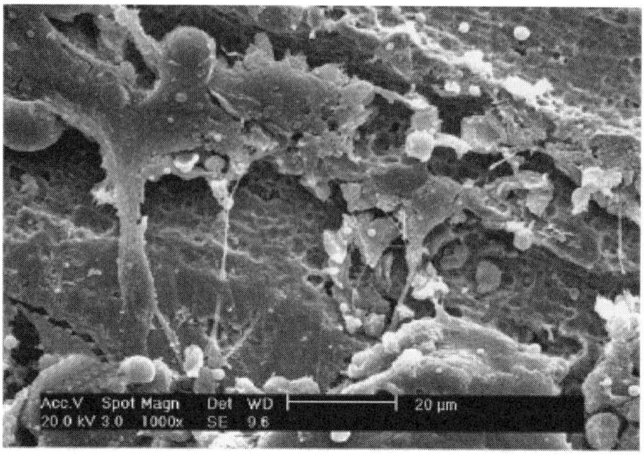

(a)

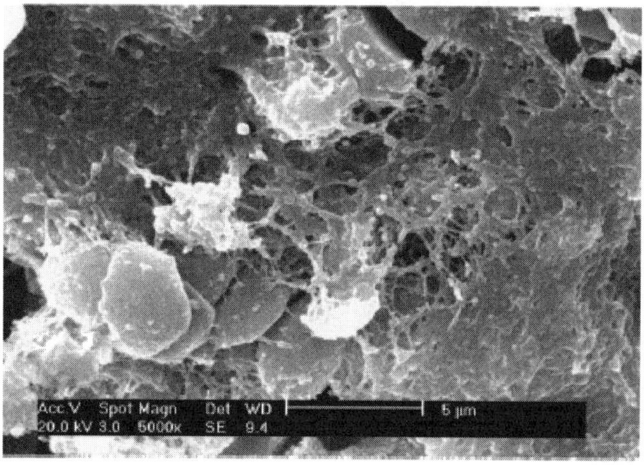

(b)

Figure 19.14 The morphologies of surface of implants after 13 weeks of implantation: (a) a little bulged cell and collagen fibrous network on uncoated NiTi implant; (b) a large number of bone cells and collagen fibrous structures on coated NiTi implant.[42]

In their work, Chen et al.[42] prepared a bioactive surface by immersing a shape memory alloy in SBF. It was found that a HAp layer can easily form on the surface of the NiTi alloy after immersing 48 hours in SBF. The surface-modified and as-received NiTi cylindrical implants were implanted in rabbit femurs. It was reported that a HAp coating can facilitate fast proliferation of osteoblasts. After 13 weeks of implantation, the interface between the coated implant and the natural bone revealed osteobonding. In contrast, the interface between the uncoated implant and the bone has gaps, showing a weak bone–implant interface (Fig. 19.14).

In a different work, Choubey et al.[43] reported the tribological behavior, under fretting contacts, of Co–Cr–Mo alloys coated by chemical vapor deposition of diamondlike carbon (CVD-DLC). They performed the experiments on coated materials in Hank's balanced salt solution to assess the *in vitro* performance in SBF (physiological) solution. In SBF solution, DLC-coated Co–Cr–Mo alloys exhibited a low coefficient of friction (COF) of 0.07–0.10, whereas a high COF (0.4) was measured for uncoated Co–Cr–Mo, under identical fretting conditions. The wear mechanism was mainly governed by mild wear with no significant change in surface morphology in DLC-coated flat material. The obtained research results confirmed the superior tribological performance of DLC coatings compared with uncoated Co–Cr–Mo alloys. In a review article, Hanawa[44] reported the effect of metal ion release from metallic implants *in vivo*. Oxide films, which formed on the surface of metallic materials, protected the surface from releasing ions. Low concentration of dissolved oxygen, inorganic ions, proteins, and cells can potentially accelerate the release of metal ions. Also, the formation and breakdown as well as regeneration time of the oxide layer on the metallic implant surface determine the release rate. In some specific applications, ions and debris are also released due to some mechanical actions, such as wear and fretting. It was concluded that there is a small chance that metal ions in combination with biomolecules can cause cytotoxicity, allergy, and other biological influences.

19.8.2 Glass-Ceramics-Based Biomaterials

Glass-ceramics are processed using prolonged heat treatment of glass at elevated temperature (above the glass transition temperature, T_g). Depending on processing conditions, the amount of crystalline ceramic phase can vary between 50 and 99 vol%. Since 2000, several glass-ceramic systems have been researched for their biomedical—in particular, dental restoration—applications. It can be recalled here that human teeth act as mechanical devices during masticatory processes, and the tooth material gets damaged or worn away with age and, therefore, partial or total replacement of human teeth with a suitable biocompatible material is required.

Among various glass-ceramic compositions, 45S5 glass, invented by Hench and coworkers, is considered as the best bioactive glass (BG) material.[45] A typical composition[46] is SiO_2, 46.1; P_2O_5, 2.6; CaO, 26.9; Na_2O, 24.4. A study by Chevalier and coworkers[47] reported the crystallization kinetics of 45S5 glass. The results indicate that 45S5 glass undergoes a series of structural transformations, leading to the formation of $Na_2CaSi_2O_6$ crystals at 610°C. However, poor mechanical properties and lack of machinability remain a major concern for 45S5 glass. Therefore, these materials are unsuitable for dental applications or any biomedical applications requiring a complex shape. Besides 45S5 glass, Bioverit®I base glass has wider applications, including the possibility of using Bioverit II base glass as a matrix for Ti-particle-reinforced composite coatings.[48] The coatings were fabricated by a single-step vacuum plasma spray method on Ti–6Al–4V substrates. Mechanical characterization revealed a good adherence of the coatings to the substrate and a toughening effect of the dispersed Ti particles. In another study, composites of Bioverit III base glass and glass-ceramic matrix with Ti-particle reinforcement were processed using a simple pressureless sintering method.[49] The sintering process was carefully optimized using differential scanning calorimetry (DSC) and hot stage microscopy. The growth of fibroblast cells on the surface of the glass-ceramic matrix composite confirms their biocompatibility. In a different system, complex reactions between titanium and HAp have been reported to take place during the sintering of Ti/HAp/BG composites.[50] The pressureless sintered composites of Bioverit III glass-ceramic[51] matrix with yttria–partially-stabilized zirconia (Y-PSZ) particulate reinforcement also confirmed the toughening effect of the Y-PSZ particles. Similarly, highly dense (>98% of relative density) Si_3N_4–bioglass composite[52] has the potential advantage of each constituent, that is, the high fracture toughness of Si_3N_4 with the bioactivity of a bioglass. The most significant feature concerning the mechanical properties of this biocomposite is the improvement in fracture toughness (4.4 MPa $m^{1/2}$) and bending strength (383 ± 47 MPa) with respect to currently used glasses and glass-ceramics for load-bearing applications. da Rocha Barros et al.[53] reported the *in vivo* bone tissue response of another fluoride-containing canasite glass-ceramic ($0.47K_2O$–$0.94Na_2O$–$1.42CaO$–$5.67SiO_2$–$1.5CaF_2$). HAp was used as control sample for this experiment. Interestingly, the canasite formulation evaluated was not osteoconductive and was reported to degrade in the biological environment.

19.9 OUTLOOK

As a concluding note, a summary of the combination of aspects related to processing as well as physical biological properties to be considered when developing bone analogue materials is provided in Figure 19.15. Among the physical properties, the

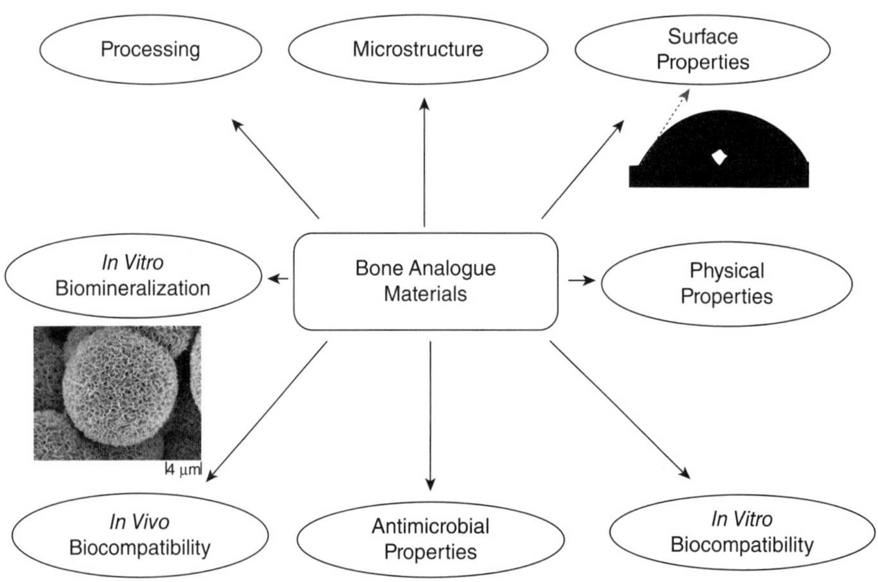

Figure 19.15 Schematic illustration of various aspects that need to be considered when developing an implant material.

E-modulus, strength, and toughness are important parameters and such properties are determined by the microstructure of the as-processed materials. Among biological properties, cellular functionality and cell fate processes as well as antimicrobial properties and *in vivo* osseointegration are important for hard-tissue replacement applications. It is impossible to optimize the array of properties in a unique material composition. This is because natural bone has a unique composition and properties, and many synthetic materials cannot even closely mimic the structure and properties of natural bone. Therefore, a synergistic approach to combining various properties in designed composite materials is possibly the only solution.

REFERENCES

1 B. D. Ratner, A. S. Hoffman, F. J. Schoen, and J. E. Lemons. *Biomaterials Science—An Introduction to Materials in Medicine*, 2nd ed. Academic Press, New York, 2004, 526.

2 J. Black. *Orthopaedic Biomaterials: Research and Practice*. Churchill Livingstone, New York, 1988.

3 D. F. Williams. *Consensus and Definitions in Biomaterials: Advances in Biomaterials.* Elsevier Publishers, Amsterdam, The Netherlands, 1988.

4 M. S. Valiathan and V. K. Krishnan. Biomaterial: An overview. Natl. Med. J. Ind. 12(6) (1999), 270–274.

5 D. F. Williams. *Definitions in Biomaterials, Progress in Biomedical Engineering*. Elsevier Publishers, Amsterdam, The Netherlands, 1987.

6 D. F. Williams. On the mechanisms of biocompatibility. Biomaterials 29 (2008), 2941–2953.

7 B. Basu, D. Katti, and A. Kumar. *Advanced Biomaterials: Fundamentals, Processing and Applications*. John Wiley & Sons, Hoboken, NJ, 2009.

8 F. H. Silver and D. L. Christiansen. *Biomaterials Science and Biocompatibility*. Springer, London, 1999.

9 M. Katsikogianni and Y. F. Missirlis. Concise review of mechanisms of bacterial adhesion to biomaterials and of techniques used in estimating bacteria material interactions. Eur. Cell Mater. 8 (2004), 37–57.

10 Y. L. Ong, A. Razatos, G. Georgiou, and M. M. Sharma. Adhesion forces between *E. coli* bacteria and biomaterial surfaces. Langmuir 15 (1999), 2719–2725.

11 L. M. Schmidt, J. J. Delfino, J. F. Preston, and G. St Laurent. Biodegradation of low aqueous concentration pentachlorophenol (PCP) contaminated groundwater. Chemosphere 38 (1999), 2897–2912.

12 D. L. Hasty, I. Ofek, H. S. Courtney, and R. J. Doyle. Multiple adhesins of streptococci. Infect. Immun. 60 (1992), 2147–2152.

13 Y. H. An, R. J. Friedman, R. A. Draughn, E. Smith, C. Qi, and J. F. John. Staphylococci adhesion to orthopaedic biomaterials. Trans. Soc. Biomater. 16 (1993), 148.

14 B. Sugarman and D. Musher. Adherence of bacteria to suture materials. Proc. Soc. Exp. Biol. Med. 167 (1981), 156–160.

15 A. G. Gristina, C. D. Hobgood, and E. Barth. Biomaterial specificity, molecular mechanisms, and clinical relevance of *S. epidermidis* and *S. aureus* infections in surgery, in *Pathogenesis and Clinical Significance of Coagulase-Negative Staphylococci*. G. Pulverer, P. G. Quie, and G. Peters (Eds.). Gustav Fischer Verlag, Stuttgart, 1987, 143–157.

16 L. W. Duran, J. A. Pietig, and J. E. Driemeyer. Prevention of microbial colonization on medical devices by photochemical immobilization of antimicrobial peptides. Trans. Soc. Biomater. 16 (1993), 35.

17 K. Merritt, J. W. Shafer, and S. A. Brown. Implant site infection rates with porous and dense materials. J. Biomed. Mater. Res. 13 (1979), 101–108.

18 E. W. McAllister, L. C. Carey, P. G. Brady, R. Heller, and S. G. Kovacs. The role of polymeric surface smoothness of biliary stents in bacterial adhesion, biofilm deposition, and stent occlusion. Gastrointest. Endosc. 39 (1993), 422–425.

19 A. S. Baker and L. W. Greenham. Release of gentamicin from acrylic bone cement: Elution and diffusion studies. J. Bone Joint Surg. 70 (1998), 1551–1557.
20 A. H. Hogt, J. Dankert, J. A. de Vries, and J. Feijen. Adhesion of coagulase-negative staphylococci to biomaterials. J. Gen. Microbiol. 129 (1983), 1959–1968.
21 M. Fletcher and G. I. Loeb. Influence of substratum characteristics on the attachment of a marine pseudomonad to solid surfaces. Appl. Environ. Microbiol. 37 (1979), 67–72.
22 N. Satou, J. Satou, H. Shintani, and K. Okuda. Adherence of streptococci to surface-modified glass. J. Gen. Microbiol. 134 (1988), 1299–1305.
23 H. J. Busscher, A. H. Weerkamp, H. C. van der Mei, A. W. J. van Pelt, H. P. de Jong, and J. Arends. Measurement of the surface free energy of bacterial cell surfaces and its relevance for adhesion. Appl. Environ. Microbiol. 48 (1984), 980–983.
24 C. Krekeler, H. Ziehr, and J. Klein. Physical methods for characterization of microbial cell surfaces. Experientia 45 (1989), 1047–1054.
25 J. Dankert, A. H. Hogt, and J. Feijen. Biomedical polymers: Bacterial adhesion, colonization, and infection. CRC Crit. Rev. Biocompat. 2 (1986), 219–301.
26 V. P. Harden and J. O. Harris. The isoelectric point of bacterial cells. J. Bacteriol. 65 (1953), 269–271.
27 P. Kuusela. Fibronectin binds to *Staphylococcus aureus*. Nature 276 (1978), 718–720.
28 P. Kuusela, T. Vartio, M. Vuento, and E. B. Myhre. Attachment of staphylococci and streptococci on fibronectin, fibronectin fragments, and fibrinogen bound to a solid phase. Infect. Immunol. 50 (1985), 77–85.
29 J. I. Flock, G. Fröman, and K. Jönsson. Cloning and expression of the gene for a fibronectin-binding protein from *Staphylococcus aureus*. EMBO J. 6 (1987), 2351–2357.
30 D. F. Mosher and R. A. Proctor. Binding and factor XIII amediated cross-linking of a 27 kilodalton fragment of fibronectin to *Staphylococcus aureus*. Science 209 (1980), 927–929.
31 R. J. Gibbons and I. Etherden. Comparative hydrophobicities of oral bacteria and their adherence to salivary pellicles. Infect. Immunol. 41 (1983), 1190–1196.
32 E. C. Reynolds and A. Wong. Effect of adsorbed protein on hydroxyapatite zeta potential and *Streptococcus mutans* adherence. Infect. Immunol. 39 (1983), 1285–1290.
33 M. Herrmann, P. E. Vaudaux, and D. Pittit. Fibronectin, fibrinogen, and laminin act as mediators of adherence of clinical staphylococci isolates to foreign material. J. Infect. Dis. 158 (1988), 693–701.
34 E. Muller, S. Takeda, D. Goldmann, and G. B. Pier. Blood proteins do not promote adherence of coagulase-negative staphylococci to biomaterials. Infect. Immunol. 59 (1991), 3323–3326.
35 P. Brokke, J. Dankert, J. Carballo, and J. Feijen. Adherence of coagulase-negative staphylococci onto polyethylene catheters *in vitro* and *in vivo*: A study on the influence of various plasma proteins. J. Biomater. Appl. 5 (1991), 204–226.
36 Y. W. Gua, K. A. Khora, and P. Cheang. *In vitro* studies of plasma-sprayed hydroxyapatite/Ti-6Al-4V composite coatings in simulated body fluid (SBF). Biomaterials 24 (2003), 1603–1611.
37 Y. PengLu, M. S. Li, S. T. Li, Z. G. Wang, and R. F. Zhu. Plasma sprayed hydroxyapatite + titania composite bond coat for hydroxyapatite coating on titanium substrate. Biomaterials 25 (2004), 4393–4403.
38 R. Godley, D. Starosvetsky, and I. Gotman. Bonelike apatite formation on niobium metal treated in aqueous NaOH. J. Mater. Sci. Mater. Med. 15 (2004), 1073–1077.
39 J. Pajamaki, S. Lindholm, O. Andersson, K. Karlsson, A. Yli-Urpo, and R. R. Happonen. Glass-ceramic-coated titanium hip endoprosthesis experimental study in rabbits. Arch. Orthop. Trauma Surg. 114 (1995), 119–122.
40 E. Munting. The contributions and limitations of hydroxyapatite coatings to implant fixation: A histomorphometric study of load bearing implants in dogs. Int. Orthop. (SICOT) 20 (1996), 1–6.
41 H. Wang, N. Eliaz, Z. Xiang, H. P. Hsu, M. Spector, and L. W. Hobbs. Early bone apposition *in vivo* on plasma-sprayed and electrochemically deposited hydroxyapatite coatings on titanium alloy. Biomaterials 27(23) (2006), 4192–4203.
42 M. F. Chen, X. J. Yang, R. X. Hu, Z. D. Cui, and H. C. Man. Bioactive NiTi shape memory alloy used as bone bonding implants. Mater. Sci. Eng. C 24 (2004), 497–502.

43 A. Choubey, A. Dorner-Reisel, and B. Basu. Friction and wear behaviour of DLC coated biomaterials in simulated body fluid solution at fretting contacts. Key Eng. Mater. 264–268(3) (2004), 2115–2118.

44 T. Hanawa. Metal ion release from metal implants. Mater. Sci. Eng. C 24 (2004), 745–752.

45 D. C. Clupper, J. E. Gough, P. M. Embanga, I. Notingher, L. L. Hench, and M. M. Hall. Bioactive evaluation of 45S5 bioactive glass fibres and preliminary study of human osteoblast attachment. J. Mater. Sci. Mater. Med. 15(7) (2004), 803–808.

46 K. D. Lobel and L. L. Hench. In-vitro protein interactions with a bioactive gel-glass. J. Sol-Gel Sci. Technol. 7 (1996), 69–76.

47 L. Lefebvre, J. Chevalier, L. Gremillard, R. Zenati, G. Thollet, D. Bernache-Assolant, and A. Govin. Structural transformations of bioactive glass 45S5 with thermal treatments. Acta Mater. 55 (2007), 3305–3313.

48 E. VerneÂ, M. Ferraris, C. Jana, and L. Paracchini. Bioverit1 I base glass/Ti particulate biocomposite: "in situ" vacuum plasma spray deposition. J. Eur. Ceram. Soc. 20 (2000), 473–479.

49 E. VerneÂ, M. Ferraris, and C. Jana. Pressureless sintering of bioverit1 III/Ti particle biocomposites. J. Eur. Ceram. Soc. 19 (1999), 2039–2047.

50 C. Q. Ninga and Y. Zhoub. On the microstructure of biocomposites sintered from Ti, HA and bioactive glass. Biomaterials 25 (2004), 3379–3387.

51 C. Fernandeza, E. Verne, J. Vogel, and G. Carlb. Optimisation of the synthesis of glass-ceramic matrix biocomposites by the "response surface methodology." J. Eur. Ceram. Soc. 23 (2003), 1031–1038.

52 M. Amarala, M. A. Lopesb, R. F. Silva, and J. D. Santos. Densification route and mechanical properties of Si_3N_4–bioglass biocomposites. Biomaterials 23 (2002), 857–862.

53 V. M. da Rocha Barros, L. A. Salata, C. E. Sverzut, S. P. Xavier, R. van Noort, A. Johnson, and P. V. Hatton. *In vivo* bone tissue response to a canasite glass-ceramic. Biomaterials 23 (2002), 2895–2900.

54 N. Saha. Development of hydroxyapatite-ZnO biocomposites with antimicrobial property. MTech thesis, Indian Institute of Technology, Kanpur, India, May, 2010.

55 S. Nath. Development of novel calcium phosphate-mullite composites for orthopedic application; PhD thesis, Indian Institute of Technology, Kanpur, India, September, 2008.

Chapter 20

Calcium Phosphate-Based Bioceramic Composites

In the biomaterials community, hydroxyapatite (HAp) and other calcium phosphates (CaPs) are being widely investigated for hard-tissue replacement applications, because HAp has close chemical proximity with the inorganic composition of natural bone and has good biocompatibility properties. Despite such advantages, three major limitations restrict the wider use of monolithic HAp in biomedical applications: (1) extreme brittleness (fracture toughness < 1 MPa $m^{1/2}$), (2) absence of antimicrobial properties, and (3) limited contact with host tissues. Therefore, the development of new HAp-based composites demands improvements in its physical properties (strength, toughness) without compromising the biocompatibility aspect. From this perspective, this chapter presents results obtained with HAp–titanium and HAp–mullite systems to demonstrate how the material composition can be optimized to enhance cellular functionality, while enhancing the physical and mechanical properties. Research results obtained with HAp–Ag and HAp–ZnO systems are also discussed to demonstrate how to optimize the cellular functionality, while enhancing the antimicrobial properties of HAp. Finally, the results of *in vivo* experiments in rabbits are presented to illustrate good histocompatibility of newly developed CaP–mullite composites

20.1 INTRODUCTION

In the quest for an ideal bone replacement material, significant efforts have been invested in developing HAp- or CaP-based bulk composites with various reinforcements.[1] From a biological-applications perspective, HAp-based composites are a reasonable choice for hard-tissue replacement, compared with bioinert ceramics, metals, or metallic alloys. Since natural bone is a hybrid composite of different minerals and organic compounds,[2] a novel approach should ideally lead to development of a multifunctional composite material, in which each element will serve a different function. Since HAp is the major inorganic component in natural bone, the

Advanced Structural Ceramics, First Edition. Bikramjit Basu, Kantesh Balani.

presence of HAp renders the composite bioactive. However, CaPs and bioactive glasses (BGs) are inherently brittle, impairing their use in load-bearing applications and making handling by the surgeon difficult. A better approach to achieve less-brittle bone substitutes is to use intrinsically tougher materials, for example, ceramic–polymer composites. Nevertheless, the CaP-based ceramics, among various bioceramics, are being investigated widely for biomedical applications, especially for hard-tissue replacement, due to their acceptable biocompatibility[3–5]—because natural bone contains HAp as a primary mineral phase. Recall that it is highly desirable that materials for bone tissue engineering should have a combination of properties, including osteoinductive (capable of promoting the differentiation of progenitor cells down an osteoblastic lineage), osteoconductive (able to support bone growth and encourage the ingrowth of surrounding bone), and osteointegration (able to integrate into surrounding bone) properties.

In designing HAp-based materials to overcome various limitations, a number of research groups have made several attempts to incorporate metals,[6] ceramics,[7–9] and polymers.[10] Ideally, any attempt to enhance physical properties (fracture toughness, strength) or to induce bactericidal effect should not compromise cellular functionality in terms of cell viability, osteoinduction, or bone mineralization properties.

Therefore, the enhancement of physical properties without compromising biocompatibility demands the use of an optimal amount of reinforcement as well as tailoring the processing parameters. Note that the processing conditions influence the combination of HAp and tricalcium phosphate (TCP) phases in a composite as well as the thermochemical interactions of CaP phases with reinforcement. This aspect needs to be pursued keeping in mind that overall biocompatibility is not compromised. Against this backdrop, an important part of research to develop new biocompatible materials also involves characterization of the *in vitro* properties of biomaterials, under scientifically controlled conditions. For example, tissue culture test methods are used as acute screening tests for biocompatibility testing of materials.[11,12]

However, HAp has the problem of poor biodegradation properties, which prevents natural bone growth for extended periods. Also, its low strength and fracture toughness have reduced the field of possible applications to only those where the implant will be subjected to low stresses. To overcome various limitations of HAp,[13–16] researchers have used mixtures of HAp–TCP materials along with second-phase reinforcement to make composites with desirable properties. In a study on HAp–TCP ceramics, MgO was used as sinter-additive; *in vitro* cell culture tests on these composites using L929 fibroblasts confirmed good biocompatibility and good biodegradation properties.[17] In another study, the addition of HAp to zirconia–alumina nanocomposites was reported to increase significantly their biocompatibility, as evident from the results of *in vitro* tests using MG63 osteoblast cells.[18] In some cases, the HAp-based composites exhibited better *in vitro* biocompatibility than pure HAp. For example, Boanini et al.[19] studied the cell fate of osteoblast-like cells on nanocomposites of HAp with aspartic acid and glutamic acid. Their results revealed that the HAp-based nanocomposites could exhibit better cell proliferation, alkaline phosphatase (ALP) activity, and osteocalcin (OC) gene expression than pure HAp.

In a different study, the response of osteoblastic cells to a thin film of poorly crystalline calcium phosphate apatite crystals was investigated *in vitro*. The osteoblastic cells exhibited high cellular activity, such as adhesion and proliferation. In fact, the bone-forming cells attached more rapidly to apatite thin film than to control dishes.[20] Studies performed on three bioactive sol–gel glasses both with and without a layer of hydroxycarbonate apatite (HCA) confirmed that the formation of the HCA layer enhanced the attachment of cells to the surface.[21] In a study to assess the *in vitro* biocompatibility of fluoridated HAp coatings, Wang et al.[22] probed the osteoblastic cell response using MG63 cell lines; while assessing the *in vitro* biocompatibility of such composites, they analyzed the cell morphology, proliferation, and differentiation (ALP activity and OC gene expression). Shu et al.[23] reported the role of HAp on the differentiation and growth of MC3T3-E1 osteoblastic cells, confirming that HAp enhances osteoblast differentiation. In contrast, Alliot-Licht et al.[24] demonstrated that, due to the phagocytosis of HAp particles, osteoblast cells exhibited reduced cell growth and ALP activity. In another study, Ogata et al.[25] compared the osteoblast responses to HAp with HAp/soluble calcium phosphate (SCaP) composites. Their results revealed that HAp/SCaP composites could exhibit greater ability in osteogenesis than HAp, by increasing collagen synthesis and calcification of the extracellular matrix.

20.2 BIOINERT CERAMICS

In the following, bioinert materials and their areas of application are briefly presented. The use of carbon and inert glass fibers, in view of their typical anisotropic properties, have some important applications.[26] These engineering fibers can be used as reinforcements in orthopedic devices, such as femoral hip stems, knee prostheses, and fracture fixation plates. Also, the carbon-based composites, used for articulating surfaces, are reported to have a tendency to cause inflammatory problems due to the loss of carbon particles.[27]

One potential bioinert ceramic, alumina (Al_2O_3), exhibited good performance *in vivo*, though its low fracture toughness (3–4 MPa m$^{1/2}$) typically restricts its use in demanding applications. In this respect, tetragonal zirconia (t-ZrO_2) ceramic (8–11 MPa m$^{1/2}$) has an edge over alumina. Various attempts are being pursued to toughen Al_2O_3 by adding monoclinic zirconia (m-ZrO_2), partially stabilized ZrO_2 (PSZ), and so on, leading to the development of zirconia-toughened alumina (ZTA).[28] It has been reported that dense ZTA has considerably better toughness[29] and wear resistance[30,31] than monolithic alumina. Although ceramic composite materials potentially can be used in load-bearing orthopedic applications, very few materials have been tested clinically so far. In an important study, Hayashi et al.[32] investigated the *in vivo* response of bioinert ceramics such as alumina ceramic (99.5% pure Al_2O_3), zirconia ceramic (5 wt% yttria-stabilized ZrO_2), and SUS316L stainless steel (Fe, 65%; Cr, 18%; Ni, 13%; MO, 2%; Mn, 2%). The *in vivo* results were compared with those of dense sintered HAp. It was concluded that the bioinert ceramics are not ideal as bone-bonding materials, but as materials for articulating surfaces. Colon et al.[33] reported the function of osteoblasts and *Staphylococcus epidermidis* on nanophase ZnO and TiO_2 inert bioceramics. It was evident from their results that nanophase ZnO and TiO_2 decreased *S. epidermidis* adhesion and increased osteoblast adhesion compared with macrophase materials.

More recent work[34,35] on Ca- and Mg-doped zirconia indicated that 8 mol% calcium oxide (CaO)-doped PSZ ceramics possessed 97.5% of theoretical density (ρ_{th}) and 16 mol% CaO-doped fully stabilized zirconia (FSZ) ceramics possessed only 91.6% ρ_{th}, whereas more than 95% ρ_{th} in both Mg-PSZ and Mg-FSZ could be obtained, when all are microwave (MW) sintered at 1585°C for 1 hour. The microstructure of Ca-PSZ ceramic exhibits a bimodal grain size distribution of coarser m-ZrO_2 grains embedded in a cubic matrix. The optimized microstructure of both Mg-PSZ and Mg-FSZ samples revealed the presence of coarser grains with size in the range of 5–10 μm. Importantly, the obtained microstructure was superior compared with the conventional densification route, which normally results in grain sizes of 20–50 μm. X-ray diffraction (XRD) analysis confirmed the presence of predominantly cubic zirconia (c-ZrO_2) phase in MW sintered Ca-FSZ, whereas m-ZrO_2 was a predominant phase in Mg-PSZ samples. The optimized Ca-PSZ and Ca-FSZ ceramics exhibited Vickers hardness of around 10 and 9 GPa, respectively, whereas the toughness was recorded as 6 MPa $m^{0.5}$ for Ca-PSZ. Similarly, Mg-PSZ ceramics, MW sintered at 1585°C, possessed a good combination of hardness (10.6 GPa) and fracture toughness (6.8 MPa $m^{0.5}$).

20.3 CALCIUM PHOSPHATE-BASED BIOMATERIALS

As mentioned earlier, HAp, other CaPs, and their composites are potential implant materials, particularly for hard-tissue replacement applications.[36] The most popular bioactive CaP material is HA or HAp (with chemical composition $Ca_{10}(PO_4)_6(OH)_2$), which has a mineral composition similar to that of natural bone and teeth. In Figure 20.1, the stability region of HAp in the $CaO–P_2O_5–H_2O$ ternary system is presented. A number of compounds with varying Ca/P ratio, belonging to the CaP family, are relevant to biomedical applications. These CaP compounds are octacalcium phosphate (OCP, Ca/P = 1.33), TCP (Ca/P = 1.5), HAp (Ca/P = 1.67), and tetracalcium phosphate (TTCP, Ca/P = 2). Also, a Ca/P ratio of less than 1.0 is not biomedically important. Although TCP and HAp are the commonly reported phases, *in vitro* or *in vivo* formation of OCP or other phases is also reported to a limited extent.[37,38] TCP exists in two polymorphs: α-TCP and β-TCP. Of these two, α-TCP forms at high temperature. It is important to note here that a number of literature reports emphasize that nonstoichiometric HAp promotes better osteoconduction.[39]

To date, several attempts have been pursued to develop HAp-based bioceramic composites in various systems, including HAp–alumina, HAp–zirconia, HAp–bioglass, HAp–HAp-whisker (HAp_w), and HAp–TiO_2 composites.[7,8,40,41] In most cases, two major common phenomena occur: (1) dissociation of HAp to TCP (α/β)

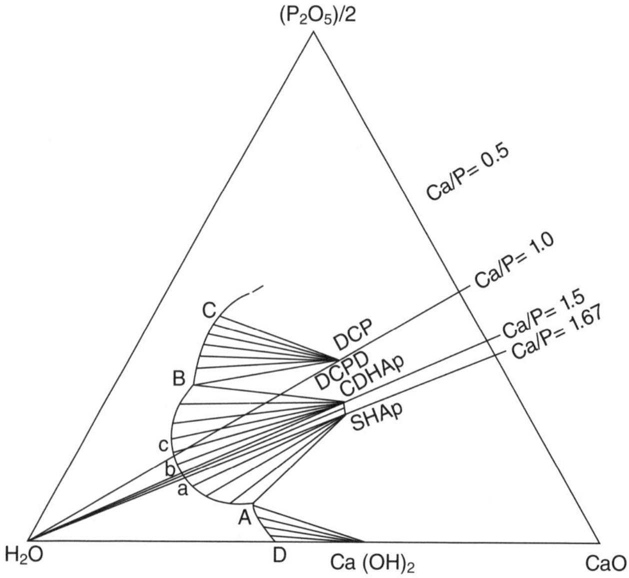

Figure 20.1 Ternary phase diagram of $CaO–P_2O_5–H_2O$ ternary system showing the stability region of different CaP phases along with hydroxyapatite.[92] SHAp, stoichiometric hydroxyapatite; DCP, dicalcium phosphate; DCPD, dicalcium phosphate dihydrate; CDHAp, calcium deficient hydroxyapatite.

and (2) interfacial reactions between HAp and the ceramic reinforcement (e.g., $CaZrO_3$ in the case of HAp–ZrO_2[42]). Besides ceramic reinforcements, limited attempts have been made to use metallic reinforcements, such as titanium, in order to fabricate biocomposites. Both Ti and TiO_2 are biocompatible and can be used as a reinforcing phase in CaP-based composites. A common problem in using Ti is oxidation at sintering temperatures, forming TiO_2 and other oxide products. It also has been reported that dissociation of HAp to other CaP phases (mainly TCP) occurs at temperatures higher than 900°C, depending on the deficiency of calcium.[43]

Various attempts also have been made to experimentally assess the cytocompatibility of CaP-based materials. For example, Suzuki et al.[44] reported that serum protein adsorbed on the surface of TCP–HAp ceramics is effective in preventing cell rupture by functioning as an intermediate layer. Chen et al.[45] reported the bone-bonding mechanism of crystalline HAp *in vivo*. Initially, a layer of amorphous HAp was formed on the HAp implant surface. After 3 months, a bone-like apatite layer had formed between the synthetic implant and natural bone tissue. After 6 months, direct bone–HAp implant contact was found to have been established and collagen fiber had entered inside the implant. As a result, the interface region exhibited good mechanical strength with the new bone apposition.

In hot pressed HAp–ZrO_2 composites,[46] zirconia did not facilitate the decomposition of the HAp matrix. In a different study, it was reported that HAp/Al_2O_3 composites with unique functionally graded structure can be fabricated by the underwater-shock compaction technique without use of any sintering-aid.[47] Hill and

Clifford found that HAp decomposes at 950°C[48] and, accordingly, sintering experiments were performed using fluorohydroxyapatite (FHA), while replacing HAp. It was found that FHA with 40% ZrO_2 can be sintered at 1400°C[49] without dissociation. Importantly, the biocompatibility of FHA is comparable with that of pure HAp.

Another study[18] reported the formation of biphasic calcium phosphate (BCP) in HA–ZrO_2–Al_2O_3 nanocomposites, when hot pressed at 1400°C. In the last two decades, BCP ceramics have attracted wider attention as an ideal bone substitute due to their controlled degradability.[50] It can be recalled here that any composite with superior physical and biocompatibility properties would be the material of choice in the near future. One report indicated that it is possible to alter the HAp/TCP ratio to form BCPs with desired properties.[50] In an earlier study, it was reported that an addition of 10 wt% phosphate glass to a HAp composite led to partial dissociation of HAp to TCP[51]. With the variation of the amount of glass added, the resorption of TCP can be controlled. Suchanek and coworkers[52] observed that the toughness of HAp–HAp_w (10% addition) could be increased to 1.1 MPa $m^{0.5}$ under optimal processing conditions (hot pressing at 1000–1100°C).

Xin et al.[53] investigated the biomineralization ability of a few bioceramic sintered porous solids, including bioglass, glass-ceramics, HAp, α-TCP, and β-TCP under both *in vitro* simulated body fluid (SBF) conditions and *in vivo* physiological environment. The presence of octacalcium phosphate was noticed in all but one (β-TCP) of the investigated bioceramic surfaces *in vitro* and *in vivo*. An important conclusion was that CaP formation on bioceramic surfaces is more difficult *in vivo* than *in vitro*. In an interesting study, Dong et al.[54] probed the bone formation of an osteoblast/porous HAp composite *in vivo*. Using scanning electron microscopy (SEM) observations (Fig. 20.2), mineralized collagenous extracellular matrix as well

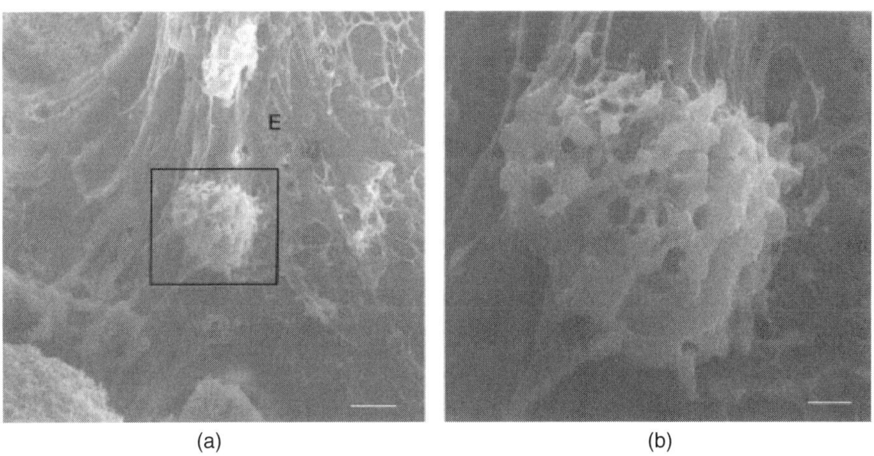

Figure 20.2 SEM images of cross sections of osteoblast–HAp composites after two weeks of implantation: (A) The pore surface is covered by round cells as well as collagenous extracellular matrix; E indicates collagenous extracellular matrix; bar = 37.5 mm. (B) Higher magnification of the large rectangular area in (A). A round cell, which seems to be an active osteoblast, can be seen on the HAp surface; bar = 5 μm.[54]

as active osteoblast was confirmed on HAp-based composite after 2 weeks of implantation. It was concluded that the application of low-pressure system to subculture bone cells on porous HAp blocks can increase bone tissue formation *in vivo*.

Gatti et al.[55] compared the bone ingrowth behavior on implantation of HAp and TCP granules. They found that the TCP granules induced total bone growth after 4 months of implantation, whereas HAp granules could not induce any new bone even after 12 months of implantation. Balcik et al.[56] performed a weight-bearing study of porous HAp/TCP (60/40) ceramics, implanted as intramedullary fixation in segmental bone defects in rabbit bone. They reported that these ceramics are limited in their ability to treat load-bearing segmental bone defects, but did not fail at the early stages of implantation. It was suggested that additional internal fixation should be used when immediate mobilization and load bearing are required. Fini et al.[57] reported the effect of pulsed electromagnetic fields (PEMFs) on the osteointegration of HAp implants in cancellous bone. They found that PEMFs could enhance early HAp osteointegration in the trabecular bone of healthy rabbits. The faster histocompatibility was associated with a higher degree of bone mineralization and maturation, which was maintained during 6- and 3-week periods of nonstimulation.

It is known that microstructure plays an important role in determining mechanical properties; therefore, a study was conducted to assess microstructural influence on implanted material *in vivo*. Okuda et al.[58] reported the effect of the microstructure of β-TCP on the metabolism of neobone. Comparing rod-shaped and globular particles, it was argued that rod-shaped β-TCP can stimulate osteoblastic activity. There have been some reports, mostly from the research of Webster et al.,[59] that the presence of increased grain boundary area promotes osteoblast cell adhesion.

Tsai et al.[60] conducted a study to investigate the bioresorption behavior of TTCP–derived CaP cement implanted in rabbit femur. Their results indicated that CaP cement did not evoke inflammatory response, necrosis, or fibrous encapsulation in the surrounding bony tissues. After 24 weeks of implantation, all the investigated CaP cement materials were resorbed. Similar to TTCP, TCP also has been reported to be a very useful biocompatible material for bone replacement applications.[61]

In the following section, research results obtained with some representative biomaterial systems are summarized.

20.4 CALCIUM PHOSPHATE–MULLITE COMPOSITES

From the preceding discussion, it is evident that HAp, TCP, and other CaP phases are bioactive. The major concern in adding a second phase to a HAp matrix is that its biocompatibility may be greatly affected.[7,8,40,52,62,63] As part of more recent research on biomaterials,[64–69] HAp–mullite ($3Al_2O_3{\cdot}2SiO_2$) composites[9] are being developed. It is known that mullite is a solid solution of alumina (Al_2O_3) and silica (SiO_2). The chemical formula for mullite solid solution is $Al_{(4+2x)}Si_{(2-2x)}O_{(10-x)}$, where $x = 0.17$–0.59. In the experimental results presented later in this section, the formula for mullite ceramic is taken as $3Al_2O_3{\cdot}2SiO_2$. Mullite is chemically inert and mainly used as a refractory material in high-temperature furnaces.[70] Also mullite has lower density (~3.05 g/cm^3) than Al_2O_3 (~3.95 g/cm^3) and ZrO_2 (~6.1 g/cm^3). It has a good combination of mechanical properties such as higher hardness (~15 GPa), higher elastic modulus (~240 GPa), and moderate fracture toughness (~3 MPa m$^{0.5}$).[71]

As far as processing is concerned, HAp powder can be synthesized using commercially available chemicals, such as CaO and phosphoric acid (H_3PO_4), following a well-established suspension–precipitation route.[72,73] To obtain a better combination of physical properties, different amounts of mullite (10–30 wt%) were mixed with HAp and the powders were conventionally sintered. The sintering of the composite pellets was carried out at 1350°C for 2 hours, while sintering of pure HAp was performed at 1200°C for 2 hours, both in a conventional pressureless sintering furnace.

A 2010 study[74] demonstrated that HAp–mullite composites with more than 90% theoretical density can be achieved using pressureless sintering at 1300–1350°C. Combining the results of XRD, transmission electron microscopy (TEM), and dilatometry, it was concluded that the densification of CaP–mullite composites occurs initially by solid-state sintering up to a temperature of 1150°C, followed by liquid-phase sintering at higher temperature. The presence of mullite needles and residual microporosity could be noticed in the sintered microstructures (see Fig. 20.3). Based on careful analysis of the shrinkage kinetics, it was confirmed that an optimal amount of mullite needs to be added to enable liquid-phase sintering of the composites. An important conclusion was that the dissociation of HAp to TCP depends on both mullite content and sintering temperature. For example, composites with higher mullite content are more sensitive toward dissociation to β- and α-TCP.

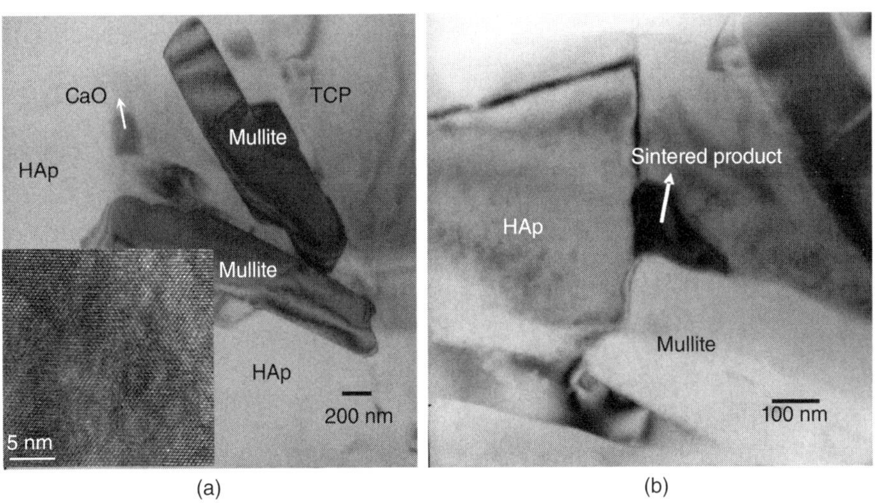

Figure 20.3 (a) Low-magnification TEM micrograph of HAp–(10 wt%)mullite sintered at 1350°C shows typical mullite needles along with HAp phase. (b) Higher magnification image showing the presence of sintered product at the mullite–HAp interface.[9]

20.4.1 Mechanical Properties

A detailed report on the mechanical properties of the CaP–mullite system is presented in a 2010 paper.[75] Elastic modulus values of sintered composites measured by impulse excitation and an instrumented indentation technique show comparable results. The elastic modulus values of the sintered CaP–mullite composites were in the range of 60–80 GPa. All the composites exhibited lower hardness values than monolithic HAp (~6 GPa) and the indentation size effect was also recorded. The hardness measured with 5-N indent load, using an instrumented indentation method, of the CaP composites with 20–30 wt% mullite content lie in the range of 4–5 GPa (Fig. 20.4). Lower hardness values were also recorded (~3.5 GPa) using Vickers indentation at 10-kg load. The nanoindentation response of CaP–mullite composites was reported in a 2009 paper.[76] The critical analysis of instrumented indentation measurements reveals that the hardness, estimated using total work done (sum of elastic recovery and plastic work), closely follows hardness values obtained using a depth penetration method (Oliver and Pharr). Fracture toughness measured by the single-edge V-notched beam (SEVNB) method provides a lower estimate than the indentation cracking method, and the maximum toughness of 1.5 MPa m$^{0.5}$ was recorded for the composites with 20–30 wt% mullite addition. Importantly, these toughness values are 2.5 times higher than for pure monolithic HAp. It was suggested that the presence of whisker-shaped mullite needles in the sintered composites is responsible for better toughness. The compressive strength of the CaP-composites (up to 350 MPa) with 30 wt% mullite addition is better by far than pure HAp. Three-

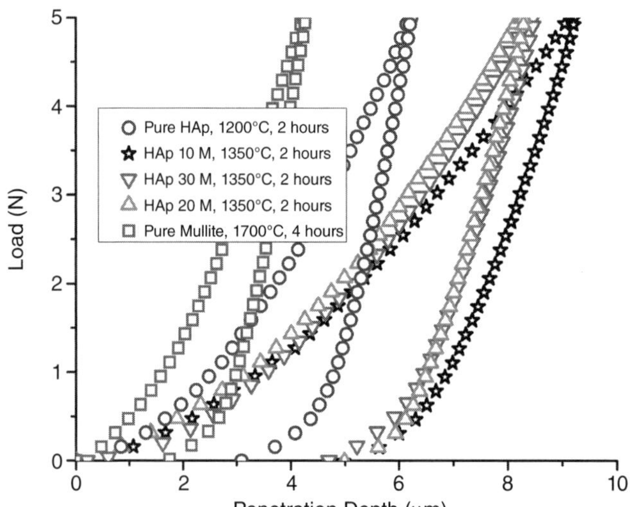

Figure 20.4 Instrumented indentation response of CaP–mullite composites. For comparison, the response of monolithic HAp and mullite are also shown.[76]

point flexural strength of 70~80 MPa was measured for CaP–(20 wt%)mullite and CaP–(30 wt%)mullite composites.

20.4.2 Biocompatibility (*In Vitro* and *In Vivo*)

The *in vitro* biocompatibility assessment using multiple biochemical assays for a series of CaP–mullite (up to 30 wt%) composites using human osteoblast-like MG63 and mouse fibroblast L929 cell lines has been reported.[77] The observation was that as-sintered CaP–mullite composites containing higher amounts of β-TCP can be used as suitable substrates for cell attachment and proliferation of functional human bone cell line (osteoblast-like MG63) and L929 mouse fibroblast cells *in vitro* (see Fig. 20.5). As far as cytotoxicity is concerned, MTT assay results with fibroblast and osteoblast-like cells did not indicate any statistically significant variation in terms of the number of metabolically active cells for the mullite-containing composites compared with pure HAp. Importantly, the results of both alkaline phosphatase and osteocalcin assay demonstrate that BCP-based composites with 20% or 30% mullite substrates have the ability to support better osteoconduction than baseline monolithic HA ceramic. The predominant presence of β-TCP phase in BCP-mullite composites results in good osteoblastic function and phenotypic expression. Based on all the *in vitro* biomedical analyses CaP-biocomposites containing 20 or 30% mullite can be considered as bone analogue materials.

In a separate study,[78] on the basis of qualitative analysis of histologic, radiographic, SEM, and atomic force microscopy (AFM) observations of the bone–implant interface after short-term implantation study (up to 12 weeks) in rabbits, it was concluded that the investigated CaP–mullite biocomposite could exhibit a consistent pattern in deposition of neobone, which appears to remodel over the survival period. It is evident that, although new bone formation is present around HAp–mullite at 12 weeks (Fig. 20.6), very few areas of direct material–bone interface were observed compared with ultrahigh-molecular-weight polyethylene (UHMWPE), which was used as control. Figure 20.6a–d displays the interface between bone and CaP–mullite composite material. No degenerative, necrotic changes or inflammation were noticed at the implant site. Neovascularization was not observed at the bone–material interface. Minimal fibrosis with fibrocytes and occasional macrophages distinguishes the new bone–material interface. Collagen with foci of chondrogenesis and osteocytes were also observed at the interface. Similar observations were made around the UHMWPE implants with direct implant–bone contact (Fig. 20.6e,f). Figure 20.7 presents a representative SEM image of bone–material interface after 4 weeks of implantation. Bone remodeling with formation of Haversian canals could be seen at the implant interface. No gap between natural bone/HAp-mullite implant was found. However, a visible gap of less than 2 μm wide was reported at the bone–Ti interface, while the bone–HAp interface was fully mineralized.[79] On the basis of the absence of inflammation as well as the presence of new bone formation and minimal collagen at the interface, the suitability of the HAp–mullite composite as a bone replacement material for orthopedic applications can be further established.

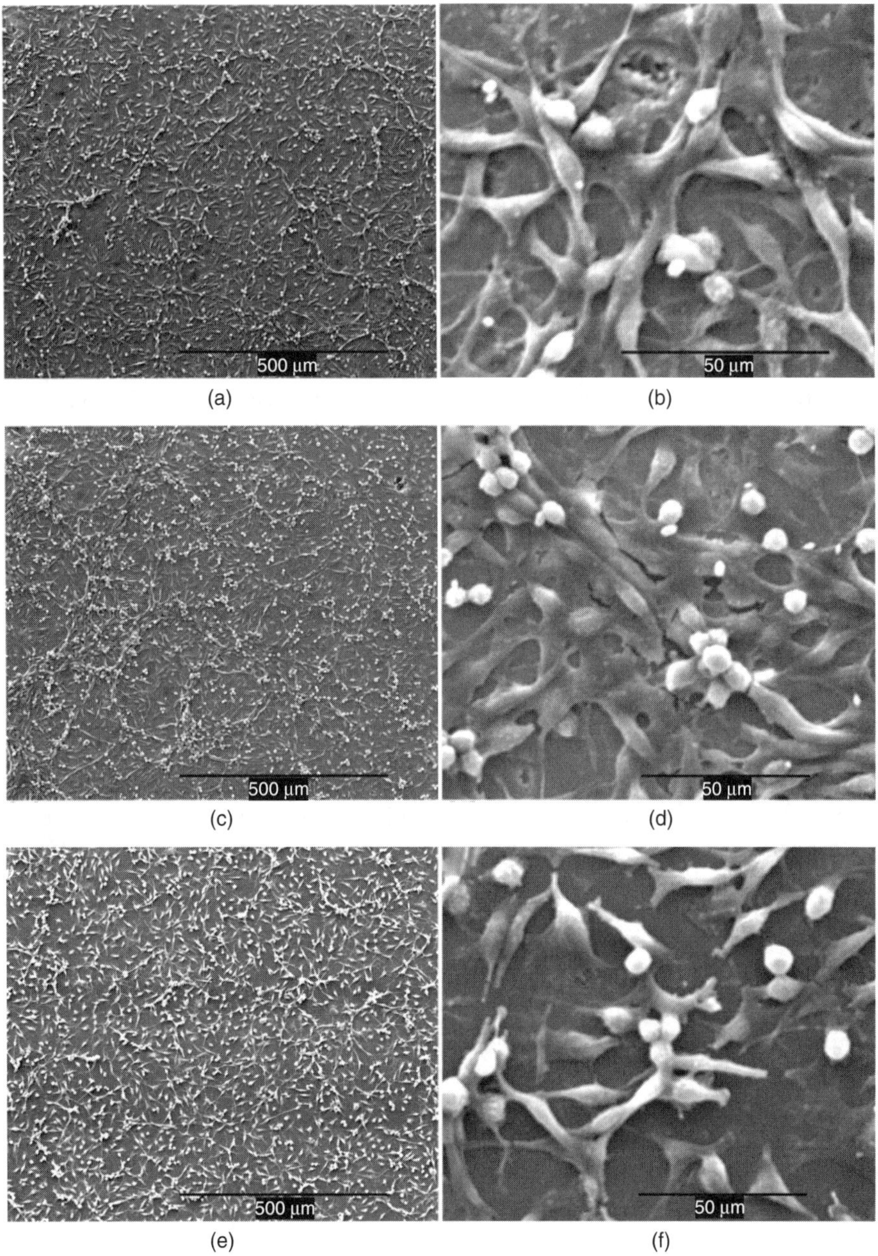

Figure 20.5 Selected SEM images illustrating the adhesion of L929 mouse fibroblast cells on various material surfaces—BCP–(20 wt%)mullite (a and b); BCP–(30 wt%)mullite (c and d); and control sample (e and f) after 24 hours of culture.[77]

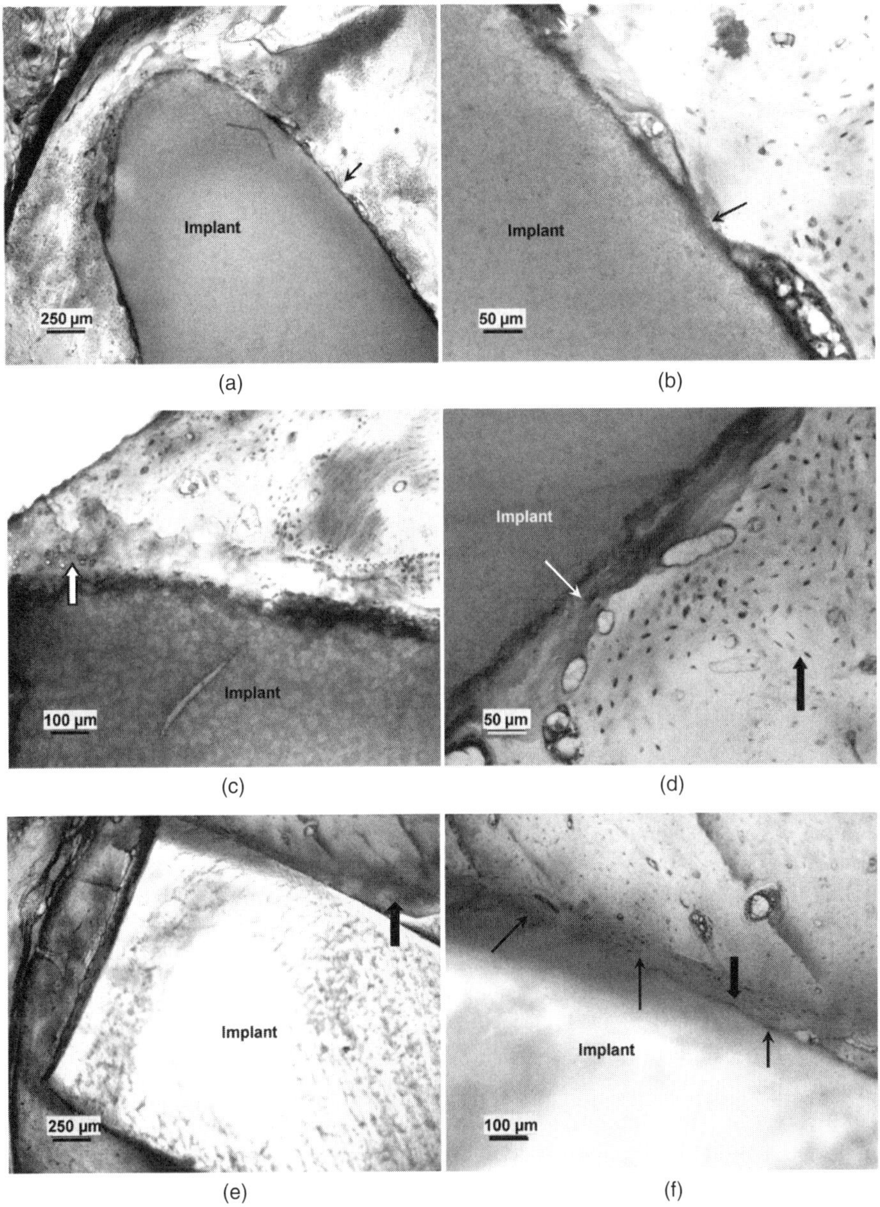

Figure 20.6 Photomicrographs of host bone–implant interface (a–d) BCP–(20 wt%)mullite and (e and f) UHMWPE after 12 weeks postimplantation: (a) bone formation around implant with focal direct contact (arrow) with implant; (b) higher magnification of focal bone–implant contact; (c) focal chondrogenesis (arrow) at interface; (d) osteocytes (white arrow) at implant–host new bone (black arrow) interface; (e) bone formation (arrow) around implant; (f) woven bone (thick arrow) at implant interface (thin arrows).[78] See color insert.

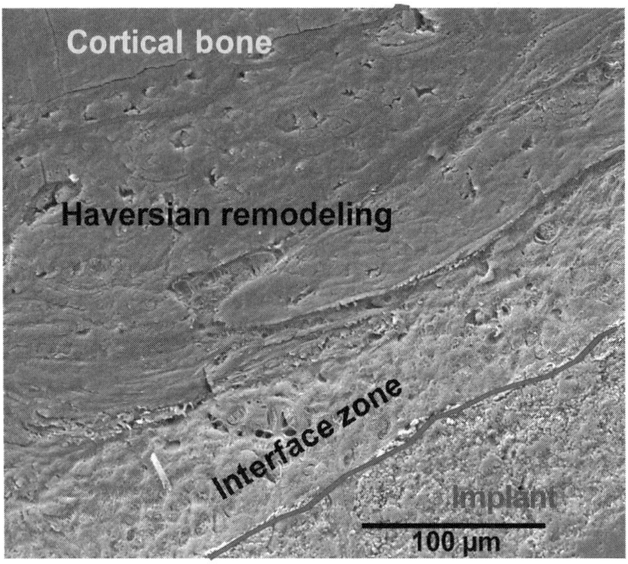

Figur 20.7 SEM image of bone–BCP–(20 wt%)mullite implant contact area showing interface zone, Haversian remodeling, and new bone formation after 4 weeks of implantation.[78]

20.5 HYDROXYAPATITE–Ti SYSTEM

Prior to presenting results on the HAp–Ti system, it can be recalled here that the important issues in developing CaP-based composites include thermochemical compatibility, dissociation, and influence on biocompatibility properties. In a 2009 study,[6] a planned set of sintering experiments was performed on different HAp–titanium starting powder mixes (10–40 wt% Ti) in the temperature range 1000–1400°C in ambient atmosphere. The results of the sintering experiments reveal that, although fully dense (~99% ρ_{th}) monolithic HAp could be obtained after sintering at 1200°C, the occurrence of extensive sintering reactions and phase dissociation of HAp resulted in TCP–TiO_2 composites with lowered density. In all the sintered samples, β-tricalcium phosphate (β-TCP) and rutile (TiO_2) are the major crystalline phases with other reaction products, such as $CaTiO_3$ and CaO (see Fig. 20.8). Cell culture experiments using L929 fibroblast cells confirm cell adhesion and cell proliferation on composites. The observation of cellular bridges, that is, cell–cell interaction as well as flattening of the cells, is indicative of good cytocompatibility of the developed composite (see Fig. 20.9).

20.6 ENHANCEMENT OF ANTIMICROBIAL PROPERTIES OF HYDROXYAPATITE

It can be reiterated now that implant-associated infection is widely considered to be a major concern in biomedical applications, and this has been the driving force for

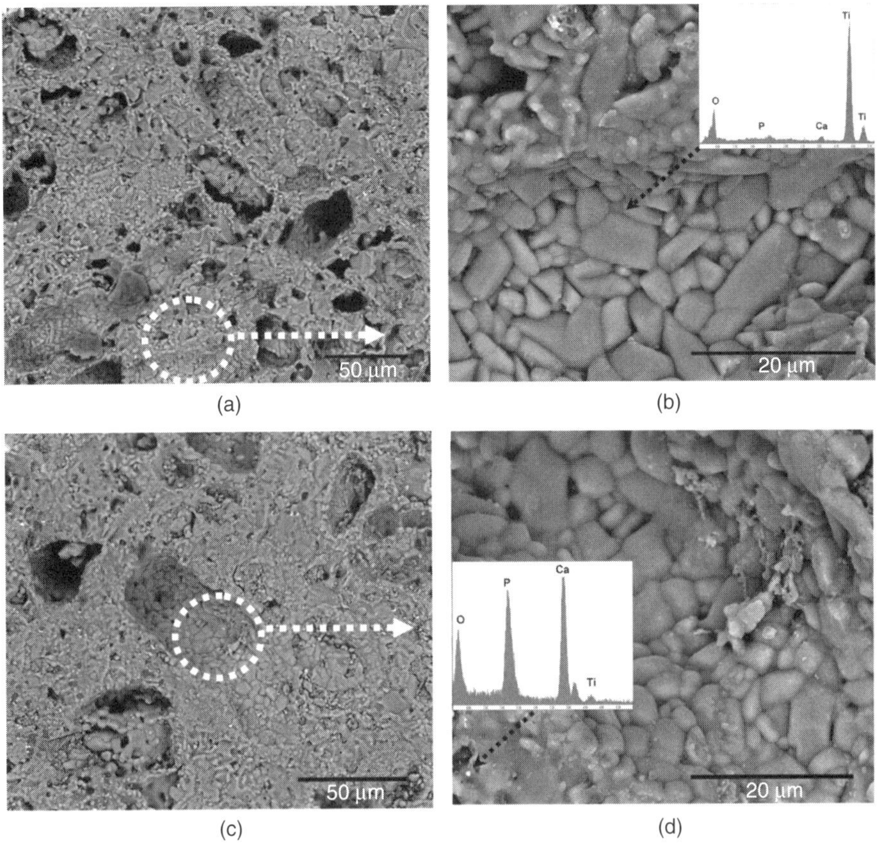

Figure 20.8 SEM images of the polished–thermally etched microstructures, after sintering at 1300°C in air atmosphere for BCP–(30%)Ti composition (a and b) and BCP–(40%)Ti compositions (c and d). The right-hand-side micrographs are captured from the circled areas indicated on the left-hand-side figures. Energy-dispersive x-ray spectrometry (EDS) compositional analysis of some microstructural region or phase (indicated by dotted arrow) is shown as an inset.[6]

developing HAp-based biomaterials with antibacterial additives for possible use in prosthetic devices. To this end, metallic (Ag) or ceramic (ZnO) particulates can be incorporated in a HAp matrix. An important advantage of silver is that it can potentially enhance the toughness and strength via a ductile metal bridging mechanism.[80,81] In addition to improvement in mechanical properties, silver provides advantageous biochemical inertness as well as antibacterial properties.[82,83] Zhang et al.[81] fabricated HAp–Ag composites from HAp and Ag_2O powder. A good combination of strength and toughness of 80 MPa and 2.45 MPa m$^{0.5}$, respectively, was measured for 30 vol% Ag, although the elastic modulus was decreased with increasing silver addition. The main toughening mechanism was crack bridging and plastic work of Ag. Chaki and Wang[82] achieved a flexural strength of ~75 MPa in HAp–(10 wt%)Ag biocomposite.

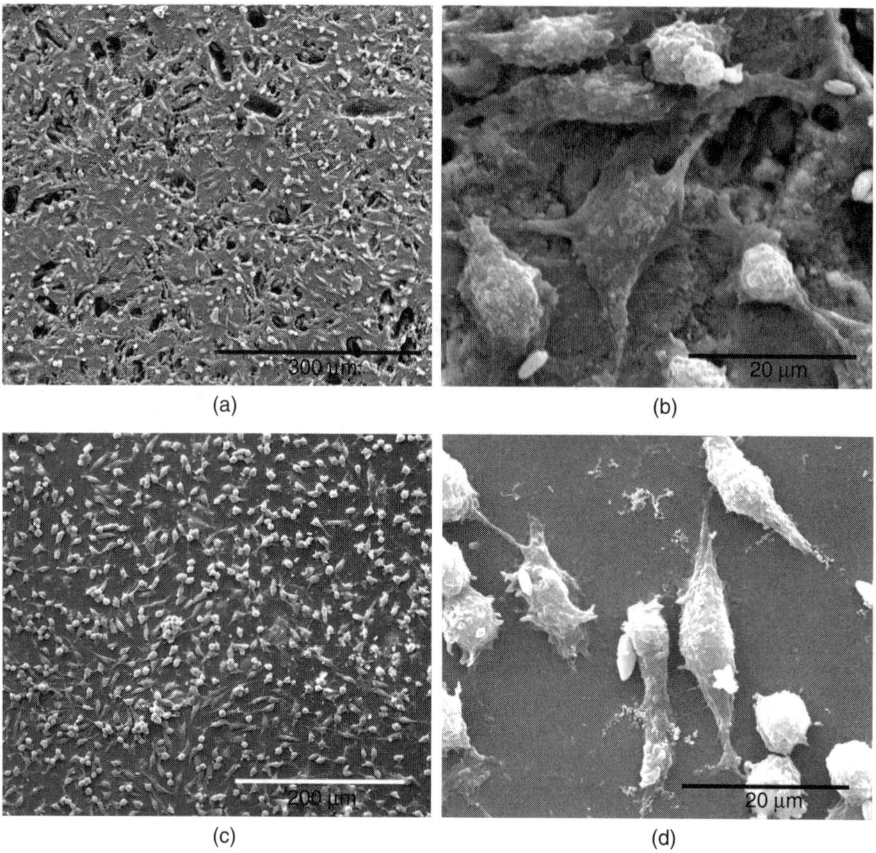

Figure 20.9 SEM images, revealing the adherence of L929 mouse fibroblast cells seeded for 24 hours on BCP–(30 wt%)Ti (a and b) and control sample (c and d).[6]

Feng et al.[84] reported the antimicrobial effect of Ag–HAp thin films on alumina substrate. They soaked the HAp-coated alumina samples in $AgNO_3$ solutions and their results reveal that the $AgNO_3$-treated samples can display antimicrobial effects against different bacterial cells, including *Escherichia coli*, *Pseudomonas aeruginosa*, *Staphylococcus aureus*, and *S. epidermidis*.

There has been a general consensus that the lack of bactericidal properties leads to prosthetic infection for many HAp-based implants. According to several studies, such an infection not only leads to the failure of arthroplasty and surgical removal of the device, but it also imposes a significant physical and psychological trauma for the patient.[85,86] While the exact mechanism of prosthetic infection remains unclear, several studies have reported that microbial biofilm formation plays a key role for most prosthetic infections. A number of researchers have pursued many different ways to prevent biofilm formation, such as by using a protective film of antibacterial agents over the implant surface or developing a new material composition. Another way to restrict biofilm growth is to use a material that has *in situ* antimicrobial properties to kill microorganisms before they grow.

In this context, several ceramics, such as CaO, CuO, MgO, and ZnO, have demonstrated considerable antibacterial activity in different studies over the last few years. Among these oxides CaO/MgO can be dissolved in growth medium, resulting in an increase in pH of the solution to 12.0–12.5. Such an increased pH level does not allow bacterial cell growth. Earlier studies showed that ZnO also can exhibit strong antibacterial behavior.[87] The main advantage of ZnO, except producing H_2O_2, is that it does not increase pH of the biological medium, which may be detrimental to osseous tissues. According to some studies, zinc oxide in water produces hydroxyl radical (OH^-), which provides an antimicrobial property by disrupting prokaryotic cell membranes.[88] In contrast, another report mentioned that hydrogen peroxide generated from the surface of ZnO is the key to inhibiting bacterial growth. In another study, Stoimenov et al. claimed that binding of particles on the bacterial surface due to electrostatic forces is the possible mechanism responsible for antimicrobial activity of ZnO. However, another study reported that that presence of Zn^{2+} ion, in solution, has no effect on the growth of bacteria cells.[89]

In the following subsections, a summary of research results obtained with HAp–Ag and HAp–ZnO systems are presented.

20.6.1 Hydroxyapatite–Ag System

A 2010 paper[90] reported that pressureless sintering of mechanically mixed powders can be employed to develop HAp–Ag bioceramic composites with superior combinations of physical (hardness, toughness), noncytotoxicity, cytocompatibility, and antimicrobial properties. An important observation is that the dissociation of HAp phase can be largely restricted when sintering HAp–(10 wt%)Ag composites at 1300°C for 2 hours. Such stability could be explained by the incorporation of Ag into the HAp lattice, and a mechanism based on defect reactions as well as the formation of nonstoichiometric HAp was put forward. Despite the addition of 10 wt% Ag particles, high hardness of around 6.5 GPa could be recorded in the HAp–(10 wt%)Ag composites, when sintered at 1200°C. Also, a modest indentation toughness of around 1 MPa m$^{1/2}$ was measured, and crack branching was identified as the only toughening mechanism. The results of the cell culture experiments with L929 mouse fibroblast cells revealed that the addition of 10% Ag to HAp matrix does not degrade the overall cell attachment, cellular bridge formation, and cell proliferation behavior, in reference to pure HAp. In fact, HAp–(10%)Ag favorably supports cell proliferation of L929 mouse fibroblast cells *in vitro* (see Fig. 20.10). As far as cytotoxicity is concerned, MTT assay results with fibroblast cells do not reveal any statistically significant difference with pure HAp. The combination of ALP activity and OC expression results indicates that HAp–(10 wt%)Ag composites can display comparable or better bone cell differentiation and better bone mineralization properties than baseline HAp ceramic. An important outcome has been that the addition of 10% Ag can provide an excellent antimicrobial property in terms of

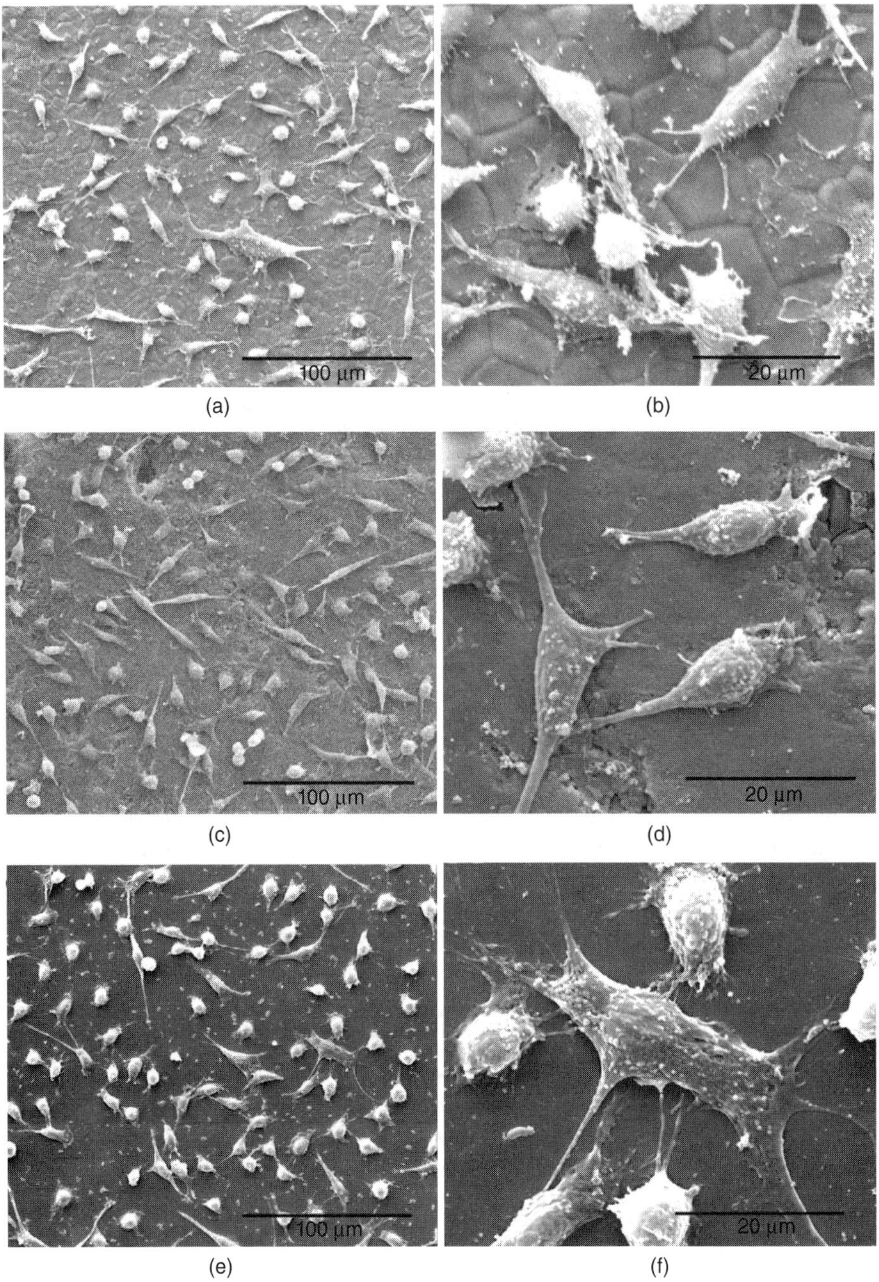

Figure 20.10 SEM images of L929 cells adhered on pure HAp (a and b), on HAp–10Ag sintered at 1200°C for 2 hours (c and d), and on negative control samples (e and f). Results were obtained after 48 hours of culture. The seeded cell density was 5×10^5 cells/mL.[90]

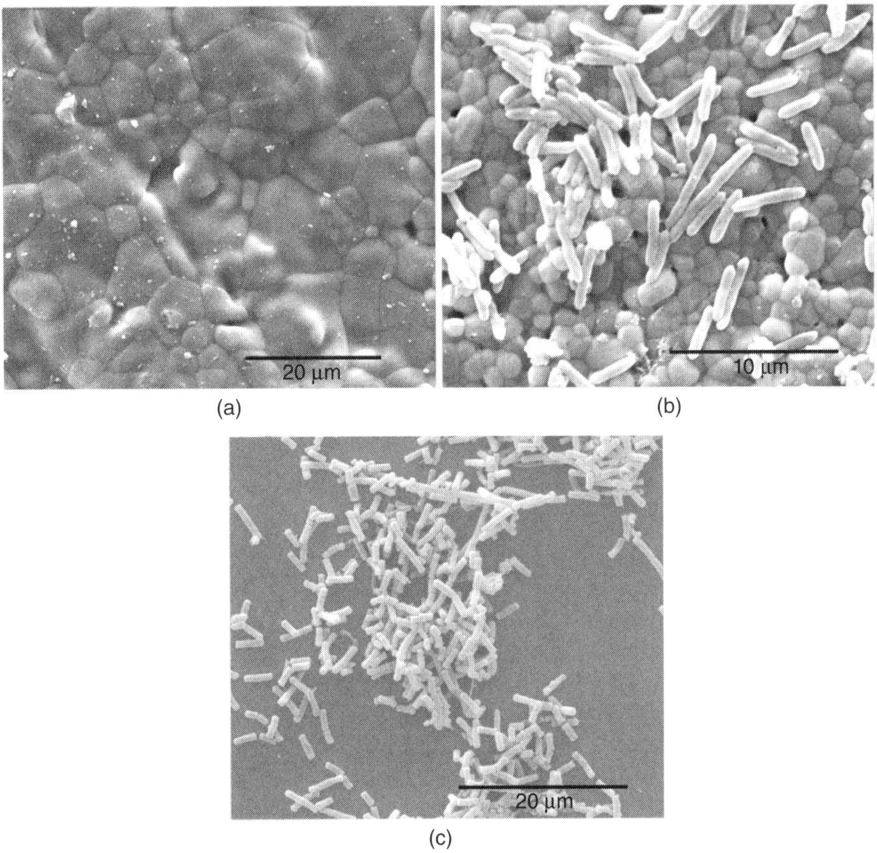

Figure 20.11 SEM images revealing *E. coli* bacterial cell adhesion on (a) HAp–(10 wt%)Ag composite (sintered at 1200°C), (b) pure HAp (sintered at 1200°C), and (c) negative control polymer disk 4 hours after inoculation.[90]

bactericidal activity against *E. coli* bacteria without compromising *in vitro* cytocompatibility of HAp (see Fig. 20.11).

20.6.2 Hydroxyapatite–ZnO System

In work published in 2010,[91] samples of HAp with different amounts of zinc oxide microrods (ZnO) were sintered at 1250°C to produce HAp–ZnO (HZ) biocomposites. Both gram-positive (*S. aureus*, *S. epidermidis*) and gram-negative (*E. coli*) bacteria were used to understand how ZnO addition (up to 30 wt%) to HAp leads to the improvement in bacteriostatic and bactericidal properties (see Figs. 20.12–20.14). SEM observation and turbidometric analysis confirmed antimicrobial behavior of HAp–ZnO composite against gram-negative *E. coli* bacteria. It was reported that increasing ZnO concentration quantitatively reduced colony formation and bacterial

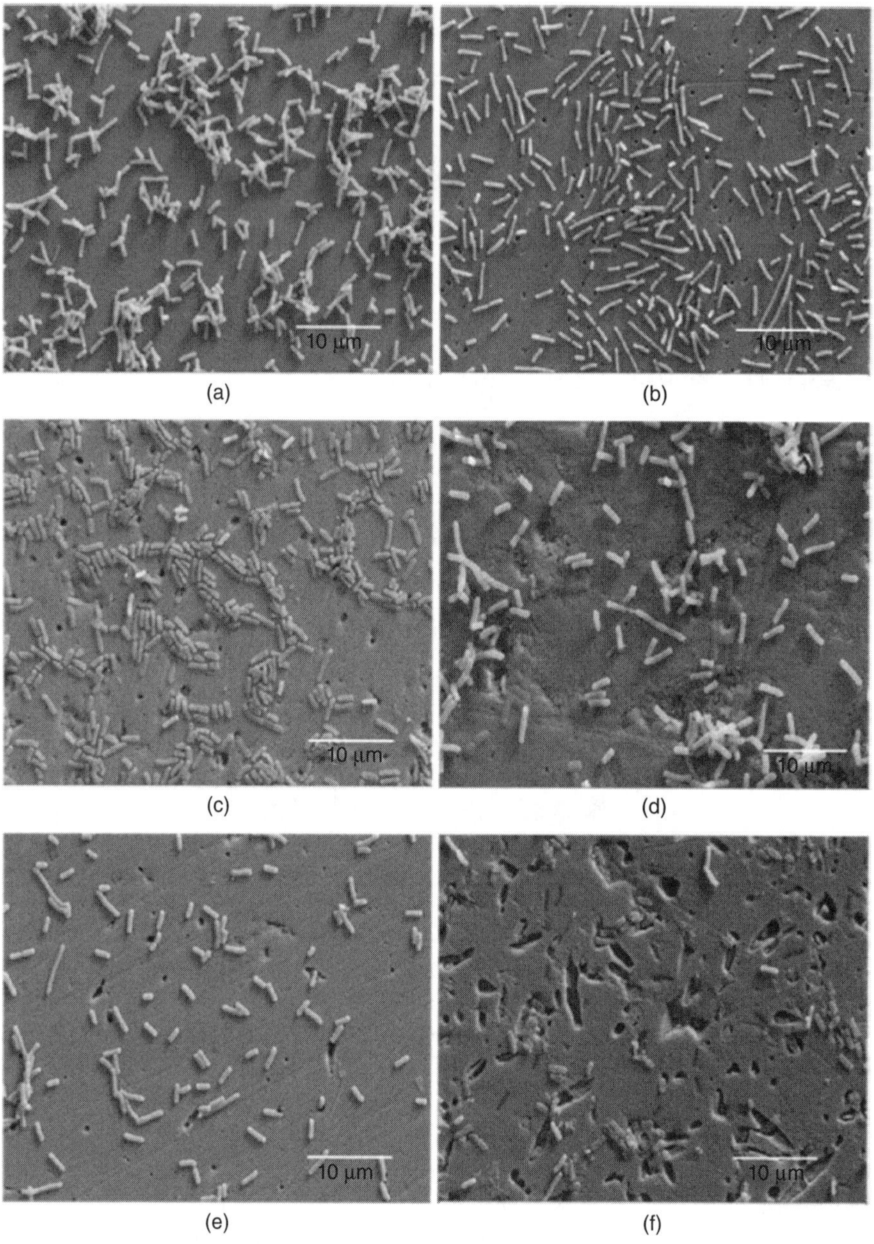

Figure 20.12 Representative SEM images of *E. coli* bacteria adhered on (a) control, (b) pure HAp, (c) HZ 5, (d) HZ 10, (e) HZ 20, and (f) HZ 30 samples.[91]

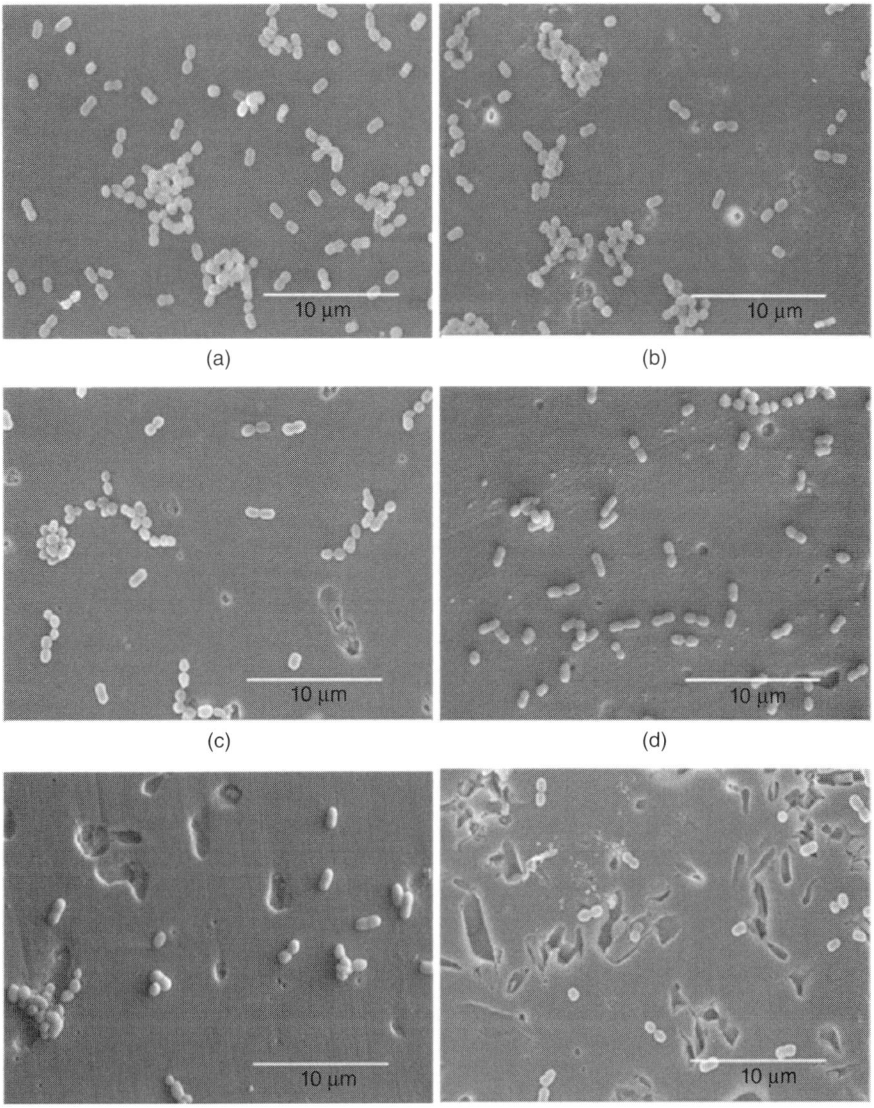

Figure 20.13 Selected SEM images of *S. epidermidis* bacteria adhered on (a) control, (b) HZ 1.5, (c) HZ 5, (d) HZ 7.5, (e) HZ 10, and (f) HZ 20 samples.[91]

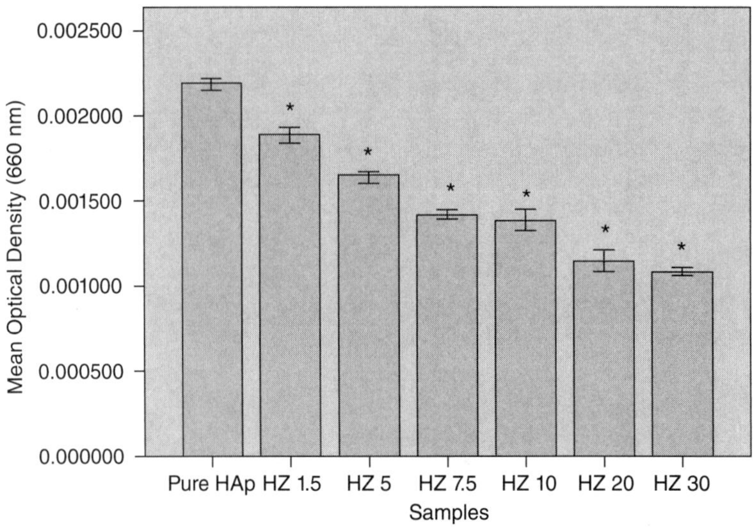

Figure 20.14 Mean optical density data after 4 hours of incubation of *E. coli* bacteria on pure HAp, HZ 1.5, HZ 5, HZ 7.5, HZ 10, HZ 20, and HZ 30 pellet samples. The initial bacterial inocula contained 1×10^7 CFU/mL. Asterisks (*) represent significant difference at $p < 0.05$ with respect to pure HAp, and error bars correspond to ±1.00 SE.[91]

Table 20.1. Number of Single Colonies Observed for Gram-Negative *E. coli* Bacteria after 24 Hours of Incubation on Various HAp–ZnO Composites[91]

Material composition	Sample designation	ZnO conc. (mg/mL)	No. of colonies (approximate)
HAp	Pure Hap	0	73
HAp–(1.5 wt%)ZnO	HZ 1.5	0.75	54
HAp–(5.0 wt%)ZnO	HZ 5	2.5	49
HAp–(7.5 wt%)ZnO	HZ 7.5	3.75	46
HAp–(10.0 wt%)ZnO	HZ 10	5.0	47
HAp–(20.0 wt%)ZnO	HZ 20	10.0	31
HAp–(30.0 wt%)ZnO	HZ 30	15.0	29

adhesion. Significant antimicrobial action was also recorded when these biocomposites were tested against gram-positive *S. aureus* and *S. epidermidis*, as significant reduction in number of bacteria over the sample surface was commonly observed with ZnO addition of more than 7.5 wt%, indicating bacteriostatic behavior (see Fig. 20.14 and Table 20.1).

Despite the aforementioned improvement in antimicrobial properties, an increase in ZnO addition demonstrated a modest influence on fracture toughness or hardness. The maximum densification and hardness for HAp–ZnO composite could be achieved for 5 wt% of ZnO addition. A maximum indentation fracture toughness up to 1.7 MPa m$^{1/2}$ and hardness of up to 6.8 GPa were measured in HAp–ZnO biocomposites. It was suggested that around 7.5–10.0 wt% of ZnO addition could exhibit a balance between mechanical and antimicrobial properties of the HAp–ZnO composite.

As a concluding note, it must be mentioned here that several attempts are being made in the biomaterials community to address specific issues of physical property (strength, toughness) enhancement in bioceramic composites, along with bactericidal effect, *in vitro* and *in vivo* biocompatibility, and so on. However, it remains a dream to develop an ideal biomaterial—one that has the desired combination of excellent biocompatibility with a good combination of physical properties as well as antibacterical properties—to mimic the properties of natural bone.

REFERENCES

1 S. Nath and B. Basu. Designing materials for hard tissue replacements. J. Kor. Ceram. Soc. 45 (2008), 1–29.
2 J. Black. *Orthopaedic Biomaterials: Research and Practice*. Churchill Livingstone, New York, 1988.
3 H. Oonishi. Orthopedic application of hydroxyapatite. Biomaterials 12 (1991), 171–178.
4 M. Wang. Developing bioactive composite materials for tissue replacement. Biomaterials 24 (2003), 2133–2151.
5 B. Basu, D. Katti, and A. Kumar. *Advanced Biomaterials: Fundamentals, Processing and Applications*. John Wiley & Sons, Hoboken, NJ, 2009.
6 S. Nath, R. Tripathi, and B. Basu. Understanding phase stability, microstructure development and biocompatibility in calcium phosphate–titania composites, synthesized from hydroxyapatite and titania powder mix. Mater. Sci. Eng. C 29 (2009), 97–207.
7 R. Rao and T. S. Kannan. Synthesis and sintering of hydroxyapatite–zirconia composites. Mater. Sci. Eng. C 20 (2002), 187–193.
8 S. Gautier, E. Champion, and D. B. Assollant. Processing, microstructure and toughness of Al_2O_3 platelet-reinforced hydroxyapatite. J. Eur. Ceram. Soc. 17 (1997), 1361–1369.
9 S. Nath, K. Biswas, and B. Basu. Phase stability and microstructure development in hydroxyapatite-mullite system. Scr. Mater. 58 (2008), 1054–1057.
10 N. Ignjatovic and D. Uskokovic. Synthesis and application of hydroxyapatite/polylactide composite biomaterial. Appl. Surf. Sci. 238 (2004), 314–319.
11 F. J. Schottler. Tissue culture methods for determining biocompatibility. ASTM STP 810 (1983), 19–24.
12 Y. M. Kong, C. J. Bae, S. H. Lee, H. W. Kim, and H. E. Kim. Improvement in biocompatibility of zirconia-alumina nanocomposite by addition of HA. Biomaterials 26 (2005), 509–517.
13 H. S. Ryu, K. S. Hong, J. K. Lee, D. J. Kim, J. H. Lee, B. S. Chang, D. H. Lee, C. K. Lee, and S. S. Chang. Magnesia-doped HA/TCP ceramics and evaluation of their biocompatibility. Biomaterials 25 (2004), 393–401.
14 E. Adolfson, P. A. Henning, and L. Hermannson. Phase analysis and thermal stability of hot isostatically pressed zirconia-hydroxyapatite composites. J. Am. Ceram. Soc. 83 (2000), 2798–2802.
15 H. W. Kim, Y. M. Kong, Y. H. Koh, and H. E. Kim. Pressureless sintering and mechanical and biological properties of flour-hydroxyapatite composites with zirconia. J. Am. Ceram. Soc. 86 (2003), 2019–2026.
16 A. Clifford and R. Hill. Apatite-mullite glass ceramics. J. Non-Crys Sol. 196 (1996), 346–351.
17 H. S. Ryu, K. S. Hong, J. K. Lee, D. J. Kim, J. H. Lee, B. S. Chang, D. H. Lee, C. K. Lee, and S. S. Chung. Magnesia-doped HA/β-TCP ceramics and evaluation of their biocompatibility. Biomaterials 25 (2004), 393–401.

18 Y. M. Kong, C. J. Bae, S. H. Lee, H. W. Kim, and H. E. Kim. Improvement in biocompatibility of ZrO_2–Al_2O_3 nano-composite by addition of HA. Biomaterials 26 (2005), 509–517.
19 E. Boanini, P. Torricelli, M. Gazzano, R. Giardino, and A. Bigi. Nanocomposites of hydroxyapatite with aspartic acid and glutamic acid and their interaction with osteoblast-like cells. Biomaterials 27 (2006), 4428–4433.
20 J. Y. Hong, Y. J. Kim, H. W. Lee, W. K. Lee, J. S. Ko, and H. M. Kim. Osteoblastic cell response to thin film of poorly crystalline calcium phosphate apatite formed at low temperatures. Biomaterials 24 (2003), 2977–2984.
21 N. Olmo, A. I. Martin, A. J. Salinas, J. Turnay, M. V. Regi, and M. A. Lizarbe. Bioactive sol-gel glasses with and without a hydroxycarbonate apatite layer as substrates for osteoblast cell adhesion and proliferation. Biomaterials 24 (2003), 3383–3393.
22 Y. Wang, S. Zhang, X. Zeng, L. L. Ma, W. Weng, W. Yan, and M. Qian. Osteoblastic cell response on fluoridated hydroxyapatite coatings. Acta Biomater. 3 (2007), 191–197.
23 R. Shu, R. McMullen, M. J. Baumann, and L. R. McCabe. Hydroxyapatite accelerates differentiation and suppresses growth of MC3T3-E1 osteoblasts. J. Biomed. Mater. Res. A. 67 (2003), 1196–1204.
24 B. Alliot-Licht, M. Gregoire, I. Orly, and J. Menanteau. Cellular activity of osteoblasts in the presence of hydroxyapatite: An *in vitro* experiment. Biomaterials 12 (1991), 752–756.
25 K. Ogata, S. Imazato, A. Ehara, S. Ebisu, Y. Kinomoto, T. Nakano, and Y. Umakoshi. Comparison of osteoblast responses to hydroxyapatite and hydroxyapatite/soluble calcium phosphate composites. J. Biomed. Mater. Res. A 72 (2005), 127–135.
26 G. Zhang, R. A. Latour Jr., J. M. Kennedy, H. D. Schutte Jr., and R. J. Friedman. Long-term compressive property durability of carbon fibre-reinforced polyetheretherketone composite in physiological saline. Biomaterials 17(8) (1996), 781–789.
27 T. R. Martin, S. W. Meyer, and D. R. Luchtel. An evaluation of the toxicity of carbon fiber composites for lung cells *in vitro* and *in vivo*. Environ. Res. 49 (1989), 246–261.
28 J. Wang and R. Stevens. Zirconia-toughened alumina (ZTA) ceramics. J. Mater. Sci. 24 (1989), 3421–3440.
29 A. H. D. Aza, J. Chevalier, G. Fantozzi, M. Schehl, and R. Torrecillas. Crack growth resistance of alumina, zirconia and zirconia toughened alumina ceramics for joint prostheses. Biomaterials 23(3) (2002), 937–945.
30 B. Kerkwijk, L. Winnubst, E. J. Mulder, and H. Verweij. Processing of homogeneous zirconia-toughened alumina ceramics with high dry-sliding wear resistance. J. Am. Ceram. Soc. 82(8) (1999), 2087–2093.
31 C. He, Y. S. Wang, J. S. Wallace, and S. M. Hsu. Effect of microstructure on the wear transition of zirconia-toughened alumina. Wear 162–164 (1993), A 314–A 321.
32 K. Hayashi, T. Inadome, H. Tsumura, T. Mashima, and Y. Sugioka. Bone-implant interface mechanics of in vivo bio-inert ceramics. Biomaterials 14 (1993), 1173–1179.
33 G. Colon, B. C. Ward, and T. J. Webster. Increased osteoblast and decreased *Staphylococcus epidermidis* functions on nanophase ZnO and TiO_2. J. Biomed. Mater. Res. A 78A(3) (2006), 595–604.
34 S. Nath, N. Sinha, and B. Basu. Microstructure, mechanical and tribological properties microwave sintered of Ca-stabilised ZrO_2 ceramics. Ceram. Int. 34 (2008), 1509.
35 S. Nath, S. Bajaj, and B. Basu. Microwave-sintered MgO-doped zirconia with improved mechanical and tribological properties. Int. J. Appl. Ceram. Technol. 5(1) (2008), 49–62.
36 S. Yu, K. P. Hariram, R. Kumar, P. Cheang, and K. K. Aik. In vitro apatite formation and its growth kinetics on hydroxyapatite/polyetheretherketone biocomposites. Biomaterials 26 (2005), 2343–2352.
37 R. Z. LeGeros. Preparation of octacalcium phosphate (OCP): A direct fast method. Calcif. Tissue Int. 37 (1985), 194–197.
38 W. E. Brown, N. Eidelman, and B. Tomazic. Octacalcium phosphate as a precursor in biomineral formation. Adv. Dent. Res. l(2) (1987), 306–313.
39 F. Betts, N. C. Blumenthal, and A. S. Posner. Bone mineralization. J. Crys. Grow. 53 (1981), 63–73.

40 J. Li, B. Fartash, and B. L. Hermansson. Hydroxyapatite–alumina composites and bone-bonding. Biomaterials. 16 (1995), 417–422.
41 V. V. Silva, R. Z. Domíngues, and F. S. Lameiras. Microstructural and mechanical study of zirconia-hydroxyapatite (ZH) composite ceramics for biomedical applications. Composites Science and Technology. 61 (2001), 301–310.
42 H. W. Kim, Y. J. Noh, Y. H. Koh, H. E. Kim, and H. M. Kim. Effect of CaF_2 on densification and properties of hydroxyapatite–zirconia composites for biomedical applications. Biomaterials 23(20) (2002), 4113–4121.
43 N. Y. Mostafa. Characterization, thermal stability and sintering of hydroxyapatite powders prepared by different routes. Mat. Chem. Phys. 94 (2005), 333–341.
44 T. Suzuki, R. Ohashi, Y. Yokogawa, K. Nishizawa, F. Nagata, Y. Kawamoto, T. Kameyama, and M. Toriyama. Initial anchoring and proliferation of fibroblast L-929 cells on unstable surface of calcium phosphate ceramics. J. Biosci. Bioeng. 87(3) (1999), 320–327.
45 Q. Z. Chen, C. T. Wong, W. W. Lu, K. M. C. Cheung, J. C. Y. Leong, and K. D. K. Luk. Strengthening mechanisms of bone bonding to crystalline hydroxyapatite *in vivo*. Biomaterials 25 (2004), 4243–4254.
46 A. Rapacz-Kmita, A. Slosarczyk, and Z. Paszkiewicz. Mechanical properties of HAp–ZrO_2 composites. J. Eur. Ceram. Soc. 26(8) (2006), 1481–1488.
47 A. Chiba, S. Kimura, K. Raghukandan, and Y. Morizono. Effect of alumina addition on hydroxyapatite biocomposites fabricated by underwater-shock compaction. Mater. Sci. Eng. A 350 (2003), 179–183.
48 R. Hill and A. Clifford. Apatite-mullite glass-ceramics. J. Non-Cryst. Sol. 196 (1996), 346–351.
49 C. M. Gorman and R. G. Hill. Heat-pressed ionomer glass-ceramics. Part I: An investigation of flow and microstructure. Dent. Mater. 19(4) (2003), 320–326.
50 I. Manjubala and M. Sivakumar. In-situ synthesis of biphasic calcium phosphate Ceramics using microwave irradiation. Mater. Chem. Phys. 71 (2001), 272–278.
51 D. C. Tancred, B. A. O. McCormack, and A. J. Carr. A quantitative study of the sintering and mechanical properties of hydroxyapatite/phosphate glass composites. Biomaterials 19(19) (1998), 1735–1743.
52 W. Suchanek, M. Yashima, M. Kakihana, and M. Yoshimura. Hydroxyapatite/hydroxyapatite-whisker composites without sintering additives: Mechanical properties and microstructural evolution. J. Am. Ceram. Soc. 80(11) (1997), 2805–2813.
53 R. Xin, Y. Leng, J. Chen, and Q. Zhang. A comparative study of calcium phosphate formation on bioceramics *in vitro* and *in vivo*. Biomaterials 26 (2005), 6477–6486.
54 J. Dong, T. Uemura, H. Kojima, M. Kikuchi, J. Tanaka, and T. Tateishi. Application of low-pressure system to sustain *in vivo* bone formation in osteoblast-porous hydroxyapatite composite. Mater. Sci. Eng. C 17 (2001), 37–43.
55 A. M. Gatti, D. Zaffe, and G. P. Poli. Behaviour of tricalcium phosphate and hydroxyapatite granules in sheep bone defects. Biomaterials 11 (1990), 513–517.
56 C. Balcik, T. Tokdemir, A. Senkoylu, N. Koc, M. Timucin, S. Akin, P. Korkusuz, and F. Korkusuz. Early weight bearing of porous HA/TCP (60/40) ceramics *in vivo*: A longitudinal study in a segmental bone defect model of rabbit. Acta Biomat. 3 (2007), 985–996.
57 M. Fini, R. Cadossi, V. Cane, F. Cavani, G. Giavaresi, A. Krajewski, L. Martini, N. N. Aldin, A. Ravaglioli, L. Rimondini, P. Torricelli, and R. Giardino. The effect of pulsed electromagnetic fields on the osteointegration of hydroxyapatite implants in cancellous bone: A morphologic and microstructural *in vivo* study. J. Ortho. Res. 20 (2002), 756–763.
58 T. Okuda, K. Ioku, I. Yonezawa, H. Minagi, G. Kawachic, Y. Gonda, H. Murayama, Y. Shibata, S. Minami, S. Kamihira, H. Kurosawa, and T. Ikeda. The effect of the microstructure of β tricalcium phosphate on the metabolism of subsequently formed bone tissue. Biomaterials 28 (2007), 2612–2621.
59 T. J. Webster, R. W. Siegel, and R. Bizios. Nanoceramic surface roughness enhances osteoblast and osteoclast functions for improved orthopaedic/dental implant efficacy. Scr. Mater. 44(8–9) (2001), 1639–1642.

60 C. H. Tsai, R. M. Lin, C. P. Jua, and J. H. C. Lin. Bioresorption behavior of tetracalcium phosphate-derived calcium phosphate cement implanted in femur of rabbits. Biomaterials 29 (2008), 984–993.
61 H. E. Koepp, S. Schorlemmer, S. Kessler, R. E. Brenner, L. Claes, K. P. Gunther, and A. A. Ignatius. Biocompatibility and Osseointegration of β-TCP: Histomorphological and Biomechanical Studies in a Weight-Bearing Sheep Model. J. Biomed. Mater. Res. B Appl. Biomater. 70B (2004), 209–217.
62 V. V. Silva, F. S. Lameiras, and R. Z. Domínguez. Microstructural and mechanical study of zirconia-hydroxyapatite (ZH) composite ceramics for biomedical applications. Comput. Sci. Technol. 61 (2001), 301–310.
63 G. Gollera, H. Demirkiran, F. N. Oktar, and E. Demirkesen. Processing and characterization of bioglass reinforced hydroxyapatite composites. Ceram. Int. 29 (2003), 721–724.
64 S. Nath, S. Bodhak, and B. Basu. Tribological investigation of Novel HDPE-HAp-Al_2O_3 hybrid biocomposites against steel under dry and simulated body fluid condition. J. Biomed. Mater. Res. A 83A (2007), 191–208.
65 S. Nath, N. Sinha, and B. Basu. Microstructure, mechanical and tribological properties of microwave sintered Ca-stabilized zirconia for biomedical applications. Ceram. Int. 34 (2008), 1509–1520.
66 S. Bodhak, S. Nath, and B. Basu. Understanding the fretting wear properties of HAp, Al_2O_3 containing HDPE biocomposites against ZrO_2. J. Biomed. Mater. Res. A 85A (2008), 83–98.
67 S. Nath, S. Bajaj, and B. Basu. Microstructure, mechanical and tribological properties of microwave sintered Mg-doped zirconia. Int. J. Appl. Ceram. Technol. 5 (2008), 49–62.
68 S. Nath, S. Bodhak, and B. Basu. HDPE-Al_2O_3-HAp composites for biomedical applications: Processing and characterization. J. Biomed. Mater. Res. B Appl. Biomater. 88 (2009), 1–11.
69 S. Bodhak, S. Nath, and B. Basu. Friction and wear properties of novel HDPE -HAp-Al_2O_3 composites against alumina counterface. J. Biomater. Appl. 23 (2009), 407.
70 H. Schneider, K. Okada, and J. Pask. *Mullite and Mullite Ceramics*. John Wiley & Sons, New York, 1994.
71 S. Kanzaki, H. Tabata, T. Kumazawa, and S. Ohta. Sintering and mechanical properties of stoichiometric mullite. J. Am. Ceram. Soc. 68 (1984), C6–C7.
72 W. Rathje. Zur Kentnis der phosphate: I. Über hydroxyapatite. Bodenk Pflernah. 12 (1939), 121–128.
73 M. H. Santos, M. Oliveira, L. P. F. Souza, H. S. Mansurand, and W. L. Vasconcelos. Synthesis control and characterization of hydroxyapatite prepared by wet precipitation process. Mater. Res. 7 (2004), 625–630.
74 S. Nath, B. Basu, K. Biswas, K. Wang, and R. K. Bordia. Sintering, phase stability and properties of calcium phosphate-mullite composites. J. Am. Ceram. Soc. 93(6) (2010), 1639–1649.
75 S. Nath, A. Dubey, and B. Basu. Hardness, compressive strength and toughness properties of novel hydroxyapatite-mullite composites. J. Biomed. Appl. Published online before print February 22, 2011 (doi: 10.1177/0885328210393292).
76 S. Nath, A. Dey, A. K. Mukhopadhyay, and B. Basu. Nanoindentation response of novel hydroxyapatite-mullite composites. Mater. Sci. Eng. A. 513–514 (2009), 197–201.
77 S. Nath. Development of novel calcium phosphate-mullite composites for orthopedic application. Ph.D. thesis, IIT Kanpur, India, 2008.
78 S. Nath, B. Basu, M. Mohanty, and P. V. Mohanan. *In vivo* response of novel hydroxyapatite-mullite composites: Results up to 12 weeks of implantation. J. Biomed. Mater. Res. B 90B (2009), 547–557.
79 J. Li. Behaviour of titanium and titania-based ceramics *in vitro* and *in vivo*. Biomaterials 14 (1993), 229–232.
80 P. Ducheyne and L. L. Hench. The processing and static mechanical properties of metal fibre reinforced bioglass. J. Mater. Sci. 17 (1982), 595–606.
81 X. Zhang, G. H. M. Gubbels, R. A. Terpstra, and R. Metselaar. Toughening of calcium hydroxyapatite with silver particles. J. Mater. Sci. 32 (1997), 235–243.
82 T. K. Chaki and P. E. Wang. Densification and strengthening of silver-reinforced hydroxyapatite-matrix composite prepared by sintering. J. Mater. Sci. Mater. Med. 5 (1994), 533–542.

83 S. M. F. Asmus, S. Sakakuoorand, and G. Pezzotti. Hydroxyapatite toughened by silver inclusions. J. Comput. Math. 37 (2003), 2117–2129.

84 Q. L. Feng, T. N. Kim, J. Wu, E. S. Park, J. O. Kim, D. Y. Lim, and F. Z. Cui. Antibacterial effects of Ag-HAp thin films on alumina substrates. Thin Solid Films 335 (1998), 214–219.

85 N. Saha, A. K. Dubey, and B. Basu. Cellular proliferation, cellular viability and in vivo biocompatibility of HA-ZnO composites. Journal of Biomedical Materials Research B (2011), in press.

86 S. D. Gittard, J. R. Perfect, N. A. Monteiro-Riviere, W. Wei, C. Jin, and R. J. Narayan. Assessing the antimicrobial activity of zinc oxide thin films using disk diffusion and biofilm reactor. Appl. Surf. Sci. 255 (2009), 5806–5811.

87 O. Yamamoto, M. Komatsu, J. Sawai, and Z. Nakagawa. Effect of lattice constant of zinc oxide on antibacterial characteristics. J. Mater. Sci. Mater. Med. 15 (2004), 847–851.

88 G. Applerot, A. Lipovsky, R. Dror, N. Perkas, Y. Nitzan, R. Lubart, and A. Gedanken. Enhanced antibacterial activity of nanocrystalline ZnO due to increased ROS-mediated cell injury. Advanced Functional Materials 19(6) (2009), 842–852.

89 J. Sawai. Quantitative evaluation of antibacterial activities of metallic oxide powders (ZnO, MgO and CaO) by conductimetric assay. J. Microb. Methods 54 (2003), 177–182.

90 S. Nath, S. Kalmodia, and B. Basu. Densification, phase stability and *in vitro* biocompatibility property of hydroxyapatite-10 wt% silver composites. J. Mater. Sci. Mater. Med. 21 (2010), 1273–1287.

91 N. Saha, K. Keskinbora, E. Suvaci and B. Basu. Sintering, microstructure, mechanical properties and antimicrobial property of HAp-ZnO biocomposites. Journal of Biomedical Materials Research 95B (2010), 430–440.

92 P. W. Brown. Phase relationships in the ternary system CaO-P_2O_5-H_2O at 25°C. J. Am. Ceram. Soc. 75(1) (1992), 17–22.

Chapter 21

Tribological Properties of Ceramic Biocomposites

Tribology is of paramount importance for evaluating the mechanical properties of an implant. Ideally, once an implant is surgically inserted *in vivo*, it should suffice for the remaining lifetime of the person and should not require replacement because of failure. Hence certain measures, at both subgrain and bulk scales, need to be envisaged for characterizing the tribological performance of an implant.

21.1 INTRODUCTION

The science of tribology can be described on the basis of synergistic interaction among concepts drawn from multiple disciplines, including physics, chemistry, metallurgy, materials science, and mechanical engineering (see Fig. 21.1). In essence, tribology comprises *wear*, *friction*, and *lubrication*. It can be mentioned here that several new metallic alloys, ceramics, and polymers as well as their composites have been developed in the last few decades using the processing–structure–property correlation, a fundamental concept used in materials engineering. In the discipline of mechanical engineering, multiple research groups have largely contributed to the lubrication aspect of tribology.

Broadly, the tribological response of any mating material combination depends on both surface properties and the tribological interaction with the environment, which depends on the mechanics at the tribological interface. This results in friction and wear of both of the mating surfaces. Therefore, an understanding of how two mating materials will respond at a loaded contact, while experiencing relative motion, is important in assessing the potential of newly developed materials for tribological applications.

Friction is generally described as the resistance to motion that arises from the solid surface interactions at the real area of contact. Low friction is desired for certain applications, such as hinges, rivets, bearings, and human hip joints. In contrast, applications such as brakes, clutches, and tires on roads demand much higher friction. In any case, the reality of controlling the optimal friction for a particular

Advanced Structural Ceramics, First Edition. Bikramjit Basu, Kantesh Balani.

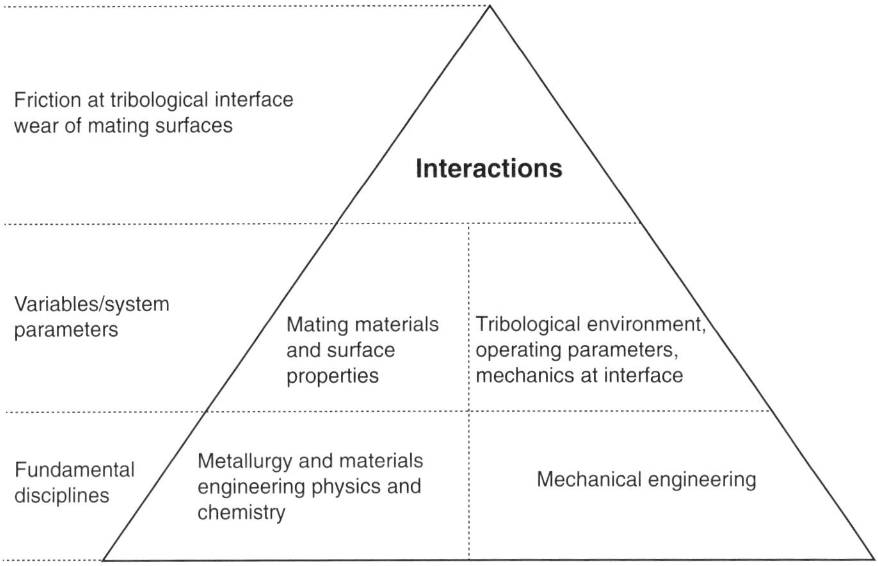

Figure 21.1 Concept triangle illustrating the interaction of basic science and engineering disciplines and multiple parameters, as involved in the science of tribology.

application depends on understanding the performance of materials and their surface properties.

The progressive loss of mating materials at contacting interfaces under relative motion is defined as wear. The variants of wear include sliding, rolling, erosion, impact, and cavitation, among others, in various atmospheric conditions. Wear can cause catastrophic failure of various implants, finally requiring replacement surgery. However, similarly as for friction, the control and prediction of wear of materials for a given application is important and is thus the key to success of every tribological application.

It is known that friction and wear are system-dependent properties.[1] Therefore, it is critical to understand and consider the nature and responses of mating materials, and this is very important for selecting suitable materials, while providing satisfactory service under the required conditions of friction and wear. To this end, the aim of material scientists is to control the properties of a specific component by incorporating suitable compositional changes at the microstructural level during the processing stage, but having in mind the tribological requirements. This is particularly important for many advanced materials, such as bioceramics and biocomposites.

21.2 TRIBOLOGY OF CERAMIC BIOCOMPOSITES

Hydroxyapatite (HAp) and various calcium phosphate (CaP)–based materials are being widely investigated as potential biomaterials, but the application of monolithic CaP materials is limited by their poor fracture toughness.[2–7] A superior CaP with mullite ($3Al_2O_3 \cdot 2SiO_2$) reinforcement has been reported that was shown to have enhanced single edge V-notched beam (SEVNB) fracture toughness ($1.5\,MPa\,m^{0.5}$), while showing compressive strength of ~300 MPa. *In vitro* and *in vivo* studies on calcium phosphate reinforced with mullite (up to 30 wt%) have shown that the biocompatibility of the composite is similar to that of pure HAp.[8,9] For overall performance assessment, it is imperative to include the tribological response of HAp–mullite composites in evaluating their potential as biomaterials.[10] Hence, their tribological response (fretting wear) is evaluated in Section 21.3 to illustrate their resistance to abrasion in contact with a mating surface.

21.3 TRIBOLOGICAL PROPERTIES OF MULLITE-REINFORCED HYDROXYAPATITE

It is often reported that biomaterials would experience a lower coefficient of friction (COF) under the lubrication conditions in simulated body fluid (SBF). Fu et al. reported that a thermally sprayed HAp coating experienced COF of 0.2–0.3 in the presence of bovine albumin lubrication, while unlubricated fretting wear yielded a COF value between 0.7 and 0.8.[11] The major wear mechanisms include delamination, fracturing, pitting, and abrasive wear. Wear-induced failure of biomaterials necessitated detailed wear and tribological studies of diverse implant materials both in dry conditions and in the wet–lubricated conditions of a simulated body fluid (SBF) environment.[12–21]

To attain superior properties, mullite-containing (10–30 wt%) CaP-composites were pressureless sintered at 1350°C for 2 hours; for comparison, pure HAp and pure mullite were sintered at 1200°C for 2 hours and 1700°C for 4 hours, respectively. The sinter density and phase assemblage are summarized in Table 21.1.

Table 21.1. The Densification, Sintering Conditions, and Phases Present after Sintering are Mentioned against the Sample Designation

Samples	Sinter density (% ρ_{th})	Sintering conditions	Phases present
Pure HAp	99.17	1200°C, 2 hours	HAp-ss
HAp–10M	94	1350°C, 2 hours	α-TCP-ss, HAp-ss, β-TCP-w, mullite-ww, CaO-ww
HAp–20M	98.1	1350°C, 2 hours	α-TCP-s, β-TCP-ss, HAp-ww, mullite-w, gehlenite-ww, CaO-ww, alumina-ww
HAp–30M	95.6	1350°C, 2 hours	β-TCP-ss, HAp-ww, mullite-s, gehlenite-ww, CaO-ww, alumina-s
Pure mullite	98.7	1700°C, 4 hours	Mullite-ss

The phases present in the sintered body are summarized based on the XRD peak intensities.
ss, very strong; s, strong; ww, very weak; w, weak.[21]

21.3.1 Materials and Experiments

Commercially available zirconia balls were used as counterbody for evaluating fretting wear (in both dry and SBF-lubricated conditions) of mullite-containing sintered HAp pellets. Fretting tests were carried out at frequency 5 Hz, swapping a linear distance of 80 μm at a normal load of 10 N, over a total of 100,000 cycles; the response was monitored *in situ* to evaluate the fretting wear resistance of the HAp–mullite composites.

After fretting, the worn surface can be three-dimensionally scanned to estimate the material loss due to fretting. A laser surface profilometer can do the job of scanning the worn zone and calculating the wear volume. The wear volume obtained for various samples can yield the comparative wear resistance. The COF for various samples is plotted in Figure 21.2. The performance of various mullite-reinforced HAp samples in dry and SBF environments are presented in Figure 21.2a,b, respectively.

From Figure 21.2a, it is evident that pure HAp exhibited a COF of 0.35, whereas the maximum COF of 0.62 was observed for pure mullite ceramic. All other HAp–mullite composites showed a COF within this range (~0.5) during steady-state conditions. However, during fretting certain fluctuations were also observed in the samples that could arise due to three-body wear, or entrapment of debris particles during fretting. Similar to the COF values observed in dry conditions (Fig. 21.2a), the lowest COF (0.3) was experienced by pure HAp samples under wet conditions (Fig. 21.2b).

21.3.2 Effect of Lubrication on the Wear Resistance of Mullite-Reinforced Hydroxyapatite

The wear rate of various mullite-reinforced HAp composites is plotted in Figure 21.3, wherein the corresponding hardness of samples does not correlate with their wear rate in dry conditions, but there is an inverse proportionality of wear rate with hardness under lubricated conditions. The maximum wear rate (8.54×10^{-6} mm^3/N·m) was measured with HAp–10M ceramic, whereas pure mullite showed the lowest wear rate (1.22×10^{-6} mm^3/N·m). Pure HAp, HAp–20M, and HAp–30M exhibited wear rates of 4.2×10^{-6}, 1.62×10^{-6}, and 1.39×10^{-6} mm^3/N·m, respectively, under dry conditions.

Similar to the dry conditions, pure mullite again showed the least wear rate (0.94×10^{-6} mm^3/N·m) in the SBF conditions. A decrease in wear rate was not significant: from 1.22×10^{-6} mm^3/N·m (under dry conditions) to 0.94×10^{-6} mm^3/N·m (under wet [SBF] conditions). Conversely, the decrease in wear rate under SBF lubricating conditions was dramatic for the pure HAp and HAp–10M samples, showing wear rates of 1.53×10^{-6} mm^3/N·m and 6.72×10^{-6} mm^3/N·m, respectively. Surprisingly, the wear rates of the HAp–20M and HAp–30M samples were higher (6.25×10^{-6} mm^3/N·m and 5.06×10^{-6} mm^3/N·m respectively) in SBF medium, compared with that of HAp–10M. The two-dimensional depth profiles at the center

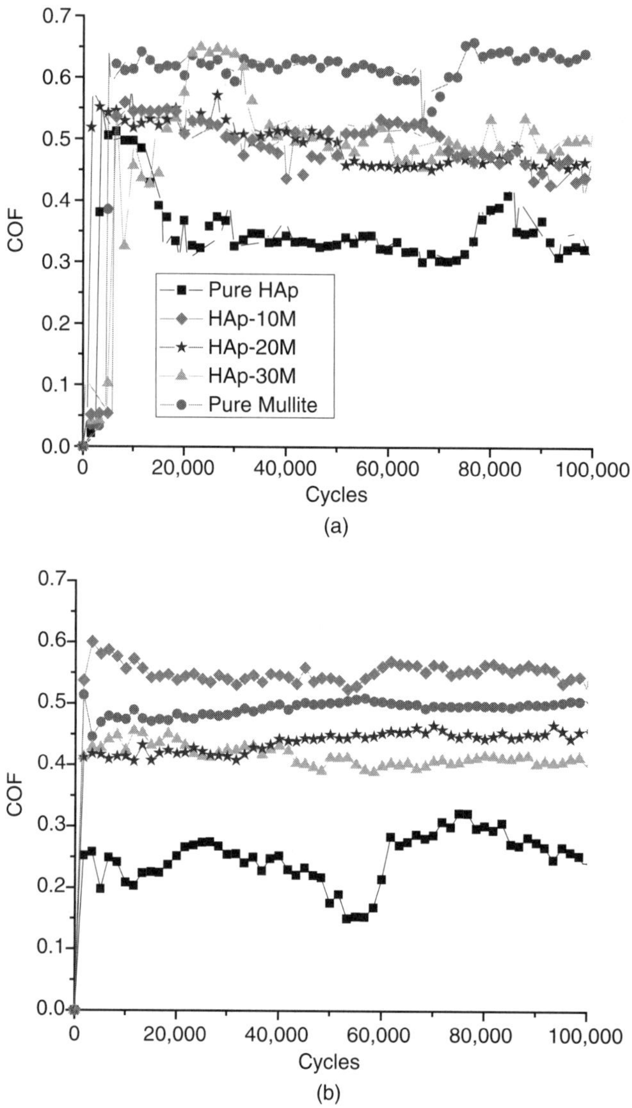

Figure 21.2 Plots of coefficient of friction (COF) vs. number of fretting cycles for HAp–mullite composites in two environments: (a) dry ambient condition; (b) simulated body fluid with albumin protein.[21]

of the wear scar were recorded using a laser surface profilometer after fretting in dry and SBF conditions (Fig. 21.4a and b, respectively).

Figure 21.4 reveals that the worn HAp surface is rough, and wear appears to be independent of the dry or lubricated conditions, whereas the addition of mullite to HAp yields a smooth surface appearance under lubricated conditions, compared with that under dry conditions. The maximum wear depth corresponds well with the

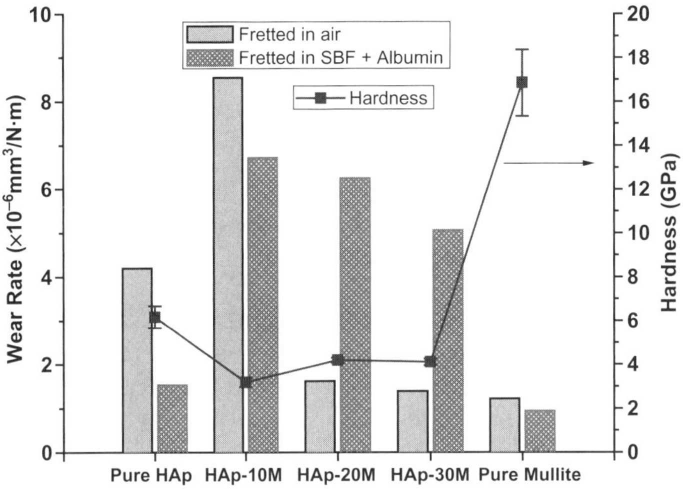

Figure 21.3 Wear rates of different samples (pure HAp, HAp–mullite composites, and pure mullite) are plotted when fretted in dry and in SBF + albumin environments. Around 10% deviation of the mean wear rate was measured in experiments.[21]

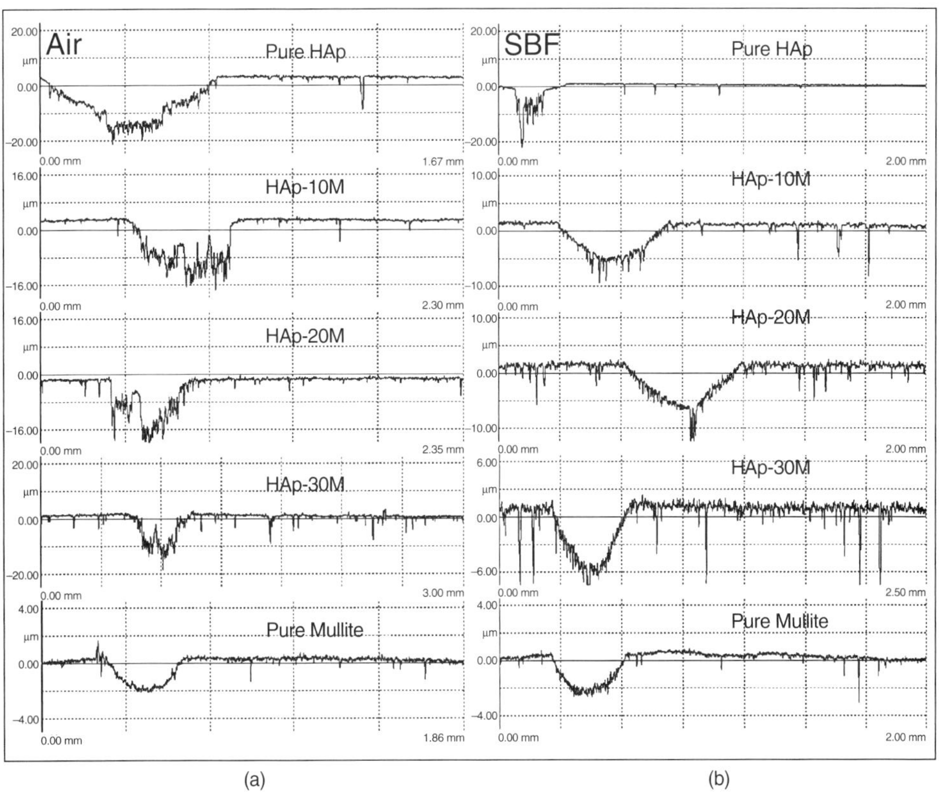

Figure 21.4 Two-dimensional wear depth profiles for pure HAp, HAp–30M, and pure mullite after fretting wear tests under (a) dry and (b) SBF conditions.[21]

fretting damage of various HAp and mullite samples. Pure HAp shows a maximum wear depth of 20 μm; the wear depth reduces to ~2.5 μm in the case of pure mullite after fretting under dry conditions. A common observation was that the fretting resulted in higher wear depths under dry conditions compared with those under SBF medium.

21.3.3 Surface Topography of Mullite-Reinforced Hydroxyapatite after Fretting Wear

The topographical features of pure HAp, HAp–20M, and HAp–30M samples are presented in Figure 21.5a–f, respectively, showing damage caused by fretting under ambient conditions. The damaged region is identified by the dashed circles in Figures 21.5a, c, and e. The delamination on the worn HAp surface layer (Fig. 21.5a) indicates severe damage due to plowing and microcrack formation (Fig. 21.5b). Fatigue cracks were observed in the HAp–20M sample due to abrasion during fretting (Fig. 21.5c,d), but a reduced propensity of cracking was observed in the HAp–30M sample (Fig. 21.5f). ZrO_2 abrasion material was absent on the flat worn surface (Fig. 21.5d,f). As for the HAp–10M sample, severe cracking was observed (not shown here), whereas only abrasive wear scratches could be noticed for the pure mullite sample.

Fretting in SBF yielded abrasion as the dominant wear mechanism. For HAp–10M and HAp–20M samples (Fig. 21.6a,b), abrasive scratches along with mild fatigue cracking were observed on the worn surface. In contrast, plowing and grain pullout were the dominant wear mechanisms in the HAp–30M composite (Fig. 21.6c).

The wear depth data, as presented in Figure 21.3, do not correlate with the wear rate trend, since wear rate depends on both depth and diameter of the wear scar. For HAp–20M and HAp–30M samples, the wear rate values were measured to be higher due to larger wear scar diameter (Table 21.2), despite exhibiting lower wear depth in SBF than in the dry environment.

A comparison of the available tribological data of HAp-based composites is provided in Table 21.3.[21–28]

An analysis of Table 21.3 reveals that the wear of HAp-based composites varies between 10^{-5} and 10^{-9} mm^3/N·m, depending on the tribological environment and the mating material. Also, mild abrasion, fracture, and fatigue wear were reported to be major wear mechanisms.

21.4 TRIBOLOGICAL PROPERTIES OF PLASMA-SPRAYED HYDROXYAPATITE REINFORCED WITH CARBON NANOTUBES

21.4.1 Bulk Wear Resistance of Hydroxyapatite Reinforced with Carbon Nanotubes

Bulk wear of HAp has shown much superior performance compared with that of a metallic Ti–6Al–4V implant. Figure 21.7 shows the wear rate and cumulative weight

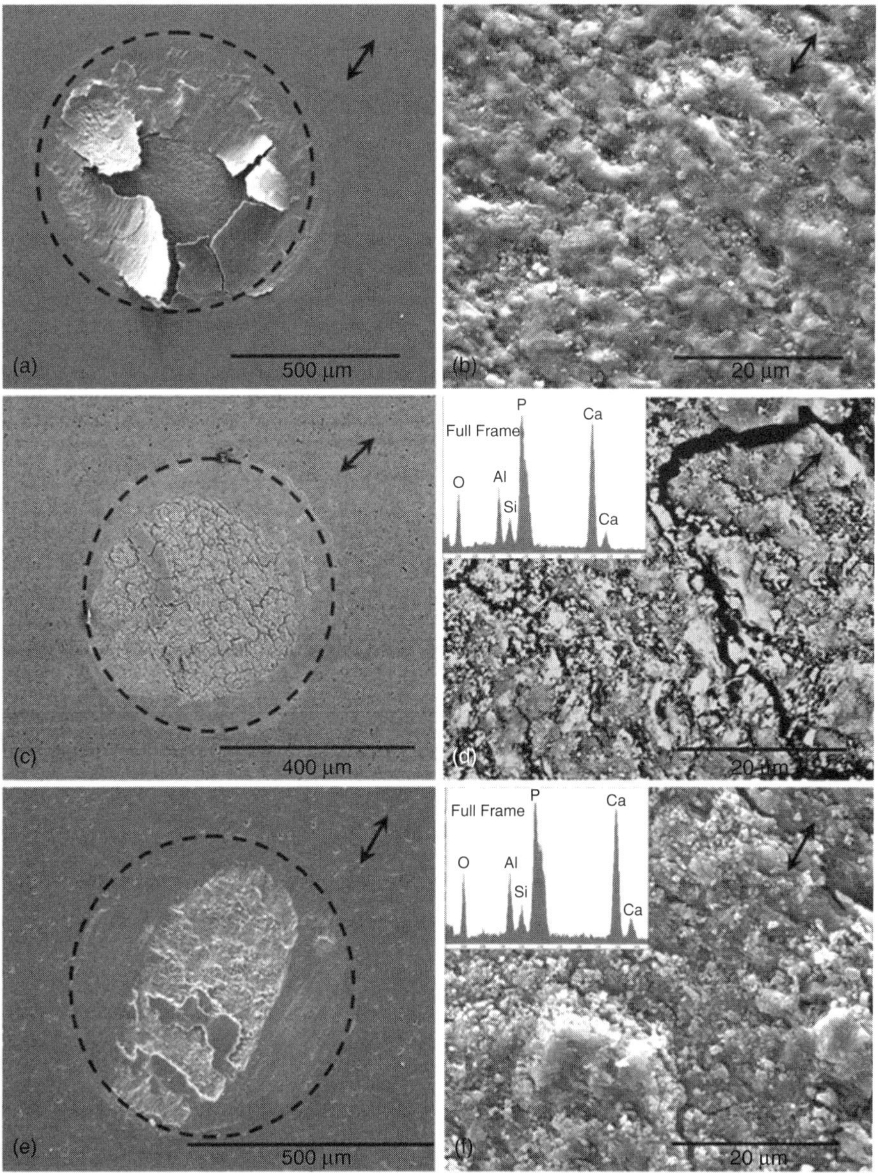

Figure 21.5 SEM images of worn surfaces of (a and b) pure HAp sintered at 1200°C, (c and d) HAp–20M sintered at 1350°C, and (e and f) HAp–30M sintered at 1350°C, after testing against zirconia in air. The double-pointed arrow indicates fretting direction.[21]

loss of the Ti–6Al–4V substrate (base) compared with the HAp and carbon nanotube (CNT)–reinforced HAp coatings. Improvement in wear resistance by up to 1.5 times has been reported with plasma-sprayed HAp coating on a Ti–6Al–4V surface.[29] Further, CNT-reinforced HAp coating has shown a reduction in the cumulative weight loss by 2.3 times compared with that of the Ti–6Al–4V substrate. This

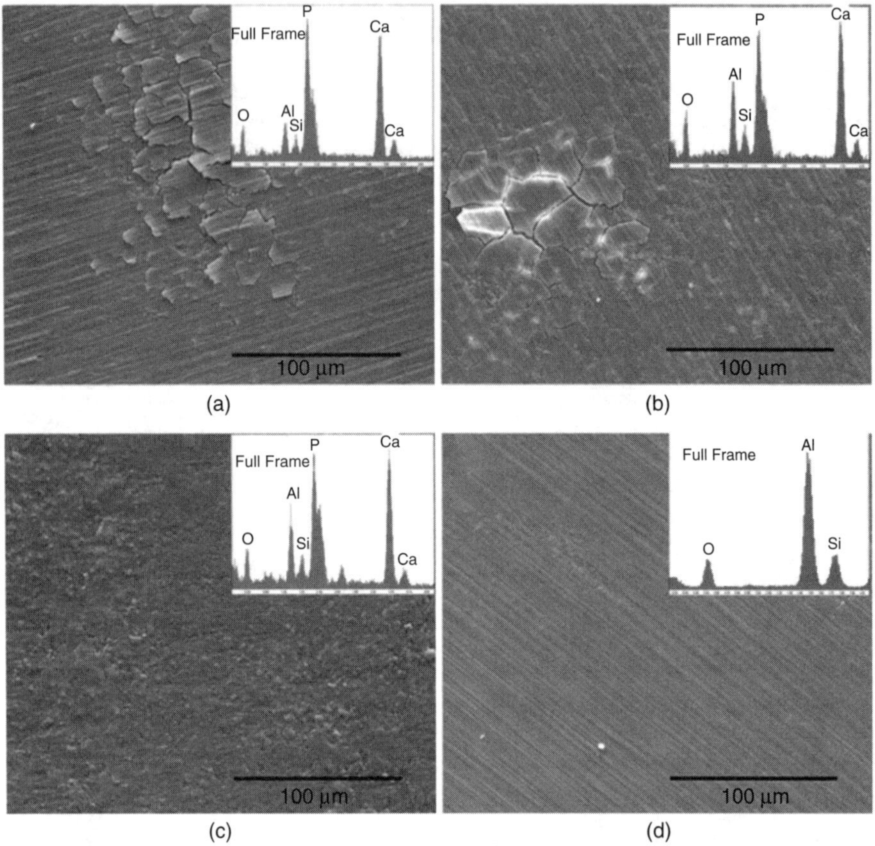

Figure 21.6 SEM images of worn surfaces of (a) HAp–10M sintered at 1350°C, (b) HAp–20M sintered at 1350°C, (c) HAp–30M sintered at 1350°C, and (d) pure mullite sintered at 1700°C, after testing against zirconia in SBF medium containing albumin. The double-pointed arrow indicates fretting direction.[21]

dramatic improvement in wear resistance is observed while retaining biocompatibility.[23]

The initial improvement in wear resistance obtained by coating with HAp is simply due to adding the ceramic coating onto a metallic implant surface. Since ceramics possess superior abrasion resistance, consequently, the material loss represented by wear rate also decreases. The enhancement in the wear resistance of CNT-reinforced HAp is attributed to anchoring by the CNTs, which keep holding the splats together. In addition, the graphitic carbon assists lubrication and also reduces the COF. Hence, the overall performance of CNT-reinforced HAp has shown superior tribological properties.

It must be noted that the interaction of bulk damage starts at a localized subgrain level; that is, upon contact, it is a few surface atoms that first encompass the damage and lead to cracking of the matrix.[24] Hence, it becomes essential to peep into the

Table 21.2. Maximum and Mean Hertzian Contact Pressures as Well as the Initial Hertzian Contact Diameter for the Investigated Fretting Couple (Containing Pure HAp, Pure Mullite, and HAp–Mullite Composites) are Presented along with Measured Wear Scar Diameter[21]

Material	Medium	Maximum Hertzian contact pressure (initial) (MPa)	Mean Hertzian contact pressure (initial) (MPa)	Initial Hertzian contact diameter (μm)	Measured wear scar diameter (μm)	*E*-modulus (GPa)
Pure HAp	Dry	809	540	153.7	767	120
	SBF	809	540	153.7	377	
HAp–10M	Dry	521	347	191.6	801	45
	SBF	521	347	191.6	671	
HAp–20M	Dry	621	414	175.4	564	72
	SBF	621	414	175.4	636	
HAp–30M	Dry	597	398	178.9	563	68
	SBF	597	398	178.9	614	
Pure mullite	Dry	1038	692	135.7	530	240
	SBF	1038	692	135.7	344	

contact damage as a nanoindenter tip interacts with the material. It has been pointed out by researchers that CNT dispersion is also one of the key parameters in resisting wear since a homogenous microstructure obtained via uniformly distributed CNTs imparts a strong reinforcing effect.[24]

CNTs tend to densify the matrix upon interaction with the abrading pin (Fig. 21.8a) and increase the resistance of the material via compaction.[24] Additionally, the graphitic layers provide lubrication (Fig. 21.8b) and tend to reduce the friction with the abrading pin, thus minimizing the surface damage.[24] As observed earlier, CNT content and dispersion hold a key to achieving superior tribological properties of bioceramics.

21.4.2 Nanomechanical Properties of Hydroxyapatite Reinforced with Carbon Nanotubes

At the nanoscale, interaction of the matrix with the indenter tip (~100-nm tip radius) becomes dependent on the (1) microstructure and (2) surface topography affecting the normal depth attained as the tip contacts the surface.[24] At atomic contact levels, van der Waals forces become dominant, and thereby these attractive forces negate the lubrication effects of CNTs. Therefore the CNT lubrication effects can become negligible even with varying CNT content and dispersion.[24]

Further, the cytocompatibility of HAp reinforced with Al_2O_3 and CNTs has been confirmed while showing enhanced mechanical properties.[22] The fracture toughness

Table 21.3. Summary of Literature Results on Various HAp-Based Materials Undergoing Wear Damage in Different Media

Material	Counterbody	Medium of testing	Experimental conditions	COF	Wear rate	Mechanism of wear	Ref.
HAp–PSZ	UHMWPE	Human plasma	Pin-on-disk Load: 4.3 N	0.04–0.06	$5.9 \pm 0.7 \times 10^{-9}$ mm^3/N·m	Fatigue wear	28
HAp–collagen with 10% hyaluronic acid	Stainless steel	Carboxymethyl cellulose Solution	Pin-on-disk Load: 10–70 N	0.01–0.04	4.1×10^{-5} mm^3/N·m	Deformation and fracture of HAp layer, transfer layer buildup	26
		Bovine serum		0.01–0.03	–		
HAp–CNT	ZrO_2	*c*-SBF	Pin-on-disk Load: 8.8 N	–	38.9 g/m^2·10^4 revolutions	Abrasion, fracture, plastic deformation fragmentation, chipping, and plowing	29
HAp–NFSS	Unknown	Air	Pin-on-disk Load: 15, 45 N	0.5–0.85 (15 N) 0.35–0.75 (45 N)	–	Abrasive wear at low load; abrasive, adhesive, and third body mechanism at higher load	27
HDPE–HAp–Al_2O_3	Stainless Steel	Hank's balanced salt solution (HBSS)	Ball-on-flat (fretting) Load: 10 N	0.07–0.11	$2.3 \pm 0.7 \times 10^{-6}$ mm^3/N·m	Mild abrasive wear and limited plastic deformation	16
HDPE–HAp–Al_2O_3	Al_2O_3	SBF	Ball-on-flat (fretting) Load: 10 N	0.11	$5.9 \pm 0.9 \times 10^{-7}$ mm^3/N·m	Mild abrasive wear and limited plastic deformation	20
HDPE–HAp–Al_2O_3	ZrO_2	SBF	Ball-on-flat (fretting) Load: 10 N	0.05	$1.1 \pm 0.7 \times 10^{-6}$ mm^3/N·m	Mild abrasive wear and limited plastic deformation	18
HAp–(30 wt%) mullite	ZrO_2	SBF	Ball-on-flat (fretting) Load: 10 N	0.4	5×10^{-6} mm^3/N·m	Mild plowing	21

The testing conditions are also indicated.

UHMWPE, ultra-high-molecular-weight polyethylene; NFSS, nickel-free stainless steel; CNT, carbon nanotube; PSZ, partially stabilized ZrO_2.

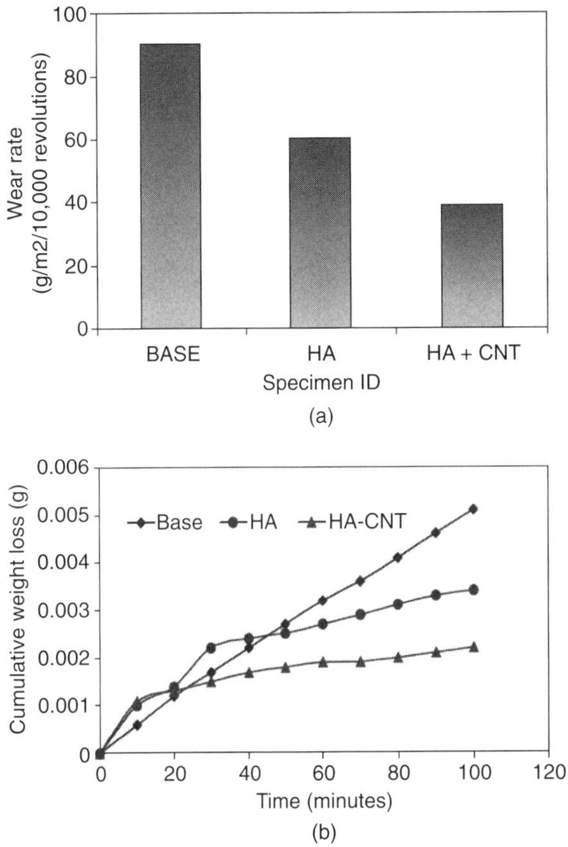

Figure 21.7 (a) Wear rate and (b) cumulative wear loss of Ti–6Al–4V (base) and coated samples. Coating of hydroxyapatite (HA) and HAp reinforced with carbon nanotubes (HA-CNT) shows superior wear resistance. Reproduced with permission from Reference 29.

was enhanced by 1.5 times (to 1.8 MPa m$^{1/2}$, compared with 0.7 MPa m$^{1/2}$ for pure HAp) with addition of 20 wt% Al_2O_3, whereas fracture toughness increased by a factor of 3 (to 2.9, compared with 0.7 MPa m$^{1/2}$) with the addition of 18.4 wt% Al_2O_3 and 1.6 wt% CNTs in the HAp matrix.[22] This study reinforced the retention of cytocompatibility while enhancing the mechanical properties of HAp biocomposites. Correspondingly, the tribological properties of HAp (HA), HAp–Al_2O_3 (HA–A), and HAp–Al_2O_3–CNT (HA–A16C) were evaluated using nanoscratching.[25] A typical load–depth curve (*P–h* curve), shown in Figure 21.9, was obtained using nanoindentation, and the mechanical properties (elastic modulus and hardness) of the samples were evaluated.[25] The superior mechanical properties of HA–A16C coatings were attributed to Al_2O_3 and CNT reinforcements while rendering a high plasticity index. Table 21.4 shows that the mechanical properties were enhanced with the Al_2O_3 reinforcement in HAp matrix (hardness increased from 2.52 to 3.52 GPa, whereas elastic modulus increased from 68.1 to 94.1 GPa). Further CNT-reinforcement

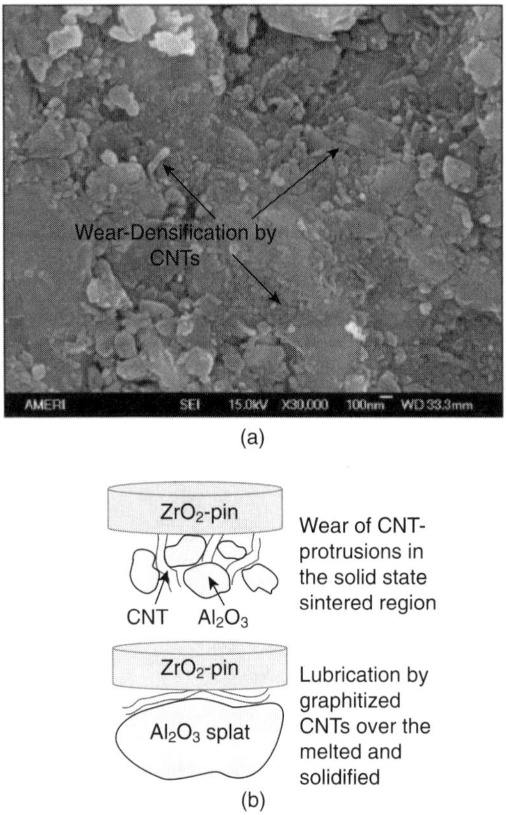

Figure 21.8 (a) Densification of CNTs upon interaction with abrading pin; (b) schematic of lubrication provided by CNTs. Reprinted with permission from Reference 24.

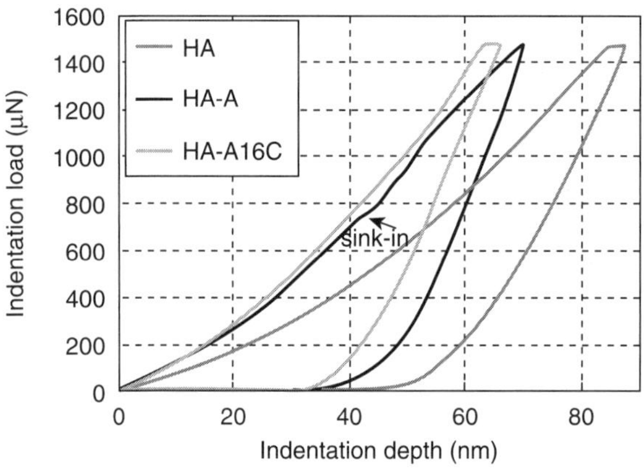

Figure 21.9 Load–depth curve of monolithic HAp (HA), HAp–Al_2O_3 (HA–A), and HAp–Al_2O_3–CNT (HA–A16C) coatings. Reprinted with permission from Reference 25. See color insert.

Table 21.4. Mechanical Properties (Hardness, Farcture Toughness, and Elastic Modulus) of Plasma-Sprayed Monolithic HAp (HA), HAp–Al_2O_3 (HA–A), and HAp–Al_2O_3–CNT (HA–16C) Coatings (Reprinted with Permission from Reference 25)

Coatings	Hardness (GPa)	Fracture toughness (MPa $m^{1/2}$)	Elastic modulus (GPa)	Indentation depth (nm)	Plasticity index
Monolithic HAp (HA)	2.52 ± 0.40	0.71 ± 0.19	68.1 ± 6.3	86.7 ± 6.8	0.49
HAp–Al_2O_3(HA–A)	3.52 ± 0.33	1.83 ± 0.11	94.1 ± 2.1	70.8 ± 2.4	0.51
HAp–Al_2O_3–CNT(HA–A16C)	4.17 ± 0.40	2.92 ± 0.23	102.4 ± 7.4	65.2 ± 1.9	0.54

(in HAp–Al_2O_3–CNT (HA–A16C) supplemented the enhancement in hardness (from 3.52 to 4.17 GPa) and elastic modulus (from 94.1 to 102.4 GPa) compared with that of the HAp–A coating.[25]

21.4.3 Nanoscratching of Hydroxyapatite Reinforced with Carbon Nanotubes

The monolithic HAp (HA), HAp–Al_2O_3 (HA–A), and HAp–Al_2O_3–CNT (HA–A16C) coatings were scratched using a Berkovich 100-nm diamond tip at a ramp load of 1000 μN (at 33.3 μN/s). Typical topographical profiles of the nanoscratched surfaces are shown in Figure 21.10. The inhomogeneity in the surface roughness wear volume (W_v) of the plasma-sprayed coatings is evaluated using the following equation[24]:

$$W_v = \frac{1}{2}\cos(70.3)d_n^2 \cdot l, \tag{21.1}$$

where 70.3° is the included half-angle of the Berkovich tip, d_n is the wear track depth of the Berkovich tip raster at ramping load, and l is the length of the wear track. The wear volume of the monolithic HAp (HA), HAp–Al_2O_3 (HA–A), and HAp–Al_2O_3–CNT (HA–A16C) coatings evaluated via nanoscratching is presented in Table 21.5.

Enhanced wear resistance of more than 13 times was observed with reinforcement with harder and tougher Al_2O_3 ceramic in the HAp matrix. Supplementing this with the lubrication and reinforcing effects of CNTs dramatically enhanced the wear resistance improvement by more than 65 times in the HAp–Al_2O_3–CNT (HA–A16C) coating.

21.5 LASER SURFACE TREATMENT OF CALCIUM PHOSPHATE BIOCOMPOSITES

Tailored calcium phosphate composites were deposited using a Nd–YAG laser on a titanium substrate (Ti–6Al–4V). Consequently, the resulting differences in the surface morphology, surface roughness, and mechanical–tribological properties indicate that the laser surface treatment can be engineered to manipulate calcium

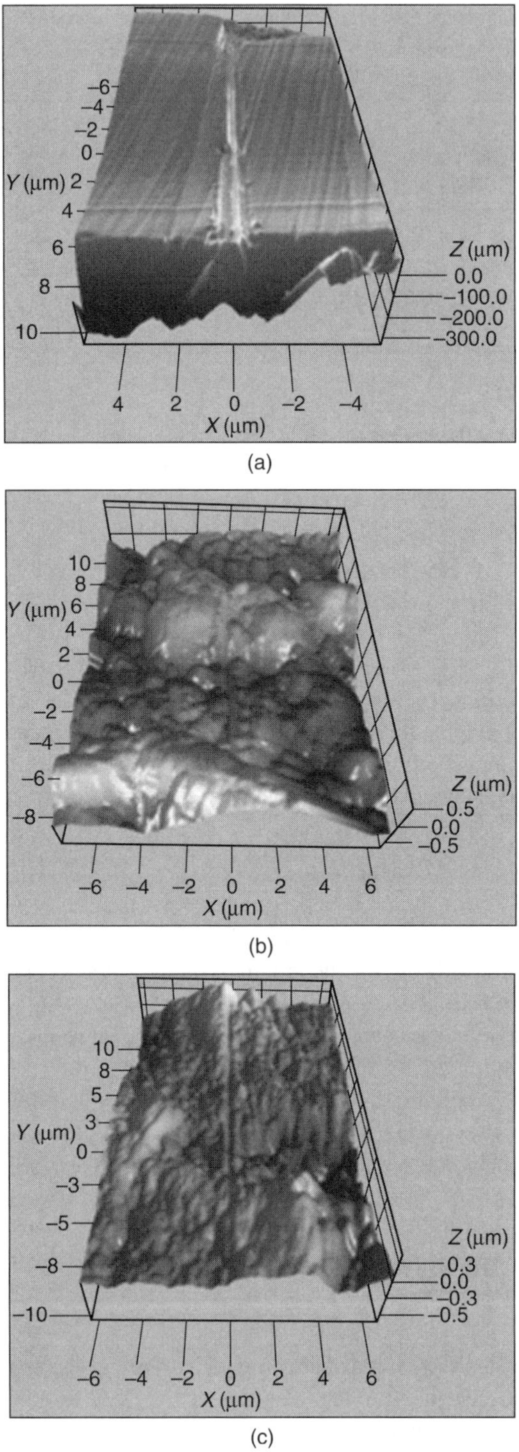

Figure 21.10 Nanoscratched surface topography of plasma-sprayed (a) monolithic HAp (HA), (b) HAp–Al_2O_3 (HA–A), and (c) HAp–Al_2O_3–CNT (HA–A16C) coatings. Reprinted with permission from Reference 25. See color insert.

Table 21.5. Effect of Surface Roughness on the Coefficient of Friction and Wear Resistance Improvement of Monolithic HAp (HA), HAp–Al_2O_3 (HA–A), and HAp–Al_2O_3–CNT (HA–16C) Coatings (Reprinted with Permission from Reference 25)

Coatings	Surface roughness, R_a (nm)	Coefficient of friction	Maximum scratch depth (nm)	Wear volume ($10^{-21}\,m^3$)	Wear resistance improvement
HAp	5	0.12 ± 0.01	644.2	177.7 ± 1.1	1
HAp–A	26	0.18 ± 0.11	101.9	13.1 ± 0.2	13.5 times
HAp–A16C	35	0.13 ± 0.02	67.3	2.6 ± 0.02	68.3 times

Table 21.6. Laser Parameters for Deposition of Ca-P Coatings on Ti–6Al–4V Substrate[30]

Pulse width	0.5 ms
Pulse energy	4 J
Pulse repetition rate	20 Hz
Average power	80 W
Laser scan speed	36, 48, 66, 78, 102 cm/min
Focus position	0.8 mm above the surface of the sample
Spot diameter on the surface	900 μm
Input energy density	1887, 1415, 1029, 871, 666 J/cm^2
Pulse height	94% of the maximum lamp current
Pulse shape	Rectangular

phosphate coatings. Laser parameters used are presented in Table 21.6. The variation in the laser speed was designed to alter the residence time of the laser spot, thereby changing the energy density available for coating deposition.

The tribology of various laser-treated depositions was tested in an SBF environment using a pin-on-disk setup. A 3-mm-diameter alumina pin was used as a pinhead (Fig. 21.11) since the calcium phosphate coating is a potential material for use on a femoral head replacement and would come in contact with a ceramic acetabular component, which is often made of alumina ceramic. Wear tests were done with a normal load of 8.8 N, at a rotating speed of 100 rpm (or a linear speed of 32 m/min), with weight loss being measured every 10 minutes.

The resultant coatings displayed an adherent interface with the substrate, with coating thickness of ~600 μm as observed under the cross-sectional image shown in Figure 21.12a. However, the top surface showed a rough topography with wide pore distribution (Fig. 21.12b). It was observed that increasing the laser scan speed resulted in a decrease in the overall average roughness (R_a, from ~29 μm at laser scan speed of 36 cm/min to ~20 μm at laser scan speed of 102 cm/min), as shown in Figure 21.13. Concurrently, maximum roughness (R_{max}) of the coatings decreased from ~38 μm at laser scan speed of 36 cm/min to ~33 μm at laser scan speed of 102 cm/min (Fig. 21.13).

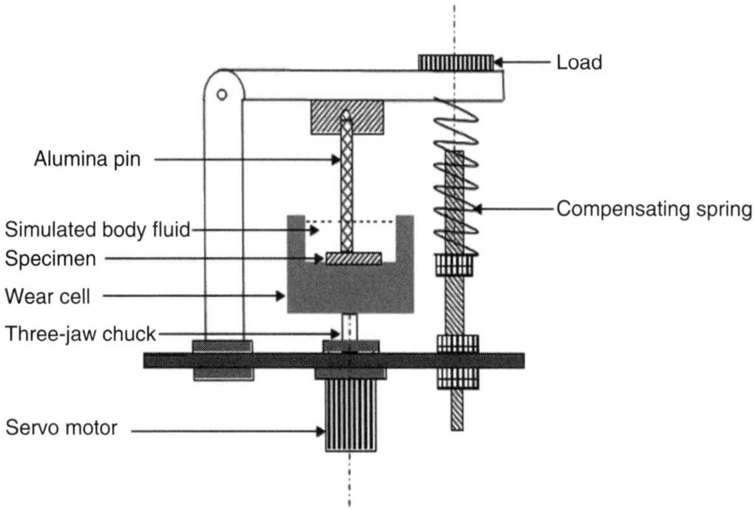

Figure 21.11 Schematic of a pin-on-disk wear tester. Adapted from Reference 29.

Phase characterization of laser-processed calcium phosphate coatings is evinced in the x-ray diffraction (XRD) spectra presented as Figure 21.14. It is observed that α-TCP, TiO_2, Ti, and Al are the major phases present. The characteristic peaks of Ti and Al appear due to laser penetration and substrate melting along with calcium phosphate surface powder and consequent solidification. Presence of TiO_2 is attributed to the oxidation of Ti substrate. TiO_2 is a chemically stable phase with enhanced resistance to corrosion and wear. Formation of α-TCP results due to dehydration of HAp during laser processing. Further, degradation of HAp and interaction with TiO_2 at high laser speeds (102 cm/min) also resulted in the formation of $CaTiO_3$ phase. With reduced laser speed, the overall thermal exposure is enhanced, due to which dehydroxylation forms a barrier to the nucleation of crystalline phases and causes a reduction of crystalline phases at high energy density.

Nanoindentation of the laser-processed calcium phosphate coatings demonstrated enhanced elastic modulus (~145 ± 20 GPa) and hardness (~4.4 ± 0.59 GPa) compared with those of the substrate material (elastic modulus of 120 GPa and hardness 2.20 GPa). Their typical *P–h* (load–depth) curves are presented in Figure 21.15.

Since the bioceramic is expected to resist wear in addition to resisting the severe body fluid environment, the pin-on-disk wear tests were done on the samples immersed in SBF solution. Evaluation of wear testing under SBF solution allows the resistance of the coating to osteolysis to be analyzed and implant loosening to be limited by selecting an optimized coating. The pin-on-disk wear response of various laser-processed calcium phosphate coatings is presented in Figure 21.16. It can be observed that Ti–6Al–4V substrate showed the highest cumulative weight loss (0.0035 g, or minimum wear resistance). Correspondingly, laser-processed calcium phosphate coatings showed 3–10 times enhanced wear resistance (weight loss between 0.00034 and 0.00098 g) compared with those of uncoated substrate.

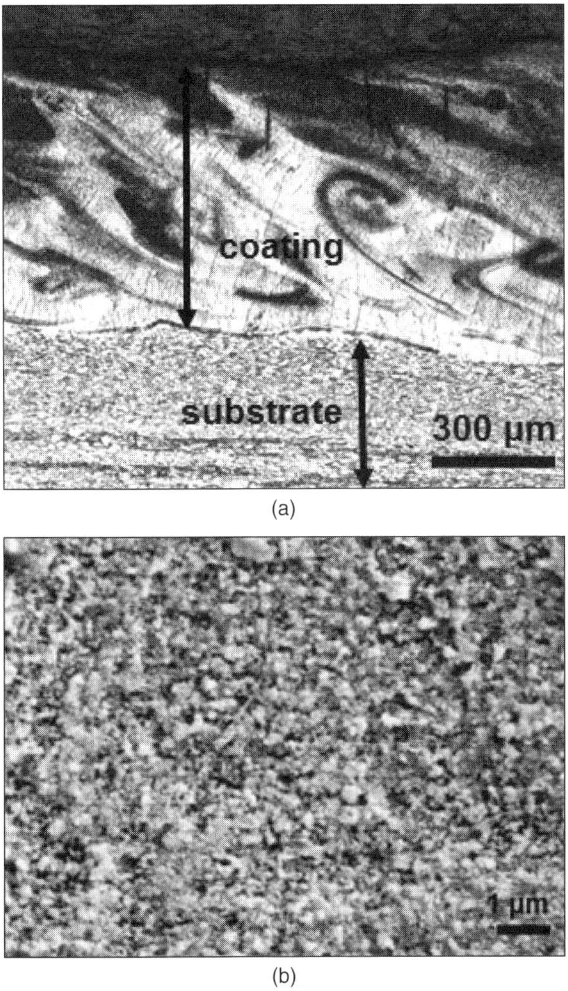

Figure 21.12 (a) Cross-sectional image and (b) top surface of laser-processed calcium phosphate coating at laser speed of 36 cm/min.[30]

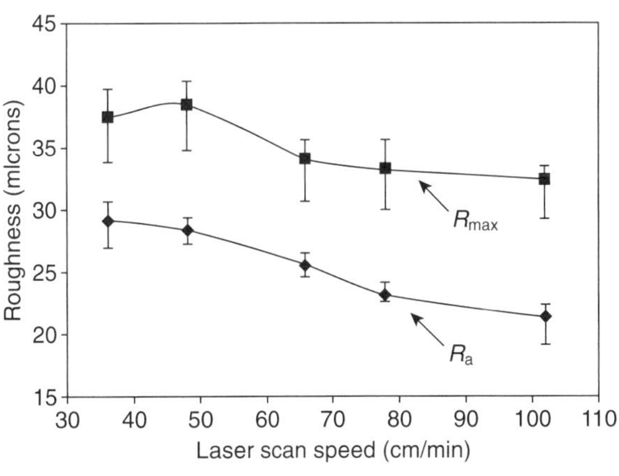

Figure 21.13 Average roughness (R_a) and maximum roughness (R_{max}) of the laser-synthesized calcium phosphate coatings.[30]

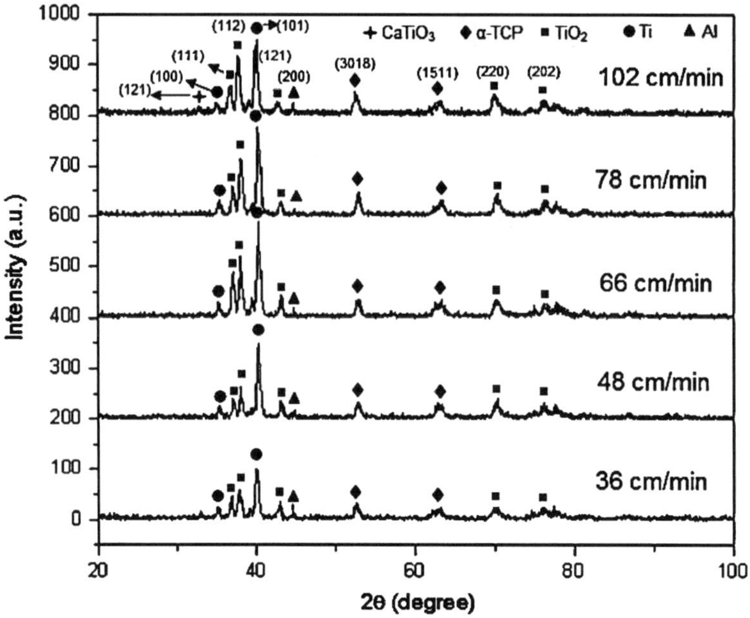

Figure 21.14 X-ray diffraction spectra of the laser-synthesized calcium phosphate coatings at various scan speeds.[30]

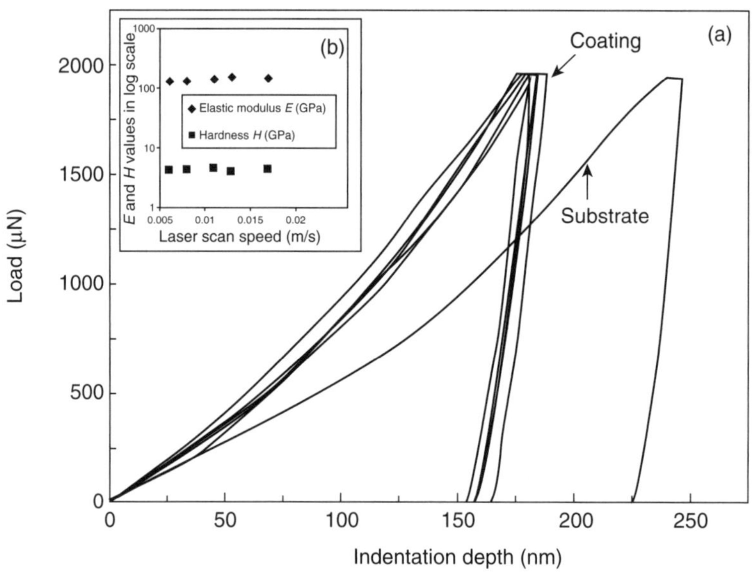

Figure 21.15 (a) Load–indentation (P–h) curves and (b) elastic modulus and hardness variation of laser-processed calcium phosphate coatings and Ti-substrate.[30]

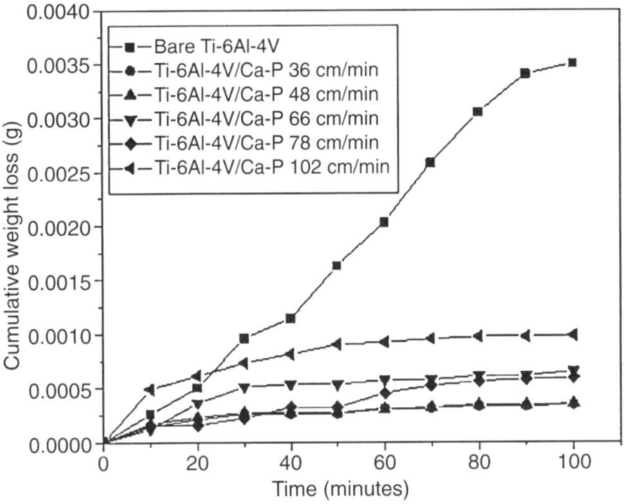

Figure 21.16 Pin-on-disk wear response of various laser-processed calcium phosphate coatings.[30]

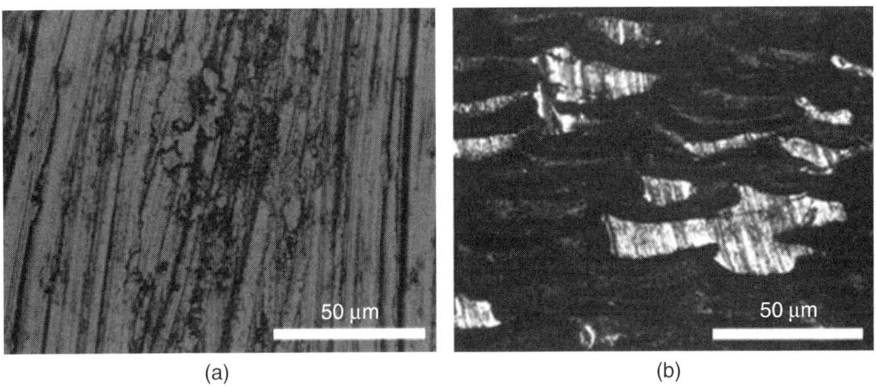

Figure 21.17 Pin-on-disk wear track surface of (a) Ti–6Al–4V substrate and (b) CaP coating processed at a laser scan speed of 102 cm/min.[30]

Pin-on-disk wear tracks on the Ti-alloy (Fig. 21.17a) show that excessive material loss has occurred compared with laser-processed calcium phosphate coating (Fig. 21.17b). A deteriorating combination of abrasive and adhesive wear on the Ti–6Al–4V substrate is the reason for the poor wear resistance. Also, the dominant influence of plowing by the transfer of adhesively bound particles on the pin surface resulted in enhanced material removal. However, the wear surface of calcium phosphate coatings shows a typical abrading surface—the hard ceramic phase resists wear upon mating with another ceramic surface (abrading pin).

Formation of an apatite-like layer was inherent in all the laser-processed surfaces (Fig. 21.18). A successive evolution of the apatite precipitation at various times

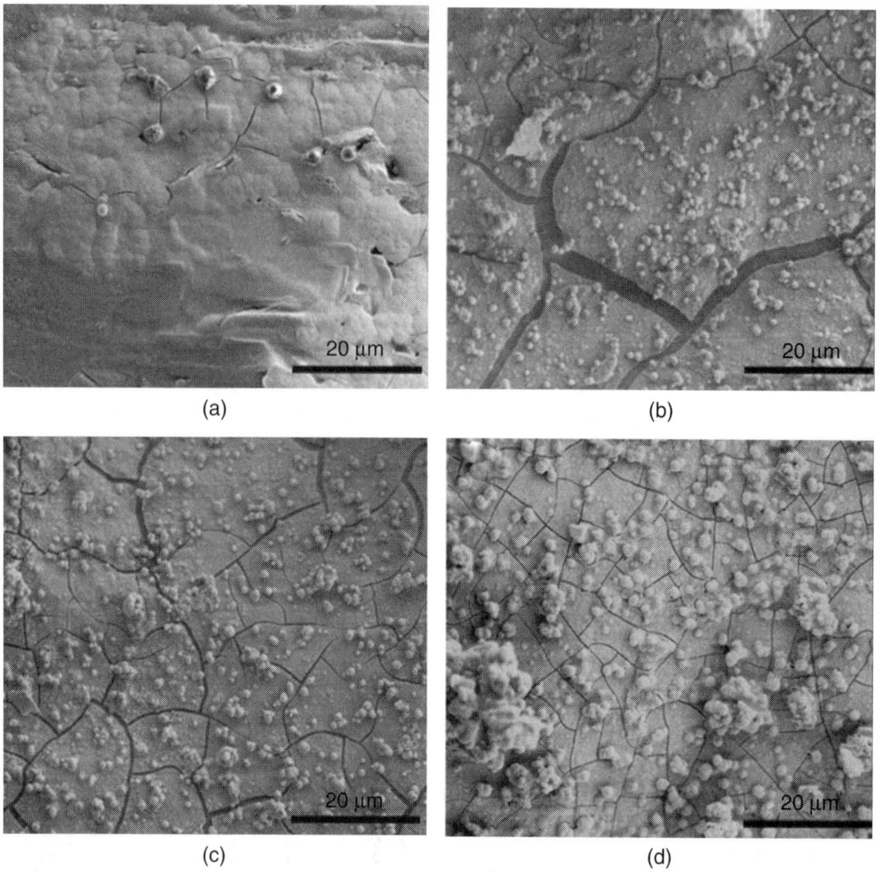

Figure 21.18 SEM micrograph of apatite-like layer after immersion in SBF of laser-processed coating (at laser scan speed of 102 cm/min) at (a) 24 hours, (b) 48 hours, (c) 72 hours, and (d) 96 hours.[30]

(24–96 hours) shows the cytocompatible nature of the laser-processed calcium phosphate surface. Cracking of the surface layer is attributed to drying, which removes moisture from the surface. Further, XRD spectra of the biomineralized surface confirm the precipitation of apatite on the surface of calcium phosphate coatings (Fig. 21.19). The comparison of the normalized integrated intensity of HAp to that of Ti (Fig. 21.19) also confirmed enhanced apatite precipitation with increased duration of immersion of the sample in SBF.

To quantify the biocompatibility of the laser-processed calcium phosphate samples, bioactivity was evaluated via biomineralization, which was quantified by the precipitation of apatite after immersion of samples in SBF for the stipulated time intervals. The solution was maintained at pH 7.4 and temperature 37°C, and the SBF was refreshed every 24 hours. Correspondingly, the biomineralization rate (*BR*) was quantified as

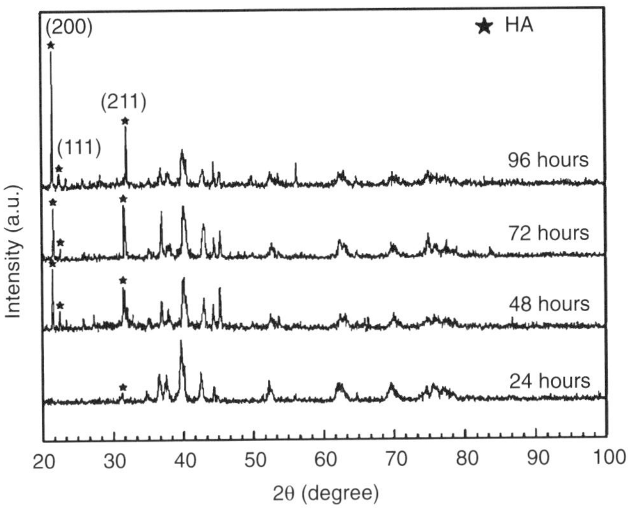

Figure 21.19 X-ray diffraction of laser-processed (scan rate of 102 cm/min) calcium phosphate coating at various time intervals of immersion under SBF.[30]

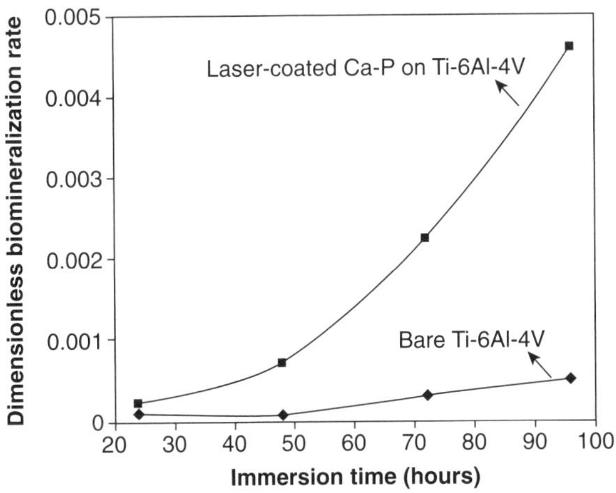

Figure 21.20 Biomineralization of laser-processed calcium phosphate coating (scan rate of 102 cm/min) compared with that of Ti–6Al–4V substrate.[30]

$$BR = \frac{W_f - W_i}{W_i},$$

where W_f is the final weight after immersion in SBF, and W_i is the initial weight before immersion in SBF. The biomineralization rates of Ti–6Al–4V and of laser-processed calcium phosphate coating (at scan rate 102 cm/min) are presented in Figure 21.20 for a duration of 24, 48, 72, and 96 hours. It is clearly indicated by the enhancement in biomineralization rate that apatite precipitation dominates (approximately eightfold increase) in the laser-processed calcium phosphate coatings compared with Ti–6Al–4V substrate.

In summary, tribological properties dictate the lifetime and response to the environment a body implant exhibits *in vivo*. Thus, the evaluation of wear loss (weight or volume) with loading becomes essential. In addition, the COF also decides the normal reaction of the adjoining interfaces. Thus, engineering the mating surfaces is highly critical in designing bone implants. Since the functionality of biocomposites strongly depends on their structural performance together with the cytocompatibility of the surfaces, merging the two requirements requires optimization. Hence the balance of matching the mechanical strength of an implant with that of the original part and obtaining the required function from the implant requires engineering. Further, the release of wear debris particles (arising due to loss of implant material *in vivo*) into the bloodstream may cause severe problems, which also need to be considered for its safe use. Hence, the evaluation of tribological properties is followed with *in vitro* analysis of the cytocompatibility of the implant to foresee its further potential for clinical trials before it can be utilized *in vivo*.

REFERENCES

1 K.-H. Zum Gahr. *Microstructure and Wear of Materials*. Elsevier, Oxford, 1987.

2 S. Gautier, E. Champion, and D. B. Assollant. Processing, microstructure and toughness of Al_2O_3 platelet-reinforced hydroxyapatite. J. Eur. Ceram. Soc. 17 (1997), 1361–1369.

3 J. Li, B. Fartash, and L. Hermansson. Hydroxyapatite–alumina composites and bone-bonding. Biomaterials 16 (1995), 417–422.

4 R. R. Rao and T. S. Kannan. Synthesis and sintering of hydroxyapatite–zirconia composites. Mater. Sci. Eng. C 20 (2002), 187–193.

5 V. V. Silva, F. S. Lameiras, and R. Z. Domínguez. Microstructural and mechanical study of zirconia-hydroxyapatite (ZH) composite ceramics for biomedical applications. Compost. Sci. Technol. 61 (2001), 301–310.

6 G. Gollera, H. Demirkiran, F. N. Oktar, and E. Demirkesen. Processing and characterization of bioglass reinforced hydroxyapatite composites. Ceram. Int. 29 (2003), 721–724.

7 W. Suchanek, M. Yashima, M. Kakihana, and M. Yoshimura. Hydroxyapatite/hydroxyapatite-whisker composites without sintering additives: Mechanical properties and microstructural evolution. J. Am. Ceram. Soc. 80 (1997), 2805–2813.

8 S. Nath. Development of novel calcium phosphate-mullite composites for orthopedic applications. Ph.D. thesis, IIT Kanpur, India, 2008.

9 S. Nath, B. Basu, M. Mohanty, and P. V. Mohanan. *In vivo* response of novel hydroxyapatite–mullite composites: Results up to 12 weeks of implantation. J. Biomed. Mater. Res. B Appl. Biomat. 90B (2009), 547–557.

10 P. Schaaff. The role of fretting damage in total hip arthroplasty with modular design hip joints—Evaluation of retrieval studies and experimental simulation methods. J. Appl. Biomater. Biomech. 2 (2004), 121–135.

11 Y. Fu, A. W. Batchelor, and K. A. Khor. Fretting wear behavior of thermal sprayed hydroxyapatite coating lubricated with bovine albumin. Wear 230 (1999), 98–102.

12 A. Choubey, B. Basu, and R. Balasubramaniam. Tribological behavior of Ti-based alloys in simulated body fluid solution at fretting contacts. Mater. Sci. Eng. A 379 (2004), 234–239.

13 D. Sheeja, B. K. Tay, S. P. Lau, and L. N. Nung. Tribological characterization of diamond-like carbon coatings on Co–Cr–Mo alloy for orthopedic applications. Surf. Coat. Technol. 146–147 (2001), 410–416.

14 D. Sheeja, B. K. Tay, and L. N. Nung. Tribological characterization of surface modified UHMWPE against DLC-coated Co–Cr–Mo. Surf. Coat. Techol. 190 (2005), 231–237.

15 R. Hauert and U. Müller. An overview on tailored tribological and biological behavior of diamond-like carbon. Diamond Relat. Mater. 12 (2003), 171–177.

16 S. Nath, S. Bodhak, and B. Basu. Tribological investigation of novel HDPE-HAp-Al_2O_3 hybrid biocomposites against steel under dry and simulated body fluid condition. J. Biomed. Mater. Res. A 83 (2007), 191–208.

17 S. Nath, N. Sinha, and B. Basu. Microstructure, mechanical and tribological properties of microwave sintered Ca-stabilized zirconia for biomedical applications. Ceram Int. 34 (2008), 1509–1520.

18 S. Bodhak, S. Nath, and B. Basu. Understanding the fretting wear properties of HAp, Al_2O_3 containing HDPE biocomposites against ZrO_2. J. Biomed. Mater. Res. A 85A (2008), 83–98.

19 S. Nath, S. Bajaj, and B. Basu. Microstructure, mechanical and tribological properties of microwave sintered Mg-doped zirconia. Int. J. Appl. Ceram. Technol. 5 (2008), 49–62.

20 S. Bodhak, S. Nath, and B. Basu. Friction and wear properties of novel HDPE-HAp-Al_2O_3 composites against alumina counterface. J. Biomater. Appl. 23 (2009), 407–433.

21 S. Nath, U. Raghunandan, and B. Basu. Fretting wear behavior of calcium phosphate-mullite composites in dry and albumin containing simulated body fluid conditions. J. Mater. Sci. Mater. Med. 21 (2010), 1151–1161.

22 J. E. Tercero, S. Namin, D. Lahiri, K. Balani, N. Tsoukias, and A. Agarwal. Effect of carbon nanotube and aluminum oxide addition on plasma-sprayed hydroxyapatite coating's mechanical properties and biocompatibility. Mater. Sci. Eng. C 29 (2009), 2195–2202.

23 K. Balani, R. Anderson, T. Laha, M. Andara, J. Tercero, E. Crumpler, and A. Agarwal. Plasma-sprayed carbon nanotube reinforced hydroxyapatite coatings and their interaction with human osteoblasts in vitro. Biomaterials 28 (2007), 618–624.

24 K. Balani, S. P. Harimkar, A. Keshri, Y. Chen, N. B. Dahotre, and A. Agarwal. Multiscale wear of plasma sprayed carbon nanocomposites reinforced aluminum oxide nanocomposite coating. Acta Mater. 56 (2008), 5984–5994.

25 K. Balani, D. Lahiri, A. K. Keshri, S. R. Bakshi, J. E. Tercero, and A. Agarwal. The nanoscratch behavior of biocompatible hydroxyapatite reinforced with aluminum oxide and carbon nanotubes. J. Miner. Met. Mater. 61 (2009), 63–66.

26 S. K. R. Chowdhury, A. C. Kulkarni, A. Basak, and S. K. Roy. Wear characteristic and biocompatibility of some hydroxyapatite-collagen composite acetabular cups. Wear 262 (2007), 1387–1398.

27 M. Younesi and M. E. Bahrololoom. Optimizations of wear resistance and toughness of HA/NFSS new bio-composites for using in total joint replacement. Mater. Des. 31 (2009), 234–243.

28 Q. Wang, S. Ge, and D. Zhang. Nano-mechanical properties and biotribological behaviors of nanosized HA/partially-stabilized zirconia composites. Wear 259 (2005), 952–957.

29 K. Balani, Y. Chen, S. P. Harimkar, N. B. Dahotre, and A. Agarwal. Tribological behaviour of plasma-sprayed carbon nanotube-reinforced hydroxyapatite coating in physiological solution. Acta Biomater. 3 (2007), 944–951.

30 S. R. Paital, K. Balani, A. Agarwal, and N. B. Dahotre. Fabrication and evaluation of a pulse laser-induced Ca–P coating on a Ti alloy for bioapplication. Biomed. Mater. 4 (2009), 015009.

Index

Advanced Structural Ceramics, First Edition. Bikramjit Basu, Kantesh Balani.